Focus on Computer Graphics

Tutorials and Perspectives in Computer Graphics

Springer
Berlin
Heidelberg
New York
Barcelona
Budapest
Hong Kong
London
Milan
Paris
Santa Clara
Singapore
Tokyo

Wolfgang Strasser Reinhard Klein
René Rau (Eds.)

Geometric Modeling: Theory and Practice

The State of the Art

With 236 Figures and 12 Tables

 Springer

Volume Editors

Prof. Dr. Ing. Wolfgang Strasser
Dr. Reinhard Klein
Dr. René Rau

Wilhelm-Schickard-Institut für Informatik
Auf der Morgenstelle 10, D-72076 Tübingen, Germany

CR Subject Classification (1991): I.3.5, D.2.10, G.1-2

Library of Congress Cataloging-in-publication Data

Geometric modeling: theory and practice: the state of the art / Wolfgang Strasser, Reinhard Klein, René
Rau (Eds.). p. cm. – (Focus on computer graphics). Proceedings of a conference held Oct. 14–18, 1996
in Blaubeuren, Germany. Includes bibliographical references and index.
ISBN-13:978-3-540-61883-6   e-ISBN-13:978-3-642-60607-6
DOI: 10.1007/978-3-642-60607-6

1. Computer graphics–Congresses. 2. CAD/CAM systems–Congresses. 3. Geometry–Data processing–Congresses. I. Strasser, Wolfgang, 1941– . II. Klein, Reinhard, 1961– . III. Rau, René, 1961– .
IV. Series. T385.G4644 1997 620'.0042'028551–dc21 97-18146 CIP

ISBN-13:978-3-540-61883-6

The use of general descriptive names, registered names, trademarks, etc. in this publication does not
imply, even in the absence of a specific statement, that such names are exempt from the relevant protec-
tive laws and regulations and therefore free for general use.

Cover Design: Künkel + Lopka, Heidelberg
Typesetting: Camera ready by authors
SPIN 10552091   45/3142 – 5 4 3 2 1 0 – Printed on acid-free paper

Preface

The Blaubeuren Conference "Theory and Practice of Geometric Modeling" has become a meeting place for leading experts from industrial and academic research institutions, CAD system developers and experienced users to exchange new ideas and to discuss new concepts and future directions in geometric modeling. The relaxed and calm atmosphere of the Heinrich-Fabri-Institute in Blaubeuren provides the appropriate environment for profound and engaged discussions that are not equally possible on other occasions. Real problems from current industrial projects as well as theoretical issues are addressed on a high scientific level.

This book is the result of the lectures and discussions during the conference which took place from October 14th to 18th, 1996. The contents is structured in 4 parts:

- Mathematical Tools
- Representations
- Systems
- Automated Assembly.

The editors express their sincere appreciation to the contributing authors, and to the members of the program committee for their cooperation, the careful reviewing and their active participation that made the conference and this book a success.

Tübingen, April 1997

Wolfgang Straßer
Reinhard Klein
René Rau

Organization

Conference Chair: Wolfgang Straßer, University of Tübingen, Germany
Organizing Chairs: Reinhard Klein, University of Tübingen
                       René T. Rau, University of Tübingen

Program Committee

W. Boehm, Technical University of Braunschweig, Germany
B. Falcidieno, University of Genova, Italy
G. Farin, Arizona State University Tempe, USA
E. Gschwind, Hewlett Packard Böblingen, Germany
H. Hagen, University of Kaiserslautern, Germany
Ch. Hoffmann, Purdue University, USA
J. Hoschek, Technical University of Darmstadt, Germany
R. Klass, Mercedes-Benz, Germany
M. Mäntylä, Helsinki University of Technology, Finnland
H. Nowacki, Technical University of Berlin, Germany
M. Pratt, Rensselaer/NIST, USA
A. Rappoport, Hebrew University of Jerusalem, Israel
J. Rossignac, IBM Watson Research Center, USA
T. Sederberg, Brigham Young University, USA
H. P. Seidel, University of Erlangen, Germany
T. Strotman, Structural Dynamics Research Corporation SDRC, USA
T. Varady, Hungarian Academy of Science, Hungary

Table of Contents

III   Systems

IV   Automated Assembly

Part I
Mathematical Tools

Variational Design with Boundary Conditions and Parameter Optimized Surface Fitting

Hans Hagen[1], Siegfried Heinz[2], and Alexa Nawotki[1]

[1] Universität Kaiserslautern, FB Informatik, Germany
[2] TransCAT GmbH, Karlsruhe, Germany

Abstract. Computer Aided Geometric Design has emerged from the need for freeform surfaces in CAD/CAM technologies; it has become a major research topic in computer science with direct applications for all engineering sciences. A major topic is the generation of smooth curves and surfaces which can be immediately supplied to the NC-Process.
The fundamental idea of the so called variational design methods is the use of modeling tools (curves or surfaces) which minimize certain functionals which can be interpreted in the sense of physics or geometry.
The purpose of this paper is to present a method to include the parametrization as an additional parameter in the variational design process and to include boundary conditions (given curves and neighbouring surfaces) in the design process.

Introduction

In many applications, only point data are available for surface fitting, and it is difficult to estimate additional information such as tangents or curvatures. Furthermore measurement of inaccuracies which occur, for example when digitizing an original model with worn out parts, must be considered.

One of the most successful and widely used variational design method is the combination of weighted least square fitting and jerk minimization along the parameterlines of a designed surface (Hagen-Santarelli '92) for a survey on variation design techniques (see [Bru93]).

In this paper we extend this method by using reparametrization as an additional optimization parameter and by using given boundary information for a reduction of the linear systen of equations.

1 Bézier and B-Spline Surfaces

The curves and surfaces now known as Bézier curves and surfaces were independently developed by P. de Casteljau and P. Bézier. The underlying mathematical theory based on the concept of Bernstein's polynomials was first introduced by R. Forrest. The fundamental idea of this approach is to evaluate and manipulate the curves and surfaces by a (small) number of control points. We first consider

Bézier curves as segmented curves. The segments $X_l(u); l = 0, \cdots, k$ of a Bézier curve of degree m over the parameter interval $u_l \le u \le u_{l+1}$ are

$$X_l(u) := \sum_{i=0}^{m} b_{lm+i} B_i^m \left(\frac{u - u_l}{u_{l+1} - u_l} \right). \tag{1}$$

The Bernstein polynomials $B_i^m(t) := \binom{m}{i}(1-t)^{m-i}t^i, 0 \le t \le 1$ are used as blending functions.

Bernstein polynomials are special degenerated B-splines. If we use B-splines as blending functions instead of Bernstein polynomials, we can generalize the whole concept to so-called B-spline curves and surfaces. B-spline curves are similar to Bézier curves in that a set of blending functions combine the effect of $n + 1$ control points

$$Y(u) := \sum_{j=0}^{n} d_j N_j^M(u). \tag{2}$$

The most important difference is the local support property of the B-spline blending functions $N_j^k(u)$. Both curve types have the convex hull and variation diminishing property.

A Bézier surface is a segmented surface. The segments $X_{pq}(u, v); p = 0, \cdots, k; q = 0, \cdots, r$ of Béziers surface of degree m, n over the rectangular parameter domain $u_p \le u \le u_{p+1}; v_q \le v \le v_{q+1}$ are

$$X_{pq}(, v) := \sum_{i=0}^{m} \sum_{j=0}^{n} b_{b \cdot m + i/q \cdot n + j} \cdot B_i^m \left(\frac{u - u_p}{u_{p+1} - u_p} \right) B_j^n \left(\frac{v - v_q}{v_{q+1} - vq} \right). \tag{3}$$

Instead of a control polygon a Bézier surface (segment) has a control polyhedron.

The definition of a B-spline surface over a rectangular parameter domain directly follows the same pattern,

$$Y(u, v) := \sum_{i=0}^{m} \sum_{j=0}^{n} d_{ij} N_i^M(u) N_j^N(v). \tag{4}$$

The so-called tensor-product surfaces can be easily generated by applying the de Casteljau algorithm or the de Boor algorithm twice.

2   Surface Generation from a Net of Curves

It is important for surface generation, in which form the input data is given. Is it a net of curves (plus ev. some further informations) or is it point data only?

The variational problem is straight forward, if a net is given. The one and only choice is made with the minimizing functional. Note, that the higher the degree of the derivatives in the functional the more bounday input is required.

A good selection is

$$F := \int A \cdot \left(\frac{\partial^2 X}{\partial u^2} + \frac{\partial^2 X}{\partial v^2} \right)^2 - B \left(\frac{\partial^2 X}{\partial u^2} \cdot \frac{\partial^2 X}{\partial v^2} - \frac{\partial^2 X}{\partial u \partial v} \right) du dv \qquad (5)$$

for some constants A, B because this expression describes the energy of a thin plate.

For solving the corresponding optimization problem $F \to min$, the boundary curves and tangents are necessary. The result depends on the parametrization and it cannot be changed after the computation.

We found an intuitive way for getting an interaction parameter on the surface: The constants A, B in equation (5) are related to material parameters,

$$A = \frac{h^3}{24} \frac{E}{1 - \nu^2} \; , \; B = \frac{h^3}{12}(1 - \nu)$$

where h is the thickness of the plate, E is the elastic modulus and ν is the Poisson's ratio. E and ν together are describing the way, a specific material deforms under external forces. Now it is possible to influence a surface systematically by changing the material parameters A, B or E, ν locally. That does not interfer with the nice computation properties of our variational problem.

Another choice for the functional is

$$F = \int \subset (\kappa_1^2 + \kappa_2^2) = 2D\kappa_1\kappa_2 \; du dv \qquad (6)$$

where κ_1, κ_2 are the principal curvatures.

This functional does not depend on the parametrization, but the cost is much more computation time. (In fact, the integral (5) can be deduced from (6)).

3   Variational Design of B-spline Surfaces

In this section we leave the classical approach of constructing a smooth net of curves and adding the surface patches smoothly into the net and present a direct method to construct a technically smooth B-spline surface, which uses only point data and refrains from determining a net.

The construction algorithm combines a weighted least square approximation with automatic surface smoothing. The smoothing criterion is the approximate minimization of the curvature variation. This technique presented here aims at constructing tangent-plane continuous B-spline surfaces. The following mathematical model serves as variation principle:

$$(1 - ws) \left\{ \sum_{k=1}^{n_p} w_{pk}[X(u_k, v_k) - P_k]^2 \right\}$$

$$+ ws \left\{ \sum_{i=1}^{n} \sum_{j=1}^{m} w_{3u} \int_{v_j}^{v_{j+1}} \int_{u_i}^{u_{i+1}} w_{3u_{ij}} \left\| \frac{\partial^3 X(u, v)}{\partial u^3} \right\|^2 du \, dv \right. \tag{7}$$

$$\left. + w_{3v} \int_{v_j}^{v_{j+1}} \int_{u_i}^{u_{i+1}} w_{3v_{ij}} \left\| \frac{\partial^3 X(u, v)}{\partial v^3} \right\|^2 du \, dv \right\} \rightarrow \quad min.$$

$X(u, v)$ is the representation of the surface; $(u, v) \in [u_1, u_{n+1}] \times [v_1, v_{m+1}]$ is the parameter value; and n, m are the number of segments in a u and v direction. P_k are the points to be approximated and n_p is the number of these points. The weight coefficients $ws, w_{3u}, w_{3v}, w_{3v_{ij}}$ are valid in the interval $[0, 1]$ and fulfill the constraints $\sum_{i=1}^{m} \sum_{j=1}^{n} w_{3u_{ij}} = 1$ and $\sum_{i=1}^{m} \sum_{j=1}^{n} w_{3v_{ij}} = 1$.

We apply this variation principle to biquintic B-spline surfaces,

$$X(u, v) = \sum_{i=1}^{4n+2} \sum_{j=1}^{4m+2} d_{ij} N_i^5(u) N_j^5(v), \tag{8}$$

with the knot-vectors

$$U := \left\{ \underbrace{u_1, \cdots, u_1}_{6*}, \underbrace{u_2, u_2, u_2 u_2}_{4*}, \cdots, \underbrace{u_n, u_n, u_n, u_n}_{4*}, \underbrace{u_{n+1}, \cdots, u_{n+1}}_{6*} \right\}$$

and

$$V := \left\{ \underbrace{v_1, \cdots, v_1}_{6*}, \underbrace{v_2, v_2, v_2, v_2}_{4*}, \cdots, \underbrace{v_m.v_m, v_m, v_m}_{4*}, \underbrace{v_{m+1}, \cdots, v_{m+1}}_{6*} \right\}.$$

This set of knot vectors guarantees the C^1-continuity of the surface.

We can now use the control points $d_{ij}; i \in \{1, \cdots, 4n+2\}, j \in \{1, \cdots, 4m+2\}$ as parameters for the calculus of variation approach.

Applying the variation principle (7) is a three-step process.

Step 1. **Least square fitting.**

$$LS := \sum_{k=1}^{n_p} w_{pk}[X(u_k, v_k) - P_k]^2 \rightarrow \quad min, \tag{9}$$

or in B-spline representation

$$\sum_{k=1}^{n_p} w_{pk} \left[\sum_{i=1}^{4n+2} \sum_{j=1}^{4m+2} d_{ij} N_i^5(u_k) N_j^5(v_k) - P_k \right]^2 \rightarrow \quad min. \tag{10}$$

The necessary condition $\frac{\partial LS}{\partial d_{lr}} = 0$ lead to a linear system of equations:

$$\sum_{i=1}^{4n+2} \sum_{j=1}^{4m+2} \left\{ 2 \sum_{k=1}^{n_p} w_{pk} \cdot N_i^5(u_k) N_j^5(v_k) N_l^5(v_k) N_r^5(v_k) \right\} d_{ij} \qquad (11)$$

$$= 2 \sum_{k=1}^{n_p} w_{pk} \cdot P_k N_l^5(u_k) N_r^5(v_k).$$

This unique solution of this system is the best point fitting in the least square sense of (9)

Step 2. **Automated smoothing process.**   As a fairness criterion we use

$$\sum_{i=1}^{n} \sum_{j=1}^{m} \int_{v_j}^{v_{j+1}} \int_{u_i}^{u_{i+1}} w_{3u} \cdot w_{3u_{ij}} \left\| \frac{\partial^3 X(u,v)}{\partial u13} \right\|^2 \qquad (12)$$

$$+ w_{3v} \cdot w_{3v_{ij}} \left\| \frac{\partial^3 X(u,v)}{\partial v^3} \right\|^2 du \; dv \rightarrow min.$$

A calculus of variation approach leads again to a linear system of equations.

Step 3. **Merging.**
We now combine the weighted least square fitting with the automated smoothing process:

$$(I - ws)A + ws \; B = 0 \; ; \quad ws \in [0,1] \qquad (13)$$

A symbolizes (11) and B symbolizes the equations of step 2. (for more details see ([HS92])).

   This method is in the meantime a standard technique used by Hella, BMW, Bosch and it is part of the CAD/CAM-Systems CATIA and EUCLID.

4   Multi Patch Approach

For the design process of complex surfaces it is in some cases necessary to create curvature continuous multipatch surfaces of arbitrary degree. Furthermore, the segmentation of the surface is a design and optimization parameter. This yields to a user input of the form

   - the degree in u, v direction
   - the continuity in u, v direction
   - the segmentation in u, v direction
     (i.e. determine the u, v parameter of the segment boundaries by indicating the segment boundaries on the surface for parametrization of the points).

   From this input we calculate the knot vectors in u and v of our new surface representation. The mathematical model (7) of Chapter 3 serves as variational principle.

$X(u, w)$ is the representation of the surface; $(u, v) \in [u_1, u_{n+1}] \times [v_1, v_{m+1}]$ is the parameter value and n, m are the number of segments in a u and v direction. P_k are the points to be approximated and n_p is the number of these points.

The weight coefficients $w_s, w_{3u}w_{3v}, w_{3u_{ij}}, w_{3v_{ij}}$ are valid in the interval $[0, 1]$ and fulfill the constraints

$$\sum_{i=1}^{m} \sum_{j=1}^{n} w_{3u_{ij}} = 1 \text{ and } \sum_{i=1}^{n} \sum_{j=1}^{n} w_{3v_{ij}} = 1.$$

c_u is the continuity of the surface patches in u direction and c_v is the continuity in v direction with $c_u, c_v \geq 0$. p is the order of the surface patches in u direction and q is the order in v direction with $p, q \geq 1$.

We apply the variation principle to the B-spline surface

$$X(u, v) = \sum_{i=1}^{ncu} \sum_{j=1}^{ncv} d_{ij} N_i^p(u) N_j^q(v)$$

with the knot-vectors

$$U := \left\{ \underbrace{u_1, \cdots, u_1}_{p*}, \underbrace{u_2, \cdots, u_2}_{l_u*}, \cdots, \underbrace{u_n, \cdots, u_n}_{l_u*}, \underbrace{u_{n+1}, \cdots, u_{n+1}}_{p*} \right\}$$

and

$$V := \left\{ \underbrace{v_1, \cdots, v_1}_{q*}, \underbrace{v_2, \cdots, v_2}_{l_v*}, \cdots, \underbrace{v_m, \cdots, v_m}_{l_v*}, \underbrace{v_{m+1}, \cdots, v_{m+1}}_{q*} \right\}$$

where

$$l_u = p - c_u - 1$$
$$l_v = q - c_v - 1$$

and

$$ncu = ((p - 1) - c_u \cdot n + c_u + 1$$
$$ncv = ((q - 1) - c_v) \cdot m + c_v + 1.$$

Now we can use the control points $d_{ij}; i \in \{1, \cdots, ncu\}, j \in \{1, \cdots, ncv\}$ as parameters for the variation approach.

Step 1. **Least square fitting.**

$$LS := \sum_{k=1}^{n_p} w_{pk}[X(u_k, v_k) - P_k]^2 \rightarrow min.$$

In B-spline representation

$$\sum_{k=1}^{n_p} w_{pk} \left[\sum_{i=1}^{ncu} \sum_{j=1}^{ncv} d_{ij} N_i^p(u_k) N_j^q(v_k) - P_k \right]^2 \rightarrow min.$$

The necessary conditions $\frac{\partial LS}{\partial d_{lr}} = 0$ lead to a linear system of equations:

$$\sum_{i=1}^{ncu} \sum_{j=1}^{ncv} \left\{ \sum_{k=1}^{n_p} w_{pk} N_i^p(u_k) N_j^q(v_k) N_l^p(u_k) N_r^q(v_k] \right\} d_{ij}$$

$$= \sum_{k=1}^{n_p} w_{pk} P_k N_l^p(u_k) N_r^q(v_k).$$

Step 2. **Automated smoothing process.** As fairness criterion we use the same criterion as in Chapter 3.

Step 3. **Merging.** The same combination of the weighted least square fitting with the smoothing process as in Chapter 3.

5   Boundary Conditions

In the whole process of surface reconstruction it is in some cases necessary to create surfaces through a cloud of points with given boundary curves and along these boundaries tangent to neighbouring surfaces.

To include such boundary curves in the design process we calculate points along the given curves. The number of points depend on the degree and the number of segments of the curves. With the user input of the degree and the continuity in u and ν direction and the same parametrization as for the surface calculation we start an automated iterative subdivision method for the number of segments. Each segment with a greater point deviation then a given tolerance is subdivided into two segments. The mathematical model

$$(1 - ws) \left\{ \sum_{k=1}^{np} w_{pk}[x(u_k) - P_k]^2 \right\}$$

$$+ ws \left\{ \sum_{i=1}^{n} w_{3i} \int_{u_i}^{u_{i+1}} \|\frac{\partial^3 X(u)}{\partial u^3}\|^2 \, du \right\} \rightarrow min$$

of Chater 3 for one direction serves as variational principle. For these curve approximation we use a small smoothing weight ws to get a "best" approximation (We do not want to smooth the curve). The resulting control points of the curve are used to reduce the linear system of equations from Chapter 4 for the surface calculation.

To include neighbouring surfaces in the design process we calculate the normals from these surfaces in the points along the boundary curves. Additional to that tangent plane information we need the tangent direction and the tangent length at these points. For this we calculate first a surface without tangent conditions, then we calculate the tangents from the surface along the boundaries in these points and project the tangents into the tangent plane. Now we have the discrete tangent information and calculate the tangent curves with the mathematical model.

The resulting control points of the tangent curves are also used to reduce the linear system of equations. Every curve (two boundary curves, two tangent curves) for one parameter direction has to be computed with the same knot vector. That means we use a global automated subdivision method. Every curve of one parameter direction has to be subdivided into several segments at the same parameter value. That defines the knot vector for the surface calculation in that parameter direction.

6   Parameter Optimization

Absolutely necessary for a successful variational design technique is an appropriate parametrization of the given points. A first step is a best plane fitting. After projecting the points onto this plane a first parametrization can be calculated. This leads to a first surface fitting. In some cases this approximation is not good enough. In a second step we project the points onto this "first approximation surface" and a new parametrization is then calculated by approximated curve lengths of the isoparametrics passing through these points.

Now we include the parametrization as an additional parameter in the variational design process and determine the parameter and the control points in one step.

A reparametrization

$$
\left. \begin{array}{l} u_k = u_{k_0} + \Delta u_k \\ v_k = v_{k_6} + \Delta v_k \end{array} \right\} k = 1, \cdots , n_p
$$

included in the variational design process leads to the additional necessary conditions:

$$
\begin{aligned}
\frac{\partial LS}{\partial \Delta u_l} &= \frac{\partial}{\partial \Delta u_l} \sum_{k=1}^{n_p} w_{pk} [X(u_k, v_k) - P_k]^2 \\
&= \sum_{k=1}^{n_p} 2 < (X(u_k, v_k) - P_k), \frac{\partial}{\partial \Delta v_l} X(u_k, v_k) > \\
&= 2 < (X(u_l, v_l) - P_l), \frac{\partial}{\partial \Delta u_l} X(u_l, v_l) > \\
&= 0
\end{aligned}
$$

and

$$
\begin{aligned}
\frac{\partial LS}{\partial \Delta v_l} &= 2 < (X(u_l, v_l) - P_l), \frac{\partial}{\partial \Delta v_l} X(u_l, v_l) > \\
&= 0,
\end{aligned}
$$

with $l = 1, \cdots, n_p$ and $< \cdot, \cdot >$ is the scalar product of two vectors.

The nonlinear system of equations

$$\frac{\partial LS}{\partial d_{lr}} = 0 \quad l = 1, \cdots, ncu \quad r = 1, \cdots, ncv$$

$$\frac{\partial LS}{\partial \Delta u_l} = 0 \quad l = 1, \cdots, n_p$$

$$\frac{\partial LS}{\partial \Delta v_l} = 0 \quad l = 1, \cdots, n_p$$

combined with the linear system of equations from the automated smoothing process can be solved with a Newton method. The appropriate initial values for the Newton method can be calculated in a first step with a best fitting plane for the parametrization of the points.

7   Applications

We use this method to construct parts of a pipeline system:

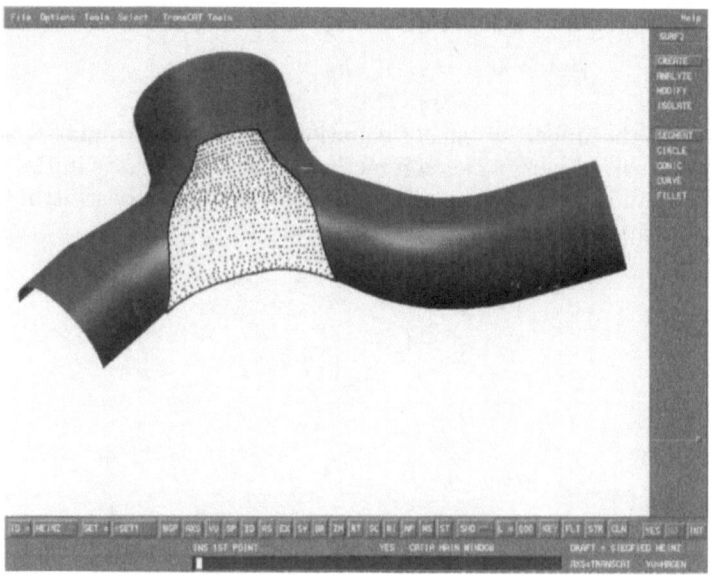

Step1: Digitizing

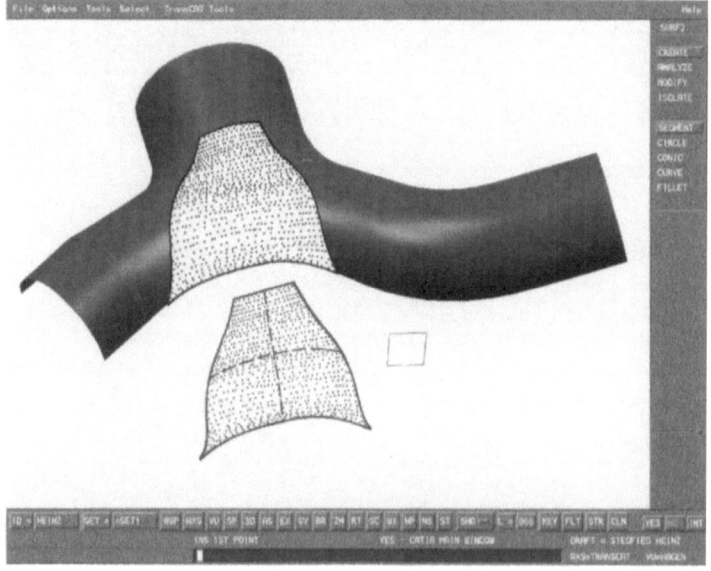

Step2: Parametrization

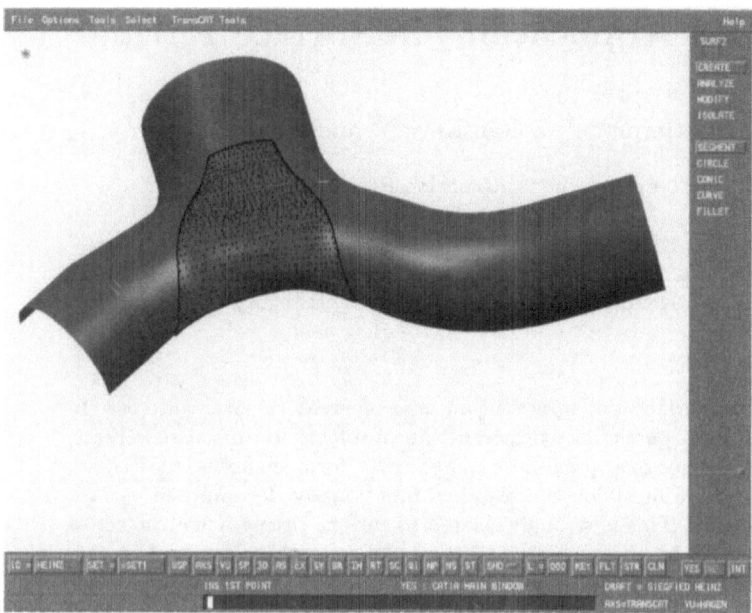

Step3: Variational Surface Design

References

[Bru93] G. Brunnett. Variational design of curves and surfaces. *Surv. Math. Ind.*, **3**: 1 – 27, 1993.

[HS92] H. Hagen and P. Santarelli. Variational Design of Smooth B-Spline Surfaces. In H. Hagen, editor, *Topics in Surface Modelling*, pages 85 – 94. SIAM, 1992.

Special Applications in Surface Fitting

Thomas Hermann[1], Zoltán Kovács[2] and Tamás Várady[1]

[1] Computer and Automation Research Institute,
Hungarian Academy of Sciences,
Kende u. 13-17, 1111 Budapest, Hungary
[2] CADMUS Consulting and Development Ltd.,
Kende u. 13-17, 1111 Budapest, Hungary

Abstract. Surface fitting is still an area of great research interest. In this paper two particular surface fitting problems are discussed. First, various strategies to approximate complex free-form shapes are described, with special emphasis on the so-called functionally decomposed surface representation. This is strongly related to surface fitting when the set of data points to be approximated is incomplete; i.e. there are areas which belong to other surfaces or which are just not accessible. In these cases we have to somehow "bridge" the unknown areas, and produce an overall pleasing surface, which smoothly connects the surface portions where data is available. The algorithm described is a combination of previous approaches to minimize hybrid objective functions, with several practical improvements. The second problem aims at fitting surfaces where not only positional data but surface normal data needs to be approximated as well. After analyzing the consistency of the data and possible objective functions to be minimized, the solution, which was found to be the most suitable is presented. Both the above problems emerged due to practical demands and have been intensively tested using real data obtained from the automobile industry. Some colour pictures illustrate the results.

1 Introduction

Surface fitting techniques are required in many application areas including variational surface design, fairing existing surfaces, approximate conversions from one representation to another, smoothing noisy data and reverse engineering of shapes based on measured data points. The problem has been intensively investigated in the past few years, an overview on the current state of the art can be obtained from the following publications by Hoschek [13], Cox [4], Dietz [6], Sarraga [20], Várady et al [22], see also the book edited by Sapidis [19].

Typically a set of various constraints are given and a smooth surface which satisfies these is created. The constraints can be specified "quantitatively", for example, a set of discrete data points needs to be approximated within a prescribed tolerance. Additionally there are "qualitative" constraints, by means of which the shape of the surface is forced to be predictable and aesthetically pleasing. It was recognized early that without a fairing term the quality of the surface is poor due to wiggles. While quantitative constraints can be formulated and tested in a relatively straightforward manner, there is no general agreement on how to characterize fairness. A great variety of objective functions has been published in the literature. The choice depends on the particular application, smoothness requirements, computational efficiency (linear contra non-linear methods), available "tuning" parameters and to some extent personal taste (see related work of Celniker [1], Dietz [6], Farin [8], von Golitschek [10], Greiner [11], Moreton [17], Nielson [18] and Terzopoulos [21] among others).

Our main interest in this paper is surface approximation with tensor product B-splines. Numerical efficiency mostly requires the use of least squares instead of maximum norm, etc., then the minimization leads to a linear system of equations. The unknown quantities to be determined are the control points of the surface. (The first papers addressing the computational efficiency of these systems were Hayes and Halliday [12] and Cox [2], [3].) The situation is more complicated when smoothness terms are added. There are various hybrid objective functions composed of weighted terms - see alternatives for example in [6]. Mostly, in order to preserve numerical efficiency, quadratic terms are preferred.

Before presenting our surface fitting algorithms which were used for two special applications, let us raise the key issues which need to be solved. First of all, what sort of smoothing terms to choose ? How to combine the quantitative and qualitative terms, how to set (automatically ?) the weights assigned to the individual terms of the objective function ? How to select an ideal number of control points and how to set the knot vectors? How to insert new knots ? How to find a good initial parametrization and how to assign surface parameter values to the data points ? How to improve parametrization ? How to sample large data sets and how to increase adaptively the number of data points in certain detailed areas ? How to enforce higher tolerance at certain places and ignore certain areas ?

The description of the following algorithms tries to provide answers for most of the above questions. Known solutions have been extended and improved. We believe that due to the lack of an overall theory of surface fitting, it is still a practical issue with several heuristic elements. The generated surfaces can prove (or disprove) the suitability of the solutions.

Our present interest partly comes from reverse engineering complex free-form shapes [22]. While the majority of related publications deals with fitting a single, standard, tensor-product surface onto a pointset within a rectangular domain, industrial parts are rarely so simple and surfaces are mostly composed by high-level geometric modelling operations, such as intersections, trimming, blending, etc. Strategies to segment large point sets with irregular topology will be analyzed in Section 2, emphasizing the advantages of a representation scheme based on decomposing the shape into functional elements. This representation is concerned with a subproblem of surface fitting with "ignore areas". Here points over certain surface areas are known, while points over other areas are either missing or belong to some another surface and for this reason need to be ignored. We still attempt to find out the shape of the original surface and assure overall smoothness. A description of the algorithm is given in Section 3. A different type of surface fitting problem is investigated in Section 4, where positional and normal vector constraints need to be satisfied simultaneously. A summary and open research topics conclude the paper.

2   Surface Fitting of Complex Free-Form Shapes

The majority of free-form objects are bounded by several surface elements; for example panels from the car industry or moulded plastic objects. Typically these are composed of relatively large primary surfaces and additional free-form surface features formed using high-level geometric operations. The features may include smooth transition surfaces between primary elements, such as blends, or various special functional parts, such as free-form steps, slots, pockets, etc. Ideally, an intelligent reverse engineering system should recognize how the part was originally constructed based on the point set and create a complex surface model accordingly. However, it seems there is a long way to go until CAD/CAM systems will be capable of doing this without user interaction.

There are various strategies for solving the above problems, though each of them has some deficiency as will be explained below. In some approaches preference is given to automatic surface generation and user interaction is minimized. In the case of applying *globally approximating surfaces* (see for example Dietz [6] or Hoschek [13], the rectangular topology is basically kept and a large four-sided surface is fitted which covers the whole object. The boundaries of the object will be represented by trimming curves within the surface if necessary. If the surface fit is not appropriate new knots are inserted. The knot vectors represent the parametrization quasi parallel to the patch boundaries, so problems may occur if there are boundaries with uneven lengths or "diagonally" placed features in the middle of the surface. At detailed parts, we may apply quadtree-type local subdivision to get more freedom for a better surface fit. A hierarchical solution was proposed in [9]. But in any case, the basic problem remains, the surface will be a composition of rectangular elements within a grid, which may approximate the data points well, but will rarely reproduce features properly (see curvature maps around features).

An alternative "automatic method" was suggested by Eck and Hoppe [7] called *arbitrary topology surfaces*. Here, based on a preliminary triangulation, an irregular topological domain with four-sided elements is created after simplifying the original triangulation. Rectangular surface patches are created which join each other smoothly and in fact interpolate a general topology free-form curve network lying on the object. The advantage of this approach is that there is no need to generate trimming curves. At the same time, it is quite demanding from a computational point of view and can assure G^1 continuity only, at present. This method can not represent features directly either, since the domain was created independently of the underlying design structure.

The next strategy seems to be the most widely applied industrial technique these days [15]. Here the user must create a patch structure, and in contrast to the previous methods, this manual segmentation requires a significant amount of user interaction. Practically, a *free-from curve network* is constructed, which is followed by interpolating the regions using the local, single patch methods. The advantage of this approach is that the user can interpret the underlying engineering structure, however there are disadvantages as well. One of them is that "drawing" on data points, such as determining the boundaries between a surface and a blend or defining proper subdividing curves, is not a well-defined task. Another problem is, that in order to create a complete net artificial edges often need to be incorporated. These may separate parts of a single functional surface into several parts and along these dividing curves the surface quality will not be as good as if the whole surface had been approximated in one piece.

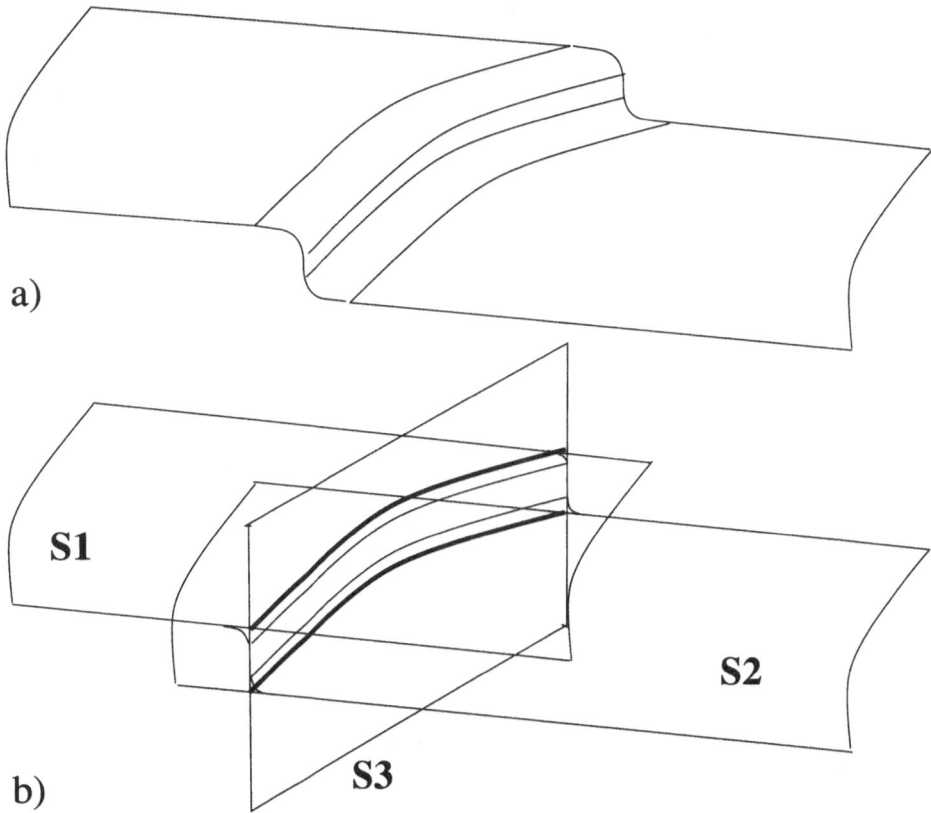

Fig. 1. Simple free-from surface

Two simple examples illustrate the above problems. In Figure 1/a two large primary surfaces are connected by a thin, blended step surface. Obviously the first two automatic surface fitting schemes can produce reasonable surface approximations with none or minimal wiggling, but will never produce evenly distributed curvatures along the blends, since the control points were not determined according to their natural structure. The curve network based approach will reflect the structure of the object in a much better way, however if the curves are not precisely drawn on the object, the corresponding surfaces will not be of high quality either. Figure 2 shows further deficiencies of the curve net approach. A relatively simple trimmed face was merged with a free-form depression. As shown, many boundary curves must be defined to construct a net of possibly four sided regions. Moreover, subpatches 1 to 8 belong to the same surface, but are represented separately, thus a global fairing for example can hardly be performed across the patch boundaries.

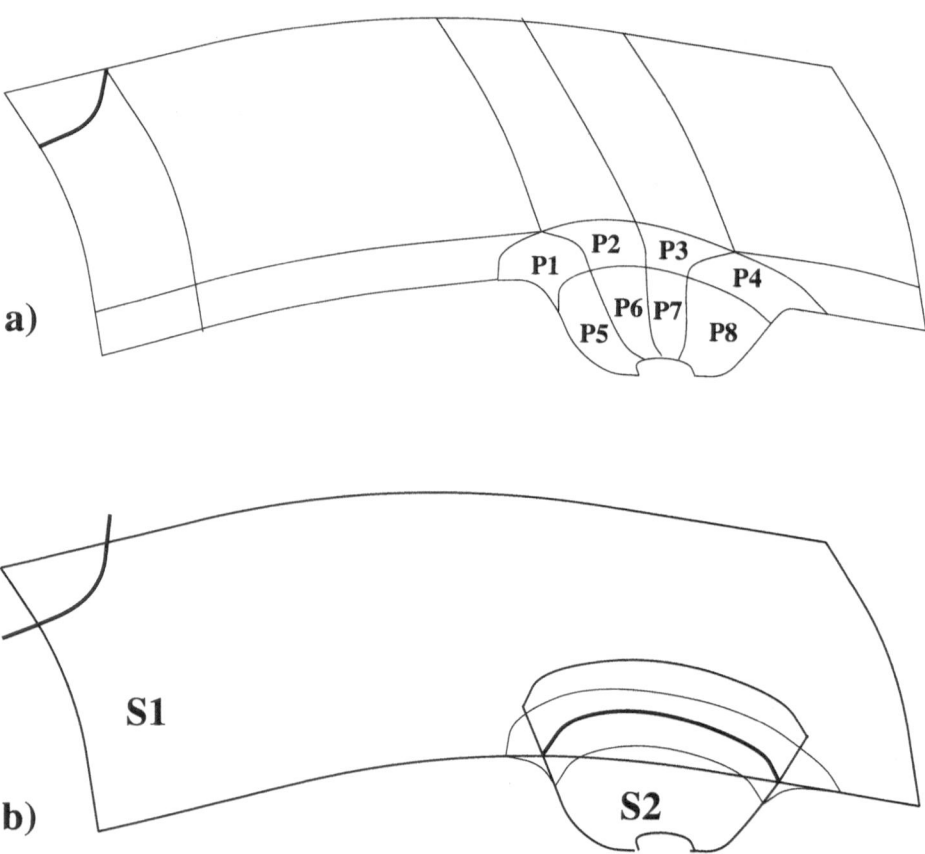

Fig. 2. Surface with depression

The above disadvantages suggest another type of segmentation strategy. The so-called *functional decomposition model* requires some a priori knowledge about the object and a relatively large amount of interaction as in curve network design. A human reverse engineer should be able to recognize the constituting elements of an object, those which are functionally important and those which depend basically on the previous primary elements. At the same time, he should understand the likely sequence of high level geometric operations how the object was created. (This information may even be stored in some history file as well.) The advantage of this approach is that it will result in a representation which many engineers prefer. The primary surfaces will be globally as fair as possible, since they will not be influenced by successive operations. It is also important that modifying the primary surfaces or changing some parameters of the features and then repeating the same sequence of operations will define a new valid model in most cases. (This can hardly be realized for the curve net model.) One disadvantage of functional decomposition is that trimmed surfaces are generated, which means that though the components will have high surface quality, they will typically join with numerical continuity only.

In order to make this strategy feasible two important issues need to be solved. The first is that we must compute primary surfaces even if there are areas over which no data is available. The other even more challenging issue is that we must be able to fit various free-form features onto the data points in a way so that they match the adjacent primary surfaces. This is not easy even for simple blending surfaces, not to mention complex free-form pockets or slots, where the best possible fit must be found related to a few characteristic curves and feature parameters. In the next section we show how surface fitting works over missing data; the insertion of free-form features is the subject of another paper being under preparation.

Now let us see how this strategy works for the previous examples. In Figure 1/b three primary surfaces are created separately; after intersecting **S1** with **S3** and **S2** with **S3** two sharp edges are created which serve as spine curves to determine the "best approximating blends". Note, if the difference between the surfaces is small, then there will not be sufficient data to fit **S3**. Then the "best step feature" needs to be found, which smoothly joins **S1** and **S2**. In Figure 2/b, it can be seen that there is a base surface **S1** and a depression surface **S2**, whose data points are mostly available, so surfaces with ignore areas can be fitted. The intersection curve between **S1** and and the extension of **S2** will drive the generation of a "best fit blending surface".

3   Surface Fitting with Ignore Areas

As explained above, in functional decomposition models we often fit rectangular surfaces which include certain areas, where there is no valid information available. For example, the points may belong to another functional surface, or there may be a hole, or a fixture may cover certain parts. In these cases, the user not only identifies the corners of the surface, but also has to define ignore areas. These are typically bounded by closed polylines, which serve to ignore data points within them. Two simple examples are shown in Figure 3 and Figure 4. The first example shows a door-panel of a car, the ignore area is located around the rounded pocket for the door handle. The other example shows a side surface of a car body panel where the middle bottom area needs to be excluded. As can be seen, these points belong to other surfaces, namely the wheel surface and the blend connecting this to the selected side surface.

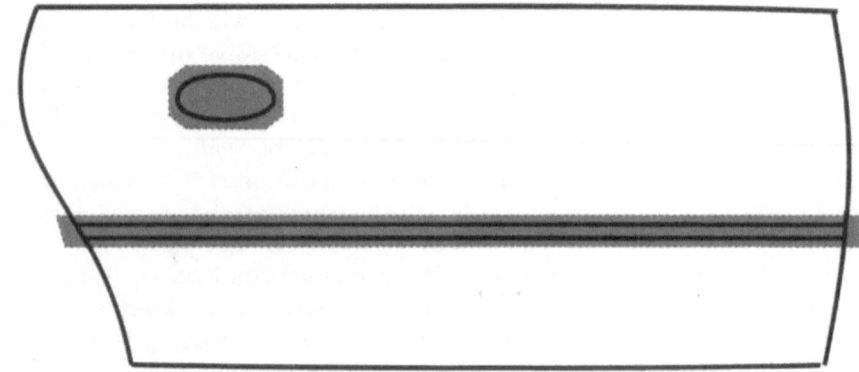

Fig. 3. Door panel

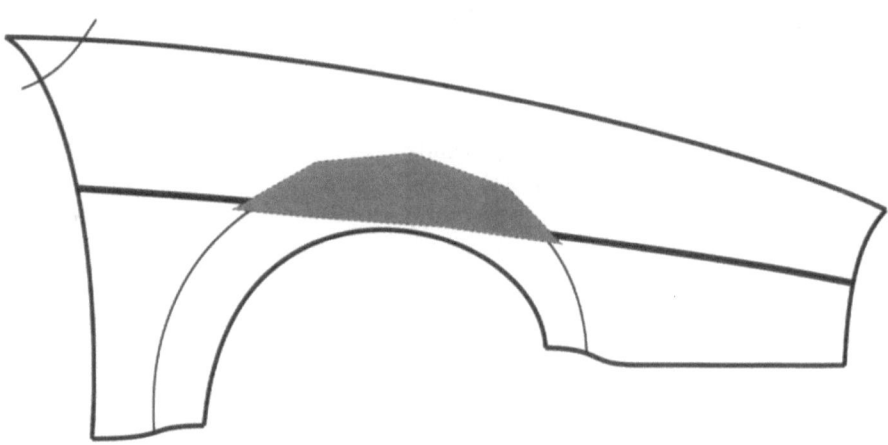

Fig. 4. Front panel

The data points are denoted:

$$\mathbf{P}_k \qquad (k = 1, \ldots, M) \ .$$

We approximate with integral tensor-product B-spline surfaces of order 4 for this application:

$$\mathbf{S}(u, v) = \sum_{i=0}^{n} \sum_{j=0}^{m} \mathbf{c}_{ij} N_i(u|U) N_j(v|V)$$

where $\mathbf{c}_{ij}$ are the de Boor points, and the N_i-s denote the B-spline basis functions. For simplicity's sake the parameter range $0 \leq u \leq 1$ and $0 \leq v \leq 1$ is considered with knot vectors $U = (u_0, u_1, ...u_{n+4}), V = (v_0, v_1, ...v_{m+4})$.

As was also suggested by other authors, we introduce two objective functions E and J to measure the deviation of the approximating surface and its smoothness, respectively:

$$E = \frac{1}{M} \sum_{k=1}^{M} w_k \left(\mathbf{S}(u_k, v_k) - \mathbf{P}_k\right)^2 \tag{1}$$

and

$$J = \int_0^1 \int_0^1 \left(\mathbf{S}_{uu}^2(u, v) + 2\mathbf{S}_{uv}^2(u, v) + \mathbf{S}_{vv}^2(u, v)\right) du dv \ . \tag{2}$$

The hybrid objective function to be minimized is a weighted sum

$$\Phi = E + \lambda J \tag{3}$$

The values w_k are non-negative weights assigned to each data point (how to set these, see later). Assuming the number of the control points, the knot vectors, the surface parameter values (u_k, v_k) belonging to each data point and the smoothness weight λ have been fixed, we can minimize the objective function Φ by solving a linear system of vector equations. The coefficients of each control point can be easily determined. (Note, the advantage of choosing this type of smoothness term is that the integrals of the related basis functions can be computed separately and directly used in the formulae.)

In fact, how to set the above quantities is one of the most difficult problems in surface fitting. Choosing only a relatively small number of control points will allow a smooth surface, but will prevent approximation within tolerance. With too many control points, computation takes a long time, and too many degrees of freedom may potentially lead to more oscillations. To set the knot vectors for a surface with large curvature variations is also a nontrivial problem. Similarly, how to find good surface parameter values (hereafter uv-values) associated with the data points is also difficult. The basic problem is that we have an *unknown* surface we still want to find related surface points which are close to the given data points. Finally, to set a proper λ is difficult, since its magnitude depends on how large the current first term is. If it is too high, the surface will be smooth, but we cannot satisfy the positional tolerance, otherwise the smoothing effect will be negligible. These arguments show that a direct solution is practically impossible, and iterative solutions need to be applied. As shown, the algorithm below guarantees that a surface with adaptively refined control net and knot vectors is generated, which remains within a prescribed positional tolerance and assures the highest possible smoothness.

The mechanism behind this iterative approach is that in the i-th step we use the parameters of the approximating surface which was generated in the $i-1$-th step. We find the minimum of the functional with the current parameters. If this approximant is satisfactory the algorithm stops, otherwise we add more degrees of freedom and continue the optimization. To simplify notations, we shall refer to quantities determined in the previous step using the symbol $\hat{\ }$, for example the previous smoothness parameter is denoted by $\hat{\lambda}$.

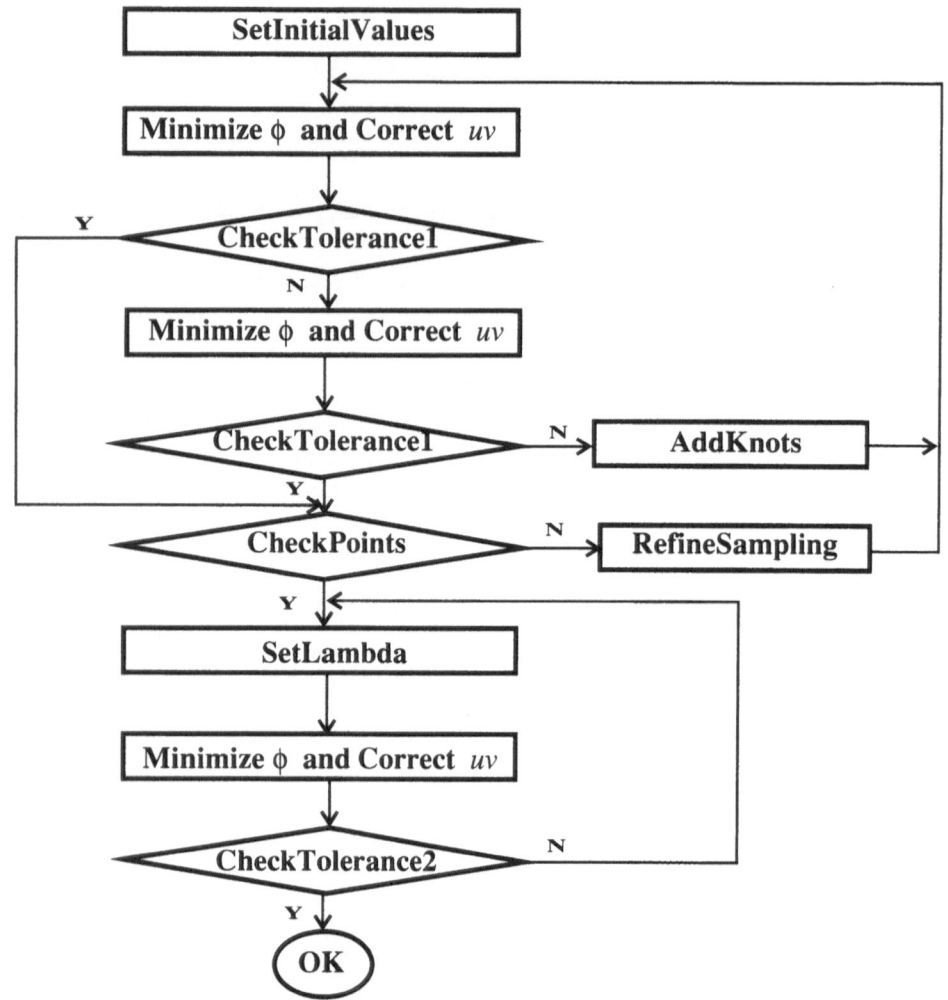

Fig. 5. Flowchart of the algorithm

The flowchart of the algorithm is depicted in Figure 5. After setting the initial values there are two parts. In the first part we adaptively approximate the data points until the largest distance will be smaller than the half of the prescribed tolerance $\epsilon/2$ to make room for the smoothing. In the second part we adaptively increase the smoothness factor λ, until we obtain the smoothest possible surface which approximates within ϵ.

SetInitialValues

Before we start iterations a coarse approximating surface needs to be created which provides an initial parametrization. First four corner points are identified. The surface can be either a parametric plane or cylinder, or if the boundaries of the patch can be roughly approximated a linear Coons patch is generated. Then the data points are projected onto the surface. Similar solutions were suggested by Ma and Kruth [16] and Dietz [6] as well. The next step is to specify the ignore areas. This can be done either in the parametric domain of the Coons patch or by projecting a 3D polyline into the Coons patch.

Once the unnecessary points are excluded an initial point set is selected. Due to efficiency reasons we deal only with a subsample of the data points which is extended later if necessary. The number of the data points may range from a few thousand to several hundred thousand or even more depending on the data acquisition process applied [22]. For example, with laser scanning, parts of a complex object are scanned in different directions. As a result, we get many point sets, within each there are parallel scanlines and often these overlap each other. To unify data a $p \times q$ uniform grid is created. In each cell there are several points, though only one, the median will be selected. In general points are weighted by 1. The boundary points play a special role, because the required accuracy is often higher there, so these are also selected and a greater "boundary weight" is assigned to them. The grid is important for refining the point sample later.

Before we start iteration we set the number of control points and the initial knot vectors U and V; uniform knots with multiplicities at the end are used. A small initial value for λ is set; in practice $\lambda = 10^{-10}, 10^{-12}$ was used. After this initial setting there is no further need to interact, the algorithm will find the most appropriate knot vectors, control points and related λ.

Minimize Φ and Correct   uv;

As explained above, the minimization of the objective function (1) actually minimizes distances between approximating surface points $\mathbf{S}(\hat{u}_k, \hat{v}_k)$ and the given points $\mathbf{P}_k$. As was also pointed out by Hoschek [14] to find a good parametrization is crucial for efficiency and surface quality. This is why the uv-parameters are modified to find new parameter values where the approximating surface is closest to the data point $\mathbf{P}_k$. For parameter correction a modification of a formula of Hoschek was used (see also Dietz [6]):

$$\begin{pmatrix} u_k \\ v_k \end{pmatrix} = \begin{pmatrix} \mathbf{S}_u^2 & \langle \mathbf{S}_u, \mathbf{S}_v \rangle \\ \langle \mathbf{S}_u, \mathbf{S}_v \rangle & \mathbf{S}_v^2 \end{pmatrix}^{-1} \begin{pmatrix} \langle \mathbf{P} - \mathbf{S}, \mathbf{S}_u \rangle \\ \langle \mathbf{P} - \mathbf{S}, \mathbf{S}_v \rangle \end{pmatrix} + \begin{pmatrix} \hat{u}_k \\ \hat{v}_k \end{pmatrix}$$

Following the parameter correction our tolerance check may be successful; if not we run the minimization again with corrected parameters and check the results again. If we fail, more degrees of freedom are needed to find a good approximant and new knots are inserted.

AddKnot

One of the knot vectors is modified. A possible strategy to insert knots is to evaluate the knot intervals by taking the sum of the squared errors divided by the corresponding points and insert a new knot halving the interval where the above sum is the largest. This solution has been found better than inserting knots at the highest deviation. (On automatic knot placement see Dierckx [5] and Cox [4]).

CheckTolerance1

As indicated before, here we check whether the largest deviation is smaller then $\epsilon/2$. If yes, another check is also performed:

CheckPoints

Each cell is subdivided into four and the distance of four additional points is checked. If for any reason the deviation is greater than $\epsilon/2$, the cell is refined. The corresponding weights of the points are divided by 4 accordingly. This illustrates why the point weights were introduced in (1). We can exclude points, we can emphasize points, for example along the boundaries, and we can also locally refine the point set where necessary. In the latter case the minimization will not be "distorted" by selecting too many points in certain areas.

SetLambda

Assume we have overcome knot insertion and sampling refinement and now the current surface approximates within $\epsilon/2$. Now we would like to gradually increase λ to get the smoothest possible surface. We simply apply

$$\lambda = const_\lambda * \hat{\lambda}$$

where $const_\lambda$ is a parameter to set the speed of change. We perform a similar minimization and parameter correction as before, the only difference is our terminating condition:

CheckTolerance2

When the largest deviation is found greater than ϵ, the iteration stops. Note, that the result surface will be the *last but one* surface, for which the above tolerance criterion was satisfied.

The above algorithm has been tested on simulated and real measured data from the automobile industry. As an example, a surface, similar to the.one schematically drawn in Figure 4 was generated with an ignore area. The number of data points is around 90000. Figure 6 - Figure 7 show that without a proper

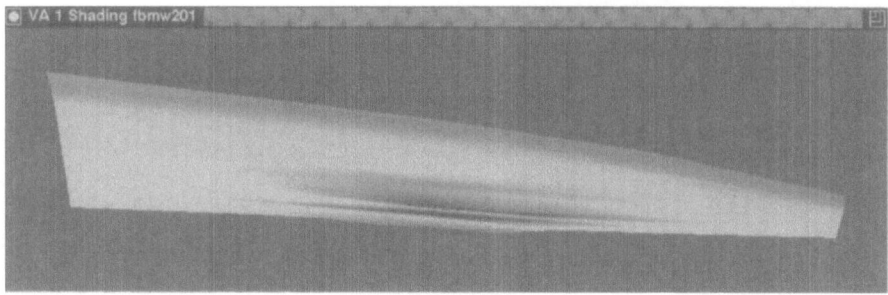

Fig. 6. Surface without smoothing: shaded image

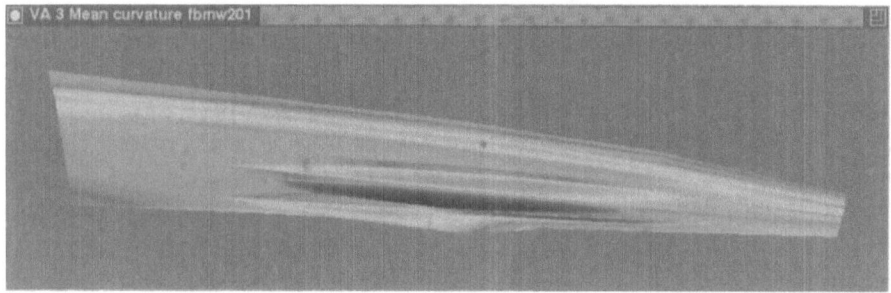

Fig. 7. Surface without smoothing: mean curvature

smoothing term we generated a surface which, while approximating the selected data points well, is horrible within the ignore area, so this would be unacceptable for functional decomposition (see mean curvature map). Figure 8 - Figure 9 illustrate, however, that applying the above algorithm the generated surface is smooth and pleasing within the ignore area as well.

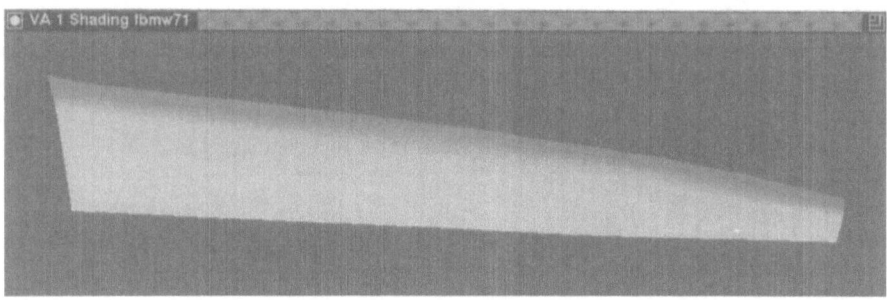

Fig. 8. Surface with smoothing: shaded image

Fig. 9. Surface with smoothing: mean curvature

4   Surface Fitting with Prescribed Positions and Normals

Our second problem is the following: given a set of data points $\mathbf{P}_k$ and a set of normal vectors $\mathbf{N}_k, (k = 1, \ldots, M)$. We would like to fit a fair surface which approximates the points within a prescribed positional tolerance ϵ and the normal vectors within an angular tolerance α. There are several objective functions to satisfy the latter constraint.

A natural aproach (this was also suggested by Dietz [6]) is the following:

$$F_1 = \sum_{k=1}^{M} \left(\langle \mathbf{S}_u(u_k, v_k), \mathbf{N}_k \rangle^2 + \langle \mathbf{S}_v(u_k, v_k), \mathbf{N}_k \rangle^2 \right)$$

This seems to be a good choice, since here the parametric derivatives are used and only a linear system of equations needs to be solved. The scalar product terms can vanish in two cases. Either the angle between the surface normals and the given normals becomes small, or the absolute value of the derivative vectors. Unfortunately, for this reason, the minimization of the above objective function may lead to uneven surface parametrization, particularly at the data points, but not necessarily between them. This may cause wiggling.

To avoid this effect we may introduce normalizing terms. The surface is unknown, but we have to normalize by related surface quantities. We have faced a similar problem when estimating the uv-parameters, and can apply the same method as before. Take the magnitudes of the derivatives at $(\hat{u}_k, \hat{v}_k)$ from the previous surface approximation and normalize accordingly, i.e.

$$F_2 = \sum_{k=1}^{M} \left(\left(\frac{\langle \mathbf{S}_u(u_k, v_k), \mathbf{N}_k \rangle}{|\mathbf{S}_u(\hat{u}_k, \hat{v}_k)|} \right)^2 + \left(\frac{\langle \mathbf{S}_v(u_k, v_k), \mathbf{N}_k \rangle}{|\mathbf{S}_v(\hat{u}_k, \hat{v}_k)|} \right)^2 \right)$$

Once we use estimates from the previous approximation, another objective function can be formalized, which is often better than F_2 according to our experience. Take the parametric derivatives of the previous iteration and project these into the plane determined by $\mathbf{N}_k$. In this way we obtain

$$\mathbf{TU}_k = \mathbf{S}_u(\hat{u}_k, \hat{v}_k) - \mathbf{N}_k \langle \mathbf{S}_u(\hat{u}_k, \hat{v}_k), \mathbf{N}_k \rangle$$
$$\mathbf{TV}_k = \mathbf{S}_v(\hat{u}_k, \hat{v}_k) - \mathbf{N}_k \langle \mathbf{S}_v(\hat{u}_k, \hat{v}_k), \mathbf{N}_k \rangle$$

With these estimated first derivative vectors we use the following simple objective function:

$$F_3 = \sum_{k=1}^{M} \left((\mathbf{S}_u(u_k, v_k) - \mathbf{TU}_k)^2 + (\mathbf{S}_v(u_k, v_k) - \mathbf{TV}_k)^2 \right)$$

Therefore a possible hybrid objective function is

$$\Phi = E + \mu F$$

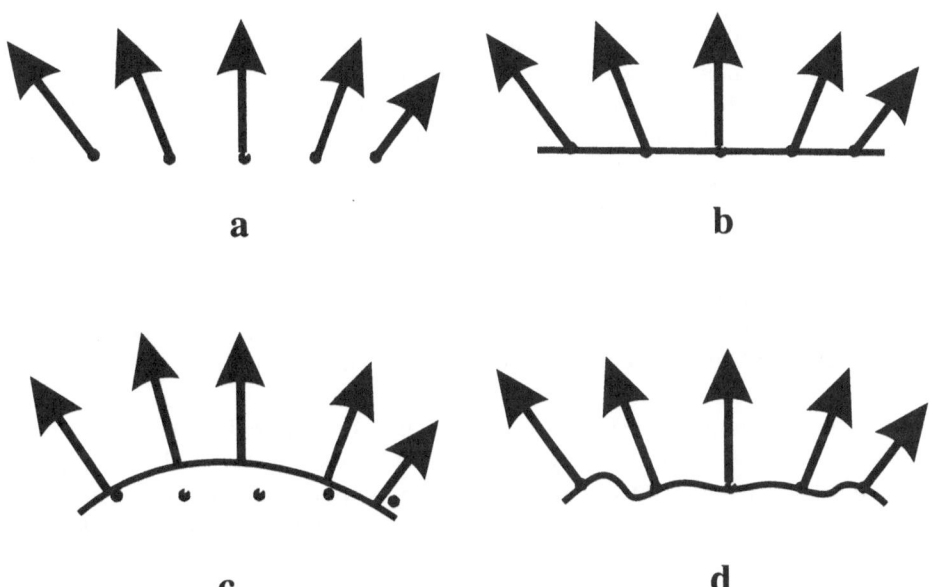

Fig. 10. Inconsistency of positions and normals

The simultaneous approximation of positions and normal vector raises the problem of what is consistent data. Clearly, with arbitrary given data and pre-scribed ϵ and α positional and angular tolerances, an acceptable solution does not necessarily exist. In Figure 10/a a simple example is shown. If the positional tolerance is tight, then we obtain Figure 10/b. If the angular tolerance is tight, we obtain Figure 10/c. If both of them are tight, a nonsense wiggling surface is obtained, which is unacceptable from an engineering point of view. These arguments force us to include a smoothing term like in the previous section.

$$\Phi = E + \mu F + \lambda J$$

The algorithm presented in Figure 5 needs to be modified in a straightforward manner. The module

`Minimize` Φ `and Correct uv`

is extended, and the first derivatives will be estimated there as well. In

`CheckTolerance1`

both the E and F terms need to be checked.

This problem emerges in several practical situations. Our investigations were motivated mainly by the design of light technical devices. For free-form reflector design the light distribution on a detector surface is often specified. There is a coarse initial reflector surface and some initial light distribution which needs to be improved. The filament of the bulb is modeled by a cylinder. By generating a sequence of filament images and repositioning them interactively to the "right" place on the detector, one can accumulate light densities approximately and thus "design" a light distribution. In this way a sequence of modified points and normal vectors are generated, and a new surface, which somehow approximates these constraints needs to be computed.

Figure 11/a shows a prescribed light distribution. Minimizing an objective function of the above type a new reflector surface was generated, Figure 11/b shows the corresponding light distribution. Different colours indicate the light intensity as shown by the colour bar. The isophotes of a reflector surface component and its Gaussian curvature map are shown in Figure 12 and Figure 13.

Similar techniques need to be applied to improve the reflection properties of aesthetic surfaces such as car bodies. Normal vectors at sampled positions are adjusted locally and then the whole surface is reapproximated. Here, as before, loosening positional tolerances and tightening angular tolerances will lead to surfaces with high quality.

5   Conclusion

Surface fitting is still a problem of practical importance. Key issues identified in this paper include the segmentation strategies of complex free-form shapes and the combination of different objective functions to obtain good approximations and fair shapes. The advantages of segmenting according to functional decomposition was emphasized, since this - although with a lot of interactive help - provides a geometric model, which reflects the underlying design intent, i.e. how the surface was possibly constructed.

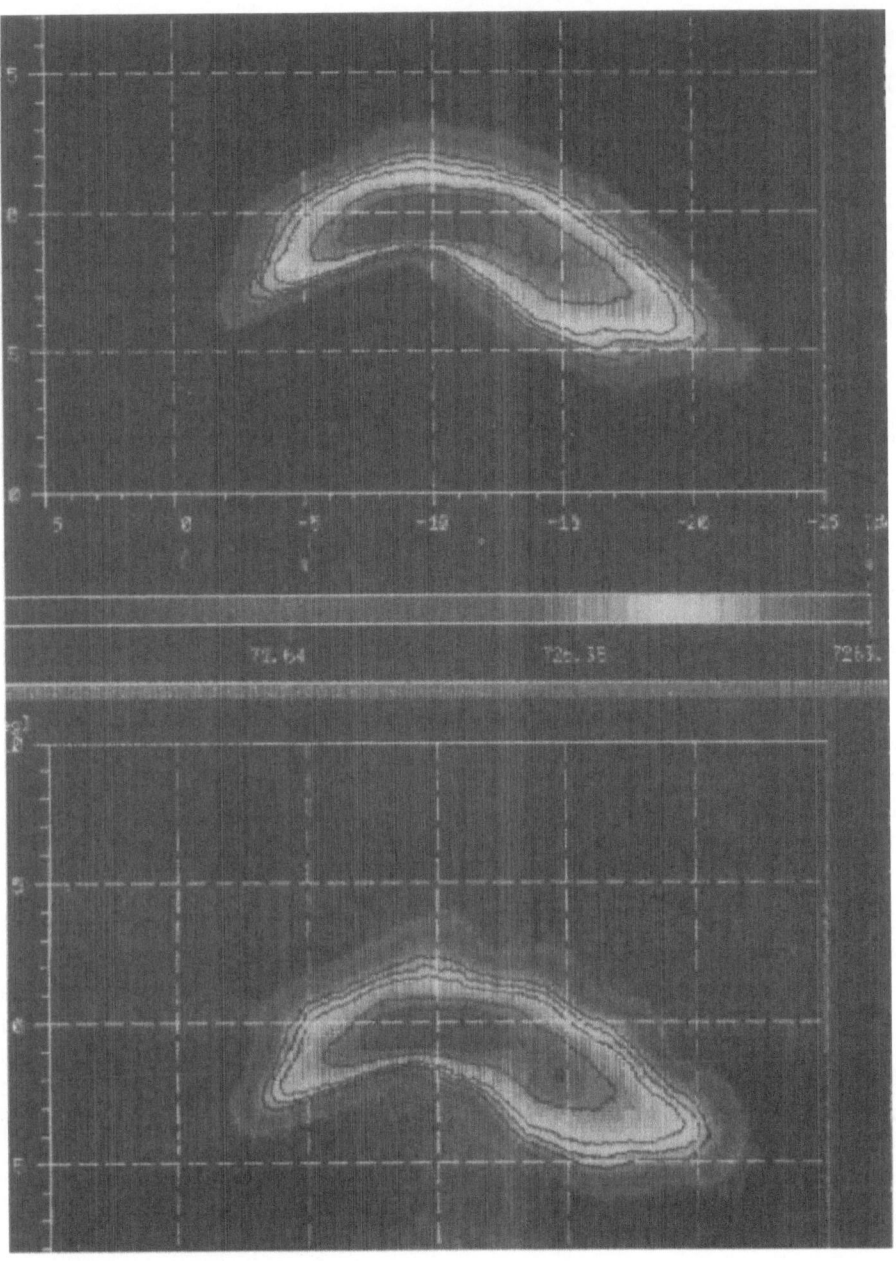

Fig. 11. Light distributions a) reference b) actual

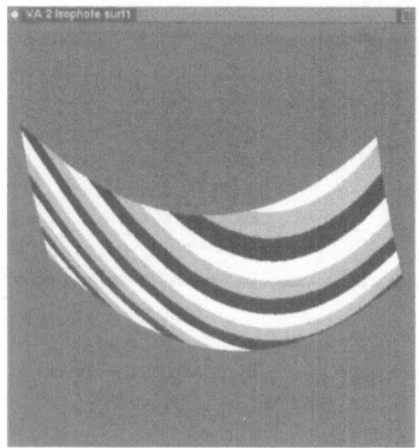
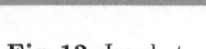

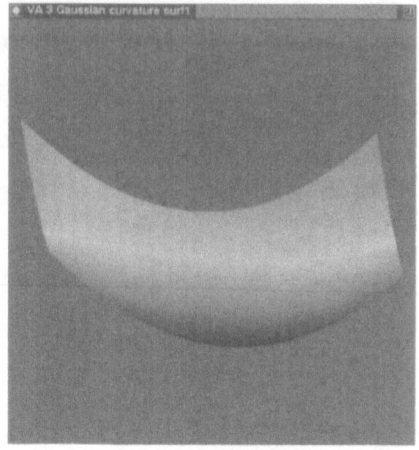

Fig. 12. Isophotes   **Fig. 13.** Gaussian curvature

Two special applications were investigated, the first surface fitting with ig-nore areas, the second simultaneous approximation of points and normal vectors. Several practical computational issues were raised while describing these algo-rithms.

The main stream in this research area is moving towards minimizing the amount of manual segmentation and automating the whole process as much as possible. To recognize connected or disconnected primary surfaces within a com-plex shape and fit free-form features between these are exciting open problems to be investigated further.

6   Acknowledgments

This research was supported by a COPERNICUS grant, RECCAD 94/1068 and a grant by the National Science Foundation of the Hungarian Academy of Sci-ences, No. 16420. Special thanks are due to Dr. Burkard Wordenweber (Hella KG Hueck & Co, Lippstadt, Germany), who directed our attention to the reflec-tor design problem. The light distributions in Figure 11 were generated by the Helios program of Hella.

References

1. G. Celniker and D. Gossard, Deformable Curve and Surface Finite Elements for Free-Form Shape Design, Computer Graphics (Procs of SIGGRAPH'91), **25** 4 (1991), 257–266.
2. M.G. Cox, The least-squares solution of overdetermined linear equations having band or augmented band structure, IMA Journal of Num. Anal., **1** (1981), 3–22.
3. M. G. Cox, Direct versus iterative methods of solution for multivariate spline-fitting problems, IMA Journal of Num. Anal., **2** (1982), 73–81.

4. M.G. Cox, Algorithms for spline curves and surfaces, in: L. Piegl (ed.), *Fundamental Developments of Computer-Aided Geometric Modeling*, Academic Press, (1993), 51–76.

5. P. Dierckx, An algorithm for surface-fitting with spline functions, IMA Journal of Num. Anal., **1** (1981), 267–283.

6. U. Dietz, Erzeugung glatter Flächen aus Meßpunkten, Teschnische Hochschule Darmstadt, Preprint-Nr. 1717, (1995).

7. M. Eck and H. Hoppe, Automatic Reconstruction of B-spline Surfaces of Arbitrary Topological Type, to appear in Procs. of SIGGRAPH'96, (1996).

8. G. Farin and N. Sapidis, Curvature and fairness of curves and surfaces, IEEE CGA, **9** (1989), 52–57.

9. D.R. Forsey and R.H. Bartels, Surface fitting with hierarchical splines, ACM Transactions on Graphics, **14** 2 (1995), 134–161.

10. M. von Golitschek and L. L. Schumaker, Data fitting by penalized least squares. in: J.C. Mason and M. G. Cox (eds.), *Algorithms for Approximation II*, Chapman and Hall, London, (1990), 210–227.

11. G. Greiner, Surface construction based on variational principles. in: P. J. Laurent, A. Le Méhauté, L. L. Schumaker (eds.), *Wavelets, Images, and Surface Fitting*, A K Peters, (1994), 277–286.

12. J. G. Hayes and J. Halliday, The least-squares fitting of cubic spline surfaces to general data sets, J. Inst. Maths. Applics., **14** (1974), 89–103.

13. J. Hoschek and D. Lasser, *Fundamentals of Computer Aided Geometric Design*, A K Peters, (1991).

14. J. Hoschek, Intrinsic parametrization for approximation, CAGD, **5** (1988), 27–31

15. *Reverse Engineering*, Ed.: J. Hoschek and W. Dankwort, Teubner, Stuttgart, 1996.

16. W. Ma and J. P. Kruth, Parametrization of randomly measured points for least squares fitting of B-spline curves and surfaces, CAD **27**(9) (1995) 663–675.

17. H.P. Moreton and C. Sequin, Functional optimization for fair surface design, ACM Computer Graphics (Procs of SIGGRAPH'92), **26** (1992), 167–176.

18. G. M. Nielson, Multivariate smoothing and interpolating splines. SIAM J. Numer. Anal., **11**(2) (1974), 435–446.

19. *Designing Fair Curves and Surfaces, Shape Quality in Geometric Modeling and Computer-Aided Design*, Ed.: N. S. Sapidis, SIAM, 1994.

20. R.F. Sarraga, Recent methods for shape optimization, General Motors, Research Publication 8477, February 1996.

21. D. Terzopoulos and H. Qin, Dynamic NURBS with Geometric Constraints for Interactive Sculpting, ACM Transactions on Graphics, **13** 2 (1994), 103–136

22. T. Várady, R.R. Martin and J. Cox, Reverse Engineering of Geometric Models - An Introduction, Geometric Modelling Studies, GML 1996/4, Computer and Automation Research Institute, Budapest, (to appear in Computer-Aided Design), 1996.

A Synthesis Process for Fair Free-Form Surfaces

Horst Nowacki[1], Feng Jin[2], Xiuzi Ye[3]

[1] Techn. University of Berlin, Division of Ship Design,
Salzufer 17–19, D-10587 Berlin, Germany
[2] SFE GmbH, Ackerstr. 71–76, D-13355 Berlin, Germany
[3] SolidWorks Corp., 150 Baker Ave. Extension, Concord, MA 01742, USA

Abstract. This paper describes a stepwise synthesis process for surface generation which leads from a set of prefaired curves, which form a regular or irregular mesh topology, via a mesh fairing process to a free-form surface that interpolates the mesh curves with G^1 or G^2 continuity and is subsequently faired based on some choice of variational fairness criterion. After briefly reviewing earlier work on the construction of a fair curve mesh the paper concentrates on two main ideas:

1. The interpolation of the given mesh topologies for given continuity and compatibility conditions with minimal polynomial degrees by local interpolants in Bézier form. Irregular, n-sided mesh cells are decomposed into assemblies of quadrilaterals.
2. A fairing process is applied to each patch. The free interior Bézier control points and twist vectors are determined by minimizing a chosen fairness measure. The effects of different variational fairness criteria on the shape and quality of the surface are explored.

The methods for surface construction are illustrated by representative examples.

1 Introduction

The industrial surface design process often proceeds from initially given representative point data via characteristic, faired curves and curve meshes to a free-form surface representation of the best possible fairness quality. This multistep procedure is usually favoured in industry because the initially given point and curve data are as relevant to the designer for the desired shape character as are the overall shape quality features of the resulting surface. The characteristic properties of the initial point and curve data must thus be retained when the surface is refined by a fairing process at the final stage. This stepwise process is a typical working procedure, e.g., in the automotive, aircraft and shipbuilding industries.

In this situation it is of advantage to combine interpolation and fairing methodologies in a concerted way. Interpolation is used to meet the hard constraints of the geometric design task ensuring the required continuities, fairing is applied to exploit the remaining freedom to improve certain quality properties of the shape. Fairing in our context means the improvement of some explicit fairness measure in face of given constraints.

In this article three stages of the synthesis process will be chiefly examined:

1. The construction of a faired curve mesh from the initial data. The topology of the mesh may be regular (with orthogonal mesh curves) or irregular (with arbitrary, n-sided mesh cells).
2. The interpolation of the mesh lines by a parametric surface of minimal degrees meeting either G^1 or G^2 continuity conditions at the patch boundaries as well as compatibility conditions at the nodes. A local construction with degrees 5×5 for G^1 and 9×9 for G^2 will be presented. This part is primarily based on Ye's work [1].
3. The interpolating patches still possess freedoms in terms of their interior control points to which a local fairing process is applied. The method will be illustrated by fairing bisextic G^1 continuous Bézier surfaces. The surface fairness measures are special cases of higher order variational criteria of quadratic form (Kallay and Ravani [2])which are varied in order to explore the effects of different functionals upon local and directional fairness properties. This part reflects the results from Jin's work [3].

The primary interest in this article will be on the new results presented on the topics (2) and (3) in Sect. 3.2 and 3.3. The stepwise synthesis process will be illustrated by examples from ship surface design.

2   Problem Formulation and Approach

A shape generation process serves to provide a shape definition for a curve or surface from given input data and some specified mathematical representation (Fig. 1). The process itself is described by a criterion or objective function and the given constraints. In this section we apply this unified type of problem formulation to all processes of curve or surface generation by interpolation or by approximation and fairing.

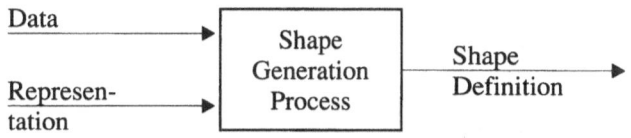

Fig. 1. Shape Generation Process

Consider the example of an interpolation process for a curve with given interpolation points and, say, a cubic parametric spline representation. It is well known that for such a spline curve an interpolant of second derivative continuity (C^2) at the segment boundaries can be constructed uniquely only if two further vector-valued conditions are applied. These maybe obtained from end conditions or, if desired, by minimizing the strain energy in the cubic spline or some other functional. The cubic spline interpolation process is thus slightly underdeterminate which may be exploited in order to improve the curve quality in terms of some chosen variational criterion.

More generally for a spline of n segments ($n+1$ interpolation points), degree d and desired continuity C^k at the knots the number of vector-valued freedoms f is

$$f = (d - k - 1)n + k, \qquad d \geq k + 1 \qquad (1)$$

That is, by raising the polynomial degree without simultaneously increasing the required continuity a sufficient number of freedoms can be obtained to allow fairing criteria to be applied to interpolating curves. For illustrations of this potential, see [4].

The same principle can be applied to surface construction. Thereby a flexible combination of interpolation and fairing requirements can be met. This enables the designer to achieve compromises between hard requirements and softer constraints. The resulting process may be characterized as 'interpolation with freedom'. It is a guiding motivation in this paper.

The steps in the synthesis process are summarized in Fig. 2. The surface construction proceeds from given point data to faired curves which are assembled to form a mesh. The curve mesh system is faired again. The resulting mesh lines are then interpolated ensuring the desired surface continuities across boundaries and compatibilities at the knots. If desired, the polynomial degrees of the surface representations can be raised beyond the minimum required for continuity. The extra freedoms are used in a surface fairing process by which the surface qualities in the patch interior can be improved without affecting the interpolated surface data.

Details about this stepwise process are given in the following section. Overall this synthesis process is destined to retain as much of the original raw data character as possible while still allowing consistent improvements in surface quality.

3   Methods and Results

3.1   Mesh Construction

Let us assume that the original point data set is initially transformed into a set of curves by some conventional interpolation or approximation process. Curve intersection points are interpolated and thus designated as mesh knots. The resulting curve set configuration will form a regular mesh of orthogonal lines or an irregular mesh with non-quadrilateral mesh cells. We will here concentrate on the regular mesh construction. The irregular mesh can be approached analogously at the mesh fairing stage; the subsequent surface interpolation differs as discussed in Sect. 3.2.

Given an initial regular curve mesh consisting of orthogonal lines $\overline{Q}_i(u)$ and $\overline{Q}_j(v)$, Fig. 3. According to Hosaka's mesh fairing method [5] this mesh can be regarded as a configuration of elastic splines with springs attached to the mesh

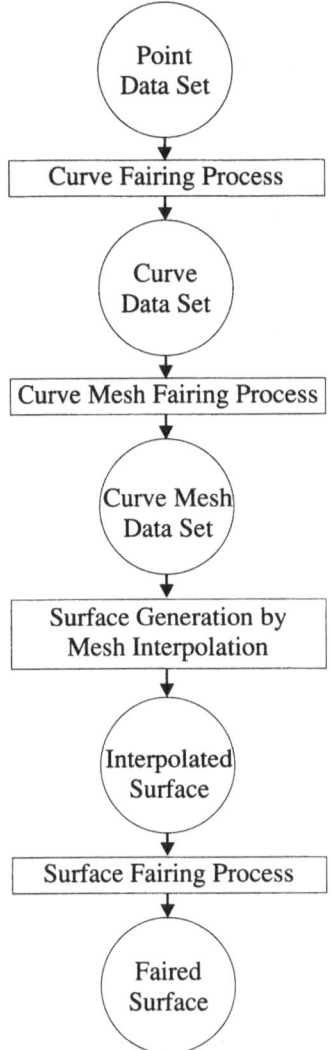

Fig. 2. Stepwise Surface Synthesis Process

knots. Then the strain energy in the spline system is equivalent to

$$L_2 = \frac{EI}{2} \left(\sum_i \int\limits_{\overline{Q}_i} \left(d^2 \overline{Q}_i(u)/du^2 \right)^2 du + \sum_j \int\limits_{\overline{Q}_j} \left(d^2 \overline{Q}_j(v)/dv^2 \right)^2 dv \right) \qquad (2)$$

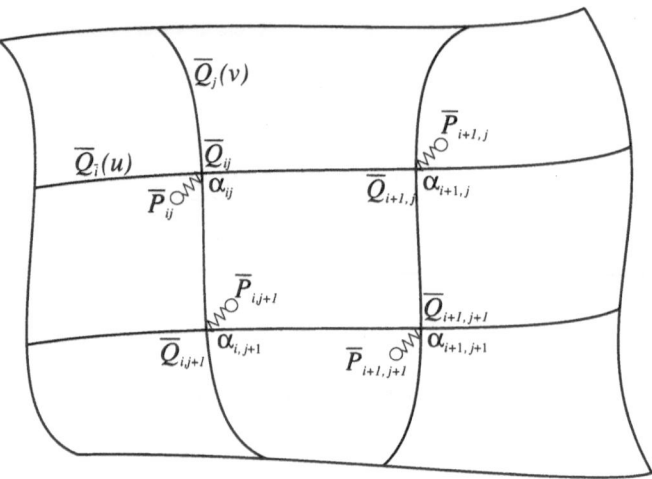

Fig. 3. Regular Curve Mesh

The potential energy resulting from the deformation of the springs of stiffness α_{ij} at knot $\overline{P}_{ij}$ is

$$U = \frac{1}{2} \sum \sum \alpha_{ij} \left(\overline{Q}_{ij} - \overline{P}_{ij} \right)^2 \tag{3}$$

In the mesh fairing problem, L_2 is to be minimized subject to the constraint

$$U \leq I_0, \quad I_0 = \text{tolerance on square error sum}$$

or $\quad A = U - I_0 \leq 0 \tag{4}$

The equivalent free variational form is

$$I = L_2 + \lambda(A - d^2) \qquad = \min \tag{5}$$

with $\quad \lambda = \text{Lagrange multiplier}$
$$d^2 = \text{slack variable}$$

The solutions to (5) yield a relaxed, prefaired curve mesh in which neighbouring curves are reasonably harmonized with each other. The process may be repeated if desired. Experiences are reported in [6].

3.2   Mesh Interpolation

Problem Statement. When a curvature continuous curve mesh has been generated, a smooth surface needs to be interpolated through the mesh curves. The desired continuity is either such that adjoining patches in the mesh will be joint with tangent plane continuity (G^1) or with curvature or equivalently Dupin indicatrix continuity (G^2). The G^1 requirement is mandatory in most applications to avoid knuckles for aesthetic and functional reasons. In some cases the higher quality of G^2 surface continuity is also required for physical or technical purposes.

The problem of mesh interpolation is a classical problem in computer aided geometric design. A good overview of the literature is given by Hoschek and Lasser [7]. The approach presented here is based on the work by Ye [1]. It differs from previous methods mainly by the surface construction, the low polynomial degrees achieved and the generality of the mesh topology for which the resulting Bézier surface patches are realized.

We are interested in nonsingular parametric surface representations. This is why the well-known solutions for Gregory patch interpolants, i.e., compatibly corrected Coons-Boolean sum interpolants, are not pursued here. Gregory proposed such surface interpolants for C^1 [8] and C^2 [9] continuous surfaces, however the resulting rational surface representations are singular at the patch corners. We rather favour constructions that result in nonsingular polynomial (or rational) Bézier patch surface topologies. The construction will be approached by first generating an auxiliary Coons-Boolean sum 'basic surface' that meets the continuity requirements along patch boundaries and compatibility conditions at mesh knots. This is then converted to a Bézier representation filling in any open choices left at that stage.

The development will begin with the special case of regular mesh topologies. The irregular mesh case will be addressed below.

More detailed information is found in Ye [1], Ye and Nowacki [10], Ye [12], Ye and Nowacki [14], Ye et al. [17].

Continuity and Compatibility. The condition for two surface patches $\overline{P}(u,v)$, $(0 \leq u,v \leq 1)$ and $\overline{Q}(u,v)$, $(0 \leq u,v \leq 1)$ to be G^1 continuous along their common boundary curve $\overline{C}(v) = \overline{P}(1,v) = \overline{Q}(0,v)$ is that there exist scalar functions $p(v)(> 0)$ and $q(v)$ such that (see Hoschek and Lasser [7], Ye et al. [11]) the first order cross boundary derivatives are related by:

$$\frac{\partial \overline{Q}(0,v)}{\partial u} = p(v)\frac{\partial \overline{P}(1,v)}{\partial u} + q(v)\overline{C}'(v) \tag{6}$$

where $'$ denotes the first order derivative.

The conditions for two surface patches $\overline{P}(u,v)$, $(0 \leq u,v \leq 1)$ and $\overline{Q}(u,v)$, $(0 \leq u,v \leq 1)$, to be G^2 along their common boundary curve $\overline{C}(v)$, as above, are that, in addition to (6), there exist scalar functions $pp(v)(> 0)$ and $qq(v)$ such that (see Hoschek and Lasser [7], Ye et al. [11]).

$$\frac{\partial^2 \overline{Q}(0,v)}{\partial u^2} = p^2(v)\frac{\partial^2 \overline{P}(1,v)}{\partial u^2} + 2p(v)q(v)(\frac{\partial \overline{P}(1,v)}{\partial u})' + \tag{7}$$
$$+ q^2(v)\overline{C}''(v) + pp(v)\frac{\partial \overline{P}(1,v)}{\partial u} + qq(v)\overline{C}'(v)$$

The 'weighting factor' functions $p(v), q(v), pp(v)$ and $qq(v)$, which are related to the coordinate transformations between parametrizations in adjoining patches (Hoschek and Lasser [7], Ye et.al [11]), cannot be arbitrarily chosen. They must comply with the compatibility conditions at the mesh knots where mesh curves meet. Mathematically, compatibility of the mixed partial derivatives (MPD) of the surface means independence of the sequence of differentiation. For G^1 surface constructions the twist vectors of the surface patches at the mesh knots must be made compatible (e.g. $P_{uv}(0,0)$). For G^2 surface constructions the compatibility of the (1,2), (2,1) and (2,2) mixed partial derivatives is also required. This imposes corresponding constraints on the vector-valued cross boundary derivative functions of a surface patch along its four boundary curves. These are taken into account in the construction of the weighting factor functions (see below and in Ye [1], Ye and Nowacki [14], Ye [12]).

G^1 Surface Construction. Mesh interpolants for regular meshes for G^1 continuity have been constructed by Sarraga [13], Ye [1],Ye and Nowacki [14] and others. They must meet the G^1 continuity and compatibility condition requirements as stated above. The compatibility condition results in a constraint on the functions $p(v)$ and $q(v)$, which in Ye's work was made explicit (Ye [1]). The construction may differ in the choice of weighting functions, hence different compatible solutions exist.

Sarraga [13] chooses a cubic polynomial $p(v)$ with

$$p'(0) = p'(1) = 0; \quad \text{and lets } q(v) = 0 \tag{8}$$

The vector-valued cross boundary derivative functions are set to be cubic Hermite interpolants of the endpoint tangents and twist vectors so that the resulting Bézier patches are bisextic. The twist vectors at the mesh knots are free. They are derived from the Coons-Boolean sum patch interpolating only boundary curves (Sarraga [13]). They might also be chosen so as to minimize a surface energy functional (Hagen et al. [15]).

Ye [1], Ye and Nowacki [14] follow a different approach which results in a biquintic G^1 continuous, compatible Bézier surface. The steps are briefly as follows:

1. The given piecewise cubic curve mesh is temporarily replaced by a basic C^1 cubic curve mesh which interpolates the mesh knots and preserves the slopes, though not the tangent vector magnitudes, at the knots.
2. The twist vectors at the mesh knots are estimated either by taking an average of the twist calculated from the assumed four Coons patches surrounding the knot or by minimizing a potential energy expression at the knot (Ye [1]).
3. Locally basic surface patches are constructed as bicubic Hermite patches interpolating the basic curves and the basic twist vectors. This basic surface is G^1 continuous.
4. The basic patches are now reparametrized to match the tangent vectors of the original curve mesh. From the transformations one derives the weighting

factor functions

$$p(v) = p_0(1 - v) + p_1 v \tag{9}$$

and $q(v) = $ cubic Hermite function with zero end values

$p_0, p_1, q'(0)$ and $q'(1)$ result from known end derivatives of the basic mesh curves and the transformations.

5. The first order cross derivative functions for the final surface are then obtained from (6). They are quintic vector-valued functions.
6. The final biquintic Bézier patches are then constructed as follows:
   - The given cubic boundary curves are degree elevated and converted into quintic Bézier curves.
   - the first order cross boundary derivative functions are also converted to quintic Bézier curves and serve to compute the first layer of interior Bézier points.
   - The remaining 2×2 central interior Bézier points remain free. They can be determined from a Coons-Boolean sum interpolation of the patch (Ye [1], Ye [14]) or by function minimization as in Sect. 3.3.

G^2 Surface Construction. The construction of G^2 continuous mesh interpolants takes into account the additional G^2 continuity and compatibility conditions stated above. The solution can be approached in analogous steps as for G^1 though allowing for the higher degrees of requirements.

In the spirit of Sarraga's G^1 method Nowacki, Kaklis and Weber [6] extended the solution to G^2 by letting $q(v) = 0$, $qq(v) = 0$ and satisfying:

$$p'(0) = p''(0) = p'(1) = p''(1) = 0 \tag{10}$$
$$pp'(0) = pp''(0) = pp'(1) = pp''(1) = 0$$

This scheme results in polynomial Bézier patches of degrees 15 by 15, a rather prohibitive magnitude. Only by skillful arrangements of patches in a chess-board pattern could the degrees be lowered to 5×15.

In order to provide a lower polynomial degree G^2 continuous interpolant Ye extended his G^1 approach using the following quite analogous steps (Ye [1], Ye [12]):

1. Construct basic C^2 quintic curve mesh, retaining slopes and curvatures at end points of original curves.
2. Estimate twist vector and higher order MPDs for mesh knots.
3. Construct basic surface patches as C^2 biquintic Hermite interpolants which are C^2 continuous.
4. Derive transformations betwen basic and original surface and construct cubic $pp(v)$ and quintic $qq(v)$ functions, keeping $p(v)$ and $q(v)$ the same as in the G^1 method.
5. Calculate first and second order cross derivative functions, which are of degree seven and nine, respectively.

6. Convert 9×9 Bézier form, constructing 4×4 interior Bézier points from additional assumptions.

Examples of the biquintic mesh and interpolated G^2 surface are shown in Figs. 4 and 5 for a ship hull surface.

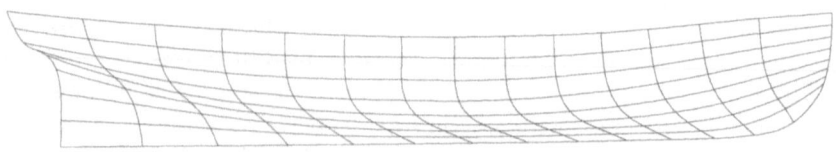

Fig. 4. A quintic curvature continuous curve mesh of a ship hull. Each mesh curve has its own chord-length parametrization.

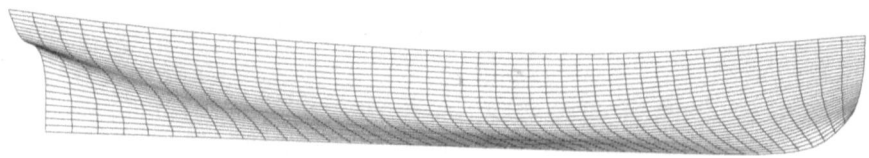

Fig. 5. The 9×9 curvature continuous ship hull generated from the curve mesh in Fig. 4 using Ye's local surface construction method

Irregular Mesh Interpolation. The approach developed by Ye [1] is equally applicable to irregular mesh topologies because the continuity conditions pertain to any boundary curve and the compatibility constraints can be posed so that they will apply to any number of mesh curves arriving at a knot. This characterizes an irregular mesh.

The essential idea adopted here is based on the fact that an n-sided patch (or hole) can be decomposed into n four-sided cells S_i by introducing a star point $\overline{P}$ in the interior that is connected with the (parametric) center points M_i of the sides (Fig. 6). The location of the star point is arbitrary, but must be judiciously chosen or adjusted because it has much effect on the resulting surface. The star lines l_i can be represented by cubic or quintic Hermite functions. Their derivatives at the midpoints are known from the cross boundary derivatives there. At the star point a suitable construction must be made to ensure that the patches meet there with the desired continuity and compatibility (G^1 or G^2). For G^1 surface construction, a solution can be achieved if the star lines meet at P in such a way that their end tangents are coplanar (for G^1) and their end normal curvatures are such that the curves are curvature continuous with the surface at $\overline{P}$. The star lines must therefore be estimated or corrected to meet this condition. A method for correcting the star lines to meet these conditions is presented by Ye and Nowacki [16]. For G^2 surface construction, the third order continuity of the star lines with the surface at $\overline{P}$ is required. A method to achieve this condition is presented by Ye and Nowacki [10].

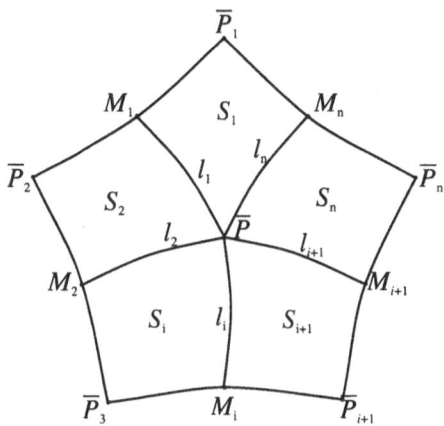

Fig. 6. Subdivision of n-sided patch into n four-sided patches S_i with star point P and star lines l_i

A general approach for dealing with n-sided patches in irregular meshes is developed by Ye [1]. The case of GC^1 multisided Bézier surface constructions is treated in Ye et al. [17].

The case of irregular meshes is thus reduced to dealing with four-sided Bézier patches within n-sided holes for which the continuity and compatibility conditions can be met.

Overview of Results. A general overview of nonsingular compatible mesh interpolating surface constructions is given in Tab. 1.

A summary of the main interpolants developed by Ye in [1] is shown in Tab. 2. It deals with G^1 and G^2 continuous, non-singular interpolants of regular and irregular meshes of individually parametrized curves achieving very low polynomial degrees and local patch constructions.

3.3 Surface Fairing

The continuous, compatible surface interpolated through a pre-faired curve mesh often still leaves room for improvement by a fairing process. The free variables at this stage may consist of the twist vectors and higher order mixed partial derivatives. In Bézier representation these freedoms are reflected in the twist vectors associated with the mesh knots and the free interior Bézier points of each patch.

Table 1. Surface Construction Interpolants

Method	Date	Degrees	Continuity	CBD	MPD Scheme
Sarraga[7]	87	6×6	G^1	(3, 6)	CBS
Nowacki, Kaklis, Weber [6]	92	15×15 or 5×15	G^2	(5, 15) or (15, 5)	free
Jin [3]	94	6×6	G^1	(6, 1) or (1, 6)	min. fairness measure
Ye [12]	96	9×9	G^2	(9, 9) or (5, 9)	CBS

Abbreviations:
CBD = cross boundary derivative function
CBS = Coons-Boolean sum construction
MPD = mixed partial derivatives

Table 2. Overview of Ye's Interpolants

Mesh Type	Surface Continuity	Polynomial Degrees	Integral (I) Rational (R)
Regular	G^1	5×5	I, R
Regular	G^2	9×9	I, R
Regular	Approximate G^2	7×7	R
Irregular	G^1, n-sided	n, 5×5	I
Irregular	G^2, n-sided	n, 9×9	I

There have been several approaches proposed for solving this post-fairing problem of G^1 continuous surface interpolants. Walter [18] first applied a curvature integral minimization to bicubic Coons surfaces. Nowacki and Reese [19] have proposed a strain energy based fairing process for fairing biquintic ship surfaces. Hagen and Schulze [15] developed a twist vector optimization also derived for a curvature minimization functional. Later Brunnett, Hagen and Santarelli [20] worked with the minimization of curvature variation, i.e., a third order derivative norm. Kallay and Ravani [2] resumed the twist optimization from an energy minimization viewpoint and also proposed a quadratic form of fairness measure which can be generalized. Table 3 gives an overview of earlier work on surface fairing.

Jin's work [3], on which this section is based, is an extension and generalization of the earlier methods. It pursues the following main thoughts:

– The surface representation must allow sufficient freedom for fairing. This is why 6×6 degree patches are favoured over lower degrees.
– The choice of fairness criterion should not be limited to second or third order derivative functionals. He explores the effects of criteria up to fourth order.

– In fact the fairness measures may be desired to have a directional orientation to permit stronger fairing in one direction while retaining the shape character in another.

Table 3. Surface Fairing Methods for G^1 Continuous Surface Networks

Method	Date	Degree	Free Variable	Order of Criterion	Direc- tional
Walter [18]	71	3×3	Twist	2nd	No
Nowacki, Rese [19]	83	5×5	Twist	2nd	No
Hagen, Schulze [15]	87	5×5	Twist	2nd	No
Brunnett, Hagen Santorelli [20]	93	5×5	Twist	3rd	No
Kallay, Ravani [2]	90	3×3	Twist	2nd	No
Jin [3]	94	6×6	Twist, MPDs	2nd to 4th	Yes

With these requirements a method was developed for fairing a G^1 bisextic Bézier surface in terms of the remaining freedoms after meeting the interpolation requirements of the given curve mesh. The approach is based on the following main thoughts:

A surface $\overline{Q}(u, v)$ is to be constructed interpolating a given regular mesh of G^2 quintic curves by a G^1 composite bisextic Bézier surface. The free variables in the surface representation are to be optimized based on some fairness functional J_k of order k:

$$J_k = \int \int \left(\overline{X}_k^T(u, v) \underline{\underline{M}}_k \overline{X}_k(u, v) \right) du dv \tag{11}$$

where
$\overline{X}_k^T$ = vector of k-th order partial derivatives of the surface $\overline{Q}(u, v)$
$\underline{\underline{M}}_k$ = coefficient matrix of quadratic form.

For example:

$$\overline{X}_2^T = (\overline{Q}_{uu}, \overline{Q}_{uv}, \overline{Q}_{vv})$$
$$\overline{X}_3^T = (\overline{Q}_{uuu}, \overline{Q}_{uuv}, \overline{Q}_{uvv}, \overline{Q}_{vvv})$$
$$\overline{X}_4^T = (\overline{Q}_{uuuu}, \overline{Q}_{uuuv}, \overline{Q}_{uuvv}, \overline{Q}_{uvvv}, \overline{Q}_{vvvv})$$

The matrix $\underline{\underline{M}}_2$ serves to select those terms in the quadratic form of the integrand and their coefficients which are chosen for a particular fairing purpose. In Jin's work [3] the following fairness criteria were explored as special cases of the quadratic form fairness functional used by Kallay and Ravani [2], for example:

$$J_2^{U2V0} = \int\int \overline{Q}_{uu}^2(u,v)\,dudv$$

$$J_2^{U2V2} = \int\int \left(\overline{Q}_{uu}^2 + 2\overline{Q}_{uv}^2 + \overline{Q}_{vv}^2\right)dudv \tag{12}$$

$$J_4^{U4V4} = \int\int \left(\overline{Q}_{uuuu}^2 + \overline{Q}_{uuuv}^2 + \cdots + \overline{Q}_{vvvv}^2\right)dudv$$

The first of these criteria is an example of a unidirectional fairness measure that is intended to fair the surface primarily in the u-direction.

The variables of surface representation can be decomposed into:

$$\overline{V} = \left(\overline{V}^{POS}, \overline{V}^{TAN}, \overline{V}^{INT}\right)$$

where
$\overline{V}^{POS}$ = positional information along curve mesh lines, given input
$\overline{V}^{TAN}$ = tangent vector cross boundary first derivative information, related to first interior layer of Bézier control points
$\overline{V}^{INT}$ = interior patch information, related to interior Bézier control points

Since the tangent vector information can be expressed in terms of the initially unknown twist vectors $\overline{V}^{TW}$ at the mesh knots, the free variables of the fairing problem are $\overline{V}^{TW}$ at all mesh knots and $\overline{V}^{INT}$ in each of the surface patches. This yields the fairing conditions

$$\frac{\partial J_k}{\partial \overline{V}^{TW}} = 0 \text{ for all mesh knots and} \tag{13}$$

$$\frac{\partial J_k}{\partial \overline{V}^{INT}} = 0 \text{ for all patches}$$

In a curve mesh there are $(M \times N)$ mesh knots and $(M-1)\cdot(N-1)$ patches. A bisextic Bézier patch has 3×3 interior control points that are not determined by positional and tangency constraints, hence free to be optimized. For large values of M and N it is therefore not attractive to solve the fairing problem simultaneously from (13). Rather it is desirable to separate the simultaneous equations into several sets of smaller systems. This is done by a mixed global and local approach for the twist vector set and the local interior control points of each patch, respectively.

To prepare the solution let us first consider a local patch (ij) and construct a G^1 transition between two adjoining patches, say, along a u-direction mesh line. We introduce a vector function, called reference tangent strip, interpolating the end cross tangents by a set of cubic Hermite function $\overline{H}^3(u)$

$$\overline{D}_{ij}^{REF}(u) = \left[\overline{P}_{0_{ij}}, \overline{t}_{0_{ij}}, \overline{P}_{1_{ij}}, \overline{t}_{1_{ij}}\right] \cdot \overline{H}^3(u)^T \tag{14}$$

where $\overline{t}_{0_{ij}}, \overline{t}_{1_{ij}}$ the initially unknown twist vectors.

This reference tangents strip is stretched in the v-direction by means of two scalar functions $\beta_{ij}^+(u)$ and $\beta_{ij}^-(u)$ for the upper and lower adjoining patch along u so that two G^1 tangent strips result

$$\overline{D}_{ij}^+(u) = \beta_{ij}^+(u)\overline{D}_{ij}^{REF}(u) \tag{15}$$

$$\overline{D}_{ij}^-(u) = \beta_{ij}^-(u)\overline{D}_{ij}^{REF}(u)$$

The two strips, which are tangent to the adjoining surfaces, ensure G^1 continuity. The scalar weighting functions $\beta_{ij}^{+/-}(u)$ are again cubic Hermite interpolants constructed to meet the compatibility constraint and to achieve interpolation of the end cross tangents of the curve mesh. This construction establishes a relationship between the tangent vector variables of the strip and the twist variables associated with the patch

$$\overline{V}_{ij}^{TAN} = h(\overline{V}_{ij}^{POS}, \overline{V}_{ij}^{TW}) \tag{16}$$

Let us now develop an approach for a global optimization of the twist vector set. For a knot (ij) all (four) surrounding patches contribute to the fairness functional J_k. The minimization condition is

$$\frac{\partial J_k}{\partial \overline{V}_{ij}^{TW}} = 0 = f_k\left(\overline{V}_{ij}^{POS}, \overline{V}_{ij}^{INT}, \overline{V}_{ij}^{TW}\right) \tag{17}$$

$\overline{V}_{ij}^{POS}$ and $\overline{V}_{ij}^{INT}$ refer to the four surrounding patches and $\overline{V}_{ij}^{TW}$ to the twist vectors at knot (ij) and its eight immediate neighbors.

The set of interior variables $\overline{V}_{ij}^{INT}$ is also unknown and shall be derived from the minimization condition

$$\frac{\partial J_k}{\partial \overline{V}_{ij}^{INT}} = 0, \qquad J_k = J_k\left(\overline{V}_{ij}^{POS}, \overline{V}_{ij}^{TAN}, \overline{V}_{ij}^{INT}\right) \tag{18}$$

This system of equations can be formally inverted to yield a relationship between the tangent vector variables and the locally optimized interior variables (interior 3×3 Bézier points):

$$\overline{V}_{ij}^{INT} = g_k\left(\overline{V}_{ij}^{POS}, \overline{V}_{ij}^{TAN}\right) \tag{19}$$

Substituting (16) $\overline{V}_{ij}^{TAN}$ can be replaced by the twist vector set:

$$\overline{V}_{ij}^{INT} = \overline{g}_k\left(\overline{V}_{ij}^{POS}, \overline{V}_{ij}^{TW}\right) \tag{20}$$

Equations of this type are obtained for all knots (ij). When combined the system

$$\frac{\partial J_k}{\partial \overline{V}^{TW}} = 0 = f_k\left(\overline{V}_{ij}^{POS}, \overline{V}_{ij}^{TW}\right) \tag{21}$$

results in a $(M \times N)$ linear equation system for the $\overline{V}_{ij}^{TW}$ with a tridiagonal coefficient matrix. This system serves for the global twist vector set optimization.

Obviously the formal derivations and inversions described in the foregoing are very complex. In practice they were performed by the symbolic formula solver system MATHEMATICA. This has provided explicit solutions to all functional relationships cited above [3]. Clearly the results decisively depend on the choice of fairness measure.

The algorithm for surface fairing through a given curve mesh based on these results proceeds as follows:

1. Degree elevate the given mesh curves to degree 6.
2. Optimize the twist vectors for the tangent strips based on (21).
3. Evaluate the tangent vector, first layer Bézier control points from (16).
4. Calculate the optimal interior Bézier points from (20) for each patch.

This results essentially in one $(M \times N)$ equation system in step 2 and the evaluation of several, already inverted systems in steps 3 and 4. The local systems in step 4 involve $3 \times 3 = 9$ Bézier control points in each patch.

4   Fairing Examples

Figure 7 shows a prefaired curve mesh (a) for a G^1 ship hull surface and the interpolated and faired surface (b) based on the criterion J_4^{U4V4}. The following figures visualize results for a segment of this ship for the same curve mesh, though with different fairness measures. Each figure set compares the null twist vector solution with results based on fairness measures J_2^{U2V0}, J_2^{U2V2}, J_4^{U4V4}.

Figure 8 visualizes isolines of normal vector z-components. In Fig. 9 the mean curvature distributions are illustrated. In both figures the null twist solution is clearly of lowest quality. In Fig. 9 for the mean curvatures it also shows patterns of diagonal wrinkles in several patches which are known from certain commercial CAD systems using null twists. The three other solutions are improved by fairing. The $U4V4$ case appears to be best in this example. This encourages the use of higher order fairness norms. The unidirectional fairing does not differ much from $U2V2$ in this case. Apparently the global twist vector set optimization creates essential improvements, the fairing of interior control points contributes further minor refinements. Further exploration of the multitudinous options of this scheme is much desired.

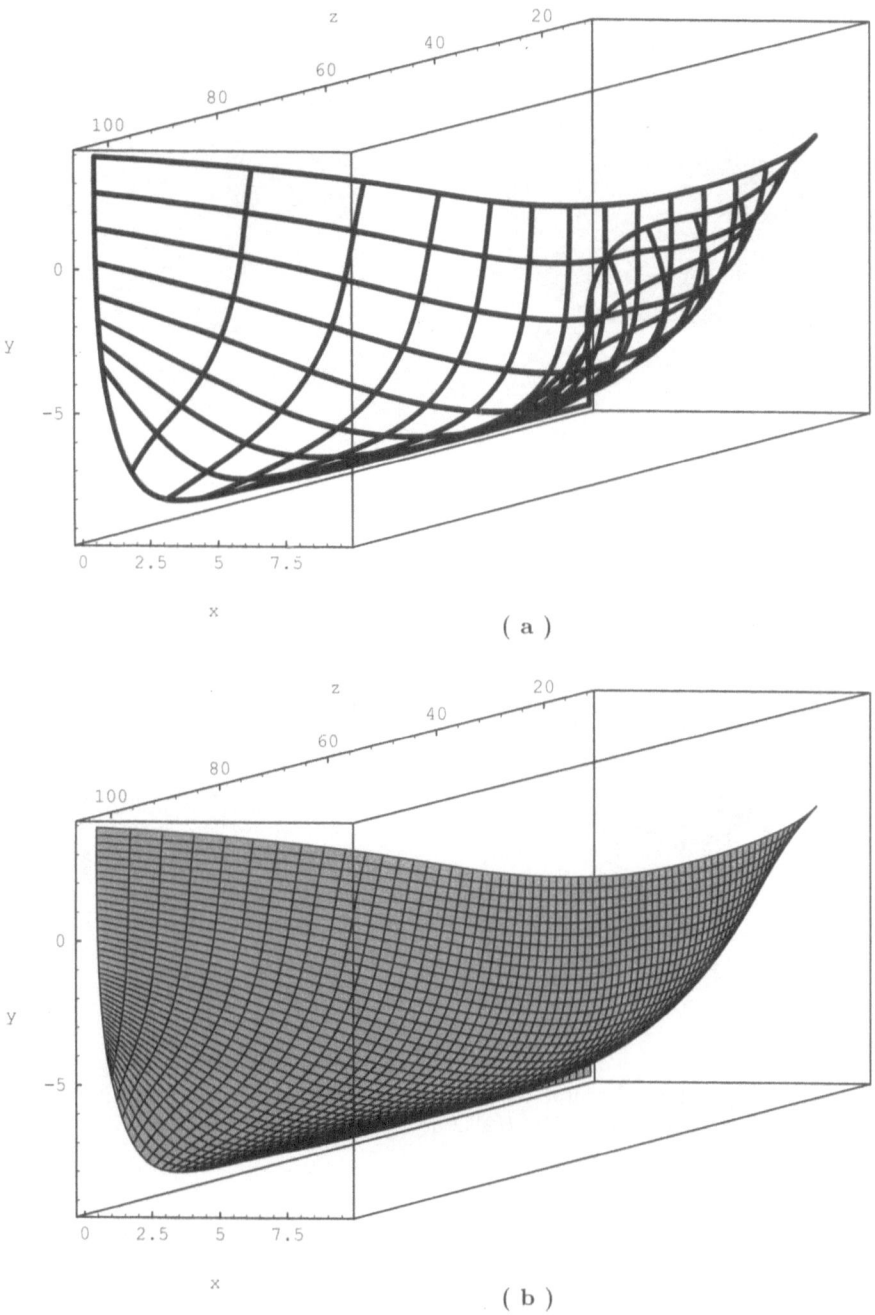

Fig. 7. Ship Surface Generation:
(a) Initial, prefaired curve mesh, (b) Faired surface, $U4V4$ norm.

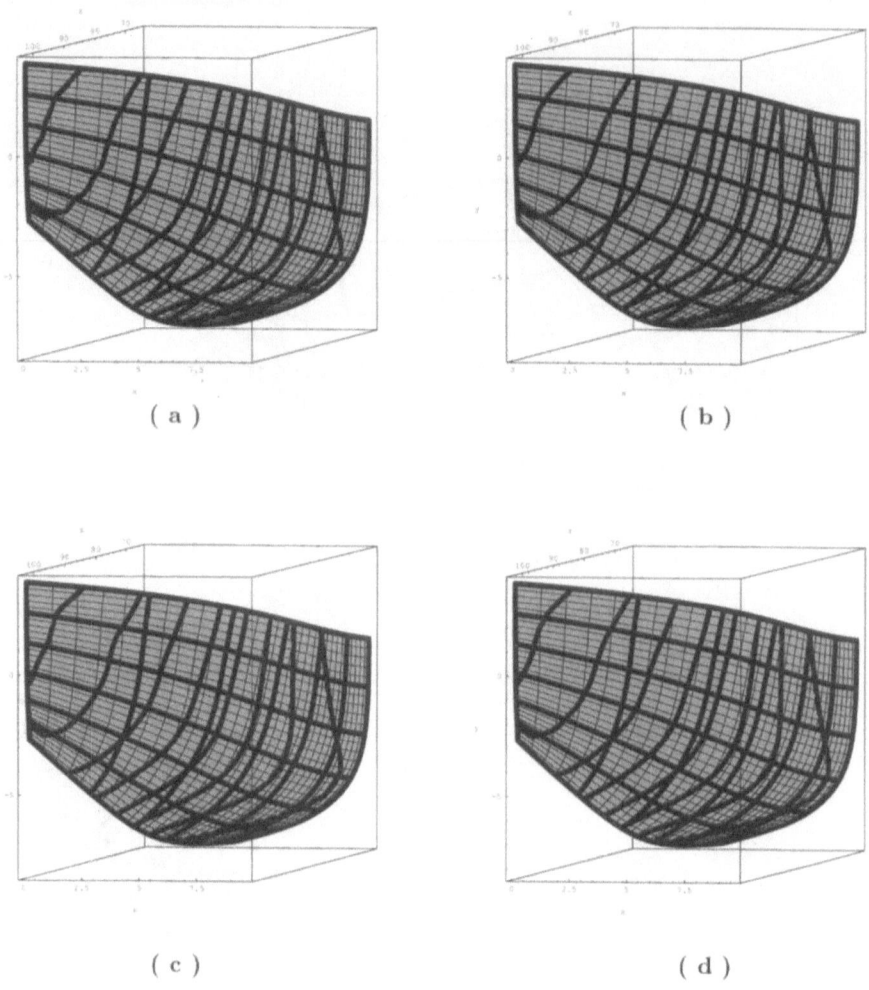

Fig. 8. Isolines of normal vector z-component of ship hull surface region: (a) Nulltwist, (b) $U2V0$ norm, (c) $U2V2$ norm, (d) $U4V4$ norm.

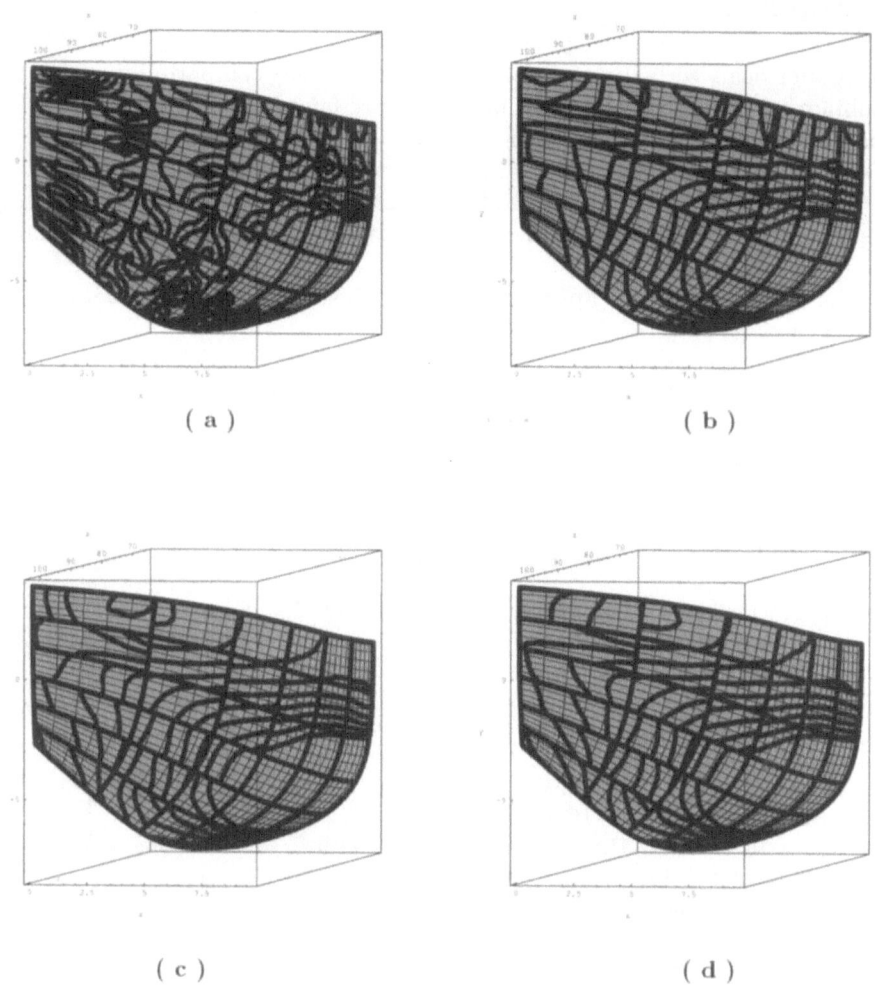

Fig. 9. Isolines of mean curvature:
(a) Nulltwist, (b) $U2V0$ norm, (c) $U2V2$ norm, (d) $U4V4$ norm.

5   Summary

This paper deals with the construction of nonsingular, compatible surfaces by a combination of interpolation and fairing. A pre-faired curve mesh is first interpolated continuously and compatibly at G^1 or G^2 levels. Low polynomial degree interpolants are presented. Subsequently, for the G^1 bisextic composite Bézier surface case, a post-fairing process globally for the twist vector set and locally for the interior Bézier control points of each patch is developed. Both methodologies are illustrated by examples.

The two stages of this approach combined ensure a good compromise in two objectives:

1. To achieve a high fidelity to the initial data set with secure surface continuity.
2. To realize a high level of fairness quality in the higher order derivative distributions with or without directional priorities.

References

1. Ye, X.: Construction and Verification of Smooth Free-Form Surfaces Generated by Compatible Interpolation of Arbitrary Meshes. Dr. Köster Publ., Berlin (1994)
2. Kallay, M., Ravani, B.: Optimal Twist Vectors as a Tool for Interpolating a Network of Curves with a Minimum Energy Surface. Computer Aided Geometric Design **7** (1990) 465–473
3. Jin, F.: A Knowledge Based System for the Design and Fairing of Free-Form Surfaces with Tangent Plane Continuity. In German, dissertation, Technical University of Berlin **D83** (1994)
4. Meier, H., Nowacki, H.: Interpolating Curves with Gradual Changes in Curvature. Computer Aided Geometric Design **4** (1987)
5. Hosaka, M.: Theory of Curve and Surface Synthesis and Their Smooth Fitting. Information Proc. Japan **9** (1969) 60–68
6. Nowacki, H., Kaklis, P.D., Weber, J.: Curve Mesh Fairing and GC^2 Surface Interpolation. Modélisation Mathématique et Analyse Numérique, AFCET Gauthier-Villars, Paris vol. **26** (1992) 1, 113–135
7. Hoschek, J., Lasser, D.: Fundamentals of Geometric Data Processing. In German, 2nd ed., Teubner Publ., Stuttgart (1992)
8. Gregory, J.A.: Smooth Interpolation without Twist Constraints. in Barnhill, R.E., Riesenfeld, R.F. (eds.): Computer Aided Geometric Design. Academic Press, New York (1974) 71–88
9. Gregory, J.A., Hahn, H.M.: A C^2 Polygonal Patch. Computer Aided Geometric Design **6** (1989) 69–75
10. Ye, X., Nowacki, H.: Ensuring Compatibility of G^2-Continuous Surface Patches around a Nodepoint. Computer Aided Geometric Design **20** (1996) (to appear)
11. Ye, X., Liang, Y., Nowacki, H.: Geometric Continuity between Adjacent Bézier Patches and Their Constructions. Computer Aided Geometric Design **20** (1996) (to appear)
12. Ye, X.: Curvature Continuous Interpolation of Curve Meshes. Computer Aided Geometric Design **20** (1996) (to appear)

13. Sarraga, R.F.: G^1 Interpolation of Generally Unrestricted Cubic Bézier Curves. Computer Aided Geometric Design **4** (1987) 23–39
14. Ye, X., Nowacki, H.: G^1 Interpolation of Rectangle Cubic Curve Meshes Using Biquintic Bézier Patches. Sixth IMA Conf. on Mathematics of Surfaces, in Mullineux, G. (ed.): Mathematics of Surfaces VI. Oxford University Press, U.K. (1994) 429–452
15. Hagen, H., Schulze, G.: Automatic Smoothing with Geometric Surface Patches. Computer Aided Geometric Design **4** (1987) 231–236
16. Ye, X., Nowacki, H.: Optimally Tangent-Plane and Curvature Continuous Modification of Curves at a Common Nodepoint. Proc. ASME 1995 Conf. on Advances in Design Automation, vol. **1**, DE-vol. **82** (1995)
17. Ye, X., Nowacki, H., Patrikalakis, N.: GC^1 Multisided Bézier Surfaces. Engineering with Computers (1996) (to appear)
18. Walter, H.: Numerical Representation of Surfaces Using an Optimum Principle. In German, dissertation, TU Munich (1971)
19. Nowacki, H., Reese, D.: Design and Fairing of Ship Surfaces. Surfaces in CAGD, Barnhill, R.E., Boehm, W. (eds.), North Holland, Amsterdam (1983)
20. Brunnett, G., Hagen, H., Santorelli, P.: Variational Design of Curves and Surfaces. Survey on Mathematics for Industry, Springer (1993) 1–27

Computing Minimal Surfaces with Particle Systems

Bernd Eberhardt[1]

WSI/GRIS, University of Tübingen,
Auf der Morgenstelle 10, 72076 Tübingen, Germany
email: beberh@gris.uni-tuebingen.de

Abstract. We describe the use of a particle system to compute minimal surfaces. This approach calculates the dynamical behavior of the surface, supports the blending of solids with rubber-like materials and makes modeling easy.

1 Introduction

Minimal surfaces have attracted mathematicians for centuries. But not only mathematicians are fascinated by the wondrous shapes of soap-bubbles. There is hardly anyone who hasn't played himself with soap-bubbles and made experiments while sitting in the bathtub.

Many famous scientists are linked to the subject, Euler, Lagrange, Monge, or Poisson just to mention some. Some made a research project out of the subject. An endless number of experiments with soap-films was made by a Belgian professor of physics and anatomy: Joseph Plateau (1801-1883). It is astonishing however, that his experiments serve as an impetus now for more than one hundred years in a growing branch of science, from his times into the present days[1]. His name is fixed to one of the most famous problems in mathematics; and it was Lebesgue[2] who stated first **Plateau's-Problem**:

To prove that for each closed curve $C \in \mathbb{R}^3$ there exists a surface S of minimal area with C as boundary.

A mathematical proof on existence and uniqueness of this problem can be found in [3]. It was in 1930 that independently two mathematicians found a proof, Douglas and Radó. Despite of the fact that minimal surfaces are still a fascinating topic of research in mathematics (differential geometry), You may still win a prize, if you find a new one. Minimal surfaces (in some sort) may be found in every-days life. They are widely used in architecture, for instance the famous roof of the Olympic stadium of Munich, as sun-roofs of pavilions and last but not least, your umbrella at home.

The approach presented here uses a simple particle model to compute minimal surfaces. The calculation is fast and allows interaction with obstacles, it is exact and calculates the dynamical behavior of the surface.

2   The Mathematics

In calculus the first year graduate student is forced to learn about the "length" of a curve $\gamma(t)$ in space. He will recall that the number

$$l = \int \left| \frac{\partial}{\partial s} \gamma(s) \right| ds$$

is the length of such a curve. Later in physics he will meet another so-called functional — $\int |\dot{\gamma}|^2 \, dt$ — the action functional. Both in common is the fact that we associate a real number to a one-parameter function $\gamma : [0, a] \mapsto \mathbb{R}^3$ in space. Now we consider a parameterized surface, i.e. a function in space with two parameters s and t,

$$S : (s, t) \in [0, a] \times [0, b] \mapsto \mathbb{R}^3 \quad .$$

Here we will deal with yet another functional:

$$A[S] = \int_0^a \int_0^b \sqrt{1 + S_s^2 + S_t^2} \, dt \, ds \qquad (1)$$

the area functional. The functions S_s and S_t are the partial derivatives of S with respect to s and t.

A long time passed since Euler and Lagrange studied problems like this, given a functional f, i.e. $A[f] = \int f(x, y, y') \, dx$. If we have to find a minimum or maximum of A at f, then f has to fulfill the following equation:

$$f_y - \frac{d}{dx} f_{y'} = 0$$

This equation is known as the Euler-Lagrange differential equation. The usual approach solving Plateau's-problem is via finite elements. However from a modelers point of view this approach is not natural and lacks the experimental side. In finite element methods each new problem (changed boundaries, fixed points etc.) needs a new implementation of the partial differential equation. In addition dynamical simulations of highly flexible materials are difficult to make using finite elements. Like Plateau we want to make experiments, without modifying too much in our code. We want to see the dynamical behavior of our membrane. Thus we will use another method and, as you hopefully will see in the examples section, you are invited to play around with minimal surfaces as if you are sitting in the bathtub. In addition we can easily add collision detection.

3   How to Compute Soap-Like Surfaces

As we have seen, minimal surfaces are determined by their fixed boundary and the property of a locally minimal surface. The catenoid and Eneper's-surface [4] where the first found minimal surfaces. The catenoid in particular is easily

Fig. 1. Meshing of a cylinder

calculated and given by an explicit formula (see below equation (3)). Thus we can compare our results with the explicit representation of the catenoid. We start with an experiment: Let us assume we have given a cylinder. What will happen if we fix the boundaries of the cylinder and calculate the minimal surface? To do this we take a (rectangular) subdivision of the cylinder. Each crossing will represent a particle which is coupled to its neighbors via the meshing. Note, that the subdivision of the cylinder is slightly disturbed, since otherwise we will get numerical instabilities in solving the differential equation. Thus we will assume spring-forces from each particle to its neighbors of the form:

$$F(x_i) = \sum_{\substack{\text{neigh.}}}^{n_i} \frac{1}{2} m_i C_t (d_i - d_{n_i})^\alpha \qquad (2)$$

Where the C_t is the tension coefficient, d_i the distance to neighbor n_i and α a fixed coefficient ($\alpha = 1$ Hooke's Law). By this force we can model almost all rubber-like materials and in the case of soap we have to choose $\alpha = 0.5$ since the force is proportional to the area [5]. The number d_{n_i} may be the unstretched distance to neighbor n_i. To calculate minimal surfaces in particular, we will set $d_{n_i} = 0$ for most particles to assure that the distance to neighbor n_i will be small and hence the area of the mesh-element.

In [6] we described the use of a particle system to calculate the draping of textiles. We use the same setting here to calculate dynamically the minimal surfaces. I.e. we add another time variable t and get via a numerical solution of the Euler Lagrange-differential equation

$$\frac{d}{dt}\left(\frac{\partial L}{\partial v_i}\right) = \frac{\partial L}{\partial x_i} \quad , \text{ where } \quad L = E_{kin} - \sum_i^n \int_0^{x_i} F(s)\, ds$$

is the Lagrange-function, a simulation of a "minimal surface".

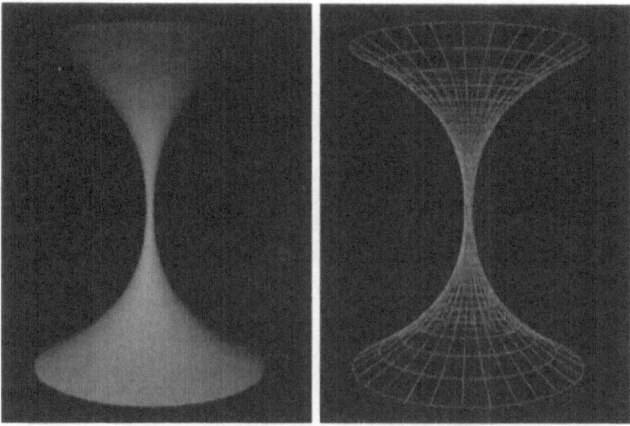

Fig. 2. Catenoid

It is clear at this point that using the force F from (2), we do not get a purely "minimal" surface since our approach minimizes a functional L different than the area functional (1). The influence of the coefficient α in equation (2) on the shape of the surface can be seen from (Fig. 3). The coefficient C_t does

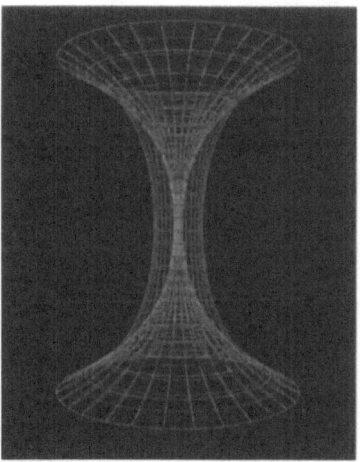

Fig. 3. Elasticity exponent $\alpha = 1$(yellow) and $\alpha = 2$(red).

not affect the shape of the surface for a fixed α (see Fig. 4), which might be astonishing on the first sight. However, if you think about it, it will become clear: Our functional L minimizes not only the area but also the used tension energy of the material, imposed by the force F from formula (2). The energy

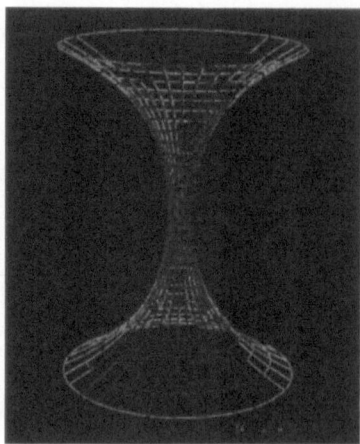

Fig. 4. Mixed C_t for $\alpha = 1$ $C = 6$(darker) and $C = 60$(brighter).

which can be "stored" in the shape of the surface is independent of the strength of the springs, i.e. our C_t.

Even in the case of Plateau's experiments, soap-films realized via dipped wireframes, will minimize the surface tension and hence are not just surfaces with minimal area. We therefore have to know the specific tension forces of the material we want to model.

In the case of soap-films we will get the well known Catenoid-Surface (cf. Fig. 5), which can be given by an explicit formula:

$$S(u, v) = (a \cosh(v) \cos(u), a \cosh(v) \sin(u), av) \tag{3}$$

where $0 \le u < 2\pi$ and a is the minimal diameter.

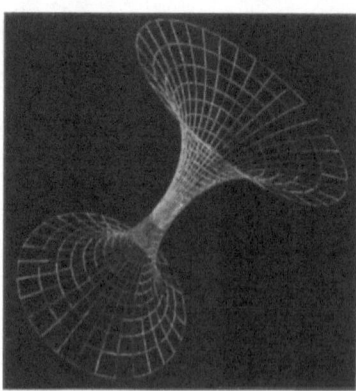

Fig. 5. The simulated Catenoid with $\alpha = 1$(darker) and the calculated explicit formula for the Catenoid (brighter).

4   More Plateau's Experiments

We are now in the situation as sitting in the middle of a playground. Like Plateau we can make numerous experiments with minimal surfaces. And we can do more: Let us first fix not only whole boundaries but just six points in space and see which minimal surface will appear. This surface can't be realized with soap-

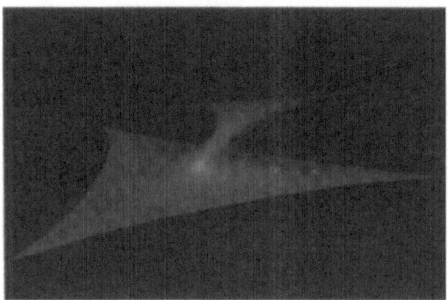

Fig. 6. An experiment.

films. Our approach is useful to blend objects with rubber-like materials in CAD systems. For this we combine the calculation module within our particlesystem with a collision detection module. This is a simple task in our implementation of a C++ particle system library. The calculation time for this last example is

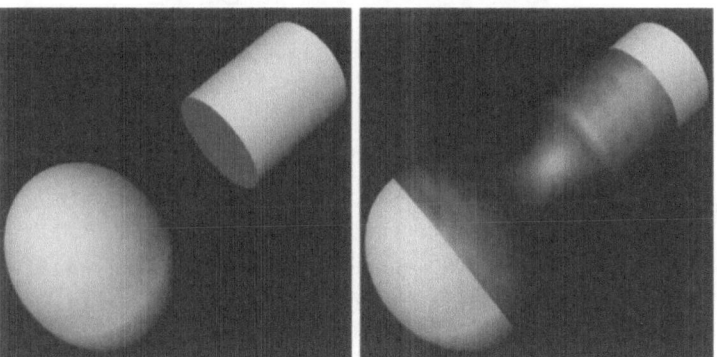

Fig. 7. Blending two obstacles.

less than 2 minutes on a SGI-Indy 134 MHZ with R4610 Floating Point Chip.

Moreover, we can calculate the tent-roof of a pavilion, or the roof of the Olympic stadium in Munich. Here we fix a set of points and use the triangulation of a given surface. The calculation is dynamic as in all of our simulations.

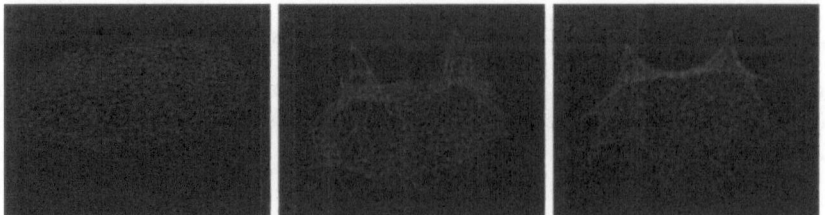

Fig. 8. Calculating the roof of a pavilion.

Last but not least you may find on our playground an object from everydays life:

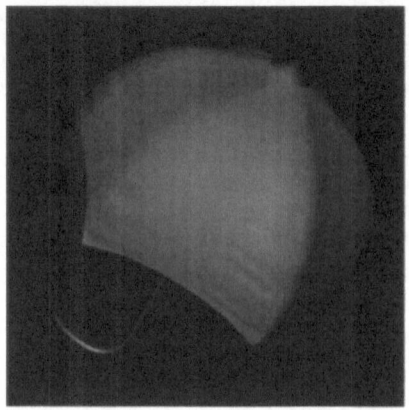

Fig. 9. An umbrella.

References

1. U. Dierkes and G. Huisken. The n-dimensional analogue of the catenary: Existence and non-existence. *Pac. J. Math. (to appear)*.
2. H. Lebesgue. Intégrale, longueur, aire. *Ann. Mat. pura appl.*, 7(3):231–359, 1902.
3. B. Lawson. *Lectures on Minimal Submanifolds*. Monographias de Mathḿatica, IMPA, Rio de Janeiro, 1973.
4. M. Spivak. *A comprehensive Introduction to Differential Geometry*, volume III. Publish or Perish, Inc., Berkeley, 1979.
5. Hermann Franke, editor. *Lexikon der Physik*. Franck'sche Verlagshandlung, Stuttgart, Germany, 1969.
6. B. Eberhardt, A. Weber, and W. Strasser. A fast, flexible, particle-system model for cloth draping. *IEEE Computer Graphics & Appl.*, 16(5):52–60, September 1996.

Orthogonality Relations for Cardinal B-Splines over Bounded Intervals

Ulrich Reif

Mathematisches Institut A, Universität Stuttgart, Pfaffenwaldring 57, D-70569
Stuttgart, Germany

Abstract. Starting from the recently discovered orthogonality relations
for cardinal B-splines over the real line we derive suitably modified bi-
linear forms providing orthonormality over bounded intervals. We conjecture that these bi-linear forms are positive definite and therefore inner
products for B-splines of arbitrary order $n \in \mathbb{N}$. For $n \leq 8$, this is verified
by explicit computation of the corresponding matrices. Further, applications to approximation theory are discussed.

1 Introduction

Let $B_{j,n}$ be the cardinal B-splines of order n supported on $[j, j+n]$. For $r \subseteq \mathbb{R}$
some interval denote by $\langle \cdot, \cdot \rangle_r$ the inner product of the Hilbert space $L^2(r)$. Let
$\mathrm{sinc}\,(y) := \sin(y/2)/(y/2)$. For positive weights $\omega_\mu(n)$ defined by

$$1/\mathrm{sinc}^{2n}(y) = \sum_{\mu=0}^{n-1} \omega_\mu(n) y^{2\mu} + O(y^{2n}) \tag{1}$$

the Sobolev space $H^{n-1}(r)$ is endowed with the inner product

$$(f,g)_{r,n} := \sum_{\mu=0}^{n-1} \omega_\mu(n) \langle \partial^\mu f, \partial^\mu g \rangle_r . \tag{2}$$

As pointed out in [Rei96], the sequence $\{B_{j,n}, j \in \mathbb{Z}\}$ is orthonormal with
respect to $(\cdot, \cdot)_{\mathbb{R},n}$. Since any linearly independent sequence is orthonormal with
respect to some inner product as a consequence of the Hahn-Banach Theorem,
the essential part of this statement is the fact that the given inner product can be
specified explicitly, and that it is quite natural. In this report, the orthogonality
relations over $\mathbb{R}$ are adapted to the case of finite intervals with integer bounds,
$r = [r_1, r_2], r_i \in \mathbb{N}, r_2 - r_1 \geq n - 1$, which is of importance for many applications.
With $f^{(n)} := [f, \partial f, \ldots, \partial^{n-2} f]$ the inner product is of type

$$(f,g)_{r,A_{1.2}(n)} := \sum_{i=1}^{2} f^{(n)}(r_i)\, A_i(n)\, g^{(n)}(r_i)^T + (f,g)_{r,n} . \tag{3}$$

The matrices $A_i(n)$ are specified explicitly and shown to be positive definite for
$n \leq 8$, what should cover most cases practical relevance. For general n, positivity
is conjectured, but could not be proven so far.

2   Definitions and Preliminary Remarks

The *Fourier transform* $\mathcal{F} : f \mapsto \hat{f}$ and its inverse are given by

$$\hat{f}(y) := \frac{1}{\sqrt{2\pi}} \langle f, \exp(-iy\cdot) \rangle_{\mathbb{R}} , \quad f(x) = \frac{1}{\sqrt{2\pi}} \langle \hat{f}, \exp(ix\cdot) \rangle_{\mathbb{R}} . \tag{4}$$

$\mathcal{F}$ is an *isometry* in $L^2(\mathbb{R})$, i.e.

$$\langle f, g \rangle_{\mathbb{R}} = \langle \hat{f}, \hat{g} \rangle . \tag{5}$$

Further, if $f(x) \leq C\,(1+|x|)^{-1-\delta}$ and $\hat{f}(y) \leq C\,(1+|y|)^{-1-\delta}$ for some constants $C, \delta > 0$, then the *Poisson summation formula* holds, [SW75], p. 252,

$$\sum_{j \in \mathbb{Z}} f(j) = \sqrt{2\pi} \sum_{k \in \mathbb{Z}} \hat{f}(2\pi k) . \tag{6}$$

With $\operatorname{sinc}(y) := \sin(y/2)/(y/2)$ the Fourier transform of $\partial^\mu B_{j,n}$ is

$$\widehat{\partial^\mu B_{j,n}}(y) = (iy)^\mu \exp(-iy(j+n/2))\,\operatorname{sinc}^n(y)/\sqrt{2\pi} , \tag{7}$$

see [Sch81], p. 139. Note that

$$\partial^\mu \operatorname{sinc}^n(y)\big|_{y=2k\pi} = 0 \quad \text{for} \quad k \in \mathbb{Z}\backslash\{0\} , \; 0 \leq \mu < n . \tag{8}$$

For f an even analytic function let $[f]_m := [f_0, \ldots, f_{m-1}]$ be the vector of the first m coefficients of the power series $f(x) = \sum_{j=0}^{\infty} f_j x^{2j}$. Denote by $* : \mathbb{R}^m \times \mathbb{R}^m \mapsto \mathbb{R}^m$ the *convolution operator* and by a^{*k} the $(k-1)$-fold convolution of a vector a with itself, i.e.

$$(a * b)_\nu := \sum_{\mu=0}^{\nu} a_{\nu-\mu} b_\mu , \quad a^{*k} := a * a^{*(k-1)} , \quad a^{*1} := a , \tag{9}$$

then

$$[fg]_m = [f]_m * [g]_m , \quad [f^k]_m = [f]_m^{*k} . \tag{10}$$

3   Orthogonality over the Real Line

In this section we recall the orthogonality relations for cardinal B-splines over the real line as introduced in [Rei96]. The main result is the following:

Theorem 1. *The sequence $\{B_{j,n}, j \in \mathbb{Z}\}$ of cardinal B-splines of order n is orthonormal with respect to the inner product $(\cdot, \cdot)_{\mathbb{R},n}$ as defined in (2). The weights are given by (1), or equivalently by*

$$\omega(n) := [\omega_0(n), \ldots, \omega_{n-1}(n)] := [1/\operatorname{sinc}]_n^{*2n} , \tag{11}$$

see Table 1.

Table 1. Weights $\omega(n)$ for $n \leq 7$.

n	$\omega_0(n)$	$\omega_1(n)$	$\omega_2(n)$	$\omega_3(n)$	$\omega_4(n)$	$\omega_5(n)$	$\omega_6(n)$
1	1						
2	1	$\frac{1}{6}$					
3	1	$\frac{1}{4}$	$\frac{1}{30}$				
4	1	$\frac{1}{3}$	$\frac{7}{120}$	$\frac{1}{140}$			
5	1	$\frac{5}{12}$	$\frac{13}{144}$	$\frac{41}{3024}$	$\frac{1}{630}$		
6	1	$\frac{1}{2}$	$\frac{31}{240}$	$\frac{139}{6048}$	$\frac{479}{151200}$	$\frac{1}{2772}$	
7	1	$\frac{7}{12}$	$\frac{7}{40}$	$\frac{311}{8640}$	$\frac{37}{6480}$	$\frac{59}{79200}$	$\frac{1}{12012}$

Proof. Since $B_{j,n}(t) = B_{0,n}(t - j)$ it suffices to consider inner products of type $(B_{0,n}, B_{j,n})_{\mathbb{R},n}, j \geq 0$. For $j \geq n$ the supports of $B_{0,n}$ and $B_{j,n}$ are disjoint, hence $(B_{0,n}, B_{j,n})_{\mathbb{R},n} = 0$. For $j = 0, \ldots, n - 1$ we obtain using (5) and (7)

$$
\begin{aligned}
(B_{0,n}, B_{j,n})_{\mathbb{R},n} &= \sum_{\mu=0}^{n-1} \omega_\mu(n) \langle \partial^\mu B_{0,n}, \partial^\mu B_{j,n} \rangle_{\mathbb{R}} \\
&= \sum_{\mu=0}^{n-1} \frac{\omega_\mu(n)}{2\pi} \int_{-\infty}^{\infty} y^{2\mu} \operatorname{sinc}^{2n}(y) \exp(ijy)\, dy \qquad (12) \\
&= \sum_{\mu=0}^{n-1} (-1)^\mu \omega_\mu(n)\, \partial^{2\mu} B_{0,2n}(j + n) .
\end{aligned}
$$

Define the column vector e by $e_j := \delta_{j,0}$ and the $n \times n$-matrices P, Q by

$$
P_{j\mu} := (-1)^\mu \partial^{2\mu} B_{0,2n}(j + n) \qquad (13)
$$

$$
Q_{\nu j} := \frac{2(-1)^\nu j^{2\nu}}{(2\nu)!} - \delta_{\nu,0}\delta_{j,0} . \qquad (14)
$$

Scaling the rows of Q appropriately yields the Vandermonde-matrix with entries $j^{2\nu}$. Thus Q is invertible, and orthonormality of $\{B_{j,n}, j \in \mathbb{Z}\}$ is equivalent to

$$
QP\omega = Qe = e . \qquad (15)
$$

For computing the product matrix $R := QP$ the summation array $j = 0, \ldots, n-1$ can be transformed to $\mathbb{Z}$ exploiting $B_{0,2n}(j+n) = B_{0,2n}(-j+n)$ and supp $B_{0,2n} = [0, 2n]$,

$$
R_{\nu\mu} = \sum_{j=0}^{n-1} Q_{\nu,j} P_{j\mu} = \frac{(-1)^{\nu+\mu}}{(2\nu)!} \sum_{j \in \mathbb{Z}} j^{2\nu} \partial^{2\mu} B_{0,2n}(j + n) . \qquad (16)
$$

So, (6) becomes applicable and we obtain using (8)

$$R_{\nu\mu} = \frac{1}{(2\nu)!} \sum_{k\in\mathbb{Z}} \partial^{2\nu}\left(y^{2\mu} \operatorname{sinc}^{2n}(y)\right)\big|_{y=2k\pi}$$

$$= \frac{1}{(2\nu)!} \sum_{k\in\mathbb{Z}} \sum_{\ell=0}^{2\nu} \binom{2\nu}{\ell} \partial^{\ell}\left(y^{2\mu}\right)\big|_{y=2k\pi} \partial^{2\nu-\ell}\left(\operatorname{sinc}^{2n}(y)\right)\big|_{y=2k\pi}$$

$$= \frac{1}{(2\nu)!} \sum_{\ell=0}^{2\nu} \binom{2\nu}{\ell} \partial^{\ell}\left(y^{2\mu}\right)\big|_{y=0} \partial^{2\nu-\ell}\left(\operatorname{sinc}^{2n}(y)\right)\big|_{y=0}$$

$$= \begin{cases} 0 & \text{for } \nu < \mu \\ \frac{1}{(2\nu-2\mu)!} \partial^{2(\nu-\mu)}\left(\operatorname{sinc}^{2n}(y)\right)\big|_{y=0} & \text{for } \nu \geq \mu. \end{cases} \tag{17}$$

Consequently, $R\omega(n) = [\operatorname{sinc}^{2n}]_n * [1/\operatorname{sinc}]_n^{*2n} = [\operatorname{sinc}^{2n}]_n * [1/\operatorname{sinc}^{2n}]_n = [1]_n = e$. The solution is positive for all $n \in \mathbb{N}$ and unique since $\det R = 1$.

By scaling, Theorem 1 can be generalized to uniform B-splines $B_{j,n,h}$ with knots $h\mathbb{Z}$ and support $[jh, (j+n)h]$.

Corollary 2. *The sequence* $\{B_{j,n,h}, j \in \mathbb{Z}\}$ *is orthonormal with respect to the inner product*

$$(f,g)_{\mathbb{R},n,h} := \sum_{\mu=0}^{n-1} \omega_\mu(n,h)\langle \partial^\mu f, \partial^\mu g\rangle_{\mathbb{R}} \tag{18}$$

with weights

$$\omega_\mu(n,h) := h^{2\mu-1}\omega_\mu(n) . \tag{19}$$

4   Orthogonality over Bounded Intervals

We start with examining orthogonality over intervals of type $r = (-\infty, r_2]$, $r_2 \in \mathbb{Z}$. By shift invariance, we may assume $r = \mathbb{R}_0^-$ without loss of generality. Since $\operatorname{supp} B_{j,n} = [j, j+n]$, the set of B-splines to be considered restricts to indices $j < 0$. Further, the inner product as specified in Theorem 1 needs to be modified only near the boundary $r_2 = 0$. More precisely, we have

$$(B_{j,n}, B_{k,n})_{\mathbb{R}_0^-,n} = \begin{cases} \delta_{j,k} & \text{if } \min\{j,k\} \leq -n \\ M_{jk} & \text{else} \end{cases} . \tag{20}$$

The values M_{jk} as defined above form an $(n-1)\times(n-1)$-matrix M, which can be computed explicitly for fixed values of n. Let A be a symmetric $(n-1)\times(n-1)$-matrix. For $f^{(n)} := [f, \partial f, \ldots, \partial^{n-2}f]$ we define the bi-linear form

$$(f,g)_{\mathbb{R}_0^-,A} := f^{(n)}(0) A g^{(n)}(0)^T + (f,g)_{\mathbb{R}_0^-,n} . \tag{21}$$

If A is positive definite then $(f,g)_{\mathbb{R}_0^-,A}$ is an inner product.

Theorem 3. *Let $D(n)$ be the $(n-1) \times (n-1)$-matrix with rows $D(n)_j :=$ $B_{j,n}^{(n)}, j = -n+1, \ldots, -1$. The sequence $\{B_{j,n}, j \in \mathbb{Z}^-\}$ is orthonormal with respect to $(\cdot, \cdot)_{\mathbb{R}_0^-, A_2(n)}$ with*

$$A_2(n) := D(n)^{-1} (Id - M) (D(n)^{-1})^T . \tag{22}$$

Proof. First, note that $A_2(n)$ is well defined, because $D(n)$ is invertible as an immediate consequence of the Schoenberg-Whitney Theorem, c.f. [dB78] p. 200. Symmetry is trivial. Now, consider $(B_{j,n}, B_{k,n})_{\mathbb{R}_0^-, A_2(n)}$. If $\min\{j, k\} \leq -n$ then $B_{\max\{j,k\},n}^{(n)} = 0$, hence $(B_{j,n}, B_{k,n})_{\mathbb{R}_0^-, A_2(n)} = (B_{j,n}, B_{k,n})_{\mathbb{R}_0^-, n} = \delta_{j,k}$. Otherwise, if $-n < j, k < 0$, we obtain

$$\begin{aligned}
(B_{j,n}, B_{k,n})_{\mathbb{R}_0^-, A_2(n)} &= B_{j,n}^{(n)} A_2(n) B_{k,n}^{(n)} + (B_{j,n}, B_{k,n})_{\mathbb{R}_0^-, n} \\
&= B_{j,n}^{(n)} D(n)^{-1} (Id - M) (B_{k,n}^{(n)} D(n)^{-1})^T + M_{jk} \\
&= \delta_{jk} .
\end{aligned} \tag{23}$$

Conjecture 4. *The matrix $A_2(n)$ is positive definite for all n.*

In Table 2 the matrices $A_2(2), \ldots, A_2(8)$ are specified explicitly. The minimal eigenvalues in the right column show that the conjecture is true for $n \leq 8$. The boundary correction as specified in Theorem 3 has to be modified slightly for a left boundary.

Lemma 5. *The sequence $\{B_{j,n}, j > -n\}$ is orthonormal with respect to the bi-linear form*

$$(f, g)_{\mathbb{R}_0^+, A_1(n)} := f^{(n)}(0) A_1(n) g^{(n)}(0)^T + (f, g)_{\mathbb{R}_0^+, n} , \tag{24}$$

where the $(n-1) \times (n-1)$-matrix $A_1(n)$ is defined by

$$A_{1,pq}(n) := (-1)^{p+q} A_{2,pq}(n) , \quad p, q = 0, \ldots, n-2 . \tag{25}$$

$A_1(n)$ is positive definite iff $A_2(n)$ is positive definite.

Proof. The statement is an immediate consequence of the symmetry of cardinal B-splines, $B_{j,n}(t) = B_{-n-j,n}(-t)$.

Combining Theorem 3 and Lemma 5 yields a bi-linear form providing orthonormality for bounded intervals of minimum length $n - 1$. The latter restriction guarantees that no B-spline is truncated simultaneously at both boundaries.

Corollary 6. *Let $r := [r_1, r_2], r_i \in \mathbb{Z}, r_2 - r_1 \geq n - 1$. The sequence $\{B_{j,n}, r_1 - n < j < r_2 - 1\}$ is orthonormal with respect to the bi-linear form*

$$(f, g)_{r, A_{1.2}(n)} := \sum_{i=1}^{2} f^{(n)}(r_i) A_i(n) g^{(n)}(r_i)^T + (f, g)_{r, n} . \tag{26}$$

Table 2. Matrices $A_2(n)$ and minimal eigenvalues for $2 \le n \le 8$.

n	$A_2(n)$	$\lambda_{\min}$
2	$\begin{bmatrix} \frac{1}{2} \end{bmatrix}$	0.500000
3	$\begin{bmatrix} 1 & \frac{1}{3} \\ \frac{1}{3} & \frac{1}{4} \end{bmatrix}$	0.123266
4	$\begin{bmatrix} \frac{3}{2} & \frac{11}{12} & \frac{1}{4} \\ \frac{11}{12} & 1 & \frac{3}{10} \\ \frac{1}{4} & \frac{3}{10} & \frac{1}{8} \end{bmatrix}$	0.030314
5	$\begin{bmatrix} 2 & \frac{7}{4} & \frac{5}{6} & \frac{1}{5} \\ \frac{7}{4} & \frac{5}{2} & \frac{119}{90} & \frac{1}{3} \\ \frac{5}{6} & \frac{119}{90} & \frac{61}{72} & \frac{19}{84} \\ \frac{1}{5} & \frac{1}{3} & \frac{19}{84} & \frac{5}{72} \end{bmatrix}$	0.007276
6	$\begin{bmatrix} \frac{5}{2} & \frac{17}{6} & \frac{15}{8} & \frac{137}{180} & \frac{1}{6} \\ \frac{17}{6} & 5 & \frac{443}{120} & \frac{19}{12} & \frac{5}{14} \\ \frac{15}{8} & \frac{443}{120} & \frac{101}{32} & \frac{3611}{2520} & \frac{1}{3} \\ \frac{137}{180} & \frac{19}{12} & \frac{3611}{2520} & \frac{101}{144} & \frac{85}{504} \\ \frac{1}{6} & \frac{5}{14} & \frac{1}{3} & \frac{85}{504} & \frac{31}{720} \end{bmatrix}$	0.001738
7	$\begin{bmatrix} 3 & \frac{25}{6} & \frac{7}{2} & \frac{29}{15} & \frac{7}{10} & \frac{1}{7} \\ \frac{25}{6} & \frac{35}{4} & \frac{1973}{240} & \frac{77}{16} & \frac{1009}{560} & \frac{3}{8} \\ \frac{7}{2} & \frac{1973}{240} & \frac{35}{4} & \frac{81863}{15120} & \frac{377}{180} & \frac{4}{9} \\ \frac{29}{15} & \frac{77}{16} & \frac{81863}{15120} & \frac{227}{64} & \frac{20389}{14400} & \frac{49}{160} \\ \frac{7}{10} & \frac{1009}{560} & \frac{377}{180} & \frac{20389}{14400} & \frac{3157}{5400} & \frac{1531}{11880} \\ \frac{1}{7} & \frac{3}{8} & \frac{4}{9} & \frac{49}{160} & \frac{1531}{11880} & \frac{7}{240} \end{bmatrix}$	0.000415
8	$\begin{bmatrix} \frac{7}{2} & \frac{23}{4} & \frac{35}{6} & \frac{967}{240} & \frac{469}{240} & \frac{363}{560} & \frac{1}{8} \\ \frac{23}{4} & 14 & \frac{2291}{144} & \frac{35}{3} & \frac{88657}{15120} & \frac{179}{90} & \frac{7}{18} \\ \frac{35}{6} & \frac{2291}{144} & \frac{182}{9} & \frac{236651}{15120} & \frac{1171}{144} & \frac{71003}{25200} & \frac{67}{120} \\ \frac{967}{240} & \frac{35}{3} & \frac{236651}{15120} & \frac{229}{18} & \frac{1544789}{226800} & \frac{65}{27} & \frac{287}{594} \\ \frac{469}{240} & \frac{88657}{15120} & \frac{1171}{144} & \frac{1544789}{226800} & \frac{107923}{28800} & \frac{8973191}{6652800} & \frac{263}{960} \\ \frac{363}{560} & \frac{179}{90} & \frac{71003}{25200} & \frac{65}{27} & \frac{8973191}{6652800} & \frac{16001}{32400} & \frac{93989}{926640} \\ \frac{1}{8} & \frac{7}{18} & \frac{67}{120} & \frac{287}{594} & \frac{263}{960} & \frac{93989}{926640} & \frac{1063}{50400} \end{bmatrix}$	0.000099

Again, the corresponding result for uniform B-splines is obtained by scaling:

Corollary 7. *Let* $r := [r_1 h, r_2 h]$, $r_i \in \mathbb{Z}$, $r_2 - r_1 \geq n-1$. *The sequence* $\{B_{j,n,h}, r_1 - n < j < r_2 - 1\}$ *is orthonormal with respect to the bi-linear form*

$$(f,g)_{r,A_{1,2}(n,h)} := \sum_{i=1}^{2} f^{(n)}(r_i) A_i(n,h) g^{(n)}(r_i)^T + (f,g)_{r,n,h} , \qquad (27)$$

where the matrices $A_i(n,h)$ *are defined by*

$$A_{i,pq}(n,h) := h^{p+q} A_{i,pq}(n) , \quad p,q = 0,\ldots,n-2 . \qquad (28)$$

The matrices $A_i(n,h)$ *are positive definite iff* $A_2(n)$ *is positive definite.*

5   Approximation of Functions

The orthogonality relations derived in the preceding section can be used for efficiently approximating functions by splines. Denote by $S_{n,h}(r)$ the space of uniform splines with knots $h\mathbb{Z}$ over the interval $r = [r_1 h, r_2 h]$. Then the B-splines $B_{j,n,h}, r_1 - n < j < r_2$ restricted to r form a basis of $S_{n,h}(r)$. Let $\tilde{\mathbb{N}} \subseteq \mathbb{N}$ be the set of orders n for which $(\cdot, \cdot)_{r,A_{1,2}(n,h)}$ is an inner product (so we know that $\{2,\ldots,8\} \subset \tilde{\mathbb{N}}$ are conjecture $\tilde{\mathbb{N}} = \mathbb{N}$) and denote by $\| \cdot \|_{r,A_{1,2}(n,h)}$ the induced norm.

Theorem 8. *For* $f \in H^{n-1}(r)$ *and* $n \in \tilde{\mathbb{N}}$ *consider the approximation problem*

$$\|f - g\|_{r,A_{1,2}(n,h)} \to \min , \quad g \in S_{n,h}(r) . \qquad (29)$$

The B-spline coefficients of the solution $Q_h f = \sum_{j=r_1-n+1}^{r_2-1} (Q_h f)_j B_{j,n,h}$ *are given by*

$$(Q_h f)_j := (f, B_{j,n,h})_{r,A_{1,2}(n,h)} . \qquad (30)$$

Thus, the projection $Q_h : H^{n-1}(r) \to S_{n,h}(r)$ *is local in the sense that* $Q_h f(x)$ *depends only on* f *restricted to* $r \cap [x - nh, x + nh]$.

Proof. Trivial.

The following estimate shows that the approximation order of Q_h is optimal as $h \to 0$.

Theorem 9. *Let* $f \in C^n(r)$. *There exists a constant* C *depending only on* n *such that for all* $k \in \{0,\ldots,n-1\}$

$$\|\partial^k(f - Q_h f)\|_\infty \leq C h^{n-k} \|\partial^n f\|_\infty . \qquad (31)$$

Proof. Let $P_h := C^n(r) \mapsto S_{n,h}(r)$ be a standard quasi interpolant of order n. Set $\Delta_h := f - P_h f$ then

$$\|\partial^k \Delta_h\|_\infty \leq C_1 \, h^{n-k} \|\partial^n f\|_\infty \tag{32}$$

with C_1 some constant depending only on n, see. [Sch81], p. 229. $S_{n,h}(r)$ is invariant under Q_h, i.e. $Q_h P_h = P_h$, so

$$\|\partial^k (f - Q_h f)\|_\infty = \|\partial^k (\Delta_h - Q_h \Delta_h)\|_\infty \leq C_1 \, h^{n-k} \|\partial^n f\|_\infty + \|\partial^k (Q_h \Delta_h)\|_\infty \; . \tag{33}$$

We obtain for the second summand

$$\|\partial^k Q_h \Delta_h\|_\infty \leq \sum_{j=r_1-n+1}^{r_2-1} |(\Delta_h, B_{j,n,h})_{r,A_{1.2}(n,h)}| \, \|\partial^k B_{j,n,h}\|_\infty$$

$$\leq h^{-k} \|\partial^k B_{0,n}\|_\infty \max_j |(Q_h \Delta_h)_j| \; . \tag{34}$$

The coefficients $(Q_h \Delta_h)_j$ can be estimated by

$$|(Q_h \Delta_h)_j| \leq \sum_{i=1}^{2} \left| \Delta_h^{(n)}(r_i) \, A_i(n,h) \, B_{j,n,h}^{(n)}(r_i)^T \right| + \left| (\Delta_h, B_{j,n,h})_{r,A_{1.2}(n,h)} \right|$$

$$\leq \sum_{i=1}^{2} \sum_{p,q=0}^{n-2} |\partial^p \Delta_h(r_i) \, A_{i,pq}(n,h) \, \partial^q B_{j,n,h}(r_i)|$$

$$+ \sum_{\nu=0}^{n-1} h^{2\nu-1} \, \omega_\nu(n) \, |(\Delta_h, B_{j,n,h})_{r,A_{1.2}(n,h)}|$$

$$\leq 2C_1 \max_{p,q \leq n-2} |A_{2,p,q}(n)| \max_{q \leq n-2} \|\partial^q B_{0,n}\|_\infty \|\partial^n f\|_\infty \, h^n$$

$$+ nC_1 \max_{\nu < n} \omega_\nu(n) \max_{\nu < n} \|\partial^\nu B_{0,n}\|_\infty \|\partial^n f\|_\infty \, h^n$$

$$\leq C_2 h^n \|\partial^n f\|_\infty \; . \tag{35}$$

The constant C_2 depends only on n, so the proof is complete.

In many applications, approximation is subject to a finite number of linear constraints, say Hermite interpolation at certain points. It turns out that the solution of such a problem is simply obtained by an orthogonal projection of the unconstrained approximant $Q_h f$ on the feasible set.

Theorem 10. *For $f \in H^{n-1}(r)$ and $n \in \tilde{\mathbb{N}}$ consider the constrained approximation problem*

$$\|f - g\|_{r,A_{1.2}(n,h)} \to \min \, , \quad \Lambda \lfloor g \rfloor = \lambda \, , \quad g \in S_{n,h}(r) \, , \tag{36}$$

where Λ is a matrix with k linearly independent rows, and $\lfloor g \rfloor$ denotes the column vector of the B-spline coefficients of g.
The solution $Q_h^\Lambda f = \sum_{j=r_1-n+1}^{r_2-1} (Q_h^\Lambda f)_j B_{j,n,h}$ is given by

$$\lfloor Q_h^\Lambda f \rfloor := \lfloor Q_h f \rfloor - \Lambda^T (\Lambda \Lambda^T)^{-1} (\Lambda \lfloor Q_h f \rfloor - \lambda) \; . \tag{37}$$

Proof. With v a vector of Lagrange multipliers, (36) is equivalent to

$$Q_h^\Lambda f + \Lambda^T v = Q_h f , \quad \Lambda Q_h^\Lambda f = \lambda . \tag{38}$$

Multiplication of the first equation by Λ and substitution yields $(\Lambda\Lambda^T)v = \Lambda Q_h f - \lambda$. The $k \times k$-matrix $(\Lambda\Lambda^T)$ is invertible, thus (37) follows.

The advantages of the given approximation scheme are evident:

- It is *explicit* — in contrast to L^2-approximation, which requires the solution of a possibly large linear system.
- It is *local*, i.e. $Q_h f(x)$ depends only on f restricted to a neighborhood of x.
- The order of convergence is optimal as h tends to zero.
- It is a *best approximation* with respect to a natural norm — in contrast to standard quasi interpolants.
- *Constrained approximation* simply splits into solving the unconstrained problem and a subsequent projection step.

6   An Example

Consider the approximation of the function $f(x) := \exp(-x^2/2)$ over the interval $r = [0,3]$ by cubic splines. As an example for constrained approximation, it is required that $[f, f', f'']$ shall be interpolated at the flex point $x = 1$. Table 3 shows the coefficients of the solutions for $h \in \{1/2, 1/4\}$. Note that only the marked entries of $Q_h^\Lambda f$ differ from $Q_h f$. The respective errors Δ and their first two derivatives are shown in Figures 6.1 through 6.4.

Table 3. B-spline coefficients for constrained and unconstrained approximation

j	$(Q_{1/2}f)_j$	$(Q_{1/4}f)_j$	$(Q_{1/2}^\Lambda f)_j$	$(Q_{1/4}^\Lambda f)_j$
-3	0.91641986	0.97914897	0.91641986	0.97914897
-2	1.04311448	1.01050758	1.04311448	1.01050758
-1	0.91167541	0.97887042	0.90979598*	0.97887042
0	0.60537521	0.88947462	0.60653065*	0.88947462
1	0.30600270	0.75827637	0.30326532*	0.75816332*
2	0.11773216	0.60645656	0.11773216	0.60653065*
3	0.03455616	0.45504127	0.03455616	0.45489799*
4	0.00760863	0.32031832	0.00760863	0.32031832
5	0.00181936	0.21154073	0.00181936	0.21154073
6		0.13106574		0.13106574
7		0.07618499		0.07618499
8		0.04154676		0.04154676
9		0.02125854		0.02125854
10		0.01019391		0.01019391
11		0.00462661		0.00462661

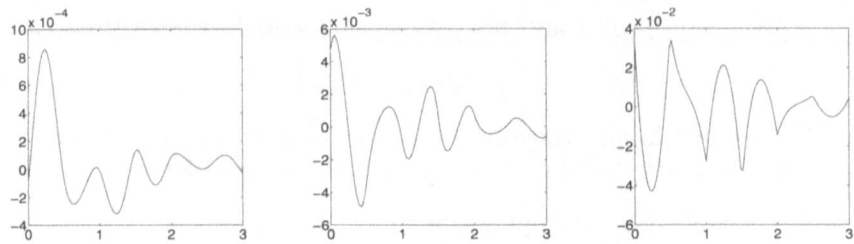

Fig. 1. Unconstrained Approximation: error $\Delta, \Delta', \Delta''$ for $h = 1/2$

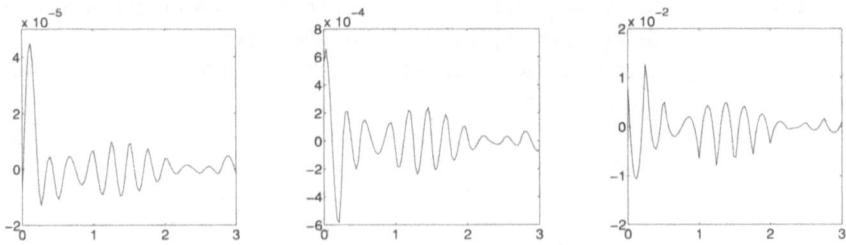

Fig. 2. Unconstrained Approximation: error $\Delta, \Delta', \Delta''$ for $h = 1/4$

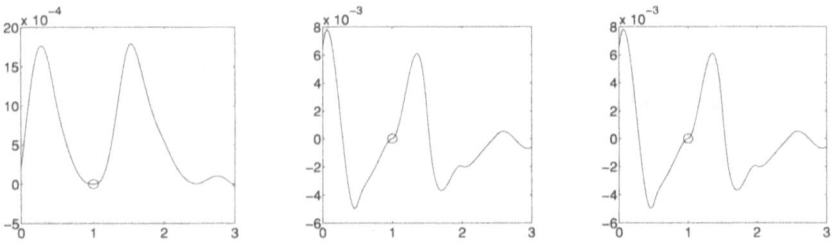

Fig. 3. Constrained Approximation: error $\Delta, \Delta', \Delta''$ for $h = 1/2$

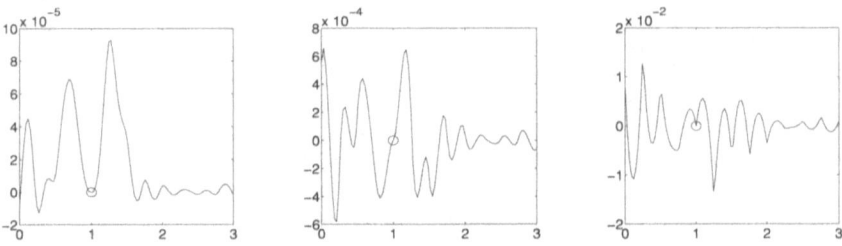

Fig. 4. Constrained Approximation: error $\Delta, \Delta', \Delta''$ for $h = 1/4$

References

[dB78]   C. de Boor. *A Practical Guide to Splines*. Applied Mathematical Sciences. Springer–Verlag, 1978.

[Rei96]   U. Reif. Orthogonality of cardinal B-splines in weighted Sobolev spaces. To appear in SIAM Journal on Mathematical Analysis, 1996.

[Sch81]   L. L. Schumaker. Spline functions: basic theory. In *Pure and Applied Mathematics. A Wiley-Interscience Publication*. John Wiley & Sons, 1981.

[SW75]   E. M. Stein and G. Weiss. *Introduction to Fourier analysis on Euclidean spaces*. Princeton University Press, 1975.

Optimizing CNC Programs Using Spline Techniques

Volker Weiß[1], Hans-Peter Seidel[1], Günther Greiner[1], Thomas Puchtler[2] and Werner Eberlein[2]

[1] Graphische Datenverarbeitung, Universität Erlangen, 91058 Erlangen, Germany
[2] Siemens AG, Industrial Automation Group, Erlangen, Germany

Abstract. While conventional CNC controlled tool-machines use linear interpolation for the geometry description, modern CNC-machines can directly deal with spline data. By converting the piecewise linear path description to a B-spline representation, the amount of data can be reduced considerably. An enormous speedup in the milling process is the consequence.

We describe design and implementation of the spline translator **reduce** which has been developed jointly with Siemens Industrial Automation in the context of the SINUMERIK 840D control unit. A series of real data examples demonstrates the efficiency of the method.

1 Introduction

Conventional CNC tool-machines use linear interpolation for geometry description. This kind of geometry representation leads to some problems. For example, to describe a curved shape accurately, a huge amount of data is necessary. Another drawback is that the speed of the milling process is strongly restricted because the single segments are very short.

Since modern CNC-machines can directly deal with spline data, there is a need for conversion of linear path descriptions to spline path descriptions. Hereby the aim is to generate splines with relatively long polynomials segments and to reduce the data as much as possible. User specified tolerances have to be met, and sharp edges have to be detected and preserved automatically.

In the following we describe design and implementation of the spline translator **reduce** which has been developed jointly with *Siemens Industrial Automation* in the context of the SINUMERIK 840D control unit. The program is based on knot removal and realizes substantial gains over previous methods.

The paper is organized as follows. In Sec. 2 we describe those aspects of a CNC-program which are important for our purposes. In Sec. 3 we discuss the basic concepts used in the new algorithm. Thereby we assume that the reader is familiar with the basic facts on (cubic) B-spline functions (e.g., in [3] the necessary information can be found). Then, in Sec. 4 a few remarks concerning the implementation are given. Finally, in Sec. 5 several examples are considered. We look at quality, compression rate and error plots of the resulting B-spline curves and compare the results with previously used methods.

2   Description of CNC Programs for Milling Machines

Numerically controlled tool-machines receive their instructions from a CNC program. Typically these programs are generated within a CAM system after the construction of the workpiece with the help of a CAD system. CNC programs are standardized by DIN66025.

2.1   Conventional CNC Programs Using Linear Interpolation

A CNC program is a sequence of blocks separated by line-feeds. Each block consists of one or more tokens. DIN66025 suggests an order of the tokens within the blocks. Each token starts with one or more letters mostly followed by a number. The letters define the token's meaning and the number determines the new value to be set.

The following list describes the tokens most important for this paper:

N*int*: Sets the block-number (optional).

F*real*: Defines the velocity valid from now on (given in $\frac{mm}{min}$).

S*int*: Defines the revolutions per second of the spindle valid from now on.

G*int*: Sets the tool-path mode:
   G0: Switch to speed-mode. Ignore given velocity and drive as fast as possible.

   G1: Switch to process-mode. Use the given velocity.

   G2/G3: Switch to circular-mode.

X*real*, Y*real*, Z*real*: Defines the coordinates of the next positions.

M*int*: Additional functions, e.g.,
   M3: Turn spindle clockwise.

   M2/M30: End of program

   M8: Coolant on.

Most tokens are modal. This means they are active until they are explicitly set to another value. The real-time control of tool-machines for free-form-surface-milling requires high computational effort. Due to this fact conventional CNC programs use linear interpolation for the description of the workpiece geometry.

The typical structure of a CNC program consists of a start-section, a geometry-description and an end-section. The start-section is used for global definitions like the selection of the right working-tool, the regulation of the coolant or the setting of the spindle's turn speed. The geometry-description is built up from a sequence of milling-paths, interrupted only by a few other tokens. Each milling-path consists of a sequence of positions which describe a three-dimensional polygon. The end-section is used to switch of the coolant and the spindle and to set the machine into a safe status.

The first lines of a conventional CNC program given below will be a useful example.

```
    ;04 (schlichten)
    . . .
N12 S15000 M3
NFFWON
F1000
N13 GO XO. YO. Z100.
N14 GO X-47.385 Y-43.973 Z2.
N15 G1 Z-0.383
N16 X-47.044 Y-44.315 Z-0.398 `
N17 X-46.337 Y-45.022 Z-0.419
N18 X-45.629 Y-45.729 Z-0.428
N19 X-44.922 Y-46.436 Z-0.418
N20 X-44.215 Y-47.143 Z-0.394
N21 X-43.973 Y-47.386 Z-0.383
N22 GO Z2.
N23 GO X-48.143 Y-43.144
N24 Z-0.383
N25 G1 X-47.715 Y-43.572 Z-0.412
N26 X-47.008 Y-44.279 Z-0.449
N27 X-46.301 Y-44.986 Z-0.469
N28 X-45.594 Y-45.693 Z-0.479
    . . .
```

Since the geometry-description is the main topic of this paper, let's take a closer look to that part of the CNC program. The description of the geometry starts at line 8 with block N13. The machine switches to speed-mode and moves the milling cutter to the start position. The next block moves the cutter to a position over the workpiece. In block N15 the machine switches to process-mode and starts the milling-path by moving the cutter with a velocity of $1 \frac{m}{min}$. The first milling-path ends at block N21. In blocks N22 and N23 the machine switches to speed-mode, raises the cutter and does the positioning for the next milling-path.

This way of geometry description leads to a set of serious problems. Curved shapes have to be approximated by polygons. To keep the approximation error small, a large amount of very short line segments has to be used. This leads to huge CNC programs, which have to be stored, transferred and processed by the machine.

An even more important problem is the restriction of the velocity by the block cycle time. The processing of each line segment requires at least one block cycle. Therefore the velocity f is bounded by the block cycle time τ and the minimal segment length l_{min}:

$$f \leq \frac{l_{min}}{\tau}$$

This means, the smaller the segment length, the smaller the attainable feed rate.

Additionally the surface quality suffers: the material cut-off on areas processed with very low feed rate is bigger than on areas with higher velocity. The resulting roughness of the surface has to be smoothed out by intensive polishing.

These difficulties can be avoided by the use of spline techniques in the geometry description of CNC programs. A new control unit which offers these abilities in form of cubic rational spline curves is the SINUMERIK 840D.

2.2 CNC Programs Using Spline Techniques

The use of splines in CNC programs requires an extension of DIN66025. The SINUMERIK 840D offers various formats to program cubic NURBS-curves. The most interesting one for our purpose is the 840D native format (see [6]).

This format allows the control of each axis by a separate cubic polynomial and a common cubic denominator polynomial. Splines are programmed as a sequence of such blocks. The token POLY is used to switch into polynomial interpolation mode. This mode stays active until the appearance of a G-token, like G0 or G1.

The programming of a polynomial has the following syntax:

$$PO[string] = (end, a_2, a_3) \{PL=t_e\},$$

where

string defines the axis to be programmed,

end defines the end position of the movement,

a_2, a_3 are the quadratic and cubic polynomial coefficients,

t_e is the optional parameter value of the end position (if omitted, it is 1).

In the standard represenation $p(t) = a_0 + a_1 t + a_2 t^2 + a_3 t^3$ the coefficient a_0 is specified implicitly by the actual position and $end = a_0 + a_1 + a_2 + a_3$. If $a_3 = 0$ it can be omitted, if $a_2 = 0 \wedge a_3 = 0$ a conventional linear block should be used. An example clarifies the way of polynomial CNC programming:

```
;04 (schlichten)
...
N12 S15000 M3
FFWON
F1000
N13 G0 X0. Y0. Z100.
N14 G0 X-47.385 Y-43.973 Z2.
N15 G1  Z-0.386
N16 POLY PO[X]=(-45.629,0.003) PO[Y]=(-45.729,0.002) \
    PO[Z]=(-0.427,-0.014,0.028)
N17 X-43.973 Y-47.386 PO[Z]=(-0.385,0.063,-0.022)
N22 G0 Z2.
N23 G0 X-48.143 Y-43.144
N24 Z-0.383
N25 G1  X-47.715 Y-43.572 Z-0.414
```

```
N26 POLY PO[X]=(-45.594,0.003,-0.001) Y-45.693 \
    PO[Z]=(-0.476,0.037,0.014)
N27 PO[X]=(-43.144,-0.001) PO[Y]=(-48.143,-0.001,0.002) \
    PO[Z]=(-0.384,0.104,-0.015)
N33 GO Z2
. . .
```

Block N16 sets the machine to polynomial interpolation mode and defines polynomials for all three axes. The polynomials are quadratic for the X- and the
Y-axis and cubic for the Z-axis. The next block only defines a polynomial for
the Z-axis. In X- and Y-direction the movement is linear. Block N22 leaves the
polynomial mode. With this spline technique the SINUMERIK 840D uses the
same curve model as modern CAD/CAM systems. If these systems are able to
produce extended CNC programs, an approximation step is saved during the
production process. But the abilities of this new control should also be available
for existing, linear CNC programs. Therefore it is necessary to transform linear
geometry description into spline format.

In a joint project with Siemens Industrial Automation Group, a program
performing this task has been implemented and tested. In this note we describe
the method, discuss the results and compare it to previously used methods.

3   The General Approach

The following problem has to be solved.

Given a piecewise linear function $P(t) \in \mathbb{R}^3$ and a tolerance $\varepsilon > 0$.
Determine a spline curve S satisfying
1. The deviation of S and P has to be less then the tolerance ε, i.e.
   $\|P(t) - S(t)\|_\infty < \varepsilon$ for all t.
2. The spline S should be "more smooth" than the polygon P. This
   means two things:
   (a) The continuity between two spline segments should be as high
       as possible (C^1 or even C^2).
   (b) The spline should be monotone on the segments, i.e. no oscilla-
       tions should occur.
3. The curve S should consist of fewer and longer segments than the
   polygon P.
4. Big jumps in the first derivative of P should be preserved by S.

We consider each component of P separately. Let $p(t)$ be such a component.

In a first step, for every vertex of the linear function p we determine the error,
occuring after removal of this vertex. Then we remove those vertices causing
the smallest errors. This will be repeated as long as the overall distance of the
reduced linear function $\tilde{p}$ is within a distance of $\varepsilon_0 < \varepsilon$ to p.

In the second step, the degree is elevated (see [3]) to transform the linear B-spline $\tilde{p}$ to a cubic B-spline (having multiple knots).

In a third step, the knot removal procedure of Lyche and Mœrken [4,5] is applied (see also [1,2]).

Again, one first checks the influence of every knot, thus establishing a *ranking list*. According to this ranking list knots are removed thus obtaining a cubic B-spline function s. Knots will be removed as long as a tolerance ε_1 is guaranteed. As an upper bound for the distance of s and $\tilde{p}$, the ℓ^∞ distance of the corresponding control points is used.

Whenever it is impossible to remove knots currently on top of the ranking list or, after 50 % of the knots have been processed already, the ranking list will be updated and the third step is repeated.

The resulting function is a cubic B-spline s with

$$\|\tilde{p}(t) - s(t)\| < \varepsilon_0 + \varepsilon_1 = \varepsilon \ .$$

Before the third step is performed for the first time, we detect the sharp edges by looking for jumps in the first and second differences of the control points. The corresponding knots then will be marked and are not considered during the knot removal, they keep multiplicity 3 during the whole procedure. The same holds for the knots at the end of the segment.

4   Implementation

The new CNC translator, named `reduce`, has been implemented in C++. As input it needs a standard linear CNC-program and a tolerance specification. It then produces an extended spline CNC-program as output. The sections of the input program which should be transformed have to be marked initially by two special tokens:

`M50`: marks the first block to be translated and

`M51`: the last block.

All process mode blocks between these two tokens are transformed to spline blocks. The translation is interrupted by any non-coordinate or non-number token. Within one CNC-program, several sections can be marked to be transformed.

Two important options of `reduce` define the handling of the end conditions of the splines:

`-(no)fix`: This option is used, to (allow)/forbid movements of the start- and end points of the spline relative to the original start- and end positions.

`-(no)cyc`: This option has to be used for closed paths. It results in a closed spline curve, which is C^2-continuous, if this is possible without breaking the given tolerance (see fig. 9).

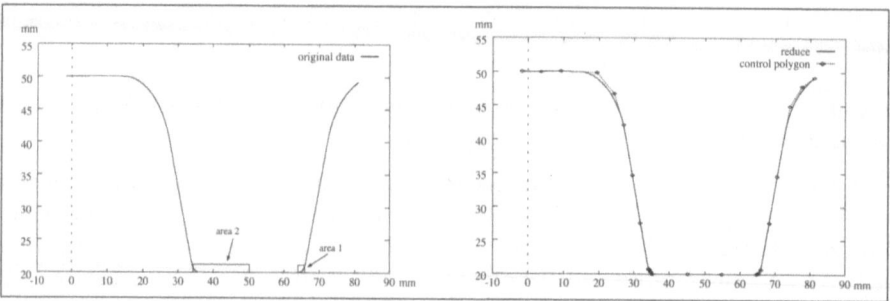

Fig. 1. Example 5.1: The original data (left) and the reduced/approximated data with control polygon (right)

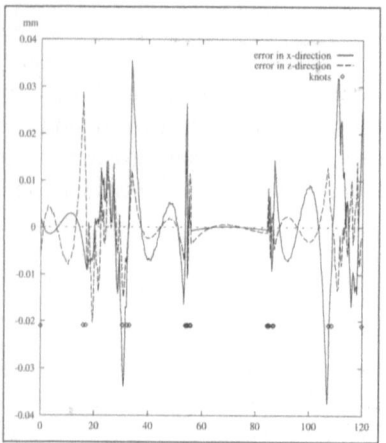

Fig. 2. Example 5.1: The approximation error and the remaining knots.

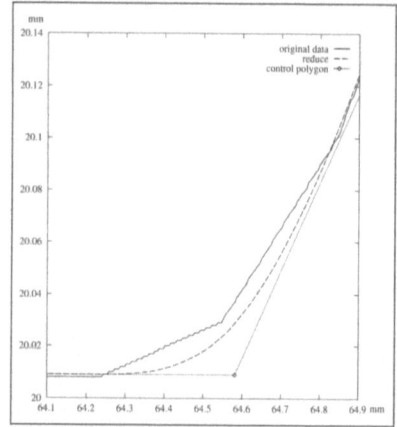

Fig. 3. Example 5.1: Enlargement of area 1 (see Fig. 1)

5 Results

This section shows the results of the CNC translator `reduce`. It is compared with three other algorithms which were previously used for this purpose. They are called `ST1`, `ST2` and `ST12`.

5.1 A Simple CNC Program

In a first test (see Fig. 1–4 and Tab. 1,2) we considered at a simple CNC program describing a C^2-continuous shape with the help of 2992 linear segments (Fig. 1). The given tolerance ε for approximation was set to $50\mu m$. Figure 1 shows the

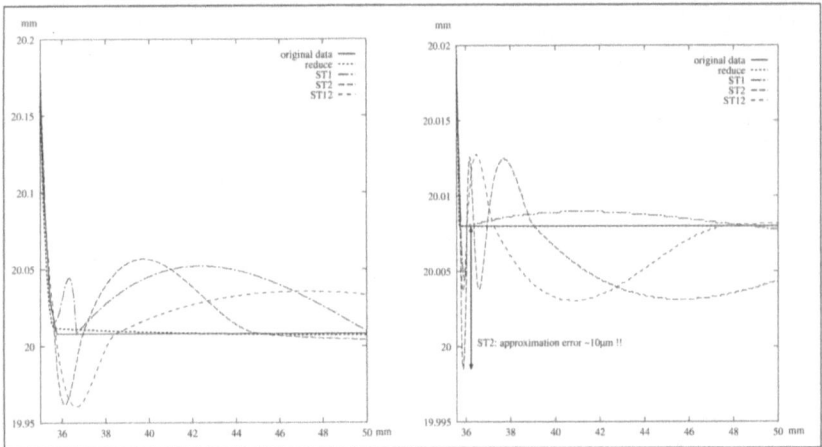

Fig. 4. Example 5.1: Enlargement of area 2 (see Fig. 1), tolerances 50μm (left) and 5μm (right)

results of **reduce**. Figure 2 shows the approximation error in x- and z-direction and the knots.

The spline is C^2-continuous, since all knots have a multiplicity of one. Figure 3 shows the enlargement of area 1. The smoothing of the original data by the spline is obvious.

The differences between the old CNC translators **ST** and **reduce** can be seen in Fig. 4 which shows the enlargement of area 2. The non-oscillating behavior of the solution of **reduce** is obvious. Figure 4 clearly shows that the old spline translator **ST2** in certain situations does not obey the given tolerance. Both figures compare the results of the different algorithms at the tolerances of 50μm and 5μm. While all "solutions" determined by the old spline translator have artificial oscillations, the result of **reduce** does not show this negative effect.

The reduction factors of the different algorithms are compared in the following two tables. The reduction rate of **reduce** is about the same as for the old programs, even slightly better. However, the quality of the curves obtained by **reduce** is much higher and, more important, the tolerances are guaranteed. This is by no means sure when the former methods are used (see the right image in Fig. 4).

5.2   CNC Programs with Jumps in the Derivative

In this example (see Fig. 5–7 and Tab. 3) the handling of data with (large) jumps of the first derivative is described. On the left side of Figure 5 the original data for this test are shown. There are four sharp corners in the lower part. The right figure contains the spline curves produced by the different algorithms with an tolerance of 20μm. The two enlargements in Figure 6 clearly illustrate the drawbacks of **ST2** and **ST12**: they are unable to recognize "sharp edge". The

Algorithm	Blocks	Reduction
ST1	20	0.67%
ST2	25	0.84%
ST12	21	0.70%
reduce	19	0.64%

Table 1. Reduction rates for example 5.1, tolerance: $50\mu m$

Algorithm	Blocks	Reduction
ST1	37	1.24%
ST2	99	3.30%
ST12	91	3.04%
reduce	85	2.84%

Table 2. Reduction rates for example 5.1, tolerance: $5\mu m$

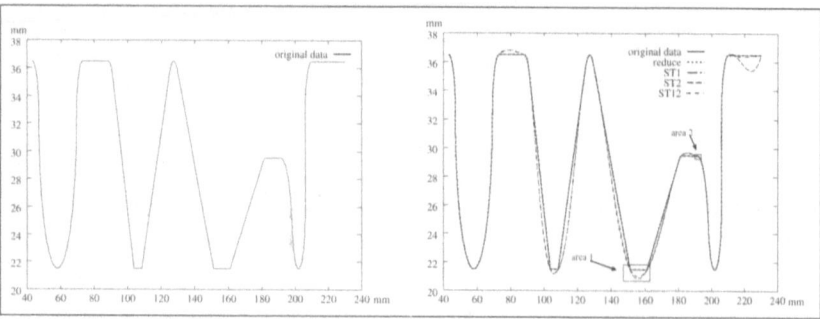

Fig. 5. Example 5.2: Original data (left) and reduced data (right)

second enlargement shows that the result of ST1 is only C^0-continuous even on areas where a C^2-continuous approximation is possible without oscillating, as the result of reduce shows.

The left part of Fig. 7 contains the approximation error of the solution calculated by reduce in both coordinate directions x and z. It is obvious, that the spline curve runs within the given tolerance of $20\mu m$. The right figure shows the first derivative of the curve with its four jumps.

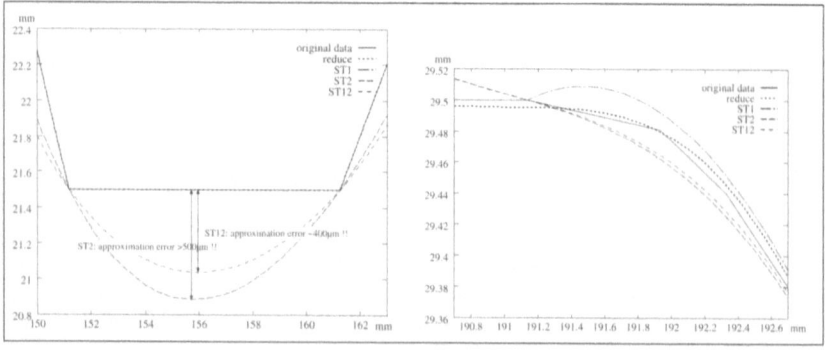

Fig. 6. Example 5.2: Enlargements of area 1 and area 2 (see Fig. 5)

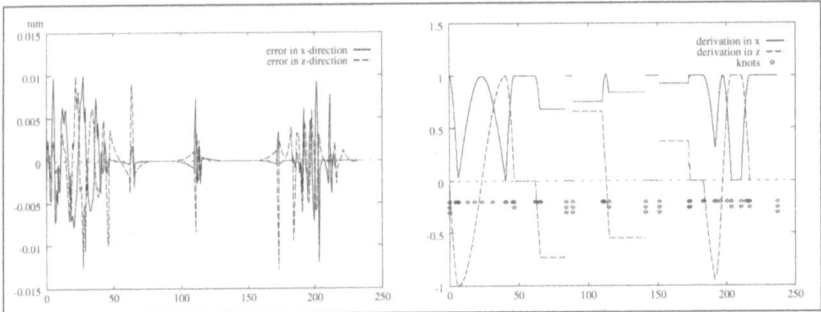

Fig. 7. The approximation error and the first derivative with remaining knots

The following table shows the reduction factors of the different algorithms. Despite the higher continuity of the result of **reduce** the reduction rate is comparable to the other algorithms. In consideration of the very high quality of the produced curve it is superior.

Algorithm	Blocks	Reduction
ST1	24	10.39%
ST2	33	14.29%
ST12	34	14.72%
reduce	42	18.18%

Table 3. Reduction rates for example 5.2, tolerance: $20\mu m$

5.3   A C^2-Continuous Closed Contour

In this example (see Fig. 8–9 and Tab. 4) a CNC program describing the grinding of a cam-shaft was optimized. It is very important for this purpose to get a C^2-continuous geometry at a very high accuracy.

The original geometry description was given by 271 linear blocks, but in polar coordinates. Since the control needs the input in a Cartesian system a conversion has to be applied. To keep accuracy the description had been translated to 36000 linear Cartesian blocks (100 points per degree) and then optimized by the algorithm. The left image of figure 8 contains the original input data, and the result of the optimization process by **reduce** is shown on the right image.

Figure 9 shows the enlargements of area 2 on the left and area 1 on the right. The left image contains the results of the new and the old CNC translation program. It is obvious that ST2 produces an intolerable output: it is only C^0-continuous, with big jumps in tangent directions. In the right image the effect of different end conditions for **reduce** is shown: Without any option for the end

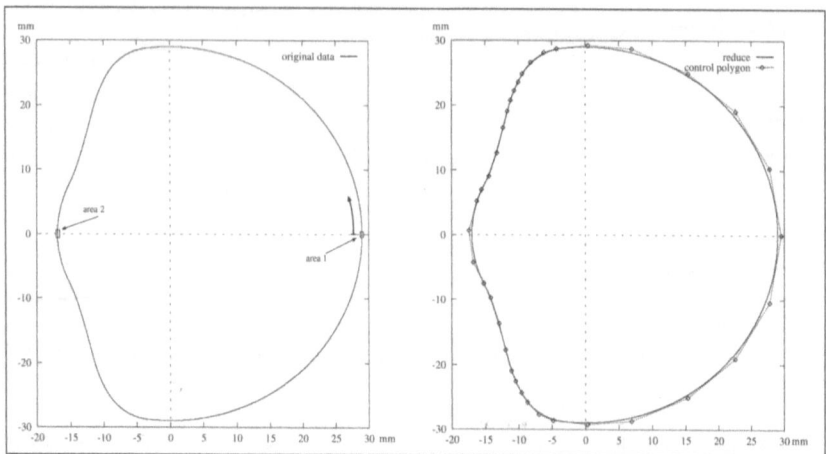

Fig. 8. Example 5.3: A cam-shaft — original and approximation

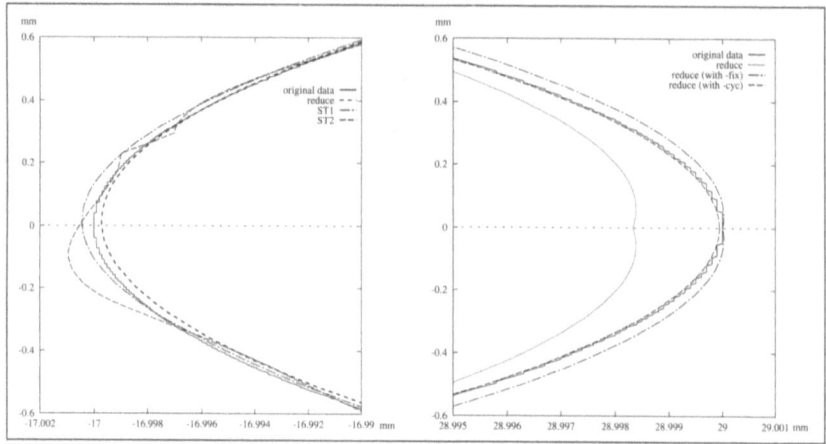

Fig. 9. Two enlargements of the cam-shaft

conditions given, reduce moves the start position and the end position of the curve. With the option -fix the start- and end positions of the original data are used for the approximation.

With the option -cyc the start- and end positions are free for optimization under the restriction that the close curve stays closed. With this option reduce produces a global C^2-continuous spline curve, even at a tolerance of only $2\mu m$.

The final table shows that the reduction factor of reduce is enormous, if the underlying original data describe a curve with high continuity.

Algorithm	Blocks	Reduction
ST1	184	0.51%
ST2	333	0.92%
ST12	–	–
reduce	37	0.10%

Table 4. Reduction rates for example 5.3, tolerance: $2\mu m$

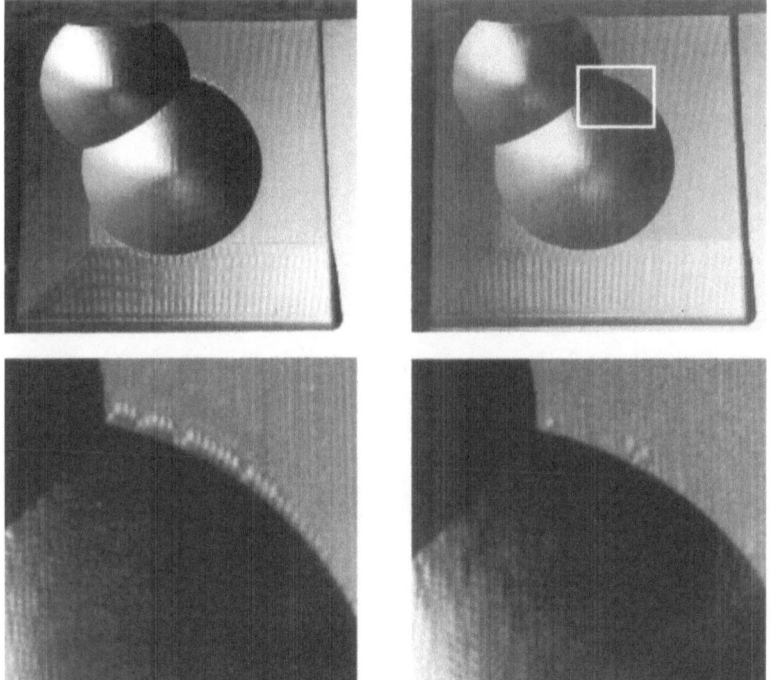

Fig. 10. A metal workpiece which has been cut by a CNC controlled milling machine which is equipped with the SINUMERIK 840D control unit.
On the right: The linear path description has been processed with reduce.
On the left: The linear path description has been processed with ST1.
In the closeups (lower figures) the quality of both methods can be compared.

References

1. Eck, M., Hadenfeld, J.: Knot removal for B-spline curves. Computer Aided Geometric Design **11** (1994).
2. Eck, M., Hadenfeld, J.: A stepwise algorithm for converting B-splines. In *P.J. Laurent, A. Le Méhauté, and L.L. Schumaker, editors,* Curves and Surfaces II. AKPeters, Wellesley 1994.
3. Hoschek, J., Lasser, D.: *Fundamentals of Computer Aided Geometric Design.* AKPeters, Wellesley 1994
4. Lyche, T., Mørken, K.: Knot removal for parametric B-spline curves and surfaces. Computer Aided Geometric Design **4** (1987) 217–230
5. Lyche, T., Mørken, K.: A data-reduction strategy for splines with applications to the approximation of functions and data. IMA Journal of Numerical Analysis **8** (1988) 185–208
6. *SINUMERIK 840D*, Interner Bericht SIEMENS AG, 1994.

New Approximation Methods for Planar Offset and Convolution Curves

In-Kwon Lee[1], Myung-Soo Kim[1] and Gershon Elber[2]

[1] Dept. of Computer Science, POSTECH, Pohang 790–784, South Korea
[2] Dept. of Computer Science, Technion, IIT, Haifa 32000, Israel

Abstract. We present new methods to approximate the offset and convolution of planar curves. These methods can be used as fundamental tools in various geometric applications such as NC machining and collision detection of planar curved objects. Using quadratic curve approximation and tangent field matching, the offset and convolution curves can be approximated by polynomial or rational curves within the tolerance of approximation error $\epsilon > 0$. We suggest three methods of offset approximation, all of which allow simple error analysis and at the same time provide high-precision approximation. Two methods of convolution approximation are also suggested that approximate convolution curves with polynomial or rational curves.

1 Introduction

Given two planar curved objects O_1 and O_2, their Minkowski sum $O_1 \oplus O_2$ is obtained by sweeping one object O_2 (with a fixed orientation) all over the other object O_1. Offset is a special case of Minkowski sum in which the sweeping object is restricted to a circular disk.

Let $C_1(t)$ and $C_2(s)$ be the boundary curves of O_1 and O_2, respectively. Then each boundary point of $O_1 \oplus O_2$ is a point of the following form:

$$C_1(t) + C_2(s), \tag{1}$$

where $C_1'(t) \parallel C_2'(s)$ and $\langle C_1'(t), C_2'(s) \rangle > 0$, for some s and t. We define the convolution curve $C_1 * C_2$ as follows:

$$(C_1 * C_2)(t) = C_1(t) + C_2(s(t)), \tag{2}$$

where $C_1'(t) \parallel C_2'(s(t))$ and $\langle C_1'(t), C_2'(s(t)) \rangle > 0$, for a reparameterization $s = s(t)$. That is, the convolution curve $C_1 * C_2$ is an envelope curve which is obtained by sweeping one curve C_1 (with a fixed orientation) along the other curve C_2 [3]. Offset curve is a special case of convolution curve in which the sweeping curve is restricted to a circle.

The boundary $\partial(O_1 * O_2)$ is a subset of $(C_1 * C_2)(t)$, in general. Moreover, when both O_1 and O_2 are convex objects, the convolution curve $C_1 * C_2$ is exactly the same as $\partial(O_1 * O_2)$. Figure 1 shows the offset and convolution of planar convex curved objects. In Fig. 1–(a), an offset is obtained by sweeping a circle along the boundary of an object and taking only the outer boundary of the envelope curve. In Fig. 1–(b), a convolution is generated by sweeping one object along the boundary of the other and similarly taking the outer envelope curve. In Fig. 1, the inner envelope curves do not contribute to the offset and convolution boundaries; their elimination is a direct consequence of the condition: $\langle C_1'(t), C_2'(s(t)) \rangle > 0$.

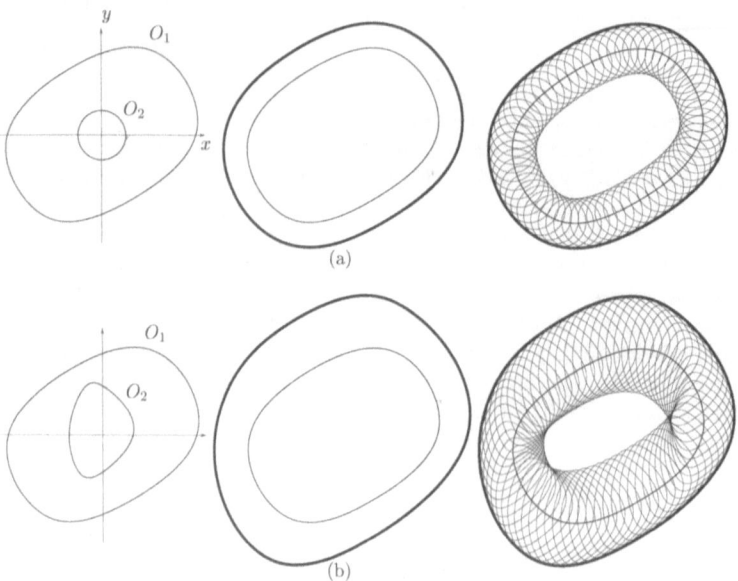

Fig. 1. (a) Offset and (b) convolution of planar objects

Offset and convolution computations are classic operations in CAD/CAM, which can be used in various interesting geometric applications such as NC machining [8,17], motion planning [3,16,27], character font and brush stroke design [13,14], blending [30], and shape transformation [21].

The exact offset and convolution curves of planar algebraic curves are algebraic, but not rational. Moreover, they have very high algebraic degree [12,18,22]. For example, the offset of a cubic Bézier curve has an algebraic degree of 10 [12]. Consequently, these fundamental limitations have led offset research to develop various offset approximation techniques [10,22,25]. Based on *quadratic curve approximation* and *tangent field matching*, we present new methods that approximate offset and convolution curves with polynomial or rational curves. Compared with other conventional approaches, our methods have many advantages: simple error analysis, high-precision approximation, and a relatively small number of curve subdivisions.

This paper is organized as follows. Section 2 presents new algorithms to approximate planar offset curves. Using similar techniques, approximation algorithms for planar convolution curves are presented in Sect. 3. Finally, in Sect. 4, we conclude this paper and suggest further research problems.

2   Offset Approximation Methods

2.1   Planar Offset Curve

Let $C(t) = (x(t), y(t))$ be a planar rational curve. Then the offset curve $C_r(t)$ (with respect to a fixed radius r) is defined by:

$$C_r(t) = C(t) + r \cdot N(t), \qquad (3)$$

where

$$N(t) = \frac{(y'(t), -x'(t))}{\sqrt{x'(t)^2 + y'(t)^2}} \qquad (4)$$

is the unit normal of $C(t)$. Due to the square root function in the denominator of $N(t)$, the exact offset curve is not rational, in general [12].

Conventional offset curve approximation techniques are focused on approximating the exact offset curve, $C(t) + r \cdot N(t)$, with low degree spline curves. In this paper, we take different approaches which approximate and/or reparameterize the offset circle $r \cdot N(t)$ by planar polynomial or rational curves.

2.2   Previous Work

Elber et al. [10] reviewed the current state-of-the-art of offset approximation methods and classified conventional approaches according to the following two categories: (i) *control polygon based approach* [5,7,8,31], and (ii) *interpolation based approach* [19,20,23,29]. Control polygon based methods are usually much simpler. In these methods, it is relatively easy to construct an offset approximation curve so that it is completely inside/outside the exact offset curve. This capability of controlling offset under/over-estimation is quite useful in NC machining for the prevention of over-cut. Interpolation based methods usually generate more precise approximation, while generating much less curve data, especially for high-precision offset approximation.

Most of the previous methods estimate offset approximation error only at finite sample points, which may inaccurately reflect the global error. Moreover, each curve subdivision is made at the midpoint of the base curve, which may not be the optimal subdivision point. Elber and Cohen [8] estimated the offset approximation error by computing the maximum value of the difference function:

$$\psi(t) = \|C_r^a(t) - C(t)\|^2 - r^2, \qquad (5)$$

where $C_r^a(t)$ is an approximation of $C_r(t)$. Curve refinement is made at the parameter value which gives the maximum error. This technique reduces the size of curve data.

Elber et al. [10] compared conventional offset approximation methods from quantitative, as well as qualitative, point of view. The difference function $\psi(t)$ of (5) was used as a criterion to measure the performance of each method. This paper also uses the function $\psi(t)$ to evaluate the performance of each suggested method.

2.3   Offset Approximation Using Circle Approximation: CAO

Let $C(t) = (x(t), y(t))$, $t_0 \le t \le t_1$, be a planar regular parametric curve. Assume $C(t)$ has a one-to-one mapping θ_C, called the *tangential angular map* of $C(t)$, defined by

$$\theta_C : t \in [t_0, t_1] \mapsto \theta_C(t) = \arctan \frac{y'(t)}{x'(t)} \in [0, 2\pi). \tag{6}$$

The unit normal vector field $N(t)$ has its curve trace as a unit circular arc.

To approximate the unit circular curve $N(t)$ (not necessarily with a uniform speed) with a rational curve, we construct a quadratic curve: $Q(s) = (x_q(s), y_q(s))$, $s_0 \le s \le s_1$, which approximates the unit circular arc between two angles, $\arctan(y'(t_0)/x'(t_0))$ and $\arctan(y'(t_1)/x'(t_1))$. Then, using a suitable rational reparameterization $s(t)$, we can approximate $N(t)$ by a rational curve $Q(s(t))$. As a result, we can construct an approximated offset curve $C_r^a(t)$ as a rational curve:

$$C_r^a(t) = C(t) + r \cdot Q(s(t)), \tag{7}$$

where $s(t)$ is a proper reparameterization of $Q(s)$ satisfying the parallel relationship: $C'(t) \parallel Q'(s(t))$, $t_0 \le t \le t_1$. In the following discussion, we show how to construct a rational reparameterization $s(t)$.

The derivative curve of $Q(s)$ is a linear curve, $Q'(s) = (x_q'(s), y_q'(s)) = (as + b, cs + d)$, where a, b, c, and d are some real constants. If $Q'(s(t)) = (as(t) + b, cs(t) + d)$ and $C'(t) = (x'(t), y'(t))$ have the same direction, we have

$$\frac{cs(t) + d}{as(t) + b} = \frac{y'(t)}{x'(t)}, \tag{8}$$

and the parameter $s(t) \in [s_0, s_1]$ is uniquely determined by

$$s(t) = \frac{by'(t) - dx'(t)}{cx'(t) - ay'(t)}. \tag{9}$$

Note that the solution $s(t)$ can be computed as a unique rational polynomial only when $Q(s)$ is a polynomial quadratic curve. The degree of the rational reparameterization $s(t)$ is $d - 1$, where d is the degree of $C(t)$. The composition $Q(s(t))$ and the offset approximation curve $C_r^a(t)$ can be computed by using symbolic computation tools for parametric curves [9,11]. The degree of $C_r^a(t)$ is $3d - 2$ for a polynomial curve $C(t)$, and $5d - 4$ for a rational curve $C(t)$. This offset approximation method is called CAO (Circle Approximation Offset).

The offset approximation error is simply r times that of approximating the unit circular arc by a quadratic curve $Q(s)$, assuming there is no local self-intersection loop in the computed offset approximation (see Reference [26] for more details). In spite of the high degree in the resulting approximation, the size of output data from CAO is smaller than those of current offset approximation methods using the control polygon based approach. Figure 2 shows the CAO approximation of a cubic B–spline curve. The CAO approximation is a rational curve of degree 7. In Fig. 2, the numbers of control points in offset

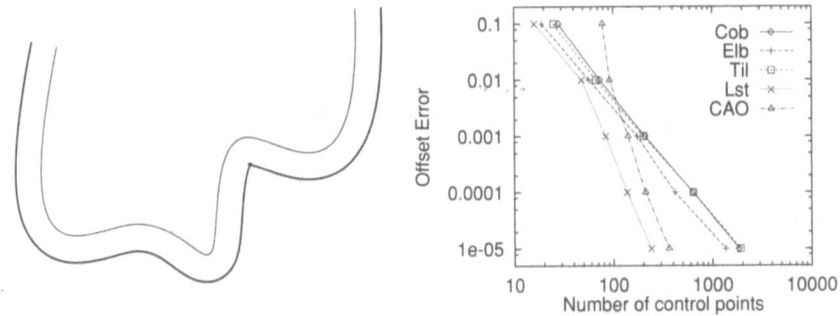

Fig. 2. CAO approximation (bold curve) of a cubic B–spline curve (light curve)

approximation curves, generated from five different methods (*Cob* [5], *Elb* [8], *Til* [31], *Lst* [20], and CAO), are compared. The CAO method outperforms the control polygon based methods (*Cob* [5], *Elb* [8], and *Til* [31]), especially for high-precision approximation. (*Lst* [20] is an interpolation based method.)

The error in a circle approximation with quadratic Bézier curves can be estimated very quickly by evaluating a simple function [28]. Since the circle approximation error is the same as the error in CAO, we can estimate the offset approximation error in an *a priori* fashion (see Reference [26] for more details). That is, we can compute the error bound even without constructing an offset approximation curve. Figure 3 shows an approximation of the unit circle with three quadratic Bézier curve segments with C^1-continuity.

Data explosion is a serious problem in offset computation, especially for high-precision approximation. Therefore, we may store only a minimum amount of computation results such as curve subdivision parameters (in subdivision based methods [5,7,19,23,29,31]), and generate all the other geometric information from these minimum data on the fly as required. In the CAO method, assuming rational curves of degree 7 are used for the offset curve approximation, this storage reduction scheme may reduce the output data size quite significantly (producing almost 14 times less data). Table 1 shows the result of this storage reduction technique applied to the examples shown in Fig. 2. The numbers of base curve subdivisions are shown for subdivision based methods, as well as

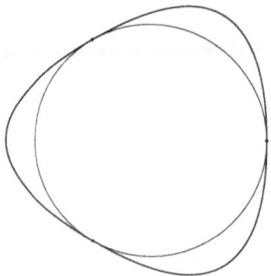

Fig. 3. Quadratic Bézier curve approximation of the unit circle

for an adapted version of least-squares based method [20] which exploits the subdivision based algorithm.

Table 1. Number of curve subdivisions in the offset approximation of Fig. 2.

Tolerance	Cob	Elb	Til	Lst	CAO
10^{-1}	7	5	6	8	10
10^{-2}	22	16	20	14	12
10^{-3}	67	61	65	29	19
10^{-4}	210	181	211	53	29
10^{-5}	613	561	637	90	51

2.4   Offset Approximation by Tangent Field Matching: MO

The basic approach of CAO suggests another way of computing offset approximation curves. That is, instead of approximating the unit circle by polynomial quadratic curves, we can use an exact rational representation $Q(s)$ of the unit circle and approximate the reparameterization function $s(t)$ for the parallel relationship: $C'(t) \parallel Q'(s(t))$. This approach is based on the concept of tangent field matching which is defined as follows.

Definition 1. Tangent Field Matching
Let $C_1(u)$, $u_0 \leq u \leq u_1$, and $C_2(v)$, $v_0 \leq v \leq v_1$, be two regular C^1 parametric curves. Assume two reparameterizations, $U : u \mapsto t$ and $V : v \mapsto t$, which map $C_1(u)$ into a curve $\hat{C}_1(t)$, $t_0 \leq t \leq t_1$, and $C_2(v)$ into a curve $\hat{C}_2(t)$, $t_0 \leq t \leq t_1$, respectively. If two unit tangent fields:

$$T_1(t) = \frac{\hat{C}_1'(t)}{\|\hat{C}_1'(t)\|} \quad \text{and} \quad T_2(t) = \frac{\hat{C}_2'(t)}{\|\hat{C}_2'(t)\|}, \tag{10}$$

are the same for all $t \in [t_0, t_1]$, the two curves $\hat{C}_1(t)$ and $\hat{C}_2(t)$ have a complete tangent field matching, and the two tangent fields of $C_1(u)$ and $C_2(v)$ are matched by the reparameterizations U and V. $\square$

When the two curves have a complete tangent field matching, the inner product $\langle T_1(t), T_2(t) \rangle = 1$, for all $t \in [t_0, t_1]$. Cohen et al. [6] presented an algorithm to approximate the matching by solving the following optimization problem:

$$\max_{v(u)} \int_{u_0}^{u_1} \left\langle \frac{C_1'(u)}{||C_1'(u)||}, \frac{C_2'(v(u))}{||C_2'(v(u))||} \right\rangle du, \tag{11}$$

where $v(u) = V(U^{-1}(u))$, $v(u_0) = v_0$, and $v(u_1) = v_1$. This optimization problem is bounded from above by $u_1 - u_0$, since the normalized inner product does not exceed one. While the optimal solution of (11) is difficult to obtain, we can instead solve an associated discrete optimization problem which may produce an arbitrarily close approximation to the solution of (11). We sample both $C_1(u)$ and $C_2(v)$ at n uniform parameter locations and compute their unit tangents as $T_{1,i}$ and $T_{2,j}$, $0 \le i, j < n$. Then, the problem is reduced to a discrete optimization problem:

$$\max_{j(i)} \sum_{i=0}^{n-1} \langle T_{1,i}, T_{2,j(i)} \rangle, \tag{12}$$

subject to

$$j(0) = 0, \ j(n-1) = n-1, \ j(i) \le j(i+1). \tag{13}$$

Cohen et al. [6] suggested a method to solve the optimization problem of (12) within $O(n^2)$ time, where n is the number of sample locations. This method employs a dynamic programming technique and provides a global optimal solution. The more sample points we use, the closer is the resulting reparametrized curve $C_2(v(u))$ to the completely matching curve with $C_1(u)$. Cohen et al. [6] also describe a method that approximates $v(u)$ from $j(i)$, by using a least-squares fitting with a B–spline curve. The composition of $C_2(v(u))$ can then be computed using symbolic computation tools [9,11], while resulting in a B–spline curve representation of the offset approximation.

The offset approximation method: $C_r^a(t) = C(t) + r \cdot Q(s(t))$, where $s(t)$ is approximated by tangent field matching, is called MO(Matching Offset). One difficulty in MO is that we cannot estimate the approximation error by using the difference function $\psi(t)$ of (5). In MO, the function $\psi(t)$ is useless since it always vanishes:

$$\psi(t) = ||C_r^a(t) - C(t)||^2 - r^2 = ||r \cdot Q(s(t))||^2 - r^2 = 0. \tag{14}$$

Because only $s(t)$ is approximated, the error in $s(t)$ results in a vector $r \cdot Q(s(t))$ which may be different from the normal direction $N(t)$ of $C(t)$. While the distance between the original curve and the offset approximation curve, for any parameter value t, remains r throughout, the error will be introduced in the form of the associated angular deviation. The curve tangent $C'(t)$ is expected to be perpendicular to $Q(s(t))$. However, since we are using $s(t)$ which is only an approximation, the orthogonality condition is not completely satisfied. Computing the deviation from the parallel condition: $N(t) \parallel Q(s(t))$, is one way to measure the error in the offset approximation of MO.

The error functional

$$\delta(t) = \frac{\langle C'(t), Q(s(t)) \rangle^2}{\|C'(t)\|^2}, \tag{15}$$

computes the value of $\cos^2 \alpha$, where α is the angle between $C'(t)$ and $Q(s(t))$ (see Fig. 4). (Note that the normal of $Q(s)$ is the same as $Q(s)$.) This angle, α, must be $\frac{\pi}{2}$ for an exact offset curve. When θ denotes the angle $\|\alpha - \frac{\pi}{2}\|$, the distance between the exact offset point $C_r(t)$ and an approximated offset point $C_r^a(t)$ is $\epsilon = 2r \sin \frac{\theta}{2} \approx r \sin \theta = r \cos \alpha$.

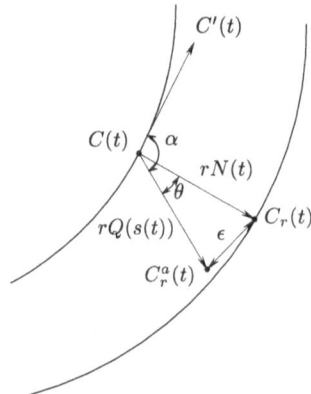

Fig. 4. Measuring the error in the offset approximation by tangent field matching

The resulting offset approximation curve is also rational. However, the degree is much lower, i.e., $d + 2$ (when we use a linear reparameterization $s(t)$), both for rational as well as polynomial base curves $C(t)$. Figure 5 shows an example of MO (Fig. 5–(a)), along with its error functional $\delta(t)$ (Fig. 5–(b)). In Fig. 6, we show MOs of a cubic Bézier curve, $C(t)$, $t_0 \le t \le t_1$, for different tolerances. Examine the line segments which connect $C(t_i)$'s and the corresponding $C_r^a(t_i)$'s, where $t_0 \le t_i \le t_1$. We can realize that the parallel relationship: $C'(t) \| C'^a_r(t)$, is enhanced as the precision of tolerance is increased.

2.5   Combined Approach: CAMO

One disadvantage of MO and CAO is that they generate a rational offset curve even for a polynomial base curve. In the case of MO, the exact circle $Q(s)$ in the

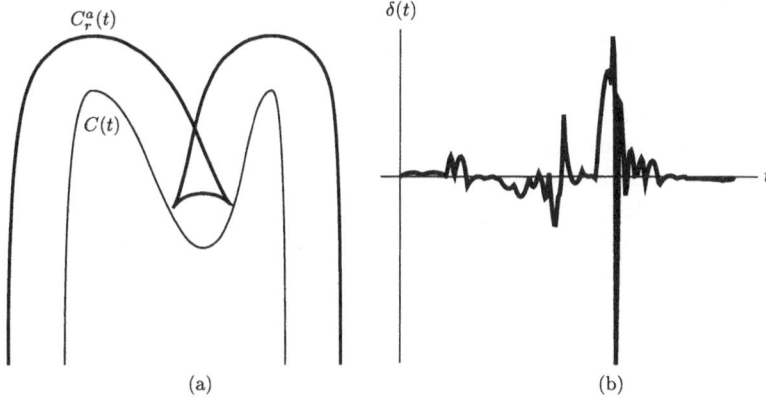

(a)   (b)

Fig. 5. MO and $\delta(t)$

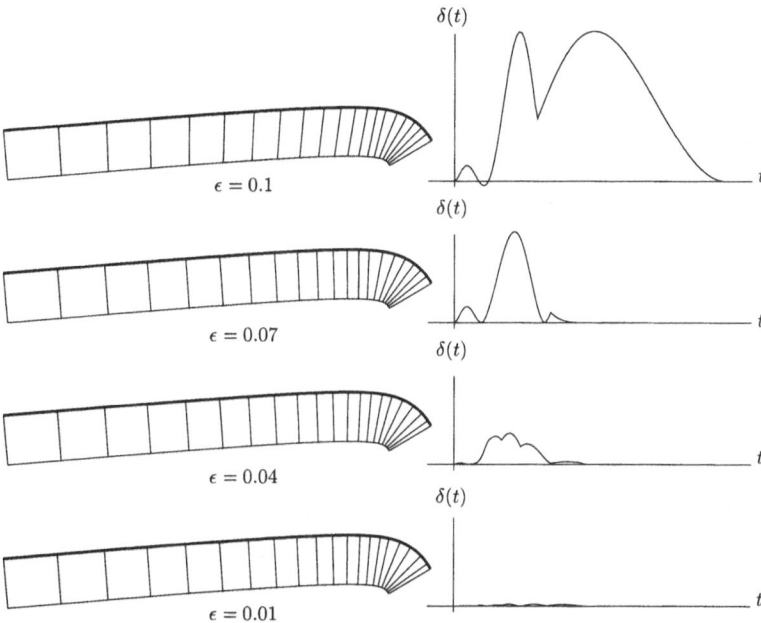

Fig. 6. MOs (bold curves) for a cubic Bézier curve for different tolerances.

offset approximation $C(t) + r \cdot Q(s(t))$ is inherently a rational quadratic curve. On the other hand, in CAO, the reparameterization $s(t)$ is a rational polynomial, while the curve $Q(s)$ is a polynomial quadratic curve.

This observation naturally suggests a new approach which adapts these two methods. That is, we use a quadratic polynomial approximation for $Q(s)$, and also use the tangent field matching technique to approximate $s(t)$ with a polynomial. Then, the resulting offset approximation curve $C_r^a(t)$ becomes a polynomial curve with the same degree as the base curve, when the base curve is a polynomial curve and the reparameterization polynomial $s(t)$ is linear. When we apply this method to a rational base curve, the degree of $C_r^a(t)$ is $d + 2$, where d is the degree of the base curve. We call this offset approximation method CAMO (Circle Approximation Matching Offset).

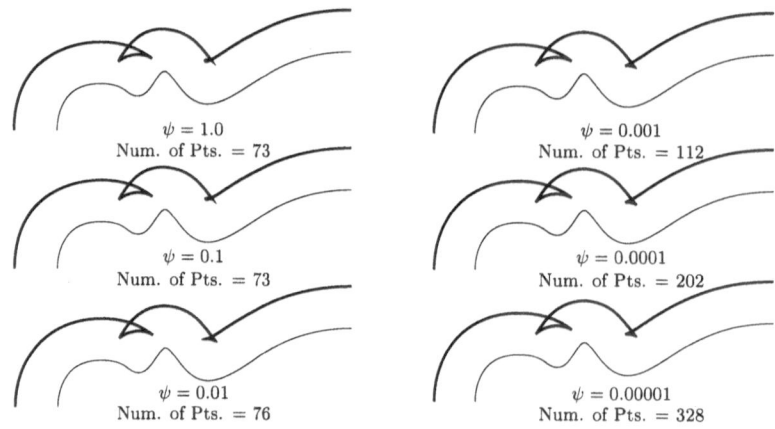

$\psi = 1.0$
Num. of Pts. $= 73$

$\psi = 0.001$
Num. of Pts. $= 112$

$\psi = 0.1$
Num. of Pts. $= 73$

$\psi = 0.0001$
Num. of Pts. $= 202$

$\psi = 0.01$
Num. of Pts. $= 76$

$\psi = 0.00001$
Num. of Pts. $= 328$

Fig. 7. Polynomial CAMOs (bold curves) of a cubic polynomial B–spline curve.

The method CAMO cannot use the error functional $\delta(t)$ of (15). This is because $Q(s(t))$ is not an exact unit circular arc. The modified error functional $\delta(t)$ is defined as follows:

$$\delta(t) = \frac{\langle C'(t), Q(s(t)) \rangle^2}{||C'(t)||^2 ||Q(s(t))||^2}. \tag{16}$$

In CAMO, we also need to consider the difference function $\psi(t)$ of (5), because CAMO does not use an exact rational quadratic curve to represent a unit circle. That is, there is no guarantee for the condition of (14). Thus, CAMO uses two tolerances of approximation error, ψ for the difference function $\psi(t)$, and ϵ for the function $\delta(t)$ of (16).

Figure 7 shows some CAMO curves of a cubic polynomial B–spline base curve for different tolerances of ψ and a fixed ϵ. All the offset approximation curves are cubic polynomial B–spline curves.

3   Convolution Approximation Methods

3.1   Planar Convolution Curve

Convolution is a classic operation which has been used as a tool for computing collision–free paths in robot motion planning [3,16,27]. Moreover, the convolution operation has applications in character font design [13,14], offset and rounding [30], and shape transformation [21].

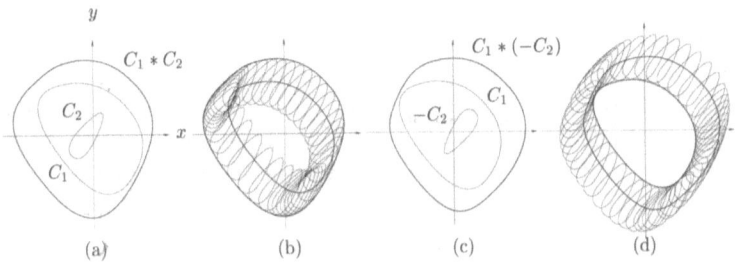

Fig. 8. Convolution and C–*space* obstacle

Given two regular parametric curves: $C_1(t) = (x_1(t), y_1(t))$, $t_0 \le t \le t_1$, and $C_2(s) = (x_2(s), y_2(s))$, $s_0 \le s \le s_1$, we define their convolution curve $(C_1 * C_2)(t)$ as follows (see [3,16]):

$$(C_1 * C_2)(t) = C_1(t) + C_2(s(t)), \tag{17}$$

where $C_1'(t) \parallel C_2'(s(t))$ and $\langle C_1'(t), C_2'(s(t)) \rangle > 0$, for some reparameterization $s = s(t)$. Figure 8–(a) shows the convolution curve $C_1 * C_2$ of two closed convex curves C_1 and C_2. The convolution curve $C_1 * C_2$ is an envelope curve which is obtained by sweeping one curve C_1 (with a fixed orientation) along the other curve C_2 [3] (see Fig. 8–(b)).

Convolution curve can be used as a tool for collision detection between two planar curved objects. Let O_1 and O_2 denote planar curved objects bounded by closed curves $C_1(t)$ and $C_2(s)$, respectively. Their *Minkowski sum* $O_1 \oplus O_2$ is defined by

$$O_1 \oplus O_2 = \{a + b \mid a \in O_1, \ b \in O_2\}. \tag{18}$$

The boundary $\partial(O_1 \oplus O_2)$ of the Minkowski sum is a subset of the convolution curve $C_1 * C_2$ [3,13,22]. When both O_1 and O_2 are convex objects, the convolution curve $C_1 * C_2$ is exactly the same as $\partial(O_1 \oplus O_2)$.

In Fig. 8–(c), we compute the convolution $C_1 * (-C_2)$, where $-C_2(s) = (-x_2(s), -y_2(s))$ is the symmetric curve of C_2 with respect to the local reference point (which is located at the origin). Similarly, we define $-O_2 = \{-b \mid b \in O_2\}$ as the reversed object of O_2 with respect to the origin. There is no collision between O_1 and O_2 as long as the reference point of O_2 does not penetrate into the Minkowski sum $O_1 \oplus (-O_2)$ which is bounded by the convolution curve $C_1 * (-C_2)$. In Fig. 8–(d), the moving object O_2 has continuous contact with the obstacle O_1 as it moves (with a fixed orientation) while its reference point follows along the convolution curve $C_1 * (-C_2)$. The object $O_1 * (-O_2)$ is called the *Configuration–space (C–space) obstacle* of the obstacle O_1 with respect to the moving object O_2. The *C–space* obstacle has been used as a tool for collision-detection in robot motion planning [3,24,27].

3.2   Previous Works

Ghosh [14,15] presented the concepts, algorithms, and data structures for various Minkowski operations in great detail. Guibas et al. [16] investigated some important properties of convolution curves. Convolution has a close relationship to other geometric entities such as common tangent and curve intersection. Lozano-Pérez [27] used the convolution technique to compute the *C–space* obstacles of polygonal objects.

Bajaj and Kim [3] discussed important issues in *C-space* generation: convolution curve generation, compatible subdivision of input curves, and the elimination of redundant convolution curve segments. They assumed exact algebraic representation of convolution curves which can be obtained by solving simultaneous algebraic equations, where input curves may be implicit and/or parametric. However, the resulting algebraic degrees are too high to be useful in practice. Ghosh and Mudur [13,15] demonstrated that (21) has a closed-form solution in some special cases. Most of his results are restricted to closed convex curves such as ellipse; moreover, the solutions are not polynomial or rational curves. Kohler and Spreng [24] suggested a method to compute the exact convolution curve in a parametric form, while providing some numerical techniques to speed up the computation; however, the parametric curves are not polynomial or rational curves, in general.

Lee and Kim [25] suggested an approximation method to compute the convolution curve. In a preprocessing step, the input curves C_1 and C_2 are approximated by a sequence of discrete points $C_1(t_j)$ and $C_2(s_j)$ $(j = 0, \ldots, k)$ at which they have predefined normal directions n_j. A simple vector addition $C_1(t_j) + C_2(s_j)$ generates a sequence of discrete points on $C_1 * C_2$ and thus constructs a piecewise linear approximation to the convolution curve $C_1 * C_2$. Kaul and Farouki [22] suggested another piecewise linear approximation. Without using any preprocessing, they generated a sequence of discrete points along the convolution curve on the fly, by computing (21) for two input curve segments. Ahn et al. [1] considered the general sweep in which the moving object may change its shape and orientation dynamically while tracing along a trajectory curve. They computed the boundary of a general sweep by approximating the sweep envelope curve with line segments.

As in the case of offset computation, it is desirable to develop efficient approximation methods for convolution curves. The previous methods either represent the convolution curve exactly by a non-rational algebraic curve or construct a piecewise linear approximation to the convolution curve. Linear approximations may not be acceptable to conventional CAD systems due to data explosion. In this paper, using the techniques of quadratic approximation and tangent field matching, we suggest two new methods to approximate convolution curves by polynomial or rational curves.

3.3   Compatible Subdivision

Let $C_1(t) = (x_1(t), y_1(t))$, $t_0 \leq t \leq t_1$, and $C_2(s) = (x_2(s), y_2(s))$, $s_0 \leq s \leq s_1$, be two regular parametric curves. Assume there is a one-to-one correspondence between $t \in [t_0, t_1]$ and $s \in [s_0, s_1]$, while satisfying the condition: $C_1'(t) \parallel C_2'(s)$ and $\langle C_1'(t), C_2'(s) \rangle > 0$; that is, $N_1(t) = N_2(s)$, where $N_1(t)$ and $N_2(s)$ denote the unit outward normal vectors (see (4)) of $C_1(t)$ and $C_2(s)$. Then the two curves $C_1(t)$ and $C_2(s)$ are called *compatible*.

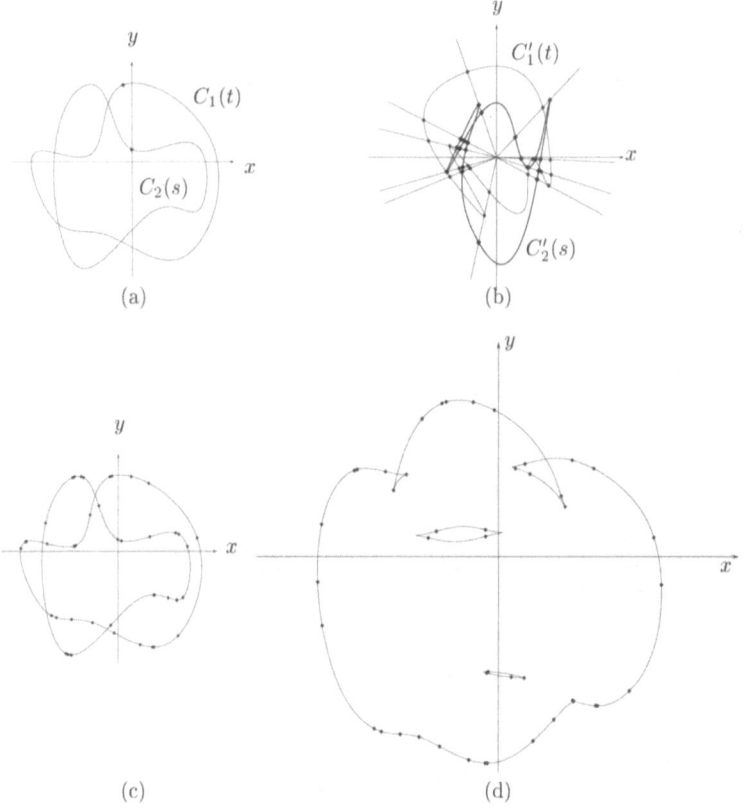

Fig. 9. Hodograph subdivision

When two input curves $C_1(t)$ and $C_2(s)$ are closed curves, but not convex, they are not compatible with each other. In this case, for the generation of convolution curves, we need to subdivide the input curves into compatible subsegments. Figure 9 shows a simple method (called *hodograph subdivision*) which subdivides $C_1(t)$ and $C_2(s)$ into compatible subsegments. Let $C_1(t_i)$ and $C_2(s_j)$ $(i = 1, \ldots, m_0, \text{ and } j = 1, \ldots, n_0)$ be all the inflection points of $C_1(t)$ and $C_2(s)$, respectively. The inflection points of a parametric curve, $C(u) = (x(u), y(u))$, can be computed by solving the following equation:

$$x'(u)y''(u) - x''(u)y'(u) = 0, \tag{19}$$

using symbolic and numeric computation tools [9]. Assume $C(u_0)$ is an inflection point. Then, in the hodograph $C'(u)$, the ray $\overline{OC'(u_0)}$ emanating from the origin O and passing through the point $C'(u_0)$ is tangent to the hodograph $C'(u)$. The set of rays:

$$\mathcal{L} = \{ \overline{OC_1'(t_i)}, \overline{OC_2'(s_j)}, \overline{OC_1'(t_0)}, \overline{OC_2'(s_0)} \mid i = 1, \ldots, m_0, \text{ and } j = 1, \ldots, n_0 \} \tag{20}$$

subdivides the plane R^2 into sector-form regions R_k $(k = 1, 2, 3, ..., l)$; see Fig. 9–(b).

After computing all the inflection points, the hodographs $C_1'(t)$ and $C_2'(s)$ are subdivided into piecewise curves $C_{1,i}'(t)$ $(i = 1, 2, 3, ..., m)$ and $C_{2,j}'(s)$ $(j = 1, 2, 3, ..., n)$ at all the intersection points with each ray in the set $\mathcal{L}$ of (20). These intersection points can also be computed using symbolic and numeric computation tools [9]. The original input curves, $C_1(t)$ and $C_2(s)$, can be subdivided into subsegments, $C_{1,i}(t)$ and $C_{2,j}(s)$ $(i = 1, 2, 3, ..., m \text{ and } j = 1, 2, 3, ..., n)$, at the same subdivision parameters for $C_{1,i}'(t)$ and $C_{2,j}'(s)$, respectively (see Fig. 9–(c)).

For a subsegment $C_{1,i}(t)$, let R_k be the sector-form region which contains the hodograph subsegments $C_{1,i}'(t)$. Then, for each $C_{2,j}'(s)$ contained in R_k, the subsegments $C_{1,i}(t)$ and $C_{2,j}(s)$ are compatible with each other. Figure 9–(d) shows a graph of convolution curve segments, in which each convolution curve segment is generated for each compatible pair of $C_{1,i}(t)$ and $C_{2,j}(s)$. In the following two subsections, we discuss how to approximate each convolution curve segment by a polynomial or rational curve.

3.4   Convolution Approximation Using Quadratic Curve Approximation: QAC

Let $C_1(t) = (x_1(t), y_1(t))$, $t_0 \leq t \leq t_1$, and $C_2(s) = (x_2(s), y_2(s))$, $s_0 \leq s \leq s_1$, be two compatible curve segments. The parallel relationship, $C_1'(t) \parallel C_2'(s)$, requires one to establish a proper reparameterization of $s = s(t)$ as a function of t. The equation representing the parallel relationship is given by:

$$y_2'(s)x_1'(t) - x_2'(s)y_1'(t) = 0. \tag{21}$$

Even if $C_1(t)$ and $C_2(s)$ are compatible, we do not have a unique algebraic representation of $s(t)$, because (21) may not have a unique closed-form solution for the curves of arbitrary degree.

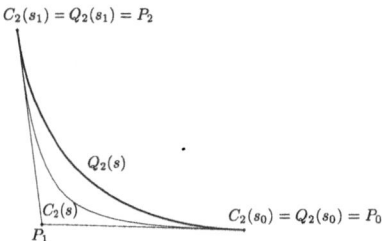

Fig. 10. Approximation of $C_2(s)$ by quadratic Bézier curve $Q_2(s)$ (bold curve)

When we approximate $C_2(s)$ by a quadratic curve, $Q(s) = (x_q(s), y_q(s))$, as in Fig. 10, we can represent $s(t)$ as a rational function of t. The approximated convolution curve $C_1(t) *^a C_2(s)$ can be represented as follows:

$$C_1(t) *^a C_2(s) = C_1(t) + Q_2(s(t)), \quad t_0 \le t \le t_1, \tag{22}$$

where $s(t)$ is the reparameterization satisfying the following equation:

$$y_q'(s)x_1'(t) - x_q'(s)y_1'(t) = 0. \tag{23}$$

The reparameterization function $s(t)$ can be computed by using a formula similar to (9), which results in a unique parametric representation of $C_1(t) + Q_2(s(t))$. When the approximation error between $Q_2(s)$ and $C_2(s)$ is ϵ, the resulting approximated convolution curve $C_1(t) + Q_2(s(t))$ also has the same approximation error ϵ. (In fact, this is true only for non-trimmed (closed or infinite) input curves and for the convolution curves after eliminating self–intersection loops; see Reference [26] for more details.) We call this method QAC (Quadratic Approximation Convolution). QAC is similar to the offset approximation method CAO in Section 2.3 which is based on the approximation of a circular arc with quadratic curves.

3.5 Convolution Approximation Using Tangent Field Matching: TMC

The other method directly approximates the reparameterization $s(t)$, rather than approximating the curve $C_2(s)$ using the tangent field matching that was introduced in Sect. 2.4.

The tangent field matching allows the computation of approximated convolution between $C_1(t)$ and $C_2(s)$, by a proper reparameterization $s(t)$ for $C_2(s(t)) = (x_2(s(t)), y_2(s(t)))$. Unfortunately, the approximation error cannot be computed with the distance function $\psi(t)$ which we use for the quadratic curve approximation. Instead of the distance function $\psi(t)$, we use the following formula which represents the value of $\cos^2 \alpha$, where α is the angle between $C_1'(t)$ and $N_2(s(t))$:

$$\delta(t) = \frac{\langle C_1'(t), N_2(s(t)) \rangle^2}{||C_1'(t)||^2}, \tag{24}$$

where

$$N_2(s(t)) = \frac{(y_2'(s(t)), -x_2'(s(t)))}{\sqrt{x_2'(s(t))^2 + y_2'(s(t))^2}} \tag{25}$$

is the unit normal vector field of $C_2(s(t))$. Then, the angle deviation can be represented as $\epsilon = \left\| \frac{\pi}{2} - \arccos(\sqrt{\max \delta(t)}) \right\|$. We call this method TMC (Tangent field Matching Convolution).

3.6   Comparisons and Examples

Each of the two approximation methods, QAC and TMC, has its own relative advantages and disadvantages. QAC is much simpler to compute than TMC. Moreover, we can guarantee that the resulting approximated convolution curve is within ϵ-distance from the exact convolution curve, where ϵ is the tolerance of quadratic curve approximation. When both $C_1(t)$ and $C_2(s)$ are polynomial

Table 2. Degrees of QAC and TMC approximations

C_1	C_2	QAC	TMC
d_1	d_2	rational $3d_1 - 2$	$\max(d_1, d_2)$
d_1	rational d_2	rational $3d_1 - 2$	rational $d_1 + d_2$
rational d_1	d_2	rational $5d_1 - 4$	rational $d_1 + d_2$
rational d_1	rational d_2	rational $5d_1 - 4$	rational $d_1 + d_2$

curves, TMC generates lower degree approximation curves than QAC (see Table 2). The degree of QAC approximation depends only on the degree of $C_1(t)$.

When we approximate $C_1(t)$ by another quadratic curve $\bar{Q}_1(t)$, the QAC approximation (applied to $\bar{Q}_1(t)$ instead of $C_1(t)$) is a rational curve of degree four. Figure 11 shows the degree and the number of control points of various QAC approximations computed from four different combinations of input curves. The QAC approximation shown in Fig. 11–(b) is computed from a cubic B–spline curve $C_1(t)$ and a rational cubic B–spline curve $C_2(s)$. Figure 11–(c) verifies the convolution curve by sweeping $C_2(s)$ (family of light curves) along $C_1(t)$.

Figure 12 shows TMC approximations computed for two cubic B–spline curves. The figure compares the number of control points that are generated for different tolerances of approximation error. Δ is the tolerance of normal deviation when we approximate $s(t)$. Figure 12–(b) verifies the convolution curve by sweeping $C_2(s)$ along $C_1(t)$.

4   Conclusion

In this paper, we have presented new methods for approximating the offset and convolution of planar curves. These methods are based on two approaches:

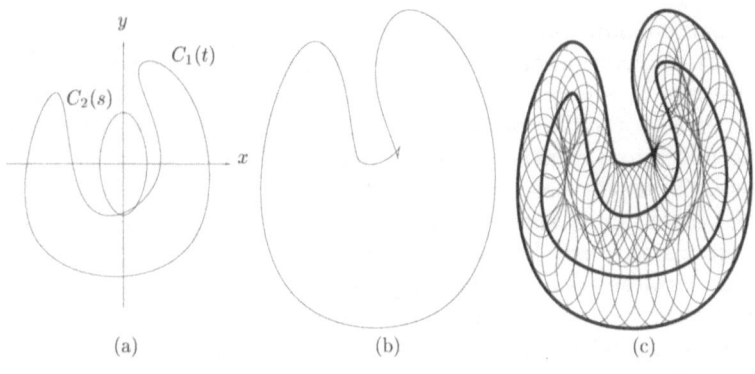

QAC	Degree	Number of control points		
		$\epsilon = 0.1$	$\epsilon = 0.01$	$\epsilon = 0.001$
$C_1(t) + Q_2(s(t))$	rational 7	211	435	1197
$C_2(s) + Q_1(t(s))$	rational 11	374	1058	3158
$\bar{Q}_1(t) + Q_2(s(t))$	rational 4	189	464	1368
$\bar{Q}_2(s) + Q_1(t(s))$	rational 4	159	472	1452

Fig. 11. Various QAC approximations computed from cubic $C_1(t)$ and rational cubic $C_2(s)$

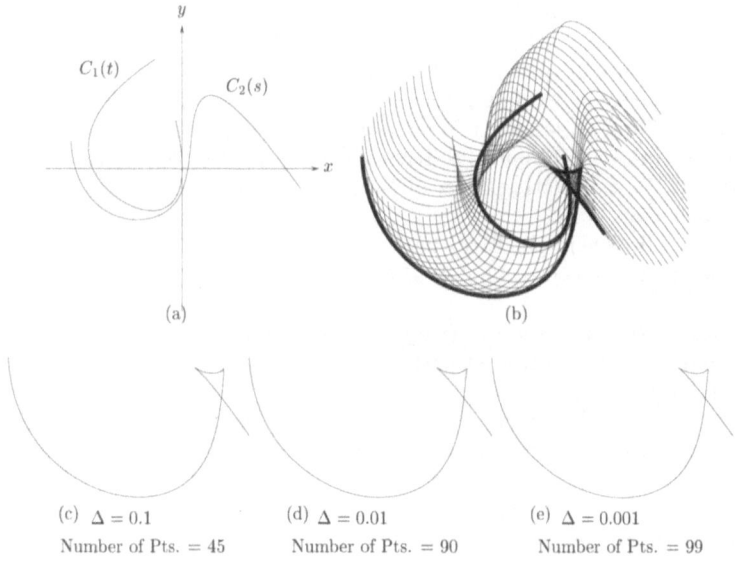

Fig. 12. TMC approximation of two cubic B–spline curves

quadratic curve approximation and *tangent field matching*. The approximations of offset and convolution curves are computed as polynomial or rational curves within the global tolerance of approximation error $\epsilon > 0$. The concepts of quadratic curve approximation and tangent field matching can be used for other geometric operations which are closely related to the normal and/or tangent directions of input curves.

The 3D extension of these approaches remains an important problem for future research. Bajaj and Kim [2,4] showed that the 3D offset and convolution of algebraic surfaces are also algebraic; however, their algebraic degrees are very high. Therefore, for their applications in practice, it is very important to approximate the 3D offset and convolution surfaces by polynomial and rational surfaces. We believe considerable research effort in CAGD should be devoted to the 3D extension of our current work as reported in this paper.

References

1. Ahn, J.-W., Kim, M.-S., and Lim, S.-B.: Approximate general sweep boundary of a 2D curved object. *CVGIP: Graphical Models and Image Processing*, **55** (1993) 98–128
2. Bajaj, C and Kim, M.-S.: Generation of configuration space obstacles: The case of a moving sphere. *IEEE J. of Robotics and Automation*, **4** (1988) 94–99
3. Bajaj, C and Kim, M.-S.: Generation of configuration space obstacles: The case of moving algebraic curves. *Algorithmica*, **4** (1989) 157–172
4. Bajaj, C. and Kim, M.-S.: Generation of configuration space obstacles: The case of moving algebraic surfaces. *The Int' J. of Robotics Research*, **9** (1990) 92–112
5. Cobb, B.: *Design of Sculptured Surface Using The B–spline Representation*. PhD thesis, University of Utah, Computer Science Department, (1984)
6. Cohen, S., Elber, G., and Bar-Yehuda, R.: Matching of freeform curves. to appear in *Computer-Aided Design*, CIS Technical Report, CIS 9527 (1995)
7. Coquillart, S.: Computing offset of B–spline curves. *Computer-Aided Design*, **19** (1987) 305–309
8. Elber, G. and Cohen, E.: Error bounded variable distance offset operator for free form curves and surfaces. *Int' J. of Computational Geometry and Applications*, **1** (1991) 67–78
9. Elber, G.: *Free Form Surface Analysis Using A Hybrid of Symbolic and Numerical Computation*. PhD thesis, Department of Computer Science, The University of Utah, (1992)
10. Elber, G., Lee, I.-K., Kim, M.-S.: Qualitative and quantitative comparisons of offset curve approximation methods. to appear in *IEEE Computer Graphics & Applications*, (1997)
11. Farouki, R.T., and Rajan, V.T.: Algorithms for polynomials in Bernsten form. *Computer Aided Geometric Design*, **5** (1988) 1–26
12. Farouki, R.T., and Neff, C.A.: Algebraic properties of plane offset curves. *Computer Aided Geometric Design*, **7** (1990) 101–127
13. Ghosh, P., and Mudur, S.P.: The brush-trajectory approach to figure specification: Some algebraic-solutions. *ACM Trans. on Graphics*, **3** (1984) 110–134
14. Ghosh, P.: A mathematical model for shape description using Minkowski operators. *Computer Vision, Graphics & Image Processing*, **44** (1988) 239–269

15. Ghosh, P.: A unified computational framework for Minkowski operations. *Computers and Graphics*, **17** (1993) 357–378

16. Guibas, L., Ramshaw, L., and Stolfi, J.: A kinetic framework for computer geometry. In *Proc. of 24th Annual Symp. on Foundations of Computer Science*, (1983) 100–111

17. Held, M.: A geometry-based investigation of the tool path generation for zigzag pocket machining. *The Visual Computer*, **7** (1991) 296–308

18. Hoffman, C.: *Geometric and Solid Modeling*. Morgan Kaufmann, (1989)

19. Hoschek, J.: Spline approximation of offset curves. *Computer Aided Geometric Design*, **5** (1988) 33–40

20. Hoschek, J., and Wissel, N.: Optimal approximation conversion of spline curves and spline approximation of offset curves. *Computer-Aided Design*, **20** (1988) 475–483

21. Kaul, A., and Rossignac, J.R.: Solid interpolating deformations: Construction and animation of pip. *Computers and Graphics*, **16** (1992) 107–115

22. Kaul, A., and Farouki, R.T.: Computing Minkowski sums of plane curves. *Int' J. of Computational Geometry & Applications*, 5 (1995) 413–432

23. Klass, R.: An offset spline approximation for plane cubic splines. *Computer-Aided Design*, **15** (1983) 297–299

24. Kohler, M., and Spreng, M.: Fast computation of the C-space of convex 2D algebraic objects. *The Int' J. of Robotics Research*, **14** (1995) 590–608

25. Lee, I.-K., and Kim, M.-S.: Primitive geometric operations on planar algebraic curves with Gaussian approximation. In T. L. Kunii, editor, *Visual Computing*, Springer-Verlag, (1992) 449–468

26. Lee, I.-K., Kim, M.-S., and Elber, G.: Planar curve offset based on circle approximation. to appear in *Computer-Aided Design*, (1996)

27. Lozano-Pérez, T.: Spatial planning : A configuration space approach. *IEEE Trans. on Computers*, **32** (1983) 108–120

28. Morken, K.: Best approximation of circle segments by quadratic Bézier curves. In P. J. Laurent, A. Le Méhauté, and L. L. Schumaker, editors, *Curves and Surfaces*, Academic Press, Boston, (1991) 331–336

29. Pham, B.: Offset approximation of uniform B–splines. *Computer-Aided Design*, **20** (1988) 471–474

30. Rossignac, J.R., and Requicha, A.A.G.: Offsetting operations in solid modeling. *Computer Aided Geometric Design*, **3** (1986) 129–148

31. Tiller, W., and Hanson, E.G.: Offsets of two dimensional profiles. *IEEE Computer Graphics & Applications*, **4** (1984) 36–46

Detection and Computation of Degenerate Normal Vectors on Tensor Product Polynomial Surfaces

Yasushi Yamaguchi

Department of Graphics and Computer Sciences, The University of Tokyo
3-8-1, Komaba, Meguro-ku, Tokyo 153, Japan
Telephone: +81-3-5454-6848, Telefax: +81-3-5454-6845
E-mail: yama@graco.c.u-tokyo.ac.jp
The author is visiting Stanford University by October, 1996.
Design Division, Department of Mechanical Engineering
Terman #540, Stanford University, Stanford, CA94305-4021, USA
Telephone: +1-415-725-0107, Telefax: +1-415-723-3521,
E-mail: yama@leland.stanford.edu

Abstract. One of the essential properties of a surface is its normal vector. Surface rendering, surface-surface intersection, and offset surface generation all require normal vectors. A normal vector at a point on a tensor product surface is usually obtained by taking a cross product of the two partial derivatives. However a normal vector can sometimes degenerate so the cross product yields a zero vector. Application programs might collapse by the degenerate normal vector. This paper is aimed at detecting any degenerate normal vectors of a tensor product Bézier surface as well as computing normal vectors at those points where they can be uniquely defined. Both the detection and the computation of degenerate normal vectors are carried out in a uniform manner based on a Bézier normal vector surface.

1 Introduction

A normal vector is essential for many applications in geometric modeling, such as surface rendering, surface-surface intersection, offset surface generation, rolling ball blending, and cutting path calculations. Most of the applications for free-form surfaces are based on normal vectors of the surfaces. Normal vectors of tensor product surfaces are usually computed by a cross product of two tangent vectors obtained as partial derivatives. However the normal vectors sometimes degenerate so that they cannot be calculated in this way. This degenerate normal vector might cause an error in the application programs. In order to avoid the difficulties, any degenerate normal vector at a point on a surface should be detected and the normal vectors at those points should be calculated in some other way. Previous studies [3,5,1] indicated that a degenerate normal vector may occur anywhere on a surface. It is not easy to detect every degenerate normal vector.

This paper proposes a new technique to detect any degenerate normal vectors of a tensor product Bézier surface as well as to compute normal vectors at those points where they can be uniquely defined. Section 2 discusses the degenerate normal vectors of a tensor product surface. Section 3 explains *Bézier normal vector surface* which represents normal vectors of a Bézier surface and an equation to calculate their control points from the original surface's control points. Section 4 proposes a method for detecting all degenerate normal vectors based on Bézier normal vector surfaces, and a method of computing unique normal vectors at those degenerate points as well. Section 5 summarizes this paper.

2 Degenerate Normal Vectors

An unnormalized normal vector $\mathbf{N}(u,v)$ at a point of a tensor product surface $\mathbf{S}(u,v)$ is defined as a cross product of two tangent vectors given as partial derivatives, $\mathbf{S}_u(u,v)$ and $\mathbf{S}_v(u,v)$:

$$\mathbf{N}(u,v) = \mathbf{S}_u(u,v) \times \mathbf{S}_v(u,v). \tag{1}$$

According to the definition of a normal vector, it is obvious that a normal vector $\mathbf{N}(u,v)$ in Equation (1) becomes zero if and only if either one of the following cases happens.

1. Zero partial derivative.
   Either one of two partial derivatives or both become zero,

   $$\mathbf{S}_u(u,v) = \mathbf{0} \quad \text{and/or} \quad \mathbf{S}_v(u,v) = \mathbf{0}.$$

2. Parallel partial derivatives.
   The partial derivatives are parallel,

   $$\mathbf{S}_u(u,v) = \lambda\mathbf{S}_v(u,v) \quad \text{where} \quad \lambda \neq 0.$$

The zero normal vectors categorized into these cases are called degenerate normal vectors. Farin pointed out some conditions of the degenerate normal vectors on the boundaries of Bézier surfaces concerning both cases in terms of the relative position of their control points [3].

A degenerate surface whose boundary edge collapses to a point results in the former case, a zero partial derivative. These kinds of surfaces are intentionally used in practice, especially when a triangular patch as shown in Figure 1 is required. The control polygon of the surface is drawn with bold dash-dotted lines. Any bold dash-dotted lines in a figure stand for a control polygon of a surface in this paper. The control points, $\mathbf{P}_{03}$ through $\mathbf{P}_{33}$, are located at the same point and the corresponding edge of the surface collapses at that point. Faux and Pratt discussed a way for computing a normal vector at the degenerate point[4]. Wolter and Tuohy proposed a method to determine curvatures at the degenerate point[6]. However this is not the only case of the degenerate normal

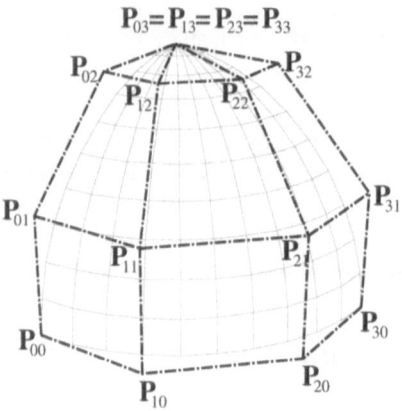

Fig. 1. A degenerate surface.

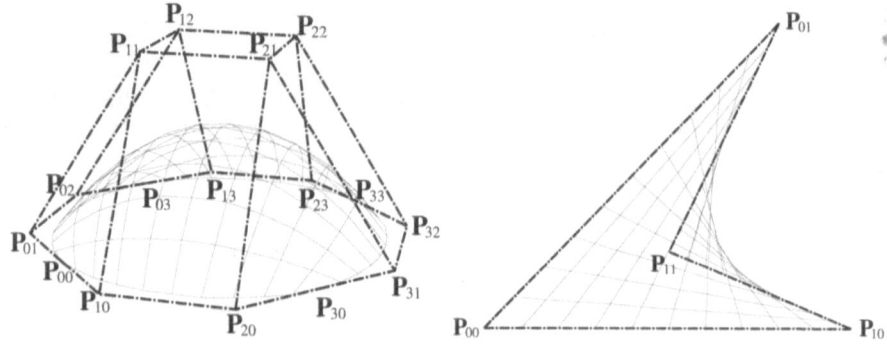

Fig. 2. Degenerate normal vectors of parallel partials.

vectors caused by a zero partial derivative. A zero partial derivative could occur even if a surface itself does not degenerate.

Figure 2 shows two examples of surfaces which have the degenerate normal vectors caused by parallel partial derivatives. The left figure illustrates a bi-cubic patch with rounded corners by locating a control point on each corner and its two neighboring boundary control points on a straight line, such as P_{01}, P_{00}, and P_{10}. Since those control points around each corner are collinear, the partial derivatives at the corner are parallel. The right figure is an example of a planar bi-linear patch specified as an example by Kim and Paralambros[5]. This patch contains degenerate normal vectors inside of the patch, because normal vectors of the surface are continuous and parallel to the Z-axis, and a normal vector at P_{00} has a positive Z-direction while a normal vector at P_{11} has a negative Z-direction. Kim and Paralambros pointed out the importance of the degenerate normal vector detection. Though they mentioned their study can cover every kind of degenerate normal vectors inside of the rectangular parametric patch,

Aumann, Rief, and Spitzmüller pointed out that an inside degenerate normal vector could occur because of zero partial derivatives[1]. Figure 3 gives an example of degenerate normal vectors caused by zero partial derivatives. Any $\mathbf{S}_u$ at $u = 0.5$ is zero and the normal vectors at those points degenerate. These kinds of degenerate normal vectors are categorized into the case 1 though they occur inside of a surface. The previous studies [3,5,1] indicated that a degenerate normal vector could occur anywhere on a surface and it is not easy to detect every degenerate normal vector only by checking the location of control points. This paper presents a method for detecting all degenerate normal vectors in a uniform manner.

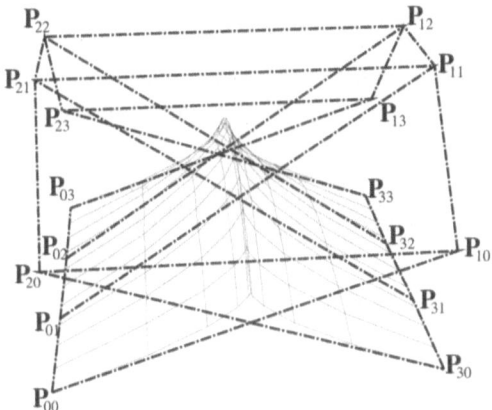

Fig. 3. Degenerate normal vectors of zero partials.

3   Bézier Normal Vector Surface

A locus of unnormalized normal vectors defined with Equation (1) constructs a surface, thus called *normal vector surface*. This section discusses the equation for calculating the control points of the Bézier normal vector surface from those of the original Bézier surface previously presented by the author[7].

A Bézier surface $\mathbf{S}(u, v)$ of degree $n \times m$ is defined by $(n + 1) \times (m + 1)$ control points $\mathbf{P}_{ij}$ as below:

$$\mathbf{S}(u,v) = \sum_{i=0}^{n} \sum_{j=0}^{m} B_i^n(u) B_j^m(v) \mathbf{P}_{ij}.$$

Here, $B_i^n(u) = \binom{n}{i} u^i (1 - u)^{n-i}$ and $B_j^m(u)$ denotes Bernstein polynomials of degree n and m, respectively, whose product is given as below (see [2]):

$$B_i^n(u) B_j^m(u) = \frac{\binom{n}{i}\binom{m}{j}}{\binom{n+m}{i+j}} B_{i+j}^{n+m}(u). \tag{2}$$

The tangent vectors in u, v-directions, namely the partial derivatives, are given as the following equations:

$$\mathbf{S}_u(u, v) = n \sum_{K=0}^{n-1} \sum_{L=0}^{m} B_K^{n-1}(u) B_L^m(v) \mathbf{Q}_{KL},$$

$$\mathbf{S}_v(u, v) = m \sum_{M=0}^{n} \sum_{N=0}^{m-1} B_M^n(u) B_N^{m-1}(v) \mathbf{R}_{MN},$$

where $\mathbf{Q}_{KL}$ and $\mathbf{R}_{MN}$ are defined as below:

$$\mathbf{Q}_{KL} = \mathbf{P}_{(K+1)\ L} - \mathbf{P}_{KL},$$

$$\mathbf{R}_{MN} = \mathbf{P}_{M\ (N+1)} - \mathbf{P}_{MN}.$$

The substitution of $\mathbf{S}_u(u, v)$ and $\mathbf{S}_v(u, v)$ in Equation (1) followed by application of the product formula (2) yields the following equation:

$$\mathbf{N}(u, v) = nm \sum_{K=0}^{n-1} \sum_{M=0}^{n} \sum_{L=0}^{m} \sum_{N=0}^{m-1} \left\{ \frac{\binom{n-1}{K}\binom{n}{M}\binom{m}{L}\binom{m-1}{N}}{\binom{2n-1}{K+M}\binom{2m-1}{L+N}} B_{K+M}^{2n-1}(u) B_{L+N}^{2m-1}(v) \right.$$

$$\left. \mathbf{Q}_{KL} \times \mathbf{R}_{MN} \right\}$$

$$= \sum_{i=0}^{2n-1} \sum_{j=0}^{2m-1} B_i^{2n-1}(u) B_j^{2m-1}(v) \mathbf{N}_{ij}.$$

This equation indicates that the normal vector $\mathbf{N}(u, v)$ of a Bézier surface $\mathbf{S}(u, v)$ of degree $n \times m$ is represented as a Bézier surface of degree $(2n-1) \times (2m-1)$. Thus this surface is named *Bézier normal vector surface*. The control points $\mathbf{N}_{ij}$ of the Bézier normal vector surface are given as below:

$$\mathbf{N}_{ij} = \frac{nm}{\binom{2n-1}{i}\binom{2m-1}{j}} \sum_{\substack{K+M=i \\ 0 \le K \le n-1 \\ 0 \le M \le n}} \sum_{\substack{L+N=j \\ 0 \le L \le m \\ 0 \le N \le m-1}} \left\{ \binom{n-1}{K}\binom{n}{M}\binom{m}{L}\binom{m-1}{N} \mathbf{Q}_{KL} \times \mathbf{R}_{MN} \right\}. \tag{3}$$

Figure 4 shows three examples of bi-cubic Bézier surfaces shown before and their bi-quintic Bézier normal surfaces. The upper left figure is the same Bézier surface shown in Figure 1 with unnormalized normal vectors. The unnormalized normal vectors around the north pole are small, because partial derivatives along latitudinal direction become small. The upper right figure shows the resulting Bézier normal vector surface and its control polygon calculated by Equation (3). The middle left figure is the Bézier surface and its unnormalized normal vectors are the same as the left surface in Figure 2. The degenerate normal vectors exist in its four corners. The middle right figure illustrates its Bézier normal vector surface with the control polygon. The lower left figure is the Bézier surface with its unnormalized normal vectors shown in Figure 3. The degenerate normal vectors occur along the line of $u = 0.5$. The lower right figure shows its Bézier normal vector surface with its control polygon obtained by Equation (3).

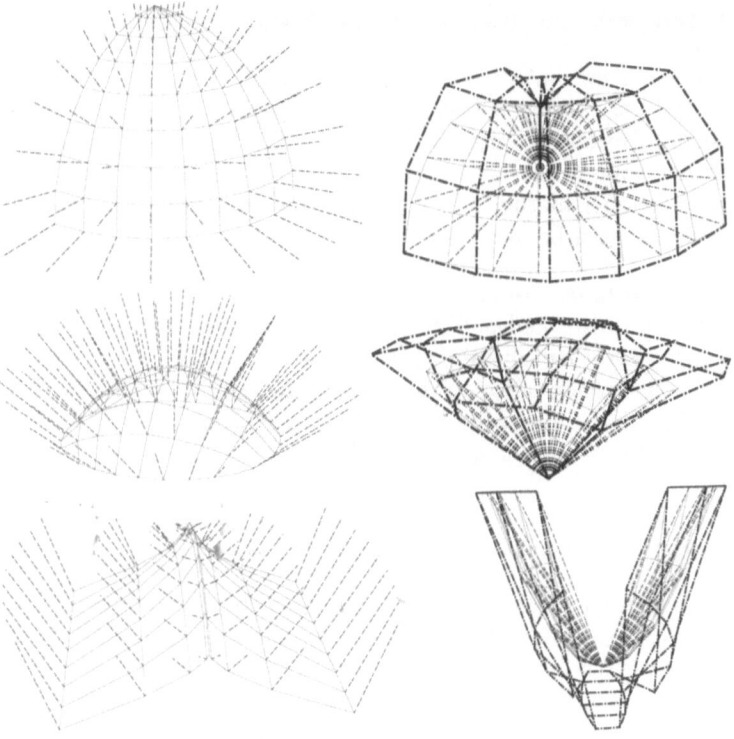

Fig. 4. Bézier surfaces and their normal vector surfaces.

Due to the nature of a Bézier normal vector surface as a Bézier surface, it has the following properties (see [3]).

1. Convex hull property.
   A Bézier normal vector surface $N(u, v)$ lies in the convex hull of the control polygon N_{ij}. Thus, a bounding box of the control polygon N_{ij} also bounds normal vectors of the surface $S(u, v)$.

2. Subdivision.
   A Bézier normal vector surface $N(u, v)$ can be subdivided at any parameter value of u and v by the de Casteljau algorithm.

4   Algorithms for Degenerate Normal Vectors

In this section a method for detecting degenerate normal vectors as well as a method of computing those degenerate vectors are proposed. Both methods are based on a Bézier normal vector surface.

4.1 Degenerate Normal Vector Detection

Since the Bézier normal vector surface is directly defined with a cross product of two partial derivatives, it coincides to the origin where the degenerate normal vector occurs. Because of the properties of a Bézier normal surface, i.e., convex hull property and subdivision, the degenerate normal vector can be detected by a divide-conquer method. The algorithm for detecting all degenerate normal vectors is illustrated below:

Degenerate normal vector detection.
Calculate a control polygon N_{ij} of the normal vector surface from the control polygon P_{ij} of the original surfaces.
while *The min-max box of the control polygon N_{ij} includes the origin.*
    if *The subdivision level is deep enough.* **then**
        Return the subsurface.
    else
        Subdivide the normal vector surface, i.e., the control polygon N_{ij}.
    { *No degenerate normal vector exists.* }

Figure 5 shows how the surfaces are recursively subdivided in the parametric space during the degenerate normal vector detection process. The upper left figure illustrates the subdivision process on the top surface in Figure 4 that is the degenerate triangular patch. In this case the boundary edge, $v = 1.0$, degenerates to a point and the point of the edge have a degenerate normal vector. The surface is subdivided along the edge into small pieces. The upper right figure corresponds to the middle surface in Figure 4 of the rounded corner patch. It has four degenerate normal vectors in its corners where the surface is actually subdivided into small pieces. The lower left figure shows the subdivision process on the planar concave bi-linear patch of the right surface in Figure 2. This figure points out that the surface has a sequence of degenerate normal vectors. The lower right one corresponds to the bottom surface in Figure 4 of the patch with zero partial derivatives. In this case partial derivatives S_u at $u = 0.5$ are all zero and the normal vectors at those points degenerate. The surface along the line, $u = 0.5$, is subdivided into small pieces.

4.2 Computation of Degenerate Normal Vectors

Some of normal vectors which cannot be calculated by Equation (1) may be uniquely defined. The concept of a Bézier normal vector surface is also useful for computing the exact normal vector at the point where the normal vector degenerates. Both the condition of a unique normal vector and the way of its computation are discussed in this section.

    There are mainly two different cases of degenerate normal vectors: the case that only the normal vector degenerates and the case that the surface itself degenerates. If the detection algorithm returned disjoint subsurfaces like the upper right case in Figure 5, the detected degenerate normal vector can be categorized

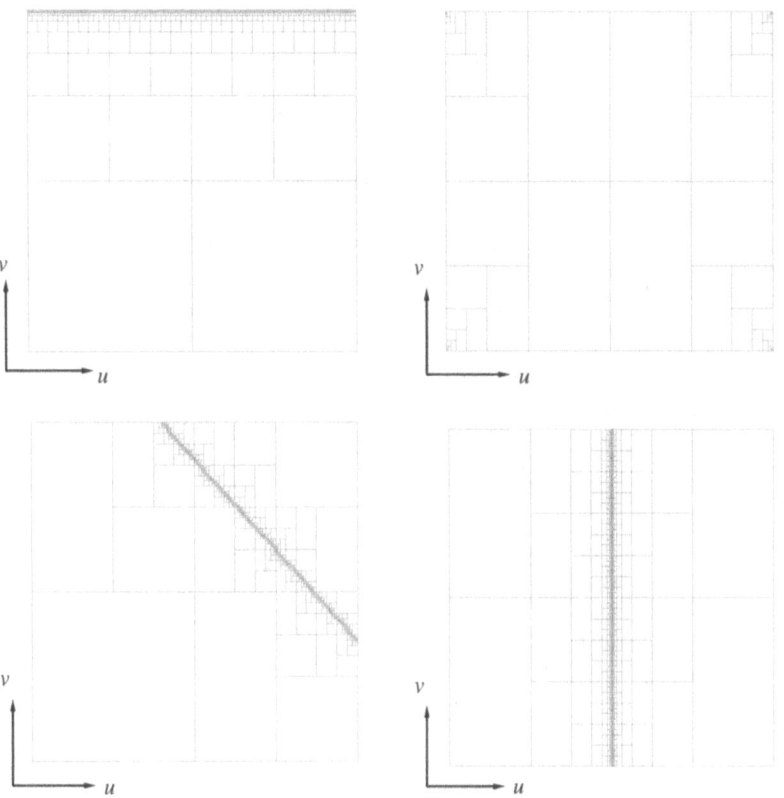

Fig. 5. Examples of degenerate normal vector detection process.

into the former case. In the latter case of a degenerate surface, the detection algorithm returns a sequence of adjacent subsurfaces and the tangent vectors along the sequence, namely the degenerate direction, are zero so that every point along the direction coincides to the same point. This degenerate direction must be one of the parameter directions, because any linear combination of two independent vectors cannot be zero except if both of the coefficients are zero. The normal vector is uniquely defined if the normal vectors along the degenerate direction are all parallel. Thus the normal vector of a degenerate surface will be computed in two steps, firstly calculate the local normal vector at each point and secondly check the normal vector consistency along the degenerate direction. In either case, the computation of a local normal vector is crucial.

The local normal vector at the degenerate point can be defined by taking the limit of the degenerate normal vector of its neighborhood. Any limits must be the same if a local normal vector is uniquely defined. By taking the Taylor series expansion of $\mathbf{N}$ up to the second order, the normal vector of the neighborhood is approximated as below:

$$\mathbf{N}(du, dv) \simeq \mathbf{N} + \mathbf{N}_u du + \mathbf{N}_v dv + \frac{1}{2} \left(\mathbf{N}_{uu} du^2 + 2\mathbf{N}_{uv} du dv + \mathbf{N}_{vv} dv^2 \right).$$

(4)

Here $\mathbf{N}$ at the degenerate point is zero due to the definition of the degenerate normal vector. The partial derivatives, $\mathbf{N}_u$, $\mathbf{N}_v$, $\mathbf{N}_{uu}$, $\mathbf{N}_{uv}$, and $\mathbf{N}_{vv}$, can be calculated either by differentiating the Bézier normal vector surface or by computing from the derivatives of the original Bézier surface $\mathbf{S}$ as follows:

$$\mathbf{N}_u = \mathbf{S}_{uu} \times \mathbf{S}_v + \mathbf{S}_u \times \mathbf{S}_{uv}$$
$$\mathbf{N}_v = \mathbf{S}_{uv} \times \mathbf{S}_v + \mathbf{S}_u \times \mathbf{S}_{vv}$$
$$\mathbf{N}_{uu} = \mathbf{S}_{uuu} \times \mathbf{S}_v + 2 \left(\mathbf{S}_{uu} \times \mathbf{S}_{uv} \right) + \mathbf{S}_u \times \mathbf{S}_{uuv}$$
$$\mathbf{N}_{uv} = \mathbf{S}_{uuv} \times \mathbf{S}_v + \mathbf{S}_{uu} \times \mathbf{S}_{vv} + \mathbf{S}_{uv} \times \mathbf{S}_{uv} + \mathbf{S}_u \times \mathbf{S}_{uvv}$$
$$\mathbf{N}_{vv} = \mathbf{S}_{uvv} \times \mathbf{S}_v + 2 \left(\mathbf{S}_{uv} \times \mathbf{S}_{vv} \right) + \mathbf{S}_u \times \mathbf{S}_{vvv}.$$

Therefore the following discussions on the normal vector computation do not depend on Bézier normal vector surfaces and are applicable to any differentiable tensor product surfaces.

According to Equation (4) the local normal vector at the degenerate point is calculated in the following way:

Local normal vector computation.
  if   *Either one or both of the first partials $\mathbf{N}_u$ and $\mathbf{N}_v$, are not zero.*
  and   *The degenerate point is on the boundary of a surface.*
  and   *The first partials, $\mathbf{N}_u$ and $\mathbf{N}_v$, are linearly dependent*
          with the same orientation in the corresponding surface domain. **then**
        The normal vector can be defined by $\mathbf{N}_u du + \mathbf{N}_u dv$ where the signs
        of du and dv are determined with the corresponding surface domain.
  else if *The second partial $\mathbf{N}_{uv}$ is zero.*
  and   *The second partials, $\mathbf{N}_{uu}$ and $\mathbf{N}_{vv}$, are linearly dependent*
          with the same orientation. **then**
        The normal vector can be defined by $\mathbf{N}_{uu} + \mathbf{N}_{vv}$.
  else if *The degenerate point is a corner of a surface.*
  and   *The second partials, $\mathbf{N}_{uu}$, $\mathbf{N}_{uv}$, and $\mathbf{N}_{vv}$, are linearly dependent*
          with the same orientation in the corresponding surface domain. **then**
        The normal vector can be defined by $\mathbf{N}_{uu} + \mathbf{N}_{vv} + \mathbf{N}_{uv} du dv$ where the
        signs of du and dv are determined with the corresponding surface domain.
  else
        The normal vector might not be uniquely defined.

Figure 6 shows the locally calculated normal vectors of the surfaces depicted in Figure 4. The degenerate normal vectors which are defined with higher order derivatives are shown in bold dotted lines.

The global check can be done by investigating higher order derivatives as well as by comparing the locally defined normal vectors along the degenerate

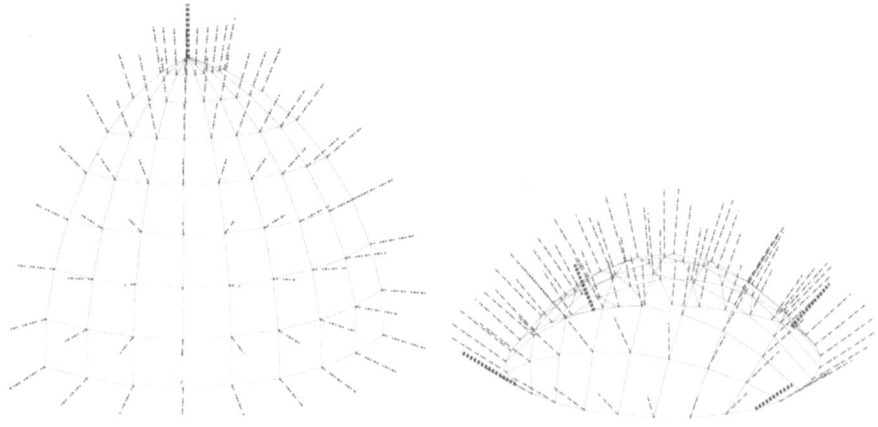

Fig. 6. Locally defined degenerate normal vectors.

direction. The degenerate direction, which is either u-direction or v-direction, is detected by computing $\mathbf{S}_u$ and $\mathbf{S}_v$. For instance $\mathbf{S}_u$, $\mathbf{S}_{uu}$ and so on are all zero if the degenerate direction coincides to the u-direction. In this case $\mathbf{N}$, $\mathbf{N}_u$, $\mathbf{N}_{uu}$, etc. are also all zero. This means the local normal vector might be defined by $\mathbf{N}_v$, $\mathbf{N}_{vv}$, or else. The global consistency of the normal vector is checked in the following way:

Global consistency check for a normal vector in u-direction.
    if $\mathbf{S}_u, \mathbf{S}_{uu}, \cdots$ *are zero.* $(\mathbf{N}, \mathbf{N}_u, \mathbf{N}_{uu}, \cdots$ *are all zero.)* **then**
        if $\mathbf{N}_v$ *dominates the local normal vector.*
          and $\mathbf{N}_{uv}, \mathbf{N}_{uuv}, \cdots$ *are all parallel to* $\mathbf{N}_v$.
        or $\mathbf{N}_{vv}$ *dominates the local normal vector.*
          and $\mathbf{N}_{uvv}, \mathbf{N}_{uuvv}, \cdots$ *are all parallel to* $\mathbf{N}_{vv}$. **then**
          The normal vectors are consistent along the degenerate direction.
        else
          The normal vectors are inconsistent along the degenerate direction.
    else
        The surface does not degenerate in u-direction.

Here the order of derivatives to be investigated depends on the order of the surface in the degenerate direction.

Figure 7 shows degenerate surfaces having non-unique normal vectors at the degenerate points. The upper surface has a degenerate point on its boundary and the lower surface has one inside it. Both at the degenerate points, normal vectors can be calculated locally and depicted in bold dotted lines in the right figures. However they are not inconsistent in a global sense. This inconsistency can be detected by investigating higher order derivatives.

Although the normal vectors at the degenerate points as seen in the examples of Figure 7 are not mathematically defined, some rendering systems might require normal vectors at those points. Since the intensity calculation is sensitive

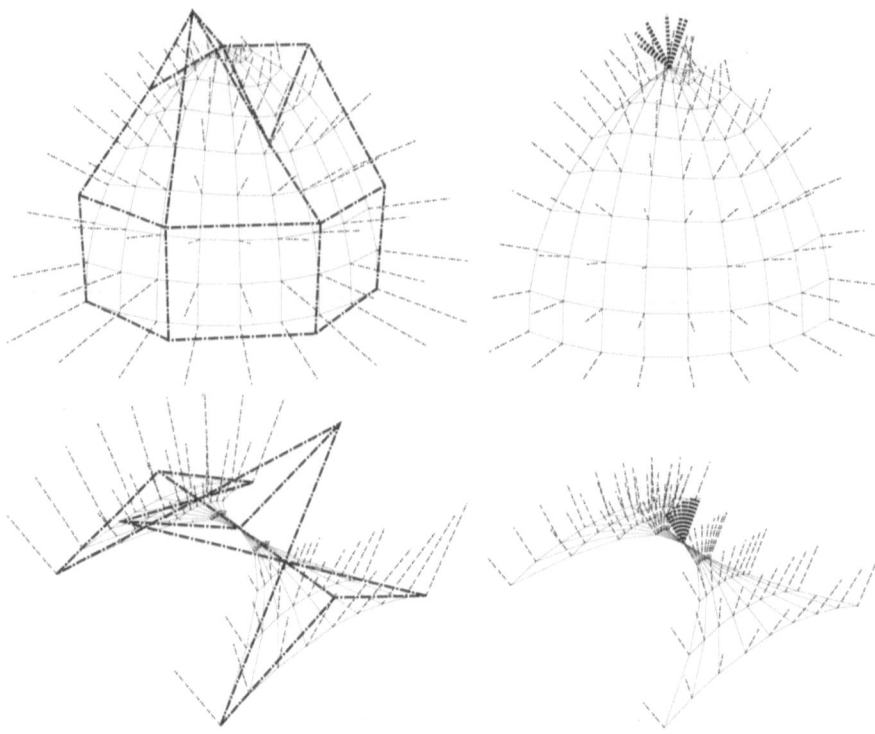

Fig. 7. Globally inconsistent normal vectors of degenerate surfaces.

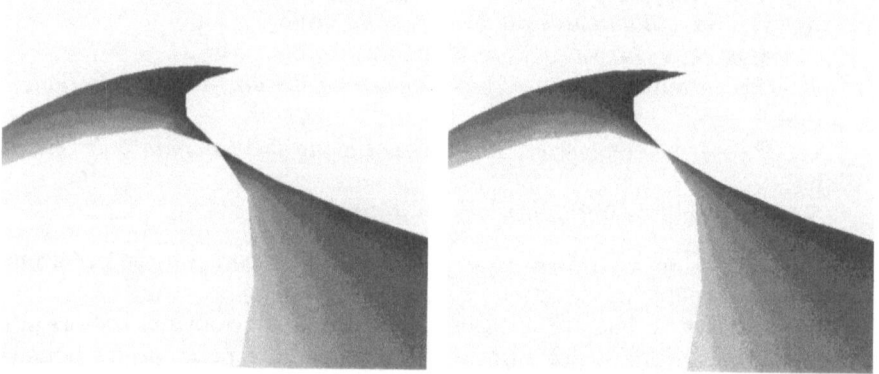

Fig. 8. Rendering with locally defined normal vectors.

to the magnitude of a normal vector, the surface could have black spots where the normal vectors are zero. The local normal vectors are useful to compromise with this problem. Figure 8 shows the effect of normal vectors. The left figure is rendered with zero normal vectors at the degenerate point and has a black spot around the point, while the right figure rendered with local normal vectors seems good.

5   Conclusions

This paper proposed a technique for solving the problems of degenerate normal vectors, i.e., degenerate normal vector detection and exact normal vector computation. The advantages of this method are as follows:

- A Bézier normal vector surface, which has good properties, is introduced.
- A method for detecting all degenerate normal vectors based on a Bézier normal vector surface is proposed.
- Methods for computing a normal vector are also discussed. These methods can calculate a normal vector in both local and global senses and applicable to any tensor product surfaces.

Acknowledgements

The author thanks F. Prinz of Stanford University for his encouragement to this study.

References

1. G. Aumann, U. Reif, and K. Spitzmüller. A counterexample to a corollary of kim et al. *Computer Aided Geometric Design*, 12(8):853–855, 1995.
2. C. de Boor. B-form basics. In G. Farin, editor, *Geometric Modeling: Algorithms and New Trends*, pages 131–148. SIAM, 1987.
3. G. E. Farin. *Curves and surfaces for computer aided geometric design*. Academic Press Inc., 1993. Third edition.
4. I. D. Faux and M. J. Pratt. *Computational Geometry for Design and Manufacture*. Ellis Horwood Ltd., 1979.
5. T.-S. Kim and P. Y. Paralambros. Detection of degenerate normal vectors on parametric surfaces: Tangent cone approach. *Computer Aided Geometric Design*, 12(3):321–327, 1995.
6. F.-E. Wolter and S. T. Tuohy. Curvature computations for degenerate surface patches. *Computer Aided Geometric Design*, 9(3):241–270, 1992.
7. Y. Yamaguchi. Surface-surface intersection with critical point detection based on Bézier normal vector surfaces. *(submitted to Computer Aided Geometric Design)*, 1996.

Surfaces Intersection for Solid Algebra: A Classification Algorithm

S. Foufou, J.M. Brun and A. Bouras

LIGIM, Bât. 710, Université Claude Bernard Lyon I
43, Bd du 11 Novembre 1918
69622 Villeurbanne Cedex, France
<foufou, bouras>@ligim.univ-lyon1.fr

Abstract. Detecting intersecting surfaces and computing their intersection curves is one of the fundamental problems in solid modeling algebra. This paper introduces a new strategy to classify pairs of surfaces according to their intersection status, where exact geometric entities are replaced by fuzzy ones. Intersecting surfaces are thus replaced by fuzzy intersecting faces and a fuzzy intersection algorithm provides a 3-State classification of pairs of surfaces: certainly intersecting, certainly non-intersecting and potentially intersecting.

In the case of certainly intersecting surface couples the fuzzy intersection algorithm provides also starting points to march along the intersection curve, which turns most of the potentially intersecting couples into certainly intersecting ones. The remaining potential intersections are then subdivided to refine their status. Statistics shows the efficiency of this global strategy.

1 Introduction

The use of free form surfaces in solid modeling changes the way one has to look at surface intersections. The detection of each and every intersection becomes a crucial problem with severe computational constraints since the solids algebra makes the comparison of faces against intersections an $O(n * m)$ problem if the solids have respectively n and m faces. Moreover the intersection of two surfaces varies from a single point (tangency) to a number of closed loops and surface portions, assuming that self intersections have been eliminated previously. A very large number of intersection computation algorithms have been proposed and a good survey of the literature in this field can be found in [6,24]. Manocha in [20] classifies surface intersection algorithms into four major categories: Subdivision, Lattice evaluation, Marching and Analytic methods.

Subdivision methods [18,13,14,2,23,19] recursively decompose the problem into similar sub-problems which are easy to solve, the surfaces being subdivided until each sub-piece is within relatively loose tolerance for flatness and edge linearity [13,5]. These methods are generally robust but very slow when a high precision is required.

In lattice evaluation methods a set of curves is computed on one of the two surfaces and their intersection points with the second surface are evaluated using a curve/surface intersection algorithm [4]. Afterwards these discrete points will be sorted and connected to form all intersection curves. Because of the discrete nature of these methods, it is often hard to find all small loops and singularities.

Marching methods [5,1,7,15,3] begin by locating at least one point on each intersection curve and using it as a starting point to step along the required curve in a direction prescribed by the curve's local geometry. The robustness of marching methods depends on two points: the detection of all intersection lines and the quality of the marching process (along tangency areas and/or in the vicinity of singularities).

Analytic methods search an analytic representation for the intersection of two low degree algebraic surfaces [26,25]. Using analytic methods to compute the intersection of two parametric surfaces requires implicitization and inversion techniques. The complexity of the representation causes analytic methods to be impractical for high degree parametric surfaces.

Mixed schemes have been proposed when one surface has a low degree algebraic description and the other one a parametric description, the intersection curve is an implicit curve in the parametric space of the parametric surface. Recently Manocha et al. [20,17] extended this approach to couples of parametric surfaces, using the implicitization algorithms pioneered by Sederberg [26]. To cope with the high degree implicit surfaces obtained, they use techniques from elimination theory such as the resultant formulation [10] to represent the intersection curve as the zero set of a determinant. While audacious, considering numerical stability problems arising from high degrees, these mixed schemes can provide an interesting alternative to all parametric approaches.

Another formulation, which represents the intersection of two surfaces as the zero set of the distance function, was proposed by Cheng and others [8,21,22]. Computing the characteristic points of the distance function helps one in this case to identify all the intersection features and to decompose it into a set of smooth branches [11,15]. This paradigm, while appealing because of its sound foundations and elegant approach, seems to contain too many intricacies to be applied widely.

This paper addresses the detection of intersecting surfaces and starting point computation in the context of solids algebra (Boolean operators) for which it is of the utmost importance to avoid unnecessary computations for non-intersecting faces. The topological intricacies of solid algebra seem to be high enough to refrain one from using and maintaining intricate solutions. For intersections, we take aim at simple as well as robust and performing solutions.

2   Detection and Classification of Surface Intersection

A classical approach to detecting surface intersections is to classify surfaces in intersecting and non-intersecting pairs. This classification is done by comparison of enclosing volumes or by tessellation of the surfaces. Since the comparison of enclosing volumes (bounding boxes) can be very efficient for distant non-intersecting faces, while tessellation can be very efficient for clearly intersecting faces, a combination of these methods is often used in practical (commercial) solid modelers.

The fact that tessellation lacks robustness in limit cases (near tangency), explains why recursive subdivision of faces and associate bounding boxes is often used in theoretic modelers to ensure that all intersections are found (at a predefined accuracy).

2.1   2-States Classification Approaches

When bounding boxes are oriented along the x, y, z axes, the comparison between two boxes:

$$b_1(xmin_1, ymin_1, zmin_1, xmax_1, ymax_1, zmax_1)$$
$$b_2(xmin_2, ymin_2, zmin_2, xmax_2, ymax_2, zmax_2)$$

is pretty simple:

If $[(xmin_1 > xmax_2)$ or $(ymin_1 > ymax_2)$ or $(zmin_1 > zmax_2)]$ or
$[(xmin_2 > xmax_1)$ or $(ymin_2 > ymax_1)$ or $(zmin_2 > zmax_1)]$
Then   boxes b_1 and b_2 are not intersecting.

Comparison between xyz boxes is often called the min-max test to remind one how it works [19]. Its efficiency is tied to its speed which comes from its simplicity, conversely this crude test often fails to detect the non-intersecting status of close by faces since the xyz boxes enclose points very far from the faces (Fig 1).

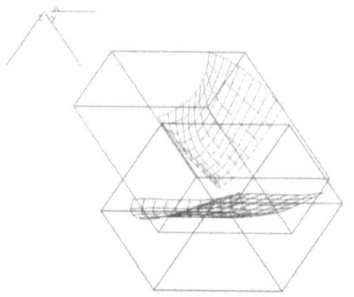

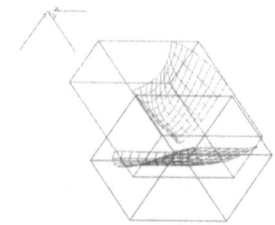

Fig. 1. xyz bounding boxes   **Fig. 2.** Oriented bounding boxes

To improve the quality of the bounding boxes, it is possible to orient the boxes along the main dimensions of the faces (Fig. 2). However the implementation of such face-oriented boxes is more costly and their intersection test ($6 * 6$ faces intersections) is pretty slow compared to the min-max test. This is why people generally prefer to subdivide surfaces and consequently reduce boxes of the resulting patches instead of computing face oriented boxes.

On one hand, subdivision is a robust and easy to implement strategy, but it is at least as costly as oriented boxes while often providing only small improvements in wasted space around the surface. This explain why CAD-CAM systems developers generally turn to a more practical scheme based on a tessellation of surfaces and computation of polyhedral intersections. Polyhedral intersections include generally a min-max test and provide starting points for marching algorithms. This strategy performs reasonably when the tessellation is not too fine but can miss nearly tangent intersections or find false ones unless the tessellation is extremely fine.

To fix ideas about the relative costs of subdivision and tessellation note that the subdivision of a Bézier patch of degree $n*m$ is of complexity $O(n*m^2)$ or more precisely $9((n+1)(m+1)(m+2n))/2$ floating operations when the subdivision is done first in the m, then in the n directions. On the other hands the tessellation need to compute $k*l$ points on the surface and the complexity of a point computation is $O(n*m^2)$ which seems equivalent, but a more precise evaluation indicates $9((m(n+1)(m+1))/2$ floating operations for the De Casteljau scheme and the global computation would be $(k-2)[(9m(m+1)/2) + (l-2)(9n(n+1)/2)]$since the n poles of an isoparametric line of k points are computed only once. Finally the subdivision of a $10 * 10$ degree patch would cost 16335 floating operations while a $7*7$ tessellation of that surface would cost 14850 floating operations only in the De Casteljau scheme. Moreover one can use the Bézier-Horner scheme to compute an isoparametric set of points and the global computation would be $(k-2)((8m+1) + (l-2)(8n+1))$ for which a $16 * 16$ tessellation would cost 17010 floating operation only. These computing cost considerations explain why a tessellation is generally chosen in place of subdivisions, mainly for high degree Bézier patches.

The classical drawback of tessellation is its lack of robustness when surfaces are in limit case position: false intersections can be found or actual ones can be missed. To avoid such errors the tessellation approximation can be taken into account and refined when surfaces are too close (within the "thickness" of the tessellation). Unfortunately this is again time consuming since minimum distance calculations between polyhedral tessellations are time consuming.

2.2   The 3-State Classification Approach

When the thickness of the tessellation is taken into account the refinement is dictated by a third state: The potentially intersecting one. A robust and practical scheme results from the 3-State classification that we formalize here under. This classification paradigm is organized around a set of tests operating as a hierarchical filtering of surface couples, classifying them as:

1. Non-intersecting.
2. Potentially intersecting.
3. Certainly intersecting along a single open line.

The coarse grain filtering is done by comparison of enclosing volume, this filter classifying couples as non-intersecting or potentially intersecting. It can be defined as a rejection test that rejects quickly the obvious non-intersecting cases from further investigation.

Enclosing volumes can be more or less close to the surface. Their quality can be defined through two criteria:

- The closeness to the surface, measured by the volume measure.
- The speed of the rejection test.

The high volume measure of xyz boxes is compensated by the high speed of the test when used as the first step in the filtering strategy. The main improvement of the coarse grain filtering is to be found in the constitution of a hierarchy of boxes, which can lower the complexity to $O(n \log n)$ from $O(n^2)$. This can be done through a space partition that list faces interfering with each space part, either voxels or octree subdivision.

Instead of a dynamic refinement of xyz boxes when the state 2 is found, which can lead to infinite subdivision when a tangency case occur, it is generally better to define a second level of treatment that switches from the subdivision scheme to a closer enclosing volume. This explains why surface oriented boxes are sometimes used.

In our approach this second level of testing is replaced by the first step of a fuzzy geometry algorithm which computes fuzzy face intersections and classifies fuzzy face couples in the above 3-State logic. The status of surface couples is closely related to their fuzzy faces one since if fuzzy faces are in:

a) State 1, there is no intersection of the surfaces.
b) State 2, there is a potential intersection that can be a closed line, a set of lines, a unique line or tangency areas (at the modeling precision level).
c) State 3, there is an intersection defined as a fuzzy line (open).

The advantages of this computation are that:

- It is much faster than the face oriented boxes with an equivalent efficiency for detection of non-intersecting faces.
- It classifies intersecting faces as such and eliminates them from further investigation.
- It selects actual limit cases for a third level treatment that can be either the simple but time consuming subdivision or a more elaborate second order algorithm.

The fuzzy geometry intersection computation is done by a fuzzy geometry algorithm, which operates on an extension of classical geometry to fuzzy geometry entities. In this approach faces, lines and points are replaced by their equivalent fuzzy entities: fuzzy face (FF), fuzzy lines (FL) and fuzzy points (FP).

3   Fuzzy Geometry Concepts

Before developing fuzzy geometry algorithm we have to define what we intend by fuzzy geometry.

Fuzzy sets were defined by Zadeh [29,30] to formalize decisions to be taken when real (continuous) data are sampled. Boolean logic operates on 2-States values 0 or 1 and the first approach to automatism and automata was to define thresholds that produce binary values out of continuous ones. Incidentally that is the way computers and logic circuits operate internally; a clock is needed in that case to synchronize all internal analogic values at values consistent with the thresholds in order to obtain effective boolean logic.

The use of pure boolean logic on non synchronized samplings of continuous data is at least hazardous. Fuzzy logic is intended to overcome these hazards and works on fuzzy data (defined as fuzzy sets) characterized by their relative "distance" to 0 and 1. Such logic proves to be less sensitive to inaccuracies in the sampling values and well suited to pattern recognition problems.

What is fundamental here is that even though fuzzy sets logic and fuzzy geometry algorithms both work on inaccurate or uncertain data, they are fundamentally different since fuzzy geometry algorithms provide two types of results:

- Geometric results that are continuous and metric.
- Topological results that have a 3-State logic (in, on, out).

For intersections the geometric result is the intersection line which may exist or not and this is the topological result. The fuzzy geometric result is a fuzzy intersection line while the fuzzy topological result is a fuzzy 3-State logic:

- No intersection
- Potential intersection
- Sure intersection (the fuzzy line)

using fuzzy sets and fuzzy logic for fuzzy geometry problems would be inappropriate and the fuzzy concepts need extensions from fuzzy sets of data to fuzzy geometric entities and from fuzzy logic to fuzzy geometry algorithms.

Note that even though interval arithmetic is also intended to cope with inaccuracies it can reflect only the metric aspects of fuzzy geometry, not the topological ones, as it will be seen later on.

4   Fuzzy Geometric Entities

The fuzzy geometric entities are computed here either from Bézier surfaces, or from Bézier sub-patches of a B-spline, or from a (precise) Bézier approximation of the surface. Other ways can be envisaged to compute the fuzzy geometric entities.

4.1   Fuzzy Points

A fuzzy point (FP) $\bar{P}$ is a ball of center c and a fuzziness value f. We use the concept of fuzzy point to express fuzzy line intersections and fuzzy line/mean plane intersections.

4.2   Fuzzy Lines

The fuzzy line (FL) concept is close to the fat line introduced by Sederberg and Nishita [28] and defined as the region between two parallel lines which bounds a Bézier curve as shown in Fig. 3.

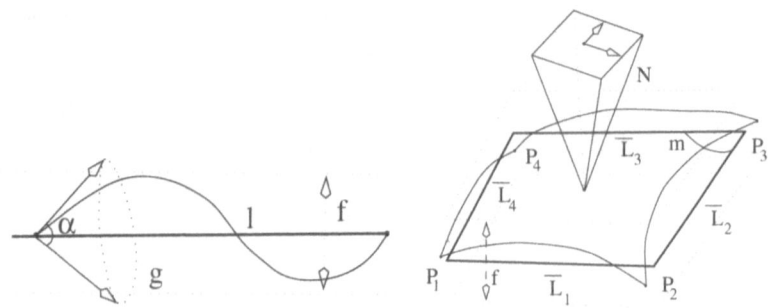

Fig. 3. A fuzzy line   **Fig. 4.** A fuzzy face

A FL $\bar{L}$ is composed of a mean line l and a fuzziness value f. In addition to the thickness of Sederberg's fat lines, a tangent envelope g is defined. The angle α (Fig.3) is called the angular fuzziness of the FL. Angular fuzziness is related to the topological fuzziness (number of potential intersections) and is the main difference from interval arithmetic.

4.3   Fuzzy Faces

We define a fuzzy face (FF) $\bar{F}$ as the region between two parallel planes (Fig. 4). It is computed from the Bézier control points of the surface and is composed of:

- A mean plane m.
- A fuzziness value f.
- An envelope of normals N.

Mean Plane. To compute the FF that bounds a surface, a plane m given by the two diagonals of the surface is first computed. Distances from Bézier control points to the plane m are used to compute the fuzziness value of the surface. Taking this fuzziness value into account, the plane m is thus translated to a mean position against the surface.

A FL $\bar{L}_i$, $i = 1, \ldots, 4$ is calculated for each boundary curve of the surface, to define the contour of the plan m.

Normal Envelope. The topology of the intersection curve between two surfaces can be derived from the interactions of their normal envelopes which can be enclosed by a more or less precise volume as is the case for the face itself. Several approaches allow the computation of a normal bounding volume N: For Sederberg and Mayer [27] N is a cone. For Kreizis et al [16] N is a pyramid with a rectangular base. For Daniel [9] it is a pyramid with a convex hull base.

In our implementation, the normal envelope is bounded by three volumes, a pyramid with a rectangular base and the enclosing/enclosed cones. These volumes allow us to define easily a 3-State logic for normal envelope interactions:

1. Certainly interacting.
2. Potentially interacting.
3. Not interacting.

5   Basic Fuzzy Geometry Algorithms

5.1   Identity of Points

The only true binary algorithm in geometry is point identity which derives from the definition of the point: A point is the portion of space such that two points are either separate or identical.

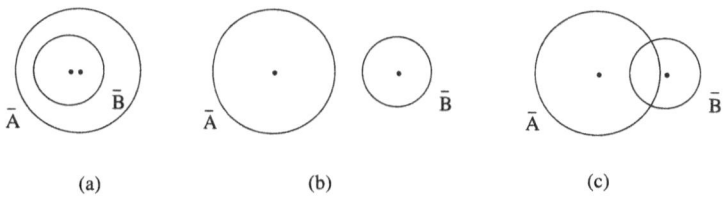

(a)   (b)   (c)

Fig. 5. Identity of fuzzy points

General point sets, continuous or not, can have some common points, in which case we say that they are intersecting.

If we introduce the notion of dimension for continuous points sets, their intersection must also be completed by the dimension parameter.

When introducing fuzzy entities, the identity of points switches to a 3-State logic (Fig.5):

 − $\bar{A} = \bar{B}$ if $d(\bar{A}, \bar{B}) \leq \min(A_f, B_f)$

- $\bar{A} \neq \bar{B}$ if $d(\bar{A}, \bar{B}) > A_f + B_f$
- $\bar{A}$? $\bar{B}$ if $\min(A_f, B_f) < d(\bar{A}, \bar{B}) \leq A_f + B_f$

where A_f, B_f are the fuzziness values of points $\bar{A}$ and $\bar{B}$ respectively.

5.2   Identity of Lines

Given two lines $L_1(A_1, \vec{V_1})$ and $L_2(A_2, \vec{V_2})$ (Fig. 6.a), the condition $L_1 = L_2$ implies that $A_1 \in L_2$, $A_2 \in L_1$ and $\vec{V_1} = \vec{V_2}$. If we try to define the corresponding fuzzy conditions, it appears that two fuzzy lines $\bar{L}_1 = \bar{A}_1 \bar{B}_1$ and $\bar{L}_2 = \bar{A}_2 \bar{B}_2$ can be considered either identical or separate (Fig. 6.b) since:

$$\left. \begin{array}{l} \bar{A}_1 \in \bar{L}_2, \ \bar{B}_1 \in \bar{L}_2 \Rightarrow \bar{L}_1 = \bar{L}_2 \\ \bar{A}_2 \in \bar{L}_1, \ \bar{B}_2 \notin \bar{L}_1 \Rightarrow \bar{L}_2 \neq \bar{L}_1 \end{array} \right\} \Rightarrow \ ?$$

If α and β are angular fuzzinesses of $\bar{L}_1$ and $\bar{L}_2$, the fuzzy equality of directions $\vec{V_1}$ and $\vec{V_2}$ would be written as:

$$\theta < (\alpha + \beta)/2 \ \Rightarrow \ ? \ldots$$

It appears from this study that identity of two lines implies a strict identity of the directions $\vec{V_1}$, $\vec{V_2}$ and the two fuzzy belongings $A_1 \in \bar{L}_2$ and $A_2 \in \bar{L}_1$ corresponding to fuzzy points identity.

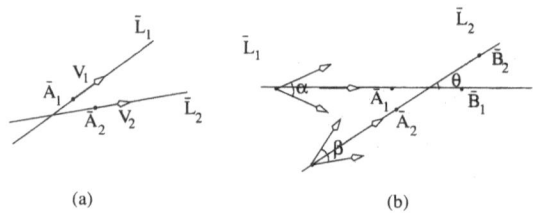

(a)   (b)

Fig. 6. Inconsistencies in fuzzy lines identity

This raises the question of the different aspects that geometric fuzziness can take and spanning limits that fuzzy geometric entities have to comply with.
Geometry in a restricted sense deals only with metric aspects of space (geometry = earth measurement), but in a broad sense it includes logical and classification aspects which can be made precise if one separates:

- Topography for pure metric aspects (or restricted sense geometry).
- Topology for logical aspects of coincidence and neighborhood.
- Morphology for classification of shapes aspect.

fuzzy geometry would thus encompass fuzzy topography and fuzzy topology as well as fuzzy morphology, consequently fuzzy line imply a spanning length (it is a line segment) and includes as fuzziness parameters:

- A thickness (for topographic aspects).
- A tangents angle (for topologic and morphologic aspects).

The identity of $\bar{L}_1$ and $\bar{L}_2$ as unlimited lines (without morphologic fuzziness) requires the definition of a spanning length which implies that within a space of length D all points of $\bar{L}_1$ pertain to $\bar{L}_2$.

This limitation to line segments rises the morphologic fuzziness aspects of fuzzy geometry in which the identity of $\bar{L}_1$ and $\bar{L}_2$ must be replaced by the notion of equivalence of $\bar{L}_1$ and $\bar{L}_2$. For some geometric computations one can consider $\bar{L}_1$ and $\bar{L}_2$ as equivalent if they provide the same fuzzy result, even though $\bar{L}_1$ is a FL used to represent a segment of line having only position fuzziness, while $\bar{L}_2$ is used to represent an arc of curve having position and shape fuzziness.

Morphologic fuzziness has an important consequence since two lines can intersect in one point at most (or be identical) whereas two curves can intersect in more that one point and remain distinct. In the three topological states of precise geometry (in, on, out), topological fuzziness replaces the state "on" by "Potentially on", which means that the number of "on" points can be $0, 1$ or > 1.

5.3   Intersection of Lines

Line intersections introduce the notion of intersection dimension since lines can be:

- Identical if they share two points (Fig. 7.a). Intersection is of dimension 1 :

$$AB \cap CD = CB$$

- Intersecting if they share one point only (Fig. 7.b). Intersection is of dimension 0:

$$AB \cap CD = I$$

- Non-intersecting if they don't share any point (Fig. 7.c). Intersection is of dimension -1:

$$AB \cap CD = \phi$$

To know if a point C is on a line implies angular computations that are explicit when a line is defined by a point and a direction $L = (A, \overrightarrow{V})$:

$$C \in L \Leftrightarrow \sin \alpha = 0 \quad \text{or} \quad \overrightarrow{V} \wedge \overrightarrow{AC} = \overrightarrow{0}$$

When one computes a fuzzy distance between a FP $\bar{C}$ and a FL $\bar{L} = \bar{A}\bar{B}$ the result $d = \|\overrightarrow{V} \wedge \overrightarrow{AC}\|$ is consistent with L_f and the condition

$$\bar{C} \in \bar{L} \Leftrightarrow d = \|\overrightarrow{V} \wedge \overrightarrow{AC}\| < \min(L_d, C_f)$$

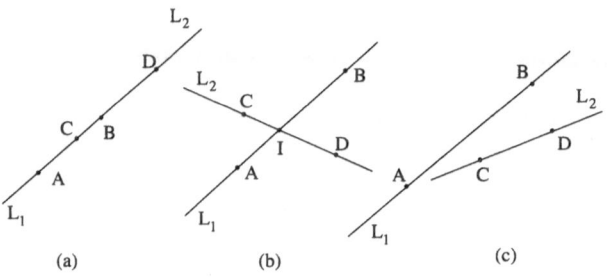

Fig. 7. Lines intersection

would apply (the direction $\vec{V}$ of $\bar{L}$ is given by $\vec{V} = \frac{\vec{AB}}{\|\vec{AB}\|}$).

The intersection of L_1 and L_2 can be computed in $2D$, and is considered as a FP $\bar{P}$ or a FL $\bar{L}$ (Fig. 8.c). If one considers $\bar{L}$ as the intersection, the fuzziness L_f can be defined as:

$$L_f = \max(L_{1f}, L_{2f})$$

while, if one considers $\bar{P}$ as the intersection, the fuzziness P_f can be computed as:

$$P_f = \sqrt{\frac{L_{1f}^2 + 2L_{1f}L_{2f}\cos\theta + L_{2f}^2}{\sin^2\theta}} \geq \frac{\max(L_{1f}, L_{2f})}{\sin\theta}$$

If one considers the intersection of two FL intersecting in a sure way the result would be "better" (less fuzzy) when considering a FL than a FP, moreover the limit points $\bar{A}_1, \bar{B}_1$ for $\bar{L}_1$ and $\bar{A}_2, \bar{B}_2$ for $\bar{L}_2$ can be inside or outside the previously defined intersection (Fig. 8.a) and can limit further the effective intersection of $\bar{L}_1$ and $\bar{L}_2$ as limited lines.

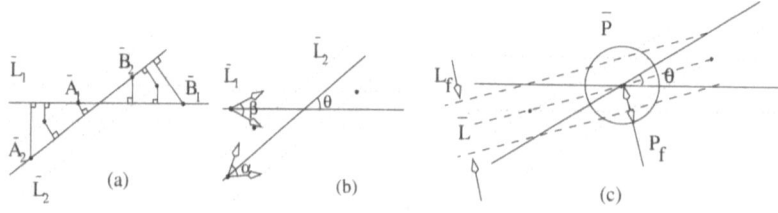

Fig. 8. Fuzzy lines intersection

When considering multi level accuracy in geometric modeling, one can envisage replacing a basic precise but complex geometry by a simpler but less precise one. If fuzzy geometry is used the result would be computed faster but be less accurate than the basic geometry would allow. Nevertheless the result can be topologically sure and it is valuable to know quickly that there is no intersection when, statistically, most of the intersections are effectively null. Moreover it can be valuable to know simultaneously that there is one and only one intersection and the domain in which this intersection lies, can it be a FP or a FL. Angular fuzziness proves useful for this purpose.

Consider two FL having a non-null intersection (Fig. 8.b). If the angular fuzzinesses α and β are such that $\theta > \frac{\alpha+\beta}{2}$, the number of intersections between the accurate curves approximated by the FL cannot exceed one point (fuzzy at the precision of the accurate curves) which is close to the FP of intersection of the fuzzy lines. This FP of intersection can be used as a starting point for the iterative algorithms generally used to compute the intersection between curves. Conversely, if the angular fuzzinesses α and β are such that $\theta \leq \frac{\alpha+\beta}{2}$ the number of intersections can exceed one point, the topological result is fuzzy and more precise geometry must be used. Since the intersection domain is known as a FL of intersection, one can restrict each curve to the portion pertaining to the fuzzy intersection then obtain a more accurate approximation and/or subdivide the curve to reduce the fuzziness (distance as well as angular fuzziness).

This fuzziness reduction process will end either in a topologically sure (0 or 1 intersection) or in a geometric fuzziness under the modeling accuracy and provides the ultimate FP of intersection.

One can note that the fuzziness reduction process can be done through a subdivision of the precise geometry or a more precise approximation (more complex than a line but simpler than the precise geometry) and the corresponding entities. These basic concepts are used in the FF intersection presented below.

6   Fuzzy Faces Intersection

The intersection of two fuzzy faces $\bar{F}_1$ and $\bar{F}_2$ is the common part $\bar{I}$ between two fuzzy lines: $\bar{I}_1$ the intersection of $\bar{F}_1$ with m_2 the mean plane of $\bar{F}_2$ and $\bar{I}_2$ the intersection of $\bar{F}_2$ with m_1 the mean plane of $\bar{F}_1$. There is no intersection between $\bar{F}_1$ and $\bar{F}_2$ if either $\bar{I}$ or $\bar{I}_1$ or $\bar{I}_2$ is null. The intersection is only potential if $\bar{I}$ is reduced to a FP.

6.1   Contours/Mean Planes Intersection

The intersection of the contour lines $\bar{L}_i$, $i = 1, \ldots, 4$. of each face with the mean plane of the other face is computed to evaluate fuzzy intersection points.

The fuzzy line/mean plane intersection is based on the computation of the Euclidean distances of the fuzzy segment extremities to the mean plane. To decide about the intersection these distances are compared to a fuzziness parameter r:

$$r = \frac{1}{2}(L_f + F_f * cos\beta)$$

where L_f and F_f are respectively fuzziness values of the FL and the FF, and β denotes the the angle between the mean line and the mean face.

When the intersection is confirmed the center c of the fuzzy intersection point is computed directly by the mean line/mean plane intersection, while its fuzziness is assumed to be the r value.

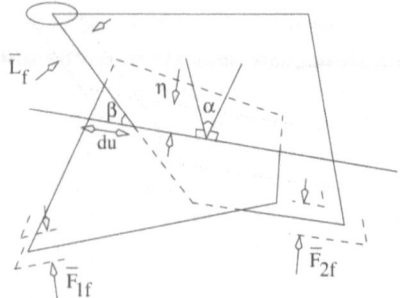

Fig. 9. Fuzzy faces intersection

6.2   Fuzzy Faces Overlapping Test

To test whether fuzzy faces overlap, we verify the overlapping of the intersection fuzzy points calculated in the above section.

For each FP of intersection $\bar{P}_i$, $i = 1, \ldots, 4$, two values are computed (Fig. 9): the parameter u_i of the point onto the FL of intersection and an overlapping tolerance $du_i = \delta du_1 + \delta du_2$ with:

$$\delta du_1 = \frac{\cos\beta}{\sin\alpha}\eta \; , \; \delta du_2 = \frac{\bar{F}_f}{2}\sin\beta \; , \; \eta = \frac{\bar{L}_f}{2\sin\alpha}$$

where F_f and L_f are respectively fuzziness of the FF and FL whose the intersection gives the FP $\bar{P}_i$. We denote by α the angle between mean planes m_1 and m_2.

> **if** $(u_3 - du_3 > u_2 + du_2) \; Or \; (u_4 + du_4 < u_1 - du_1)$
>   **then** No Intersection
> **else if** $(u_3 + du_3 < u_2 - du_2) \; And \; (u_4 - du_4 > u_1 + du_1)$
>   **then** Intersection
> **else**   Potential Intersection

7   Classification Results Analysis

The classification paradigm is organized around a set of tests as a hierarchical filtering of surface couples.

7.1   Bounding Boxes Test

This first level of rejection is the classical xyz bounding boxes.

This test classifies surfaces into potentially intersecting and non-intersecting pairs. This well known strategy has the advantage of rapidity, even if it often makes rough selections by classifying as potentially intersecting surfaces which

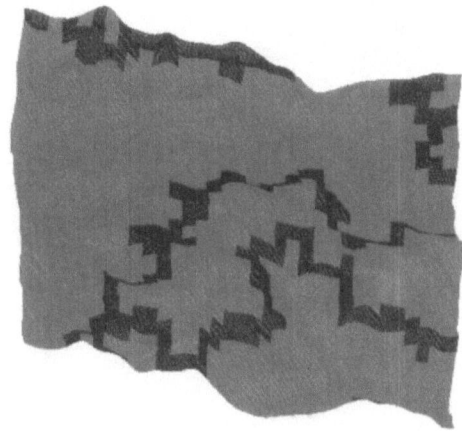

Fig. 10. Bounding boxes test classifications

are not. However, in solids algebra, the number of couples made from far away surfaces is high enough to make this test significantly efficient.

We have applied this methodology to a piece of land (mountains) modeled by a B-spline surface split down in 1089 bicubic Bézier surfaces to find regions which may intersect a horizontal plane.

This test has the advantage of clarifying the interpretation of results since face couples boil down to faces of the mountain surface while the number of faces is significant and intersection status complicated enough to test effectively the detection of intersections. Robustness and computation complexity statistics are realistically exercised through this example. The brown colored regions in Fig. 10 are classified as potentially intersecting with the plane, while the green colored ones are classified as non-intersecting by the bounding boxes test.

One can note that bounding boxes are near optimal in such cases since lands are $z = F(x, y)$ models for which xyz boxes are particularly well suited.

For general cases xyz boxes would be less adequate and fail to reject intersections more often. For solid modeling, the horizontal plane (of our example) would be replaced by the set of faces bounding a solid and the land by another set of faces. In this case, xyz boxes would have been built for all the faces of both solids. The test would have been more realistic but less easy to interpret.

7.2   Fuzzy Faces Test

When the first test returns a potential intersection status one can envisage sub-dividing faces, computing new bounding boxes and apply the bounding boxes test recursively until a size limit is reached. This provides a starting point (the bounding box center) for intersection search. Although the classical subdivision scheme is simple, it has severe performances drawbacks.

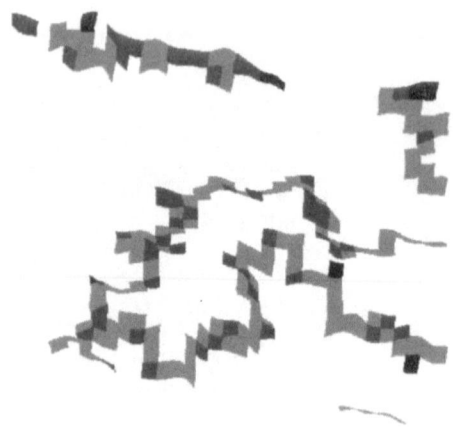

Fig. 11. Fuzzy faces test classifications

In our fuzzy geometry approach the fuzzy faces intersection provides a further classification of surfaces as:

- Intersecting: Figure 11 shows intersecting surfaces (red colored regions) among surfaces classified as potentially intersecting by the first test
- Non-intersecting: violet colored regions in Fig. 11 are classified as non-intersecting with the plane by this test
- Potentially intersecting: beige colored regions in Fig. 11 are still considered as potentially intersecting

One can note that the status of about 60% of surfaces classified as potentially intersecting by the first test changes to be certainly intersecting, the status of about 7% changes to be certainly non-intersecting, and the status of about 33% is still potentially intersecting.

For general cases one can expect to see large modifications in these percentages since xyz boxes would fail to classify correctly non-intersecting faces, which would be corrected by the fuzzy intersection.

If one doubles the number of potentially intersecting faces left by the min-max test the statistics would change to: 30% certainly intersecting, 53% non-intersecting and 17% potentially intersecting. This shows clearly the interest of the certainly intersecting status obtained at no extra cost in the comparison since it lowers significantly the number of undecided faces.

7.3   Normal Envelopes Test

The last test is based on the intersection of the normal envelopes criterion [27,12]. It verifies the intersection between normal envelopes in order to provide a complementary status for the intersection curve. Two cases can occur:

- Single and open intersection (normal envelopes don't intersect)
- Possibility of multiple or closed intersections (normal envelopes intersect)

For single and open intersection curves (for certain or possible cases) the intersecting faces connect themselves in closed loops. These sets of faces correspond to closed loops (made out of edges which are open intersections) and one can march along these intersections. The fuzzy geometry algorithm used to intersect fuzzy faces provides information about the best starting point for the marching algorithm out of the certainly intersecting faces already known.

Fig. 12. Lines of fuzzy faces intersections

Fig. 13. Intersection Curves after the marching step

Intersection segments between fuzzy faces of Bézier surfaces and the fuzzy faces of the plane of the above example are shown in Fig. 12. They are colored yellow. One can note that these Segments take roughly the form of the real intersection curve presented in Fig. 13.

When there is a possibility of multiple or closed intersection, a subdivision algorithm is to be applied for further classification. These subdivisions are stopped at a certain level in case of tangency. They also provide a tangency area between surfaces.

Finally one can note that the marching algorithm pass through faces classified as potentially intersecting (with a single intersection possibility) and modifies their status. The subdivision scheme is reserved to close or multiple intersection faces which are in very small number.

8   Conclusion

The detection and classification of surface intersections was done in a 3-State classification paradigm, through the definition of fuzzy geometry concepts and the related fuzzy geometry algorithms.

In the fuzzy geometry approach faces, lines and points are considered as fuzzy at the topological, and morphologic levels and replaced by (simpler) geometric entities: mean planes, mean lines and mean points. These new entities are afterwards qualified by fuzziness attributes such as thicknesses and normals' or tangents' envelopes. Fuzzy faces and their intersection algorithm provide results whose fuzziness is defined at the topographic and topologic levels.

A fast and robust process for the detection and computation of surfaces intersections derives from this approach.

9   Acknowledgments

This work was supported in part by a grant from the Coretech International Society.

References

1. C. Asteasu and A. Orbergozo. Parametric piecewise surface intersection. *Computer & Graphics*, 15(1):9–13, 1991.
2. N.M. Aziz, R. Bata, and S. Bhat. Bézier surface/surface intersection. *IEEE Computer Graphics and Applications*, pages 50–58, January 1990.
3. C.L. Bajaj, C.M. Hoffmann, R.E. Lynch, and J.E.H. Hopcfart. Tracing surface intersections. *Computer Aided Geometric Design*, 12:285–307, 1988.
4. R.E. Barnhill, G. Farin, M. Jordan, and B.R. Piper. Surface/surface intersection. *Computer Aided Geometric Design*, 4(1-2):3–16, 1987.
5. R.E. Barnhill and S.N. Kersey. A marching method for parametric surface/surface intersection. *Computer Aided Geometric Design*, 7(1-4):257–280, 1990.
6. E. Boender. A survey of intersection algorithms for curved surfaces. *Computer and Graphics*, 15(1):109–115, 1991.
7. J.J. Chen and T.M. Ozsoy. Predictor-corrector type of intersection algorithm for C^2 parametric surfaces. *Computer Aided Design*, 20(6):347–352, 1988.
8. K.-P. Cheng. Using plane vector fields to obtain all the intersection curves of two general surfaces. In W. Straßer and H.-P. Seidel, editors, *Proc. Theory and practice of geometric modeling*, pages 187–204, Springer-Verlag, 1989.
9. M. Daniel. Calcul d'une enveloppe des normales d'une surface definie par des pôles. Rapport de recherche no 43, Institut de Recherche en Informatique de Nantes, 93.
10. A. L. Dixon. The eliminant of three quantics in two independent variables. In *Proceedings of London Mathematical Society*, volume 6, pages 209–236, 1908.
11. Rida T. Farouki. Direct surface section evaluation. In Gerald E. Farin, editor, *Geometric Modeling: Algorithms and New Trends*, pages 319–335, Philadelphia, 1987. SIAM.
12. M.E. Hohmeyer. A surface intersection algorithm based on loop detection. *International Journal of Computational Geometry and Applications*, 1(4):473–490, 1991.

13. E.G. Houghton, R.F. Emnett, J.D. Factor, and C. Sabharwal. Implementation of a divide-and-conquer method for intersection of parametric surfaces. *Computer Aided Geometric Design*, 2:173–183, 1985.

14. P.A. Koparkar and S.P. Mudur. Generation of smooth curves resulting from operations on parametric surface patches. *Computer Aided Design*, 18(4):193–206, 1986.

15. G. A. Kriezis. *Algorithms for Rational Spline Surface Intersections*. PhD thesis, Massachusetts Institute of Technology, Cambridge, USA, March 1990.

16. G.A. Kriezis, P.V. Prakash, and N.M. Patrikalakis. Method for intersecting algebraic surfaces with rational polynomial patches. *Computer Aided Design*, 22(10):645–655, 1990.

17. Shankar Krishnan and Dinesh Manocha. Algebraic loop detection and evaluation algorithms for curve and surface interogations. Technical Report TR95-038, Department of Computer Science, University of N. Carolina, Chapel Hill, 1995.

18. J.M. Lane and R.F. Riesenfeld. A theoritical development for the computer generation and display of piecewise polynomial surfaces. *IEEE Transactions on Pattern Analysis and Machine Intelligence*, 2(1):150–159, 1980.

19. D. Lasser. Intersection of parametric surface in the Bernestein-Bézier representation. *Computer Aided Design*, 18(4):186–196, 1986.

20. D. Manocha and J.F. Canny. A new approach for surface intersection. *International Journal of Computational Geometry and Applications*, 1(4):491–516, 1991.

21. R.P. Markot and R.L. Magedson. Solutions of tangential surface and curve intersections. *Computer Aided Design*, 21(7):421–429, 1989.

22. N. M. Patrikalakis and P. V. Prakash. Surface intersections for geometric modeling. *Journal of Mechanical Design*, 112:100–107, mar 1990.

23. Q.S. Peng. An algorithm for finding the intersection between two B-spline surfaces. *Computer Aided Design*, 16(4):191–196, 1984.

24. M.J. Pratt and A.D. Geisow. Surface/surface intersection problems. In J.A. Gregory, editor, *Proc. of the Mathematics of surfaces*, pages 118–148, Oxford University Press, U.K., September 1984.

25. R. F. Sarraga. Algebraic methods for intersection. *Computer Vision, Graphics and Image Processing*, 22:222–283, 1983.

26. T. W. Sederberg. *Implicit and Parametric Curves and Surfaces*. PhD thesis, Purdue University, 1983.

27. T.W. Sederberg and R.J. Meyers. Loop detection in surface patch intersections. *Computer Aided Geometric Design*, 5(2):161–171, 1988.

28. T.W. Sederberg and T. Nishita. Curve intersection using Bézier clipping. *C.A.D,*, 22(9):538–549, November 1990.

29. L. A. Zadeh. Fuzzy sets. *Information and Control*, 8:338–353, 1965.

30. L. A. Zadeh. Fuzzy algorithms. *Information and Control*, 12:94–102, 1968.

A Degree-of-Freedom Graph Approach

Ching-yao Hsu and Beat Brüderlin

Computer Graphics Program
Technical University of Ilmenau
98693 Ilmenau, Germany
http://www.prakinf.tu-ilmenau.de/ bdb

Abstract. In this paper, we present a new graph-based approach to geometric constraint solving. Geometric primitives (points, lines, circles, planes, etc.) possess intrinsic degrees of freedom in their embedding space. Constraints reduce the degrees of freedom of a set of objects. A constraint graph represents the objects and geometric relations between them. A graph algorithm which transforms the undirected constraint graph into a directed acyclic dependency graph is developed. The dependency graph is used to derive a sequence of construction operations as a symbolic solution to the constraint problem. The approach is based on a dimension independent degree-of-freedom analysis, which, among other things, allows for a uniform handling of 2-D and 3-D constraints as well as algebraic equations between parameters. The approach handles completely constrained, as well as under- and over constrained definitions with a worst-case time complexity of O(n), where n is the number of geometric elements.

Introduction

Conventional computer aided design systems required users to construct geometric objects by a sequence of geometric construction operations. Mechanical parts designed by such a modeling system are represented as fixed geometry – the geometric design is separated from the original design criteria. It is therefore necessary for a user to determine the exact coordinates of the objects in the beginning, and it is very difficult to add information under a different view, later on. Locally changing the shape may inadvertently violate previous design decisions.

Most CAD systems, nowadays, allow users to define geometric constructions by means of parameters *(Parametric Design)*. The value of the design parameters can be changed later. A dependency propagation mechanism automatically propagates the new values to all directly and indirectly dependent parts of the object. This way, so-called families of objects can be defined. Although this increases the flexibility of CAD based design significantly, great care has to be taken to define the geometric operations in the right order, which puts an undue burden on the designer.

Geometric constraints have shown to be useful in interactive geometric design[2]. The idea is to specify shapes by relations between geometric primitives (distances, angles, etc) and use a constraint solver to derive the shape from such a specification *(Variational Design)*. Constraints can be added or deleted by the user, in any order. A clear drawback of a constraint based approach is that it is very difficult and not at all intuitive for a designer to come up with a complete and consistent set of constraints. Often we encounter over and under specified parts simultaneously, which are hard to resolve in a specification. Moreover, constraint solving (even for consistent specifications) is a very difficult problem. Following is an brief survey on different kinds of constraint solving techniques in the literature.

Constraint Solving Methods

Constraint propagation is one of the basic mechanisms used in early constraint based system for the derivation of solutions that satisfy the given constraints. Here, the system of variables and constraints are represented as an undirected graph. The nodes of the graph represent variables or constants, and the edges represent equations relating the variables and constants. The solving of constraints is done by finding an order of evaluation to satisfy all the equations from the constants progressively. Propagation methods are described, for instance, in [3,10,16,24]. The weakness of this approach is that it cannot handle cyclic dependencies, and hence the method is usually combined with numerical methods. *Cyclic dependence* refers to the situations where the dependence relationship among several variables forms a cycle. To be more specific, we define n variables, say $v_1, v_2, \ldots, v_n$. If the solution of v_i depends on the solution of v_{i+1} for $i = 1..n - 1$, and the solution of v_n is dependent on the solution of v_1, this constitutes a cyclic dependency among the n variables.

Many practical constraint based systems use numerical techniques (e.g., relaxation, Newton-Raphson iteration) which can handle cyclic dependencies, and theoretically can solve problems even if they don't have a closed form algebraic or geometric solution. In this approach, constraints are translated into a system of algebraic equations and then solved using iterative methods. Numerical approaches are described in [17,21]. Because of their generality, a lot of systems switch to numerical methods when their basic mechanisms fails. Early systems such as Sketchpad, ThingLab and Magritte used relaxation as an alternative to their propagation methods. While they are quite powerful and general, numerical techniques have convergence problems that make them very unpredictable.

Lately, another kind of constraint solvers is emerging, which satisfy the constraints using a sequence of construction steps and mostly solve problems solvable by ruler and compass construction. Some are extended to 3D, as well. The approaches can be viewed as an extension of the constraint propagation paradigm into higher dimensions. The basic principle behind the constructive approach is that an object can be evaluated when enough information about it is available. These methods differ in the method by which the order of evaluation is determined; we distinguish between rule-based approaches [19,5–7,25,22] and the graph-based approach [18,4,20,13,23]. Also, the approach described in this paper, belongs to the category of graph based approaches. The novel idea developed here is to use an abstract graph which solely uses the degree of freedom information, before applying any geometric or algebraic interpretation. A previous version of this paper was published in [14], and it is a generalization of the approach described in [13].

Overview

We develop an approach for evaluating the degrees of freedom of under constrained or completely constrained networks. We will show that with this approach, users are not forced to specify shapes completely by constraints but can add constraint definitions incrementally, and manipulate the geometric models within their degrees of freedom, which is especially useful in the early design stage.

The next section introduces the graph representation for the geometric constraint problems. Afterwards the basic principle behind the degrees of freedom analysis of a constraint graph are developed, based on the so-called 'balance equation'; The analysis phase returns an intermediate representation, the dependency graph. The next section describes an efficient algorithm for the construction of such dependency graph. Ways to evaluate the dependency graph by geometric constructions are also described. Examples for 2D, and 3D constraint solving as well as algebraic constraints with the described methods are given.

Graph Representation

A *parametric geometric model* in our approach is defined by a set of geometric objects $O = \{o_1, o_2, \dots, o_n\}$ and the set of geometric relations (constraints) $C = \{c_1, c_2, \dots, c_m\}$ between the elements of O.

Geometric objects such as points, lines, and circles (see also figure 1) *own* degrees of freedom, which allow them to vary in shape, position, size, and orientation. The set of degrees of freedom $DOF = \{dof_1, dof_2, \dots, dof_l\}$ owned by the objects in O, represent different types of domains. Each domain, represents a set of allowable values. The cross product of these domains contains the possible states of the geometric model. The number of degrees of freedoms l is the dimension of the model (also called it's total degree of freedom).

Geometric constraints such as those in figure 1 define an n-ary geometric relation among a set of n objects, $c_i = c_i(o_{i_1}, o_{i_2}, \dots, o_{i_n}, \lambda)$, where λ is the parameter of the constraint. Depending on the constraint type, the parameter may be a scalar, a vector, or empty. A constraint reduces the degrees of freedom of the model by a certain number (called the *valency* of the constraint [1]).

We also introduce a special class of constraints, so-called *local (unary) constraints*, which may *consume* all or part of the degrees of freedom owned by an object. For example, a constant x-coordinate constraint on a 2-D point fixes its x-coordinate in space (we say one degree of freedom owned by the point is consumed by that constraint).

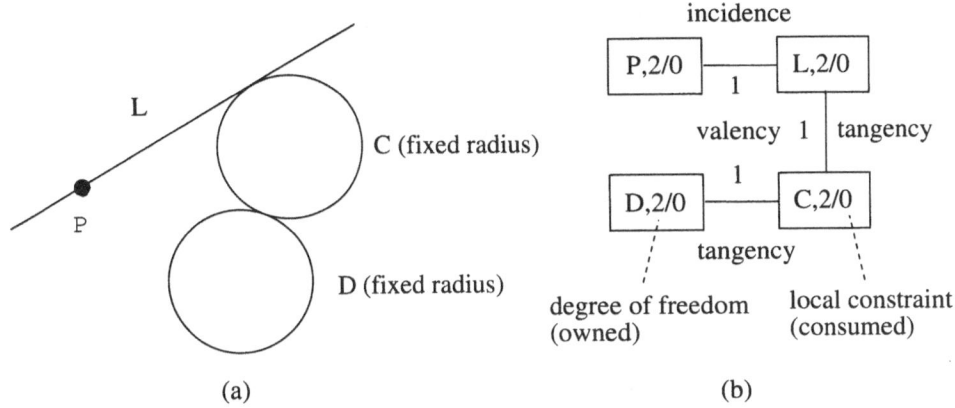

Fig. 1. The nodes and edges of the constraint graphs. (a) A parametric geometric model. (b) The graph representation.

In the following sections, we will introduce constraint graphs as the graph representation of geometric models. A *constraint graph* is an undirected graph which consists of a finite set of nodes and a finite set of arcs. A *node* in the graph represents a geometric object and is depicted as a rectangular box annotated with $DOF_{owned}/DOF_{consumed}$. An *arc* represents a constraint between two objects. We label the arc with the valency of the constraint, as shown in figure 1. In the following, we will represent n-ary constraints also as nodes (hyper arcs) in the graph, with arcs to the n connected objects.

Degrees-of-Freedom Analysis

The constraint solver presented here is composed of a planning phase and an execution phase. In the center of the planning phase is the degree of freedom analysis which is the main topic of this paper. The goal of the degree of freedom analysis is to extract from a constraint graph a connected portion which possesses a specified degree of freedom (see also figure 2). We can then manipulate the portion of the constraint graph within the degrees of freedom acquired.

The result of the degrees-of-freedom analysis is in the form of a dependency graph, which can be viewed as an annotated directed constraint graph. We then use the dependency graph in the evaluation phase which finds an embedding of the dependency graph in the Euclidean space, by a sequence of geometric construction operations. The evaluation phase is briefly described at the end of this paper. For more details see [15].

The Balance Equation

The degree of freedom analysis is based on the so-called *balance equation*, which defines the degree of freedom as the invariant χ of the system. The degrees of

Constraint graph   Specificed degree of freedom

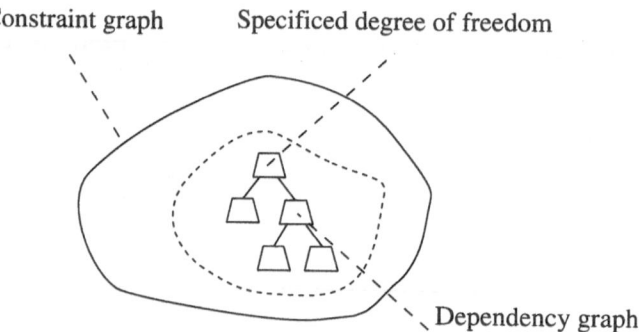

Dependency graph

Fig. 2. The overview of the constraint solving strategy.

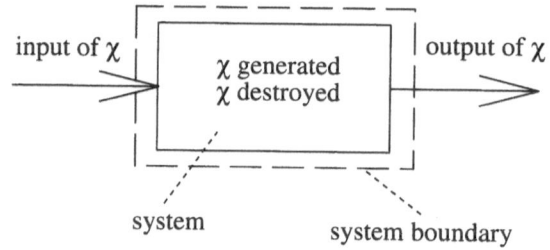

input of χ χ generated   output of χ
              χ destroyed

system   system boundary

Fig. 3. A single System operated under steady-state conditions.

freedom are transmitted from node to node. For a system which is operated under *steady-state conditions* (as shown in figure 3), the degree of freedom χ is conserved, and the balance equation can be written as:

$$[\text{output of } \chi] = [\text{input of } \chi] + [\chi \text{ generated}] - [\chi \text{ destroyed}] \qquad (1)$$

The degrees of freedom of geometric primitives are represented by positive numbers, whereas constraints (local constraints or valencies) are represented by negative degrees of freedom. The balance equation can be extended to multiple systems, connected together. By repeatedly applying equation 1, we obtain the following balance equation for connected multiple systems acting as a single system (figure 4):

$$\sum_i output_i = \sum_i input_i + \sum_i generated_i - \sum_i destroyed_i \qquad (2)$$

The Dependency Graph

A *dependency graph* is a directed, acyclic hyper-graph derived from the constraint graph. Nodes and edges in the dependency graph obey the balance equation for degrees of freedom. By connecting nodes and edges in a proper way, we can obtain the degrees of freedom from the dependency graph according to equation 2.

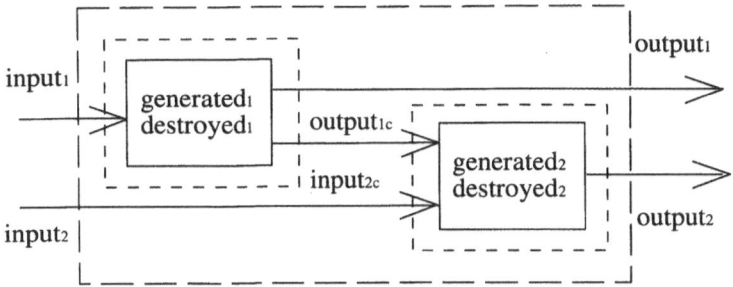

Fig. 4. Two systems connected together.

A *node* represents an object and is depicted as a trapezoidal box with the long side representing the input side and the short side representing the output side. The top row enumerates the individual degrees of freedom leaving the node to the next level up. The bottom row shows the individual degrees of freedom entering the node from the previous level down. The center row represents the balance equation for the node (see also figure 5):

$$DOF_{in} = \sum_i DOF_{in}(i)$$

$$DOF_{out} = \sum_j DOF_{out}(j)$$

$$DOF_{out} = DOF_{in} + DOF_{owned} - DOF_{consumed} \tag{3}$$

A *(hyper) edge* represents a constraint between at least one child node and a single parent node. The edge is labeled with the negative degree of freedom (valency) of the constraint and the arrow at the top shows the direction of the flow. For an edge, the balance equation is as follows (see also figure 6):

$$DOF_{in} = \sum_k DOF_{in}(k)$$

$$DOF_{out} = DOF_{in} - valency \tag{4}$$

Example 1. Figure 7 (a) shows a constraint graph with five nodes and four edges. Figure 7 (b) shows one instance of the dependency graph derived from it. In the example, there are no degrees of freedom entering the nodes at the lowest level of each branch, i.e. all the degrees of freedom owned by the nodes are consumed by local constraints. As a result, there are zero degrees of freedom at the output of these nodes.

In the next level up, one degree of freedom is offset by each constraint, which results in a negative degree of freedom entering the parent nodes. We can interpret this geometrically as follows. For instance, point D is restricted by one degree of freedom to lie on a one a-one dimensional curve (circle). We proceed upwards, until finally, we obtain one degree of freedom out of the topmost node, the root node.

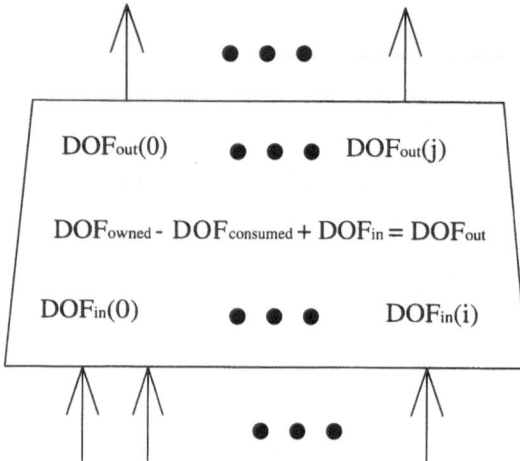

Fig. 5. The graph representation of a node

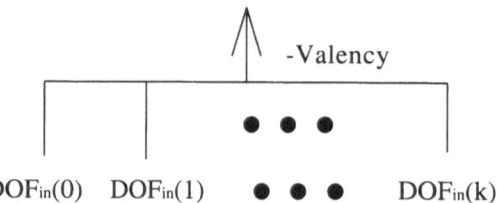

Fig. 6. The graph representation of an edge.

This example illustrates the flow of degrees of freedom for a dependency graph in a bottom-up fashion.

Constructing the Dependency Graph

The goal of the algorithm described here, is to construct a dependency graph for a given root node (top down). We start by requesting degrees of freedoms from that root node, and propagate the requests down to it's children. The geometric interpretation of this procedure is as follows: Requesting zero degrees of freedoms means to find a valid (zero dimensional) solution for the object (similarly to solving a system of equations for a chosen variable). We can also request more degrees of freedoms within which we can manipulate the object (for instance interactively with a mouse controlled cursor) and the dependency graph determines how other objects are affected as a consequence of the manipulation.

In the following sections, we will derive the algorithm step by step. First, we will describe the basic idea algorithm which produces dependency graphs observing the balance equations. Then, we introduce validity rules which restrict

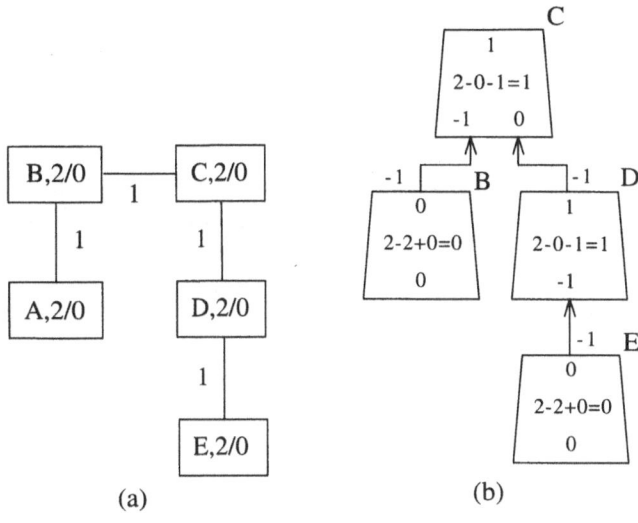

Fig. 7. An instance of the dependency graph. (a) An example constraint graph. (b) A dependency graph for dragging node C.

the kinds of dependency graphs to the ones which have a geometric interpretation in the embedding space. Finally, we present an efficient search strategy for the algorithm.

Requesting Degrees of Freedom from Objects

We start by selecting an object from the constraint graph and request degrees of freedom (called $DOF_{requested}$) from it.

We declare the selected object the *root node* and compute the total degree of freedom that has to enter the node giving the amount leaving it by means of the balance equation 3:

$$DOF_{in}(root) = DOF_{out}(root) - DOF_{owned}(root) + DOF_{consumed}(root)$$
$$(5)$$

where $DOF_{out}(root) = DOF_{requested}$. The amount DOF_{in} determines the total degree of freedom we need to request from the child nodes in order to satisfy the equation.

Suppose that m constraints, $c_1, c_2, \ldots, c_m$ are connected to the node, we then *expand* it by distributing DOF_{in} among the m constraints and creating corresponding edges for each one of them:

$$DOF_{in}(root) = DOF_{out}(c_1) + DOF_{out}(c_2) + \cdots + DOF_{out}(c_m) \qquad (6)$$

(Note that, if the node expanded is not the root node, we need to exclude from the set of the m constraints any constraints that have been used elsewhere before.)

For each edge i, we can calculate the degrees of freedom entering it by rearranging equation 4:

$$DOF_{in}(c_i) = DOF_{out}(c_i) - valency(c_i) \tag{7}$$

If the number of the fan-in of the edge is greater than one, we need to distribute $DOF_{in}(c_i)$. Let the number of the fan-in be equal to n, and the child nodes are $o_1, o_2, \ldots, o_n$. Then

$$DOF_{in}(c_i) = DOF_{out}(o_1) + DOF_{out}(o_2) + \cdots + DOF_{out}(o_n) \tag{8}$$

where $DOF_{out}(o_j)$ is the degree of freedom we need to request from the child node j.

We repeat this process in a breadth-first manner, down the branches, until either of the following termination conditions is met:

1. There are no more unused constraints, or
2. DOF_{out} of a node is less than or equal to zero.

When one of those conditions is met, we can stop recursion and make the node a leaf node. If the second condition applies, we also set $DOF_{consumed}$ equal to DOF_{owned} (i.e. to locally fix the state of the object).

Example 2. We return to the example illustrated earlier (figure 7 (b)) which shows only one instance of the dependency graphs that can be derived by requesting one degree of freedom from node C. In order to satisfy this request, we need to import minus one degree of freedom from the neighbors of node C (i.e. nodes B and D). We can choose from the following combinations to meet the requirement:

− requesting -1 from node B and 0 from node D, or
− requesting 0 from node B and -1 from node D, or
− requesting -2 from node B and 1 from node D, etc.

The figure shows the first combination is used. For node B, the degree of freedom requested is zero, which signifies a termination condition. Therefore, we set $DOF_{consumed}$ equal to DOF_{owned} for node B, and stop recursion there. By applying the algorithm in this manner, we obtain the dependency graph shown.

The degree-of-freedom analysis algorithm described here is nondeterministic in nature because there can be more than one dependency graph for a specific request and we obtain different results by distributing the degrees of freedom in equation 6 and 8 differently. However, not all the combinations resulting from the distribution are valid. They also subject to the validity rules, described in the next section.

The Validity Rules

The validity rules, introduced here, ensure that the embedding of the graph in the configuration space has a geometric meaning (e.g.the local degrees of freedoms do not exceed the dimension of the object, and also, that there will be no over-constrained subsystems).

1. If a request passed down to a node is negative, we will set it to zero. That is,

$$DOF_{out}(j) \geq 0 \text{ for all } j \tag{9}$$

   This rule guarantees that there will not be any over-constrained parts in the dependency graph. Also, positive requests will not be canceled out by negative ones.
   To maintain the balance equations, we can apply what we call the normalization process to the negative requests. Please refer to [11] for more details.
2. The total amount requested at each node should not exceed that available in the node. That is,

$$DOF_{out} \leq DOF_{owned} - DOF_{consumed} \tag{10}$$

   This rule ensures that we do not request more degrees of freedom than the node can afford. This, a node will not import positive degrees of freedom elsewhere to meet the request (which would violate rule 3).
3. No positive degrees of freedom are imported by a node. That is,

$$DOF_{in}(i) \leq 0 \text{ for all } i \tag{11}$$

   Basically all imported degrees of freedoms, stemming from other objects in the graph, are negative (i.e. they restrict the degrees of freedom of this object) or zero (i.e. there is no restriction from the related object).

With these validity rules, the possibilities of distributing the requested degree of freedom among the input nodes are now finite. The simplest distribution method for equation 6 and 8 is based on the generate-and-test paradigm. Basically, we enumerate all valid distributions in some order. Since the distribution is done blindly, there will be circumstances when a request for a degree of freedom fails (for example, there is a violation, when the total amount requested is greater than a node can afford according to second validity rule). Backtracking can be used to search for alternative dependency graphs under these circumstances. However, this amounts to an exhaustive search, and hence it is not practical.

Efficient Search Strategies

We will next describe a search algorithm based on the idea of migration of degrees of freedom. The algorithm is composed of two alternating phases: a raw construction phase and a correction phase (see also figure 8). In the construction

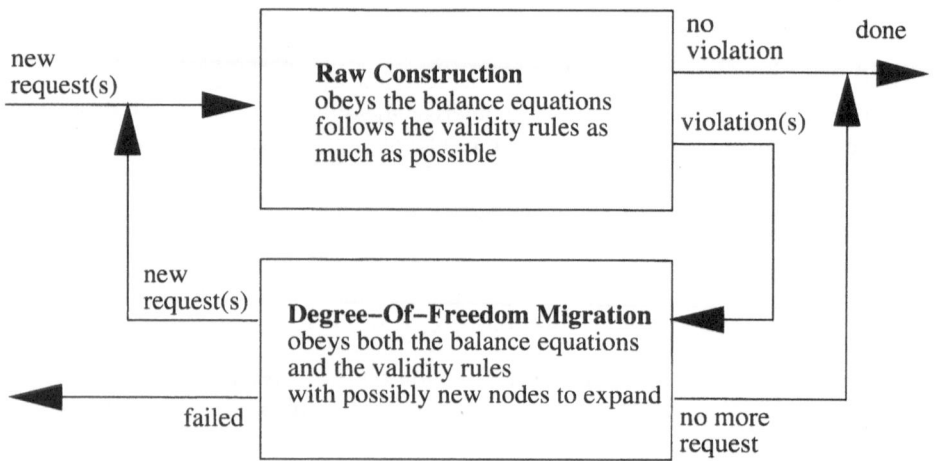

Fig. 8. The flowchart of the dependency graph construction algorithm.

phase, a raw dependency graph is built in response to requests, starting with the root node. The balance equations must be strictly observed, but the validity rules may be violated, temporarily. Any violations are marked. In the correction phase, these violations are removed by migrating degrees of freedom. The result of the migration will (possibly) produce new requests in new places. If that is the case, we return to the raw construction phase. The algorithm succeeds when a dependency graph with no violations is built or fails when there is no way to remove the violations.

The violations occurring in the raw construction phase always result from requesting a degree of freedom more than a node can afford (violation of rule 2); all other violations are eliminated directly at the source. Figure 9 (a) shows an example where the total degree of freedom requested is more than the one owned by the node. This violation can be identified when we try to expand the node. To fix it, we suspend the extra degrees of freedom, as shown in Figure 9 (b). We call the extra positive degrees of freedom as DOF_{ext}. The balance equation for a node now becomes:

$$DOF_{in} = \sum_i DOF_{in}(i)$$

$$DOF_{out} = \sum_j DOF_{out}(j)$$

$$DOF_{out} = DOF_{in} + DOF_{owned} - DOF_{consumed} + DOF_{ext} \qquad (12)$$

As a result, the balance equations are preserved and only the validity rules are violated.

The running time of the raw construction phase is similar to that of breadth-first search without backtracking. Since the final dependency graph can only be as big as the constraint graph, the time complexity of the algorithm is

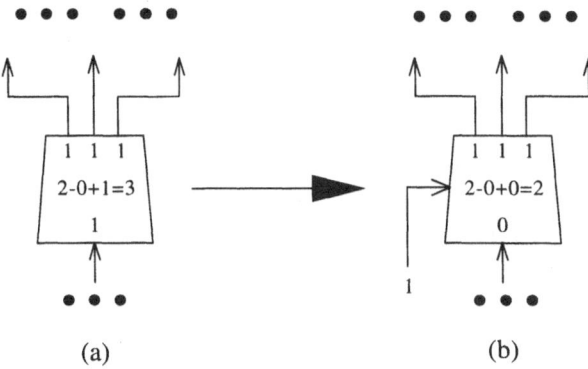

Fig. 9. An example of violation to the validity rules. (a) A violation. (b) An interpretation for (a).

$O(max(v, e))$, where v is the number of nodes, and e the number of edges in the constraint graph.

In the correction phase, we employ a migration mechanism to remove the violations while preserving the balance equations. We introduce two migration operations that alter the dependency graphs without violating the balance equations. By combining these two operations, we will be able to transform the dependency graph so as to meet the validity rules. The net effect of applying a sequence of these operations is to move an extra degree of freedom from one node to another.

In a *down migration* a degree of freedom travels down one level, and a request of that degree of freedom is posed on the child node. Therefore, we need to add one degree of freedom to the path traveled. Figure 10 (b) shows the down operation.

In an *up migration*, (see figure 10 (a) a degree of freedom travels up one level. As a consequence, we need to subtract one degree of freedom from the path traveled.

It is obvious that these operations preserve the balance equations, but they do not necessarily enforce the validity rules of the nodes the validity migrated to. To maintain the rules, we need to make sure that by applying these migration operations, the nodes will not result in states that contradict the rules. So the following restrictions are applied to the operations.

- We do not apply an up-migration to a path where the total DOF_{out} of the node is zero because by applying the up operation to such a node would make DOF_{out} of that node become negative. However, it is acceptable if DOF_{out} becomes zero. We do not make the node a leaf node in this case.
- We do not apply a down-migration to a path where its DOF_{in} is already zero because by applying the down operation, we would import a positive degree of freedom into a node.

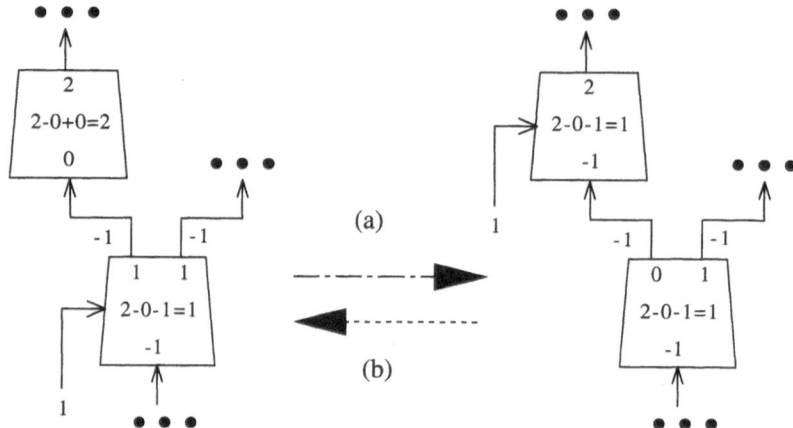

Fig. 10. The balance equation preserving operations. (a) An up operation, and (b) A down operation.

To remove the violations identified, we may have to apply a sequence of up and down operations to move the extra degrees of freedom to places allowed by the validity rules.

If the process successfully ends with an up operation, then the place where it ends must be the root node. To remove the violation when we are at the root node, we simply subtract the requested degrees of freedom by one. The interpretation of this result is that the requested degrees of freedom are not available, and only a lesser degree of freedom can be requested from the object.

On the other hand, if the sequence successfully ends with a down migration, the place where it stops will be one of the leaf nodes. Depending on the sign of the DOF_{in}, we can divide the leaf nodes into three categories (see also figure 11). A leaf node is called

1. a *zero complete node* when both DOF_{out} and DOF_{in} are equal to zero.
2. a *zero incomplete node* when DOF_{in} is equal to zero, but DOF_{out} is not.
3. a *negative incomplete node* when DOF_{in} is less than zero.

For zero complete leaf nodes and for zero incomplete leaf nodes no special treatment is necessary. When the leaf node is a negative incomplete node, the violation can be removed simply by adding the corresponding DOF_{in} to the node. The negative incoming constraints can be eliminated by defining local constraints. After this, any dependency graph will have no external degrees of freedoms. Except for the requested degrees of freedom (at the root node) the dependency graph will be a completely constrained subgraph of the constraint graph.

The migration path is generated in a depth first order. First, for each node, we try to migrate the extra degree of freedom downwards. If this is not possible, the degree of freedom is migrated up one level, and again, downward migration

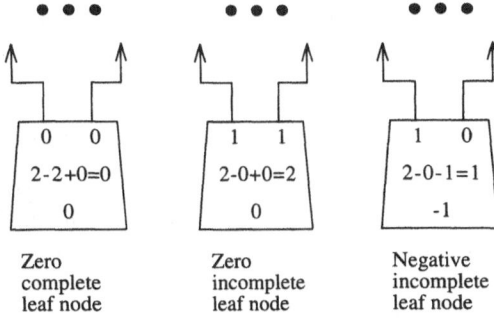

Fig. 11. Three different kinds of leaf nodes.

operations, along other paths are attempted from there. We will mark the exhausted degrees of freedoms along any path. An exhausted degree of freedom will not be used in a future search again.

The search may fail because the constraint graph is over-constrained in the neighborhood of the violation. Therefore, there will be an up migration which ending at a node with negative DOF_{out} In this case, it is impossible to build a dependency graph of any kind, because the node is locally over-constrained. To fix the problem we will have to relieve some constraints that led to the negative degree of freedom, and possible keep them as a reference, or to determine if the definition is consistently over-constrained.

The proof of correctness the algorithm is straightforward. We can argue that all the operations maintain the balance equation. Therefore, if the algorithm does produce a dependency graph, it will be one that meets the balance equations and the validity rules.

To prove termination of the algorithm, we notice that the algorithm cannot introduce new violations in the current dependency graph. Therefore, this process will eventually terminate even in the worst case, when we exhaust the constraint graph.

To determine the time complexity of the complete degree of freedom analysis algorithm, we observe that each node may violate the validity rule once. Migrating the violation downwards will generate further requests in the remaining graph. For each node, visited we mark the exhausted degrees of freedom, so it will not be available in future searches, along other paths. Exhausted degrees of freedoms are never released during a query. Either they are available, and will be reserved, or they are unavailable because of other constraints. If the arity of the constraints is limited (usually 2 or 3), then from each node we will only try to migrate along constraints attached to the node, and to the few objects related by that constraints. Once used, the constraint will not be followed, from other nodes. Thus, the overall complexity of the degree of freedom analysis will be linear to the number of objects plus the number of constraints. If the problem is not over-constrained, the maximum number of constraints is also proportionate to the number of degrees of freedom, and hence, the whole algorithm will be linear in the number of objects.

Controlling the Generation of a Dependency Graph

Because of the nondeterministic nature of the analysis algorithm, users have little control over the dependency graphs produced. To alleviate the problem, the idea of heuristics and constraint hierarchy in [3,9] can be incorporated into the searching process to help determine the proper dependency graphs to return. However, heuristics may still fail. Under these circumstances, a manual adjustment by setting temporary local constraints can be used as the last resort. This is discussed next.

Suppose we refer to the set of all possible dependency graphs for a request as the *solution space* for the analysis algorithm. Normally, the algorithm just returns the first dependency graph found in the solution space. The idea of setting temporary local constraints is to control the search by progressively "trimming" the solution space and thus steering the algorithm to the solution we want. Observe that imposing a full local constraint to a node will disconnect the constraint graph at that point in terms of passing degrees of freedom (trimming the solution space). The interactive user may, for instance, temporarily fix some objects, and force the algorithm to use up-migration to find alternative dependency graphs. Another idea is to distinguish between different types of degrees of freedoms (translational, rotational, dimensional) and, for instance, first use up translational degrees of freedom before rotational ones are used, because rotations are generally less predictable and less intuitive than translations.

In [12] we describe TWEAK an interactive toolkit which uses temporary local constraints and typed constraints to control the behavior of the constraint solver in interactive situations.

Evaluating the Symbolic Solution

After deriving the dependency graph, we will be able to evaluate the corresponding part of the constraint graph within the degrees of freedom acquired. This is achieved by deriving an evaluation plan for the dependency graph, changing the state of the root node, and reevaluating the rest of the dependency graph to maintain the constraints.

After eliminating the extra degrees of freedom at the leaf nodes, the degree of freedom of the dependency graph is equal to that requested. The requested degree of freedom is at our disposal. For instance, if we use an interactive user interface which supports locator devices, we can use them to specify these degrees of freedom interactively.

After consuming the degrees of freedom acquired, the remaining dependency graph becomes a fully-constrained system. We can evaluate the embedding of the constraint graph in Euclidean space. We have experimented with both numerical and constructive methods in our implementation of the algorithm.

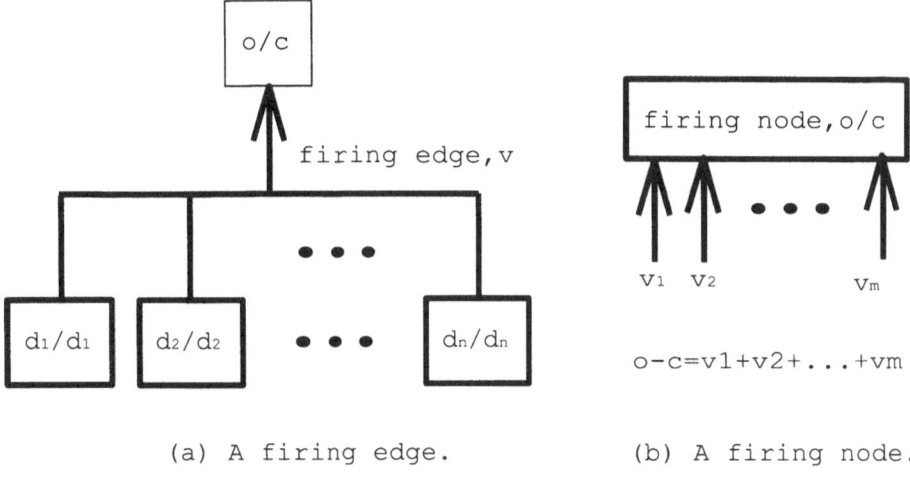

(a) A firing edge.   (b) A firing node.

Fig. 12. Firing conditions for edges and nodes. (a) A firing edge. (b) A firing node.

Here we propose a hybrid constructive method. This method tries to satisfy the constraints by geometric closed form solutions whenever possible. An iterative constructive method is used when a cyclic dependence cannot be resolved by geometric construction operations (more details can be found in [11,15]).

In the hybrid constructive method, the order of evaluation is determined by a set of firing rules. To evaluation a dependency graph, the following firing rules are used:

1. The zero complete leaf nodes are evaluated first.
2. The dependency graph is then evaluated top down, starting at the root node. After a node is evaluated, it has zero degrees of freedom.
3. When all but one node attached to an edge become fully constrained, the edge becomes a *firing edge* (see also figure 12 (a)).
4. A node becomes ready to fire (see also figure 12 (b)), when the sum of the valencies of the connected firing edges is equal to the available degrees of freedom of the node ($DOF_{owned} - DOF_{consumed}$).
5. If a node is under-constrained, but no other unevaluated nodes are attached, it can be fired.
6. Also, firing rules for cyclic dependencies exist. In these cases a hybrid iterative constructive approach is used (please refer to [11,15]).

Example 3. Figure 13 (a) Assume that we drag point B (using a mouse) requesting one degree of freedom: We can derive a dependency graph as shown in figure 13 (b), using the previously described algorithm. We can then derive an evaluation plan for the dependency graph using the firing rules above. The plan is shown below:

1. Points A and D become firing nodes according to rule 4.

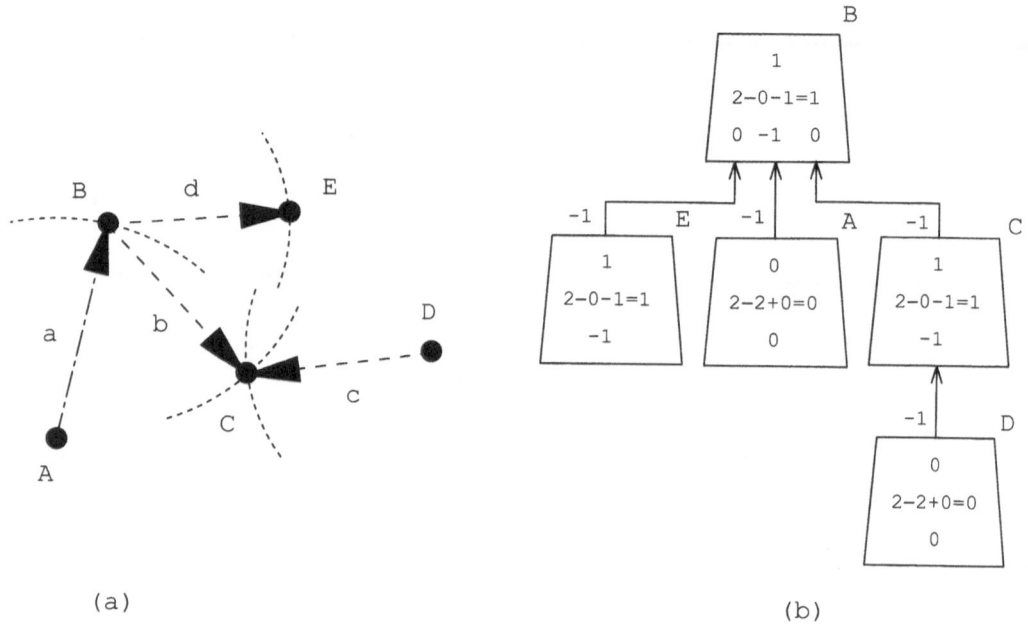

(a)   (b)

Fig. 13. Evaluating the dependency graph. (a) A geometric model. (b) A dependency graph for dragging point B.

2. Point B is fired by the second rule. To evaluate, the cursor position of the mouse is projected onto the circle centered at point A. Point B is assigned the coordinates of the projected point.
3. Point C becomes a firing node after point B and D have been fired. Point C is determined by intersecting the two circles centered at new point B and point D respectively.
4. Point E is fired by the fifth rule and can be determined by projecting the old position onto the circle centered at new point B.

Mixed-Dimensional Applications

As we mentioned initially, the graph representation of the constraint problem, as well as the degree of freedom analysis algorithm work independent of the dimension of the embedding space. We can easily define problems in two and three dimensions, even together with algebraic equations between parameters (so-called engineering equations) with the approach presented. However, the necessary geometric constructions and algebraic operations, necessary to evaluate the dependency graphs require the implementation of a large number of functions. In principle, all operations typically occurring in CAD, will be used. In our implementation, we limited ourselves to the most common constraint types and operations. In the following, we describe some examples of applications.

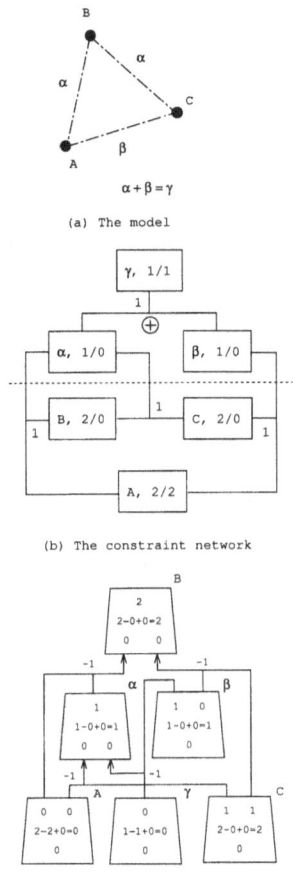

(a) The model

(b) The constraint network

(c) The dependency graph

Fig. 14. Mixed algebraic and geometric constraints. (a) A geometric model. (b) The constraint graph (c) A dependency graph for dragging point B.

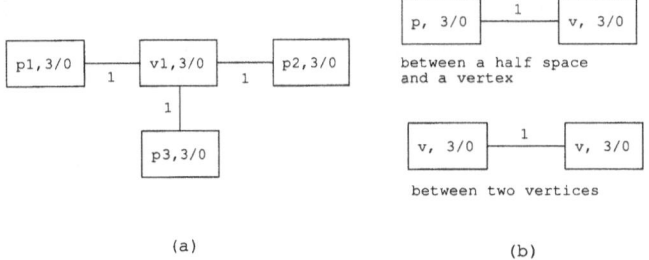

(a)

(b)

Fig. 15. 3D constraints. (a) Incidence between vertex and planes. (b) Distance between point and plane and distance between points.

Figure 14 shows a triangle with variable distances. Two sides are congruent, and the length of the third side is related to the other two by the relation $\alpha + \beta$ = const (γ). These relations are represented by the graph in 14(b). The algebraic equation is represented by the '+' relation which is a hyper arc with valency 1 in the graph. When requesting two degrees of freedom from point B, a dependency graph as shown in figure 14(c) is derived. The geometric evaluation is as follows: Point A remains where it is, and point B follows the cursor. Then α is evaluated as the new distance between points A and C. Then β becomes the difference between γ and the new value of α. Then point C is constructed by intersecting two circles with radii α and β, respectively. The example shows how algebraic operations as well as geometric constructions are used in the evaluation.

Figure 15 shows the graph definition of some 3D constraints. Figure 16 shows an example of editing a polyhedron. All vertices are shared by several faces. In the constraint graph this fact is represented by incidence relationships between the corresponding planes and points. When requesting one degree of freedom from v1, the dependency graph shown in figure 16(b) is derived. The geometric interpretation is indicated under (a), as follows: Vertices v6 and v4 remain constant, plane p3 therefore rotates around the axis defined by the two points. The evaluation is again top-down, using geometric constructions. First, the new position of v1 (the root node) is determined by projecting the cursor position onto the line defined by intersecting planes p1 and p2 (which remain constant as well). Then p3 is recalculated from v4, v6 and the new v1. Afterwards v2 and v3 are recalculated by intersecting the new p3 with p5 and v6 (which also remain constant. All other vertices that belong to the top plane p3 also have to be recalculated by corresponding intersections (this is not shown on the dependency graph, in order not to clutter the illustration). All other parts of the polygon remain constant.

Figure 17(a) shows screen dumps of the example polyhedron. The example was calculated at interactive speed on a low-end workstation or PC. We also ran examples, involving several hundred constraints at interactive speed [8], and in most practical applications, unless a high-end graphics workstation is used, the rendering seems to be the bottleneck, and not the constraint solving.

Figure 17(b) shows a modified example with additional point - point distance constraints defined. The behavior is, of course, different, due to the additional constraints. Here the calculations also involve intersecting spheres and planes.

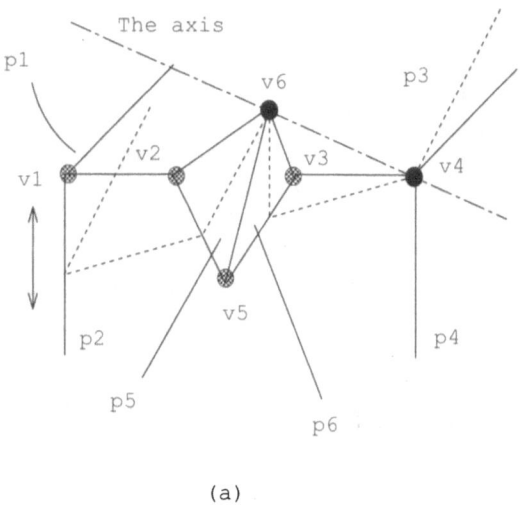

(a)

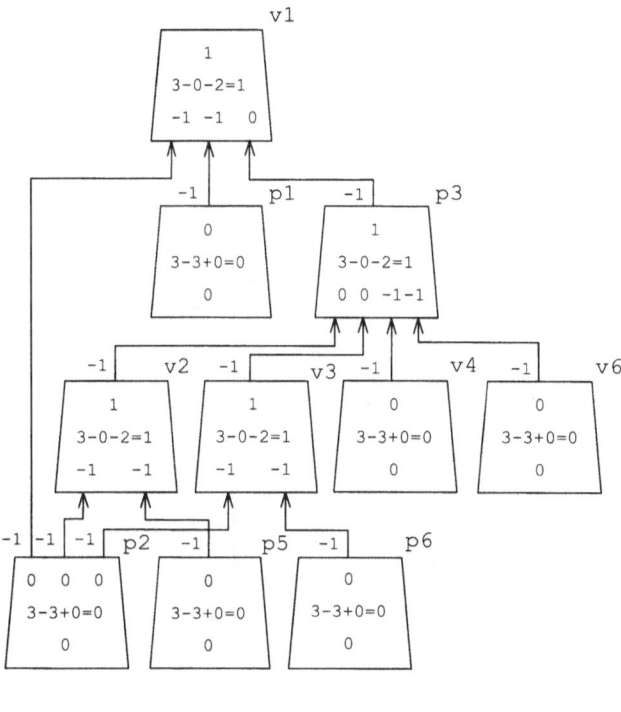

(b)

Fig. 16. Manipulating a 3D polyhedron. (a) Moving vertex v1 along an edge (b)The dependency graph for dragging point v1.

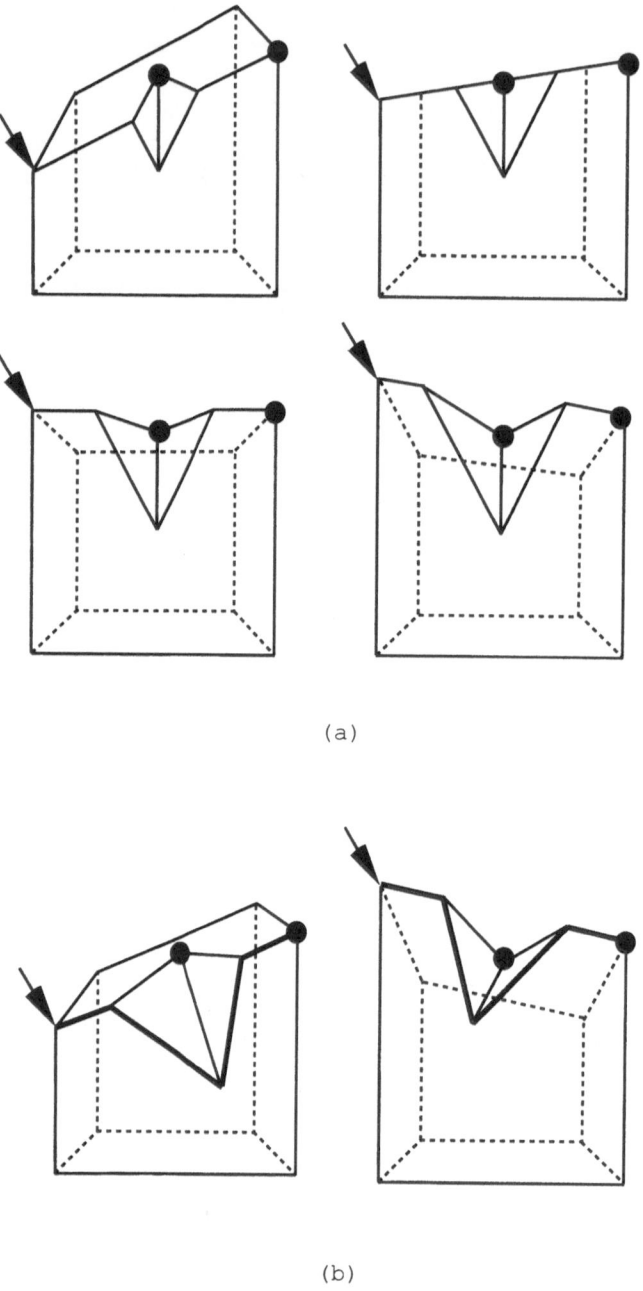

(a)

(b)

Fig. 17. Screen shots of the example. (a) Without additional constraints (b) With point - point distance constraints (thick lines).

Conclusion

The graph based algorithm presented in this paper is a very general tool for reasoning about constraint systems. The algorithm is capable of handling fully-constrained, as well as under-constrained, and consistently over-constrained problems. The degree of freedom analysis approach is dimension independent. 2D, and 3D shapes, as well as engineering equations between parameters can be handled in a uniform approach.

The degree of freedom analysis algorithm used in the analysis phase has an $O(n)$ worst case time complexity (assuming the arity of the constraints is constant). The dependency graph can then be evaluated in time O(n) (where n is the number of objects) by a sequence of geometric construction operations which satify the constraints. For this, we assume that an exact solution is found, which is not always the case. In case an iterative approximation is applied, the evaluation time is multiplied by the number of iteration steps, and possibly no solution may be found at all. However, the iterative approach may find existing solutions, even if no exact solutions exist[15]. First experience shows that the constraint solver is fast enough to even represent classical data structures, such as boundary representation, and CSG, by a more general relational representation, i.e. by geometric constraints.

Several applications of the method have been implemented, and some the results have been published (e.g., [8]) applying the approach to 3D modeling. Objects can be sketched roughly, with a pen, and later exact dimensions can be defined, and other geometric relations can be defined on demand, using the constraint solver. Shapes can be modified interactively, using an interactive manipulator tool set[12]. Thus, the approach is a first step towards a conceptual design system, in which geometric objects can first be defined by sketching simple geometric shapes, capturing the functionality. Later, a design may be refined by adding exact numeric information, and more specific shape information.

Acknowledgments

I would like to thank Ari Rappoport for helpful comments on an earlier version of this paper.

References

1. B. Aldefeld. Variation of geometries based on ageometric-reasoning method. *Computer Aided Design*, 20(3):117–126, April 1988.
2. R. Anderl and R. Mendgen. Parametric design and its impact on solid modeling applications. In *Proceedings of the Third Symposium on Solid Modeling and Applications*, Salt Lake City, 1995. ACM Press.
3. A. H. Borning. The programming language aspects of ThingLab, a constraint-oriented simulation laboratory. *ACM Transactions on Programming Languages and Systems*, 3(4):353–387, October 1981.

4. W. Bouma, I. Fudos, C.M. Hoffmann, Jiazhen Cai, and Robert Paige. A geometric constraint solver. Technical Report CSD-TR-93-054, Department of Computer Science, Purdue University, 1993.
5. B. Bruderlin. Constructing three-dimensional geometric object defined by constraints. In *Proceedings of the 1986 Workshop on Interactive 3D Graphics, ACM SIGGRAPH*, Chapel Hill, North Carolina, 1986.
6. B. Bruderlin. *Rule-Based Geometric Modelling*. PhD thesis, ETH Zürich, Switzerland, 1987.
7. B.D. Bruderlin. Using geometric rewrite rules for solving geometric problems symbolically. *Theoretical Computer Science*, 2(116):291–303, August 1993.
8. L. Eggli, C. Hsu, G. Elber, and B. Bruderlin. Inferring 3d models from freehand sketches and constraints. *Computer Aided Design*, February 1997.
9. Bjorn N. Freeman-Benson and John Maloney. The DeltaBlue algorithm: An incremental constraint hierarchy solver. Technical Report 88-11-09, Computer Science Department, University of Washington, November 1988.
10. Bjorn N. Freeman-Benson, John Maloney, and Alan Borning. An incremental constraint solver. *Communications of the ACM*, 33(1):54–63, January 1990.
11. C. Hsu. *Graph-Based Approach for Solving Constraint Problems*. PhD thesis, Computer Science Department, University of Utah, May 1996.
12. C. Hsu, G. Alt, Z. Huang, E. Beier, and B. Brüderlin. A constraint-based manipulator toolset for editing 3d objects. In *Proceedings of the 1997 ACM/SIGGRAPH Symposium on Solid Modeling Foundations and CAD/CAM Applications*, Atlanta Georgia, 1997 (submitted).
13. C. Hsu and B. Brüderlin. Constraint objects - integrating constraint definition and graphical interaction. In *Proceedings of the 1993 ACM/SIGGRAPH Symposium on Solid Modeling Foundations and CAD/CAM Applications*, Montreal, Canada, May 19-21 1993.
14. C. Hsu and B. Brüderlin. A graph-based degrees of freedom analysis algorithm to solve geometric constraint problems. In *Proceedings of Theory and Practice of Geometric Modeling (Blaubeuren II)*, Blaubeuren, October 1996.
15. C. Hsu and B. Brüderlin. A hybrid geometric constraint solver using exact and iterative geometric constructions. *To appear in "CAD Tools and Methods for Design System Development", D. Roller and P. Brunet eds, Springer-Verlag*, 1997.
16. W. Leler. *Constraint Programming Languages: Their Specification and Generation.* Addison-Wesley Publishing Company, Inc, 1988.
17. Robert Light and David Gossard. Modification of geometric models through variational geometry. *Computer Aided Design*, 14(4):209–214, July 1982.
18. J.C Owen. Algebraic solution for geometry from dimensional constraints. In *Proceedings of the 1991 ACM/SIGGRAPH Symposium on Solid Modeling Foundations and CAD/CAM Applications*, May 1991.
19. Jaroslaw P. Rossignac. Constraints in constructive solid geometry. In *Proceedings of Workshop on Interactive 3D Graphics*, pages 93–110, Chapel Hill, NC, October 23-24 1986.
20. D. Serrano. Automatic dimensioning in design for manufactoring. In *Proceedings of the First Symposium on Solid Modeling and Applications*. ACM Press, 1991.
21. D. Serrano and D.C. Gossard. Combining mathematical models and geometric models in CAE systems. In *Proc. ASME Computers in Eng. Conf.*, pages 277–284, Chicago, July 1986.
22. W. Sohrt and B.D. Brüderlin. Interaction with constraints in 3D modeling. *International Journal of Computational Geometry and Application*, 1(4):405–425, December 1991.

23. L. Solano and P. Brunet. Constructive constraint-based model for parametric cad systems. *Computer Aided Design*, 26(8), 1994.
24. I. Sutherland. *Sketchpad, a man-machine graphical communication system*. PhD thesis, MIT, January 1963.
25. A. Verroust, F. Schoneck, and D. Roller. Rule-oriented method for parameterized computer-aided design. *Computer Aided Design*, 24(10):531–540, October 1992.

Part II
Representations

Multiple-View Feature Modelling and Conversion

Willem F. Bronsvoort, Rafael Bidarra, Maurice Dohmen,
Winfried van Holland and Klaas Jan de Kraker

Faculty of Technical Mathematics and Informatics, Delft University of Technology
Zuidplantsoen 4, NL-2628 BZ Delft, The Netherlands
email: bronsvoort@twi.tudelft.nl

Abstract. A product model containing information for all product life cycle activities is central to concurrent engineering. Preferably each activity has its own view on the product model, with information relevant to that activity. In this paper, a feature modelling approach is outlined in which each view consists of a specific feature model, containing features relevant to the corresponding activity. Attention is paid to the product model, feature validation, feature conversion and specific assembly features. Feature validation is the basis for maintaining the meaning of features. Feature conversion is used to convert features from one view to other views; multiple-way conversions are possible. As an example of view-specific features, it is shown which assembly-specific features can be included in a feature model for the assembly planning view.

1 Introduction

Concurrent engineering is a systematic approach to the integrated, concurrent design of products and their related processes, in particular production processes [10]. There are two main aspects to concurrent engineering.

The first aspect is that already during the design phase of a product, criteria from downstream product life cycle phases, such as manufacturing and assembly, are taken into account by considering product properties and available resources. This is called *Design for X (DFX)*, where the X can stand for any product life cycle phase. Application of DFX can lead to better product designs, for example in the sense that they are cheaper to produce and easier to assemble, and to a reduction in product lead times by eliminating redesign iterations.

The second aspect is that certain activities for different product life cycle phases can be executed simultaneously. For example, part of process planning for manufacturing and assembly can already be executed when the detailed design has not yet been completely finished. Such simultaneous engineering can further reduce product lead times.

Both aspects require a product model containing information for all product life cycle phases. Solid modelling only deals with information about the geometry of a product. In a concurrent engineering environment this is a shortcoming, because here also non-shape information, eg functional information, is involved. This can, for example, be the function of some part of the product for the user, or information about the way some part of the product is manufactured or assembled [2]. *Feature modelling* does deal with such non-shape information in addition to shape information; both are represented in *features*.

Features can be used in several product life cycle activities. In the design of products, features can be used to model products with entities that are on a higher level, and closer to the way of thinking of a designer, than the entities used in geometric modelling; an example is a stepped hole, consisting of two concentric holes. In process planning for manufacturing, features can identify areas in a product that can be manufactured in one machining operation with one type of equipment; an example is a slot that can be milled with a particular milling machine. In assembly process planning, features can identify connections between parts.

Each activity has its own *view* of a product, ie its own way of looking at it. Each view contains the features relevant to the specific activity. An example of two views of a product is given in Figure 1. In the design view, the object is represented by a base block with a protrusion and a blind hole (a). In the manufacturing view, it is represented by a larger stock with a step and a blind hole (b).

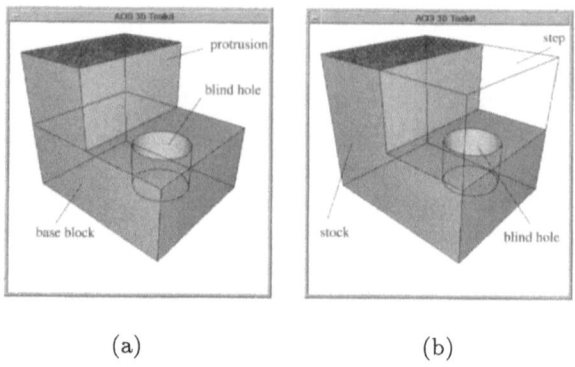

(a)   (b)

Fig. 1. Design and manufacturing view.

In this paper, an outline is given of such a feature modelling approach, with in each view a feature model specific for the corresponding activity. In Section 2, the product model is described. In Section 3, feature validation, the basis for maintaining the meaning of features, is discussed. In Section 4, a method for feature conversion from one view to other views is presented. In Section 5, a feature model specific for the assembly planning view is described. In Section 6, conclusions and future developments are discussed.

2   Product Model

The basic entity in our product model for concurrent engineering is a *feature*, defined as the representation of shape aspects of a physical product that are mappable to a generic shape and are functionally significant. A feature is characterized by a number of parameters, such as shape parameters. A parameterized description of a feature is called the *generic definition* of the feature. A generic feature can be instantiated multiple times by specifying values for its parameters. The topologic entities of a feature, such as vertices, edges, faces and volume, are called *feature elements*.

A feature has a certain *meaning*. An example is that a cylindrical blind hole has a circular top face, a cylindrical side face, and a circular bottom face, and, in addition, that the first face is not on the boundary of the modelled object, whereas the two other faces are.

In many of the current systems that are called feature modelling systems, very little support is given to maintain the meaning, or *validity*, of features when a model is edited. If, for example, a slot feature is inserted into a model with a cylindrical blind hole feature in such a way that the bottom face of the hole is no longer on the boundary of the object, then the feature is no longer a cylindrical blind hole, but has become a cylindrical through hole instead, and its functional meaning has thus changed (see Figure 2). Modelling systems that fail to notify this, are in essence only geometric modelling systems, but no feature modelling systems, because they do not maintain the meaning of features. In our feature models, each feature does have a well-defined meaning, and this meaning is maintained.

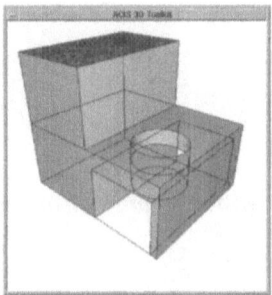

Fig. 2. Invalid blind hole.

The meaning of a generic feature is defined by its *feature validation constraints* (see Section 3). These include geometric relations within the feature, relations with other features, and validity conditions for feature elements. After feature instantiation, it is possible to specify additional relations between feature elements, possibly of different features, ie *model validation constraints*. A collection of feature instances and their relations is called a *feature model*.

Concurrent engineering requires multiple views of one product to be supported simultaneously, each view containing features relevant to its life cycle activity. Many of the activities in a concurrent engineering system will result in modifications of the product model. For example, manufacturability analysis may show that dimensions of some part need to be modified. Modifications should preferably be made in the view in which the need for them arises, and all modifications made in any view should be reflected in all other views. This requires *feature conversion* from features in one view to features in other views (see Section 4). Such feature conversion involves management of the product's shape, which is represented by a different feature model in each view, and management of the constraints, which specify validity conditions of the features.

We are currently working on a prototype implementation of a system, called SPIFF, to test our approach to feature modelling for concurrent engineering. In SPIFF, a product model contains the different views and their feature models, see Figure 3. The feature models of the different views are linked by the product geometry.

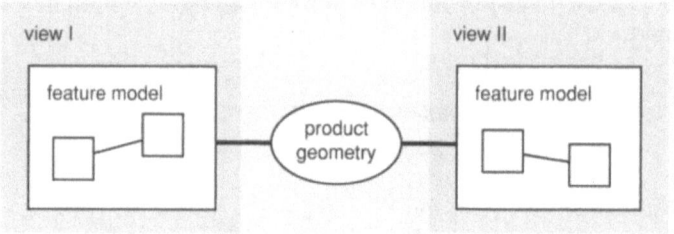

Fig. 3. The product model.

The representation of features is handled at two different levels: *specification* and *maintenance*. This separation provides a clean way of feature specification.

At the specification level, features are specified with all their properties, such as view-specific parameters and attributes and constraints. Specifications are made in an object-oriented specification language, via a graphical user interface, see Figure 4.

At the maintenance level, feature validity is maintained. A Constraint Manager and a Feature Geometry Manager store validation constraints and feature shapes, respectively, in their data structures, and they maintain these when the product model is edited, see Figure 5.

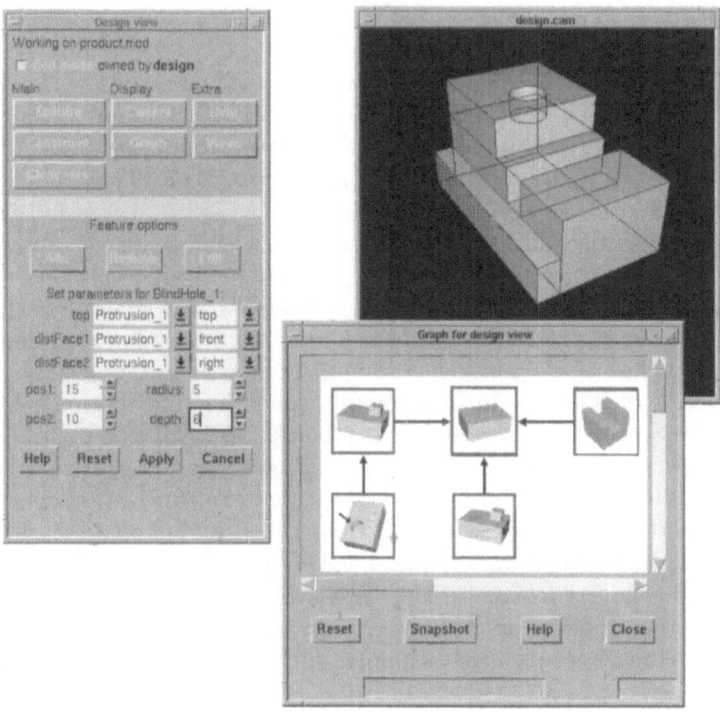

Fig. 4. The SPIFF graphical user interface.

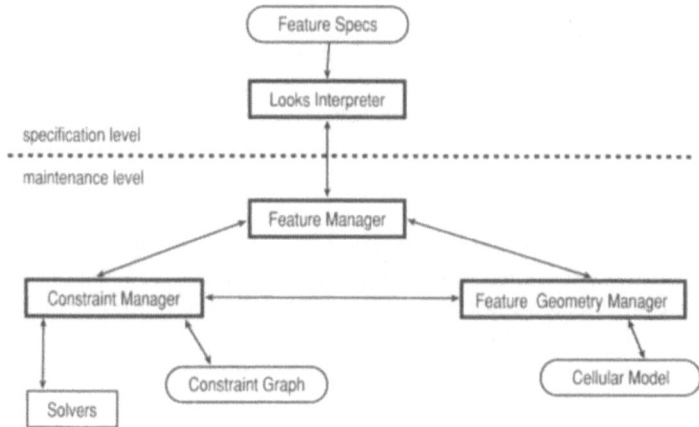

Fig. 5. Feature specification and maintenance.

The Constraint Manager stores all constraint instances in a *constraint graph*, in which two types of nodes represent constraints and variables, and edges connect constraints to associated variables. The variables are feature elements and feature parameters (see Section 3).

The Feature Geometry Manager maintains a *cellular model*, which represents the product geometry. Cells in the cellular model are volumetric; they can have overlapping boundaries, but they cannot have overlapping volumes. They reflect all feature intersections, and therefore can have an arbitrary shape.

Features as described until now, which are often called *form features*, occur in all views on the product model, although different types may occur in different views. In addition, also other, view-specific features may be useful in each view. These features add activity-specific functional information using the form features. This has been implemented for the assembly planning view, in which, for example, connection features represent connection information between form features (see Section 5). In other views, other features may be useful besides the form features. For example, in the design view these might be conceptual features at a higher abstraction level than the form features. Such a model, with form features and other, view-specific features, is called an *enhanced feature model*.

A product model with an enhanced feature model for each life cycle activity, is ideal for a concurrent engineering system, because it contains all information required in the whole product life cycle. In the sequel, several aspects of our feature modelling approach that supports such a product model are discussed.

3   Feature Validation

The basis to maintain the meaning of features is feature validation. Feature validation is performed by maintaining validation constraints. These are part of the definition of features and specify, for example, the kind of properties mentioned in Section 2 for the cylindrical blind hole. Both at the creation of a feature and at subsequent modelling steps, these constraints are checked to see whether the feature is still valid.

The following constraint types are used to specify feature and model validity. Shape, attach and semantic constraints are used only for specifying feature validity; the other constraints can be used both for specifying feature validity and for specifying model validity.

Shape constraints correspond to the type of feature shape, eg a block for a slot. They are used to control the relations between feature parameters and feature geometry.

Attach constraints specify the attachment of a feature to a feature model. A hierarchical relation is specified between two feature elements, eg faces, including the remaining degrees of freedom of the attached elements (see Figure 1a in which the top face of the cylinder is attached to the top face of the base block). These attachments are used instead of the parent-child relations between features used in many other systems. The advantage is that a feature instance does not

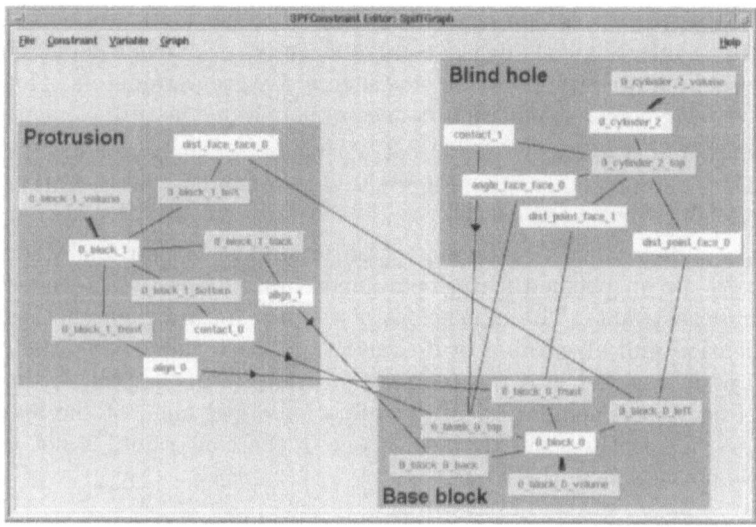

Fig. 6. Constraint graph.

necessarily have only one parent feature; it can be attached to feature elements of several features.

Semantic constraints specify topologic properties of feature elements. For a vertex, edge or face, a semantic constraint specifies the extent to which the element must lie on the product boundary. For a volume, a semantic constraint specifies the extent to which the volume is allowed to intersect other feature volumes. The cylindrical blind hole of Figure 1a is represented by a cylinder that is subtracted. Its bottom face must be completely on the product boundary. The side face, on the other hand, must be at least partly on the boundary, and the top face may not be on the boundary. This may be the semantic specification of a blind hole in a design view. If a blind hole is part of a pen-hole connection feature in an assembly view (see Section 5), a semantic constraint on its volume may declare that no additive feature instantiated later may intersect the hole.

Geometric constraints specify geometric relations, such as parallelism and distance, between feature elements. When used as a model validation constraint, geometric constraints can dimension a feature model. An example is a geometric constraint specifying a distance between two parallel slots.

Dimension constraints specify an interval for the value of a feature parameter. An example of a dimension constraint in a manufacturing view, is a constraint on the width of a slot, declaring it to be within some specified range, because there is no milling equipment available to mill slots with smaller or larger widths.

Algebraic constraints specify equations containing feature parameters. An example is an expression for the length of the protrusion in Figure 1a to be equal to half the length of its base block.

Figure 6 shows part of the constraint graph for the feature model shown in Figure 1a. Light grey icons depict constraints, dark grey icons depict constraint variables, which are either feature elements or feature parameters. The features each have a shape constraint that connects the various feature elements.

After changing a feature model, all constraints in the graph are maintained by the Constraint Manager (see Figure 5), which handles the various constraint types by calling dedicated solvers.

For example, when a feature is instantiated, the Constraint Manager first determines the unspecified feature parameters by solving its attach, shape and geometric constraints. The dimensions of the feature are then checked against the dimension and algebraic feature validation constraints. After this, the feature shape is inserted into the geometry of the product model, ie the cellular model. Finally, the semantic feature validation constraints of the feature and its intersecting features are checked. If one of the constraints is not valid, the feature is not added to the model.

Dimension, algebraic and semantic constraints are currently only checked. Dimension constraints can be checked by testing whether the feature parameter involved has a value within the specified range. For algebraic constraints, a dedicated equation solver will be used. After the model has been edited, the semantic constraints are checked of all features that have been repositioned or redimensioned, and of the features that intersect any of these before or after the change.

To maintain the feature's attach, shape and geometric validation constraints, these constraints are mapped onto a constraint graph containing *primitive* geometric constraints that restrict translational and rotational degrees of freedom [4]. Examples of primitive constraints are a parallel-z-axes constraint and a coincident-points constraint. The primitive-constraint graph is solved using degrees of freedom analysis [9].

With the feature validation approach outlined, the validity of features can be maintained after each modelling operation. If it is detected that a feature is no longer valid, the modelling operation simply can be forbidden, and the user can then take an appropriate action, for example, change the feature parameters, or even change the feature type. For an ideal feature modelling system, however, this method is too rigid. The least that is desirable is that the user gets a good explanation on what has caused a feature to be no longer valid. A further improvement would be that he gets hints on how to avoid or overcome the problem. Ideally, the system automatically adapts the model to get a valid model again in cases where this is possible and desirable, although this should probably not be done without consulting the user. In the example in Figure 2, in which the bottom of the cylindrical blind hole was removed by the insertion of a slot into the model, the blind hole might be automatically converted into a through hole, if the user permits this.

Most validity violations result from modelling operations that cause feature interactions. Feature interactions may have a very wide range of effects on a feature model. Although these may often be intended, in many situations they may affect the semantics of a feature, ranging from slight changes in parameter values to the complete suppression of its contribution to the model shape. To get more insight in this, a taxonomy for feature interactions has been developed that takes into account relevant design and technological criteria [1].

An immediate goal of our research on feature interaction management is the detection of each interaction class, as a result of monitoring each modelling operation. This is based on the analysis of the interaction extent and feature natures, the evaluation of semantic constraints, and a variety of topologic and geometric queries to the cellular model.

Our next research goals in this area include providing the user of SPIFF with a choice of reactions and suggestions, whenever one of the above interaction situations is detected (eg readjust some feature parameters, suggest a change in attachments, or perform a change in feature type).

4   Feature Conversion

In this section, our method to convert a feature model from one view to feature models in other views is presented.

Conversions proposed until now are one-way: a designer has to input a primary view, and conversion modules are available to generate other feature models for secondary views, in particular for a manufacturing process planning view [11,3]. If a modification in the product model is required on the basis of the outcome of an activity, this modification has to be entered in the primary view, after which new secondary views can be generated when needed.

To support concurrent engineering with multiple feature views, the solution with a primary view and a number of secondary views is far from ideal. If an engineer finds in some secondary view that, for example, a dimension of the designed product has to be adjusted, he has to switch from his secondary view to the primary view to make the adjustment. In this view, however, the feature model is different from the one in his own view, and it may be difficult to determine the right adjustments of the feature parameters. We have therefore developed an approach that supports all feature views simultaneously, and performs multiple-way feature conversion between these views [7].

When supporting multiple views simultaneously, two types of operations can be identified that correspond to two types of feature conversion. Firstly, it is possible to *open* a view. Feature conversion to the opened view is then performed, making all views consistent. Views are consistent if they represent the same product geometry. In the view that is opened, a new feature model is automatically built, on the basis of the product model as already specified in all other views. Secondly, in line with the design by features approach, it is possible to *edit* a view. Either parameters can be changed, or feature and constraint instances can be created or deleted. These changes are reflected in the other views by propagating the changes from the edited view to the other views.

After the product model has initially been specified in one view, another view can be opened. A generic method for opening a view has been developed that uses the cellular model and view-specific information [8]. A new feature model is automatically built by instantiating features one by one, until the opened view is consistent with the other views.

Given a set of generic features of a view and the cellular model, many interpretations in terms of features exist. One approach is to generate all possible interpretations and choose one, but, since there can be very many interpretations, this is not a satisfactory solution. Therefore, we try to find a good interpretation for each view directly.

For finding a good interpretation, we assign to each view a *strategy* for identifying feature instances in the cellular model. Such a strategy uses the view's generic features, and it reflects the view's function. Of the latter, important aspects are the structure of the feature model, and the order in which the features can occur. For example, the structure of a design feature model may be mainly constructive, but with some subtractive features. A manufacturing feature model, however, may be completely destructive. In both cases the feature order is important.

To open a view, its strategy provides a sequence of feature classes, and it is tried to identify instances of these classes in that order. This is repeated until the view is consistent. To identify an instance of a feature class, an instance of its shape must be identified. We propose to choose the largest feature shape that satisfies the feature's validation constraints. To identify a feature shape, first its attach faces are identified, and then these are used to perform a directed search to identify the other shape faces. Attach faces are sought on the boundary of the view's not yet consistent feature model. The other shape faces are sought on the product boundary. Both face types are matched with the generic shape faces using topologic and geometric relations. With the faces found, a shape is constructed. After this, values for the feature's parameters are derived. If all its validation constraints are satisfied, a valid feature instance has been created. An important advantage of this approach is that it can deal with any feature intersection.

We illustrate the opening of a view with an example. Assume that the product in Figure 7a has been created in a design view, by starting with a base block, and adding a protrusion and a blind hole (this is the same view as in Figure 1a). Now a feature model in the manufacturing view is built. First a stock is identified by adding the bounding box of the designed object, see Figure 7b. Assume that the strategy prescribes to identify a step, which is represented by a block shape. The four faces indicated in the manufacturing model in Figure 7c are selected as attach faces. The other two shape faces are sought in the model of Figure 7a. This results in the creation of the step's block shape, which is subtracted in Figure 7d.

Next the strategy prescribes to identify a blind hole. The circular face indicated in the manufacturing model in Figure 7e is selected as attach face. The side and bottom faces are sought in the model of Figure 7a. This results in the

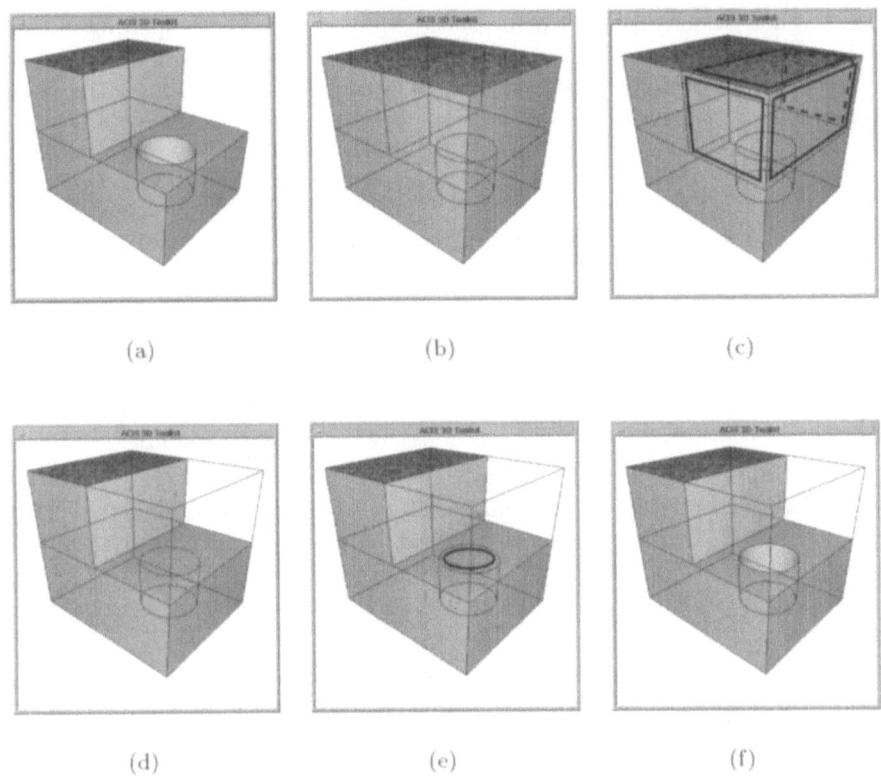

(a)   (b)   (c)

(d)   (e)   (f)

Fig. 7. Open view example.

creation of the blind hole's cylinder shape, which is subtracted in Figure 7f (this is now the same view as in Figure 1b).

Now, since the manufacturing view represents the same geometry as the design view, and thus the manufacturing view is consistent with the design view, the opening of the manufacturing view has been completed.

Editing a view requires maintaining the product model. This involves, among other things, simultaneously maintaining the validation constraints in all views.

After a view has been opened, and before a parameter can be changed, links with the other views implicitly stored in the cellular model are made explicit in geometric constraints between the views. Constraints for this are called *link constraints*, and they couple degrees of freedom of overlapping boundary feature elements of different views. These inter-view constraints are established automatically, and are considered in the constraint maintenance.

To deal with conflicting constraints, each view is assigned a priority that is used to decide whether the edit operation will be issued and propagated or not.

If the edit view conflicts with higher priority views, the change is issued only if the higher priority views can be successfully reopened. Reopening a view is efficiently done with an incremental version of the open view function. If the edit view conflicts with lower priority views, the change is issued, and it is tried to reopen the lower priority views.

5   An Assembly Feature Model

As described in Section 2, for a specific view the form feature model can be enhanced with other, view-specific features. In this section, an enhanced feature model for the assembly planning view is discussed, and it is shown how this model can profitably be used in several assembly planning modules.

Most products need to be assembled from subassemblies and parts. These components are related to each other, often with specific connections. Characteristics of these connections are stored in so-called *assembly features*. We distinguish two types: *connection* and *handling* features, storing information on connections between components, respectively information specific for handling a component.

A handling feature stores for a component the areas where it can, or where it cannot, be grasped, the way it is transported to the assembly environment, the orientation on and the geometry in contact with the fixture, and the grippers that can be used to assemble the component.

With connection features, relations between components are modelled with relations between form features on the components. The information stored in these features is, among other things, used to verify whether the contacts can be made, and whether tolerances can be managed.

The idea of connection features is that characteristics of connections can be incorporated in these features, eg insertion point, insertion path, final position, tolerances, contact faces, internal freedom of motion, attachment agent, and geometric refinements such as chamfers and rounds to ease assembly operations. The final position, or goal position, is the position and orientation of the assembled component relative to the already assembled components, called the partial assembly, after the assembly operation has been completed. The insertion point is the position and orientation relative to the final position where there is not yet contact between the assembled component and the partial assembly, and where the insertion operation is started. The insertion path is the trajectory from the insertion point to the final position. Tolerances and contact faces between assembled component and partial assembly give clues for calculation of the internal freedom of motion, ie the set of motions that can separate the component and the partial assembly. An attachment agent is a component that is needed to enforce the specific connection, eg a bolt and a nut to fasten two plates are attachment agents in the connection of the plates.

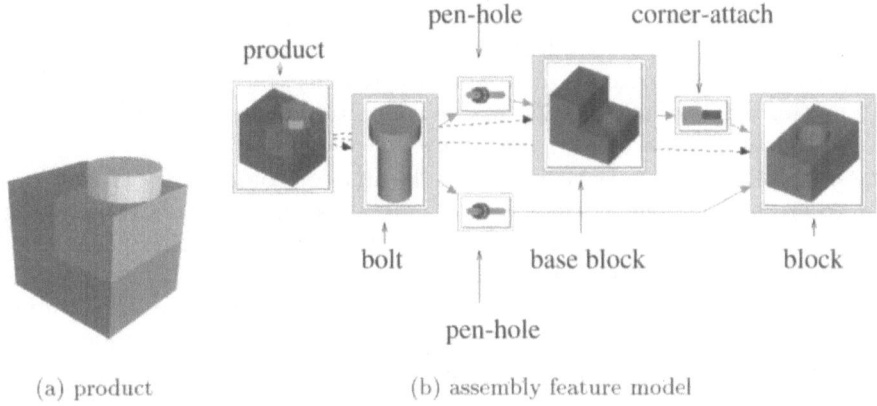

(a) product   (b) assembly feature model

Fig. 8. An assembly feature model for a product.

The assembly feature information can be added to a model either by a designer or by an assembly planning expert. In figure 8, an example is given of an assembly model, consisting of three components (with grey border) and three connection features.

It is now illustrated how this assembly feature information can be used in planning modules.

Connection features can be very useful for stability analysis, in which it has to be determined which components can move relative to each other. Each connection feature contains the internal freedom of motion between the mutually connected components. Combining the internal freedom of motions of all connection features of a component, gives information on the resulting internal freedom of motion for that component relative to other components. Because the internal freedom of motion is known for every connection feature, it is no longer necessary to calculate it from the geometry of the components, which makes stability analysis simpler and faster.

In motion planning, a path is determined to move a component from its feed position to its final position on the partial assembly. Gross motion and fine motion planning are distinguished.

During gross motion planning, a collision-free path for the assembled component from its feed position to the insertion point is searched. Because connection information is not used here, the assembly feature concept can hardly be exploited. Only the position and orientation where the gross motion must stop, and change to fine motion, can be gathered from the connection feature.

Fine motions, on the other hand, can hardly be planned without assembly features. In fine motion planning, the final stage of assembling a component onto a partial assembly is planned. Component and partial assembly are very close to each other, or even make contact. All kinds of subtle movements, including compliant movements, must be executed to make the connection. For example, for assembling a round pen-hole connection, other fine motion strategies are needed than for assembling a square pen-hole connection. Associating predefined fine-motion-strategy information with a connection feature, can result in better strategies and in a reduction of planning time.

Grasp planning is done to determine the areas on components where grippers can grasp the component. Here both handling features and connection features can be used. Handling features contain gripper information that is used to compute gripper-specific areas of a component where it either can or cannot be grasped. These areas are independent of the actual position and orientation of the component. Connection features can give additional information on areas where not to grasp, because these features specify areas that are involved in a connection. These areas are dependent on the components already assembled in the partial assembly. By combining information derived from handling features and connection features, areas can be determined where the component can be grasped. Handling features can give additional information on how to grasp the component, eg which gripper can be used and which forces the gripper should apply on the component [5].

The assembly features are also very useful for finding possible assembly sequences [6]. In assembly sequence planning, all information found in the previously presented modules is combined to determine possible assembly sequences. Finding the ideal assembly sequence, in terms of assembly time and used resources, out of all possible sequences can be very time consuming, and the assembly information stored in connection features can be used as heuristics to speed up this process. Some connection features already contain information about possible assembly sequences. A connection feature with agents for connecting two plates with a bolt and nut, for example, 'knows' that first the plates and thereafter the bolt and nut must be assembled, instead of first the bolt and nut, leaving no room for assembling the two plates.

From the above, it is clear that in many assembly planning modules it is advantageous to use the information stored in the assembly features. The presented enhanced feature model is thus useful for the assembly planning view.

6   Conclusions and Future Developments

Several new concepts and methods have been presented to apply feature modelling in concurrent engineering. Each product life cycle activity can have its own view on the product model, each view consisting of a feature model with features relevant for that view.

Feature validation is an aspect of feature modelling that erroneously is neglected in most current feature modelling systems. It becomes even more important in a concurrent engineering environment, because in different views, features can have very specific meanings. Feature interaction management is a good basis for assisting the user in creating valid models.

Feature conversion as described here, for obtaining a feature model for any view, is a very promising approach to multiple-way feature conversion. Such conversion is a prerequisite in a concurrent engineering environment, because all feature views have to be simultaneously supported in such an environment.

An enhanced feature model for a view contains, besides the form features specific for that view, also other features specific for that view. This concept has been demonstrated for the assembly planning view, in which assembly features are added to the form features. It has been shown that assembly features can be used profitably in assembly planning. For other views, other enhanced feature models are foreseen.

The work described will continue. The main goal is the development of a multiple-view enhanced feature modelling system, with good facilities for feature validation, interaction management and conversion. Such a system can adequately support the kind of product model that is ideal for concurrent engineering.

Acknowledgments

Rafael Bidarra's work is supported by the Portuguese Praxis XXI Program of JNICT.

Klaas Jan de Kraker's work is supported by the Netherlands Computer Science Research Foundation (SION), with financial support from the Netherlands Organization for Scientific Research (NWO).

We thank Bart Vergouwe for implementing the SPIFF graphical user interface, and Kees Seebregts for his support in the software development.

References

1. R Bidarra and W F Bronsvoort. Towards classification and automatic detection of feature interactions. In D Roller, editor, *Proceedings 29th International Symposium on Automotive Technology & Automation*, pages 99–108, 1996.
2. W F Bronsvoort and F W Jansen. Feature modelling and conversion - Key concepts to concurrent engineering. *Computers in Industry*, 21(1):61–86, 1993.
3. P Dave and H Sakurai. Maximal volume decomposition and its application to feature recognition. In *Proceedings of the 15th ASME International Computers in Engineering Conference*, pages 553–568, 1995.
4. M Dohmen, K J de Kraker, and W F Bronsvoort. Feature validation in a multiple-view modeling system. In J M McCarthy, editor, *CD-ROM Proceedings of the ASME 1996 Design Engineering Technical Conferences and Computers in Engineering Conference*, 1996.
5. W van Holland and W F Bronsvoort. Extracting grip areas from feature information. In J M McCarthy, editor, *CD-ROM Proceedings of the ASME 1996 Design Engineering Technical Conferences and Computers in Engineering Conference*, 1996.

6. W van Holland and W F Bronsvoort. Assembly features and sequence planning. In M Pratt, R D Sriram, and M J Wozny, editors, *Product Modelling for Computer Integrated Design and Manufacture*. Chapman & Hall, 1997. To be published.

7. K J de Kraker, M Dohmen, and W F Bronsvoort. Multiple-way feature conversion to support concurrent engineering. In C Hoffmann and J Rossignac, editors, *Solid Modeling '95, Third Symposium on Solid Modeling and Applications*, pages 105–114. ACM Press, May 1995.

8. K J de Kraker, M Dohmen, and W F Bronsvoort. Multiple-way feature conversion - opening a view. In M Pratt, R D Sriram, and M J Wozny, editors, *Product Modelling for Computer Integrated Design and Manufacture*. Chapman & Hall, 1997. To be published.

9. G A Kramer. *Solving geometric constraint systems: a case study in kinematics*. The MIT Press, Cambridge, Mass., USA, 1992.

10. H R Parsaei and W G Sullivan. *Concurrent Engineering; Contemporary Issues and Modern Design Tools*. Chapman & Hall, London, 1993.

11. J H Vandenbrande and A A G Requicha. Geometric computation for the recognition of spatially interacting machining features. In J J Shah, M Mäntylä, and D S Nau, editors, *Advances in feature based manufacturing*, pages 83–106. Elsevier Science B.V., Amsterdam, 1994.

A Knowledge Representation Perspective on Geometric Modeling

Rüdiger Klein

Daimler-Benz AG - Research Dept.,
Knowledge Based Systems Group (F3S/W)
Alt-Moabit 96a, D-10559 Berlin;
Tel.: (++49-30) 399 82 211;
fax.: (++49-30) 399 82 107;
email: klein@dbresearch-berlin.de

1 Introduction

Geometric modeling is a very complicated matter in its own. But in many applications the real challenge is the *integration* of geometry with other kinds of information. That is especially true for geometric modeling and knowledge representation (KR). Feature concepts[1] have been developed (at least in part) for this integration, but they are strongly related to current geometric modeling approaches. A *deep, conceptual integration* of knowledge representation and geometric modeling seems to be an open issue.

From the knowledge representation point of view, there is a considerable "mismatch" between knowledge representation and current geometric modeling techniques. These problems are all more or less directly related to the procedural, "history-based" way in which geometric modeling is currently done, and to the intermixed explicit/implicit way in which the various geometric entities are represented and related in current geometric modeling approaches.

The points in *geometric modeling* which are essential for knowledge representation can be summarized as follows:

- Representation and generation of geometry: Given a geometric shape we should be able to say which information (or knowledge) we can get therefrom. That's the representational aspect. Another aspect - the *generative* one - is how to generate a geometric shape which fulfills a given (or evolving) set of requirements. What does it mean to perform a certain modeling operation? In general, each such operation adds new information to the model, but as a result of the "built-in semantics" of the physical space it also changes other things. How can we adequately describe the meaning of geometric modeling operations from a knowledge representation viewpoint?
- Expressivness: Relevant knowledge may be related to various types of geometric entities: volumes, surfaces, lines, points, topologies, their parameters, etc. These entities are normally not independent - the various relations and

---

[1] for a recent review see, for instance, [20]

constraints between them should be representable. Features as well as other forms of knowledge should be related to geometric representations.

- Declarativity: The meaning of the represented knowledge should result from its structure - independent of how it is processed. A declarative (for instance, Tarsky-style) semantics may be given which has to be realized by an appropriate geometric modeling approach.

The paper is organized as follows: in the next chapter we outline what a geometric model and a geometric modeling scheme are from a KR viewpoint. Whereas this discussion is related to "static" geometric models, we extend this discussion in chapt. 3 to geometric modeling as a process, i.e., to the generative aspect of geometric modeling. In principle, every geometric modeling scheme which is unambiguous and sufficiently expressive can serve as a "geometric reasoner" in conjunction with other knowledge processing systems. But this results in very high demands to its reasoning capabilities which seem not to be a realistic option for the next future. In chapt. 4 we outline a new geometric modeling scheme (called 'G-Rep' for 'Geometric Representation' [8]) which better fits the needs of a knowledge based system due to its *expressiveness* and *declarativity*. In chapt. 5 we will compare our achievements to related works and discuss the merits and open problems.

2   The Logics of Geometric Models

In this chapter we want to start with a short discussion of knowledge representation in general, and then investigate the relation between knowledge representation and geometric modeling.

Logic-based knowledge representation

For the purpose of our discussion we will take first-order logic as a formal basis for knowledge representation. Then we have a first-order language L which allows us to represent our knowledge of that part of the world we are interested in. This language is based on a signature $\Sigma = <F, R>$ consisting of a set F of functional symbols of defined arity (including a subset C of 0-ary function symbols, the constants), and a set R of predicate symbols (of defined arity) [2]. Therefrom [3] knowledge can be represented as a set Φ of sentences (well-formed formulas, wff) in our language L. We can represent the objects in our world by naming them with constants, we can assign them attributes (as predicates relating objects with other "attribute-value" objects), we can express those relations between them we are interested in, etc. And we can formulate general knowledge of our world by using quantified variables in compound logical expressions (like implications, equivalences, etc.) which, for instance, allows us to relate more special expressions to more general ones, or to express consistency information.

---

[2] There may be functional and relational symbols in this signature with a pre-defined semantics, for instance to represent arithmetic or geometric aspects.

[3] Together with logical variables, logical connectives, punctuation symbols, and quantifiers well-formed formulas (wff) can be created from symbols in the signature S as syntactically correct expressions using the well-known syntactic rules.

The meaning of these formulas comes from an interpretation relating ("mapping") the symbols in our language to "things" and their relations in the real world. Formally, an interpretation $I = < \Delta, \cdot^I >$ is a pair which consists of a set Δ of "real-world" objects (the universe), and interpretation functions $\cdot^I$ assigning meaning to each functional and relational symbol in the signature Σ.

Given a set Φ of well-formed formulas in our language we say an interpretation is a model of this set if each formula in Φ gets true under this interpretation. In knowledge representation we typically have knowledge of two kinds: knowledge about the "things how they are" (the factual knowledge), and knowledge about "general rules". These two ingredients allow us to reason also about those aspects of our world which are not represented explicitly, but which follow from the facts and the general knowledge. This entailment relation (abbreviated '$\models$') can be described using interpretations: a set Φ of 'basic knowledge' ("facts" and "general rules") entails another assertion ϕ (wff) about our world if and only if each interpretation I which is a model of Φ is also a model of ϕ:

$$\Phi \models \phi$$

One of the key issue in knowledge representation is to *formalize* this entailment, i.e., to provide a formal apparatus (a set of *inference rules*) operating on well-formed formulas of our language L in such a way that each formula ϕ which is entailed by Φ can be inferred or derived ('$\vdash$') from Φ by a (finite) sequence of formal manipulations:

$$\Phi \models \phi \quad \text{iff} \quad \Phi \vdash \phi$$

A formal knowledge representation scheme which fulfills this condition is called sound and complete, in contrast to unsound formalisms (which allow us to deduce assertions which are not entailed) or incomplete ones (which can not deduce everything which is entailed). There is - fortunately - a calculus for classical frist-order logic which is sound and complete. It provides us with a declarative semantics (interpretations which are models), so we can get the meaning of any well-formed formula "from itself", without relation to any kind of derivation *procedures*.

Geometric modeling

How and how far can we establish an analogy between geometric modeling and knowledge representation? First let's suppose we have an unambiguous geometric representation scheme [15] where each (syntacticly correct) solid [4] represents unambiguously a valid geometry (an 'r-set'). Let's further assume that our solids are completely specified, i.e., there are no variable parameters.

The next question is which are the objects (in the sense of knowledge representation) in our "world" of geometric shapes, and what can we say about

---

[4] In order to avoid confusion we will use the term 'solid' (instead of 'geometric model') if we mean a geometric representation (a CSG tree, a B-Rep or what ever), reserving the term 'model' for logical interpretations.

them. Conceptually, geometric objects in the sense of KR are: points in the 3-dimensional physical space, one-, two- and three-dimensional sets of points (lines, surfaces, volumes), and arbitrary complex "structures" composed of them. Each of these objects can be assigned a set of attributes (geometric and non-geometric ones), and they can be summarized as classes or types. In the following we will use the notion 'geometric entity' (or simply 'entity') for the representation of such objects. Objects can be related in various ways *some of which* are typically geometric (of course, also depending on their nature): points can be *contained-in* a volume, surface, or line; a surface can form (part of) the border of a volumetric object, etc.

How can we interpret "classical" geometric modeling in this context? Given an unambiguous geometric representation scheme s and a correct solid Γ in this scheme, the semantics of this solid is the geometry ('r-set') represented by Γ under the inverse representation of s: $s^{-1}(\Gamma)$ [15]. Now we can ask queries q [14] to this solid Γ. By calculating an answer σq to such a query q the geometry modeller gives us the information (knowledge) represented in the solid [5]. So we can say that an "ideal" geometric modeller GM can be considered as an *inference engine* which infers those answers to queries from the solid which are also entailed by the geometry represented by that solid:

$$\Gamma_{GM} \vdash \sigma q \qquad \text{iff} \qquad s^{-1}(\Gamma) \models \sigma q$$

For later discussions we introduce formally the set $A_{GM}(\Gamma)$ of all assertions[6] (i.e., answers to queries) we can infer from a solid Γ (with the corresponding geometric modeller GM):

$$A_{GM}(\Gamma) = \{\sigma q \mid \Gamma \vdash_{GM} \sigma q\}$$

In current (non-ideal) geometric representation schemes there are some problems which make it difficult to follow this approach with all necessary consequences. The main problem is that some of the geometric entities and their relations are represented explicitly [7], and others are left implicit (which means

---

[5] There is no deeper formalization of this query approach in [14]. So one can not say what a query really is, how it relates to the solid, and what a correct answer to a query is. Without going into technical details, we adopt here a logical view on queries and answers. We restrict queries to be atomic expressions which may contain existentially quantified variables, and answer substitutions s contain values for those variables. A query which is not entailed (and consequently has no answer) is said to have failed.

[6] Of course, this is a completely formal concept without practical relevance. It is never meant to calculate it explicitly (which is impossible taking alone the infinitely many possible queries and their answers). A practical "approximation" could be a geometric modeling approach where all "really interesting" geometric entities and their relations are retrievable and adressable. This will be discussed in more detail in chapter 4 in conjunction with G-Rep.

[7] Though current geometry modeling systems not always support a suitable naming scheme this could in principle be built-in for the explicitly represented entities.

. that in principle they can be calculated by the geometric modeller from the explicit entities and their relations). Also the relations between geometric entities (explicit and implicit ones) are often left implicit. In CSG, for instance, volumetric entities are represented explicitly as well as their composition relations - but faces, edges, vertices, etc. and their attributes and relations must be calculated. That's not a big problem (as many hybrid modellers demonstrate which calculate a boundary representation from a CSG representation). The critical point is how to adress such "implicit" entities in queries, and (as will be discussed in the next chapter) in requirement specifications? A geometric modeling scheme which fulfills the needs of knowledge representation should allow us to represent geometric entities of *all kinds* and relations between them and to non-geometric aspects.

3   The Generative Aspect of Geometric Modeling

The interpretation of existing, complete solids is one point, another is the *generation* of solids. Up to now, geometric modeling is more or less completely done by humans in an interactive way. This left the question open what a knowledge representation view on geometry *generation* could be. The following knowledge representation approach to geometry generation is to be thought of as being part of a *design* process in which geometric shapes of objects are generated as part of the solution. Thus, our KR approach to geometry generation will be embedded in a view on design.

There are various formalizations of knowledge representation in design [7], [5]. Without going too much into detail, for our discussion we will adopt the model generation approach outlined in [7], [9]: we describe the general domain knowledge in a design domain by a set D of formluae representing definitional knowledge (which relate more abstract notions in the domain to more concrete ones), and by a set K of consistency information (constraints). A concrete design problem is then represented as a set G of formulae representing goals or requirements which the artifact to be designed should fulfill. The solution to such a design problem is then represented as a (semantical) model M which, using the definitional knowledge D, entails the goals G, and which is consistent w.r.t. the constraint set K:

$$M \models_D G \quad \text{and} \quad M \models_D K$$

where M has to fulfill some further requirements (for instance, to contain only most specific or basic expressions, or not to contain expressions which are not needed to fulfill goals or constraints)[8].

There are two main questions:

---

[8] The practicability of this KR approach to design strongly depends on the kind of knowledge which has to be represented in the domain. There are relatively simple, well-structured domains (for instance some configuration domains) where the model generation can be used more or less directly. In other domains, for instance with moving parts and complicated geometries, our approach can in the best case provide

- First, which expressiveness is needed in order to represent the general domain knowledge (the definitional knowledge D and the constraints K) in a domain and the concrete requirements G in a given design problem?
- And second, how should the geometry representation and the corresponding geometry modeller look like in order to be able to follow this approach?

Let's illustrate the first point with some examples: As mentioned above, we are interested in the representation of geometric entities of all kinds including their attributes (geometric and non-geometric), relations between them, and various forms of constraints. For instance, we want to say (as part of our definitional knowledge) that for each cylinder there are other geometric entities related to it: a cylinder surface, two (top and bottom) planes, a central axis of type straight line, etc. In first-order logic, these things could be represented as a well-formed formula:

$$\forall C \forall F \forall P \forall P' :$$

$$cyl-surf(F) \wedge plane(P) \wedge top(C,P) \wedge plane(P') \wedge bottom(C,P') \wedge \cdots \leftrightarrow cylinder(C)$$

Constraints which are relevant in this example (but not shown) are for instance that if a cylinder is related to a plane by the top (or bottom) relation then the involved objects have to satify certain geometric parameter conditions.

If we know that there is a geometric entity c1 in a given solid Γ which is a cylinder we can also conclude from the given definitions that there are other entities of corresponding type which are related to c1 by the corresponding relations. We can also conclude that there is a cylinder in a solid Γ if we know that there are entities of corresponding type which are related [9] ... And knowing about entity c1 enables us to adress its surface entity even if we don't know how it is named in the solid.

Now suppose that there is a *requirement* that a geometric entity c1 of type cylinder (with attributes ...) should exist. From the logical framework we have discussed follows that a requirement can be treated as a query q: q is fulfilled by a solid G if there is an answer sq which can be calculated from G. What a geometric modeller does with such a requirement is to modify the existing solution, i.e., to generate the corresponding object in a way which guarantees that "thereafter" this requirement and all the constraints resulting therefrom are fulfilled. For instance, it also generated the top and bottom planes and relates them to the c1 cylinder [10].

The problem is that normally things are interrelated: generating a new geometric entity may result in *modifications* to existing entities. Formally this can be describes as follows:

---

some guidance of how a knowledge based system could solve such problems in the future (and provide a feeling of how much KR know-how is still missing).

[9] which is an instance of feature recognition.

[10] Model generation is an established technique in formal logics, though the special geometric requirements need some modifications.

Suppose there is a solid Γ which has been generated in order to fulfill a set G of requirements. Then these requirements have to be satisfied by G, i.e., the set $\overline{G} = \{\sigma q | q \in G\}$ of the corresponding answers form a subset of all assertions which are true in Γ:

$$\overline{G} \subseteq A_{GM}(\Gamma)$$

Adding a set δG of new requirements to the existing ones results[11] in a *modified* solid Γ' which has to fulfill the corresponding relation

$$\overline{G} \cup \overline{\delta G} \subseteq A_{GM}(\Gamma')$$

The critical point is that the set $A_{GM}(\Gamma')$ of assertions which hold in the modified solid Γ' in general is neither a superset of those holding in the "old" one $A_{GM}(\Gamma)$, nor do the new assertions depend only on the new requirements δ G (at least some of them result only from the interaction of old and new geometry). More accurately, we can distinguish the following sets:

those which are not affected by the modifications:
$$A_{GM}^0(\Gamma, \Gamma') = A_{GM}(\Gamma) \cap A_{GM}(\Gamma')$$
those which had been true in the old solid but which are not true in the new one:
$$A_{GM}^-(\Gamma, \Gamma') = A_{GM}(\Gamma) \setminus A_{GM}(\Gamma')$$
those which result from the modifications:
$$A_{GM}^+(\Gamma, \Gamma') = A_{GM}(\Gamma') \setminus A_{GM}(\Gamma)$$
That's the classical situation of non-monotonic logics: the addition of new information (in this case the modification of a solid) makes assertions invalid which had been true before. Some aspects of this non-monotonicity of geometric modeling deserve special attention:

- The assertions in the two sets $A_{GM}^+(\Gamma, \Gamma')$ and $A_{GM}^-(\Gamma, \Gamma')$ are in conflict. A more comprehensive analysis would reveal that for each "invalidated" assertion in $A_{GM}^-(\Gamma, \Gamma')$ there is a subset of new assertions in $A_{GM}^+(\Gamma, \Gamma')$ which constitute the conflict. One approach to deal with these conflicts is to give assertions "from one side" preference over those from the other side (as, for instance, in preference logics [19]) [12].
- As outlined at the beginning of this chapter interrelations between geometric entities and relations can be represented as logical implications or equivalences. Such a logical view implies a monotonicity of the inferences - in contradiction to the observed non-monotonicity of geometry generation.

---

[11] For the moment we assume that the requirements are formulated in such a way that the modifications to the solid result therefrom deterministically. The problem of "multiple choices" will be omitted here.

[12] In geometry modeling the technical details are different from those in "classical" non-monotonic reasoning. This still has to be waorked out formally. The analogy has been used in an intuitive way to develop the G-Rep approach to geometric modeling (discussed in the next chapter). It allows one to say which geometric entity is preferred over others if a set of entities is in conflict.

Consequently, we should interpret implications and equivalences as *defaults*. For instance, a cylinder is *typically* related to a plane via a 'top' relation - if *nothing contradicts* this default "assumption". But in order to fulfill another requirement the geometric modeller may change things so that this cylinder entity has a curved surface on top of it...

– The semantics underlying geometry generation is an open world semantics. This especially means that we can not use the "meta assumption" of a closed world (CWA) which says that if something is not known its negation can be assumed. If we can assume that we know everything about a certain face, for instance, we can classify it as a rectangle, a hexagon, etc. But the addition of new geometry information may change things there. So we have to be aware of the pre-conditions under which a certain conclusion had been drawn.

– Some of those assertions which lose validity by a modification could have been used before to fulfill some of the requirements. Formally, this means that $A_{GM}^-(\Gamma, \Gamma') \cap \overline{G} \neq \emptyset$. This implies that the modification of the solid[13] is inconsistent with existing requirements.

– Especially interesting in this context is the way geometric modeling is currently done by humans. Often they do not explicitly formulate a list of requirements and then look for a geometric entity or *feature* which could be added to the solid in order to fulfill them. By a kind of heuristic reasoning they start directly with the second step (feature creation and solid modification) and then look "if it worked". Maybe a more sophisticated reasoning as outlined here would help to find other alternatives and thus avoid wasting time and efforts.

– In practice the set of requirements is far from being static: it evolves in an incremental way. At a certain point of problem solving new requirements are added, others are withdrawn, etc. From this viewpoint the currently used history based geometry modeling is inadequate. A dependency-directed management of geometry is needed.

These points can only be briefly outlined here - they need full elaboration (which also means a lot of work still to be done).

The knowledge representation view on geometric modeling given here does not conclusively determine how a geometric modeller should work. In principle, of course, we can try to extend current CSG- or Brep-oriented systems in a way which allows them to fulfill the requirements formulated here. Current efforts to develope feature and variational/parametric constraint techniques go into such a direction. But we feel that there are some principle limitations to such approaches. In the following chapter we outline a new approach to geometric modeling which better fits the knowledge representation requirements.

---

[13] or, if there are different alternatives to fulfill a requirement, the chosen alternative gets inconsistent.

4   Towards a Geometry Modeling which fits the Knowledge Representation Needs

4.1   Requirements

As a summary of the discussion in the previous two chapters we list the main requirements to geometry modeling which are essential from a KR viewpoint:

- expressiveness: All geometric entities of interest, their attributes and relations between them must be expressible (including typical geometric relations as those between volumes and its bounding surfaces etc., but also non-geometric ones). Knowledge must be formulated as generic as well as specific propositions, including the definition of classes/sorts, compound entities, variables, and constraints.
- declarativity: Knowledge should be represented in a way where its meaning comes out without any reference to the kind it is processed.
- (non-)monotonicity: One of the characteristics of geometry is that the addition of new geometric information (which in some sense is a monotonic process) may change the total geometry (which is non-monotonic). Thus a geometric modeling approach is needed which allows to monitore such situations in an appropriate way.
- classification and geometric interactions: Strongly correlated with this monotonicity problem is the classification and interaction problem of geometric modeling. In order to classify a part of the geometry all geometric information concerning this part has to be available (i.e., a closed world semantics). But the generation of geometry is based on an open world semantics. This has to be taken into account formulating the various inferences.

4.2   An Integration of Knowledge Representation and Geometric Modeling

We developed a new approach to geometric modeling (called 'G-Rep' for 'geometric representation' [8]) which resulted from the following considerations:

Whereas recent approaches to feature modeling and to the integration of KR and geometry are based on current geometric modeling techniques we go the opposite way: we start with the KR perspective on geometric modeling, discussed in the previous chapters, and then describe how these requirement can be "mapped" onto an appropriate approach to geometric modeling.

This knowledge representation perspective can be summarized as follows (fig. 1):

- First we need a knowledge representation language (usually a subset/variant of first order logic) which allows us to express all 'facts' and general 'rules' about geometric entities. This language will have a declarative semantics. Normally it will be extended "somehow" to arithmetic and geometric constraints, it also may include cardinality restrictions on relations [21] and other appropriate representational means.

- Given a set G of requirements expressed in such a logic-oriented language, from this set a geometry description has to be generated by problem solving[14].

In order to understand how this kind of geometry representation and generation works it is essential to know how it is coupled with knowledge representation and problem solving. Fig. 1 describes the principles of the interplay between knowledge representation and G-Rep' geometry modeling: on the knowledge level all the queries, requirements, general rules and constraints are processed by an appropriate problem solver (or a system of coupled modules). The main 'constraint' on this problem solver is to generate a complete (and, of course, contradiction-free) representation of the geometry (containing no variables or other forms of indefiniteness). This representation is called a G-Rep representation. Therefrom a "classical" boundary representation or what ever may be adequate can be deduced.

Three points should be mentioned:

- The G-Rep level keeps the "connections" between "pure geometry" and the other information the problem solver may be interested in (for instance, non-geometric attributes).
- The meaning of this set of expressions (the represented geometric shapes) should be independent of any procedural, history-based considerations. Thus an appropriate geometric modeling approach is needed which is declarative, non-history-based. This is also essential for the (normally encountered) incremental way of problem solving.
- The interaction between the three levels is not a one-way: for instance, the emergence of geometric interactions (not the intended, foreseen ones, but those which simply result from geometry) which can be represented as logical implications, results in the need to draw conclusions on the knowledge level from geometry.

4.3   The Basic Idea of G-Rep

In G-Rep two notions are important for the description of geometry [8]:

- basic volume and
- G-Rep solid.

A basic volume is the 'atomic unit' of geometry modeling in G-Rep. It may be a standard elementary geometry like box, cylinder, sphere, it may be composed in a pre-defined manner, or it may be "configured" dynamically during problem solving (using the logic oriented representation). Associated with each basic volume is a set of parameters (which can be variables, maybe related by

---

[14] Some practical limitations may come into this approach taking the available processing techniques into account (for instance, the limitations of current geometric constraint solving [6]).

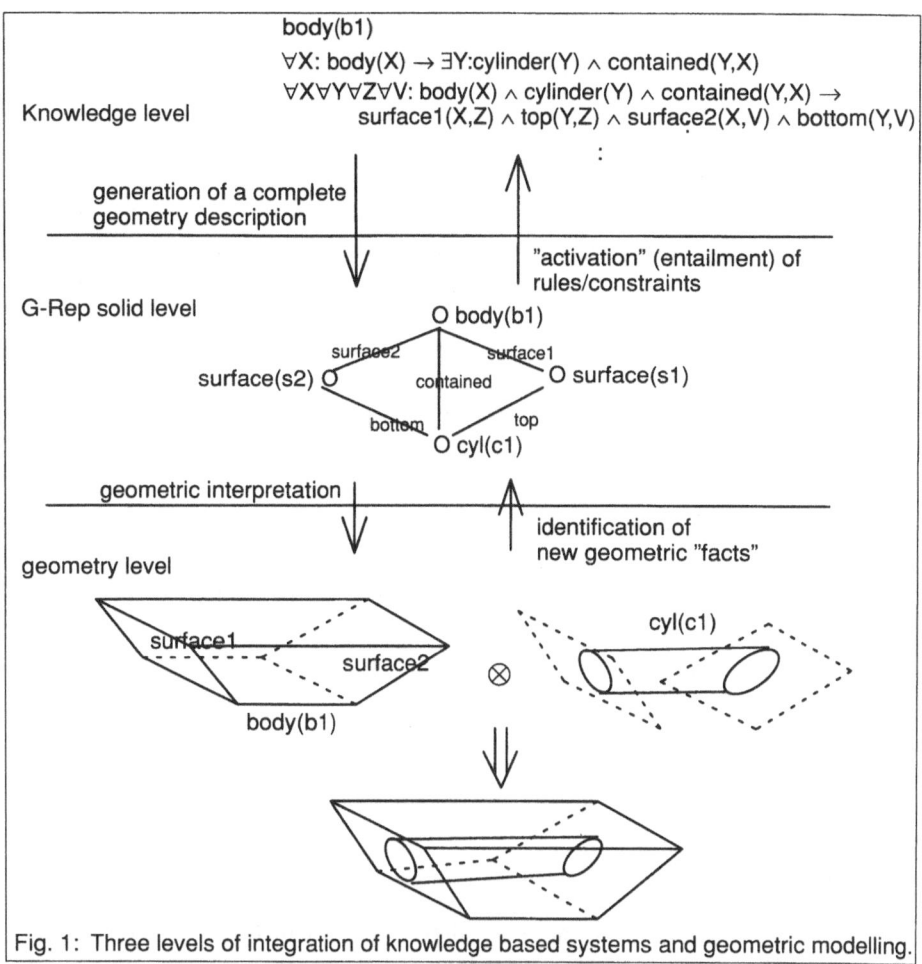

Fig. 1: Three levels of integration of knowledge based systems and geometric modelling.

constraints) and a set of surfaces which form - as in a boundary representation - a topology of faces, edges, and vertices.

More complex geometries can be described in G-Rep by composition of basic volumes. For this purpose the notion of a G-Rep solid in conjunction with G-Rep's composition operation '⊗' are introduced:

Definition: G-Rep solid Γ A G-Rep solid Γ is the integrated geometry representation. It consists either of a basic volumen V:

$$\Gamma = \{V\}$$

or it is the result of a G-Rep composition operation '⊗' applied to two G-Rep solids Γ_1 and Γ_2:

$$\text{if} \quad \Gamma_1 = \{V_1, V_2, ..., V_n\}$$
$$\Gamma_2 = \{V_1', V_2', ..., V_m'\}$$
$$\text{then} \quad \Gamma = \Gamma_1 \otimes \Gamma_2$$
$$= \{V_1, V_2, ..., V_n\} \cup \{V_1', V_2', ..., V_m'\}$$
$$= \{V_1, V_2, ..., V_n, V_1', V_2', ..., V_m'\}$$

The essential point in the G-Rep concept is that the *meaning* of a G-Rep solid (the geometry it describes) does neither depend on the sequence of the basic volumes it "contains"[15] nor on the "history" by which the solid has been built up from the basic volumes through successive application of the -composition operator:

$$\Gamma = \{V_1, V_2, ..., V_n\}$$
$$= \{V_2, V_1, ..., V_n\}$$
$$= \{V_n, V_1, ..., V_{n-1}\} \dots \text{etc., for all permutations of} \quad V_1, V_2, \dots, V_n.$$

For being *declarative*, this G-Rep $\otimes$-composition has to be commutative and associative:

$$\Gamma_1 \otimes \Gamma_2 = \Gamma_2 \otimes \Gamma_1 \text{and} (\Gamma_1 \otimes \Gamma_2) \otimes \Gamma_3 = \Gamma_1 \otimes (\Gamma_2 \otimes \Gamma_3)$$

The way this can be reached is quite simple: instead of 'sequence' we take another kind of information which 'tells' the $\otimes$-composition operator how the individual basic volumes have to be combined. By reasons of vividness we call this information 'density'.

Formally the density is defined as follows:

1. At each point p in space there is a density $\rho(p)$ which is a real number.
2. $\rho(p) > 0$ means that at this point there is some 'material' (i.e., this point is inside or on the surface of any G-Rep solid).
3. $\rho(p) \leq 0$ means that there is no material, i.e., this point is outside. The use of negative densities will get clear soon.
4. A point p not contained in any basic volume (an thus in no G-Rep solid) has per definition a density of 0.
5. If a point p is in more than one basic volume $V_1, V_2, \dots, V_n$ with the associated density values $\rho_1, \rho_2, \dots, \rho_n$ then it gets the density with the highest absolute value: $\rho(p) = max\{|\rho_1|, |\rho_2|, \dots, |\rho_n|\}$.

---

[15] That's also the reason why we use the set notation to express the relationship between a G-Rep solid and the basic volumes it "contains", though of course normally a basic volume will not stay unaffected by being put into a G-Rep solid. The point simply is, that the way it is affected depends on the other basic volumes in the same solid, but not on the sequence they have been composed.

Especially the last point makes it clear that 'density' is simply a synonym for *ranking*. Density values are only *compared*, not added[16]. They can be specified dynamically on runtime. Fig. 2 gives a simple (2-dimensional) example, how this kind of density modeling is used to replace sequence information:

Density could be positive or negative: independent of the sign of its density the stronger volume "cuts" into the softer one. It is not affected by this composition in itself, but it determines the changes the softer volume experiences therefrom. In this way a G-Rep solid is not homogeneous, but formed as a *compound* geometry (see fig. 2) consisting of "zones" of different density. In order to achieve this, some more topology information has to be remembered: the influence of the stronger to the softer volume has to be represented.

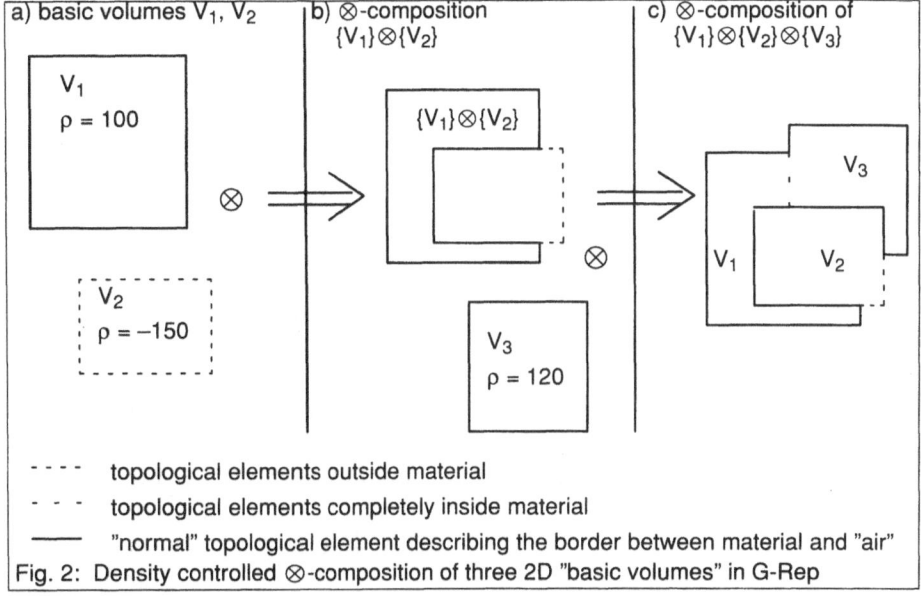

Fig. 2: Density controlled ⊗-composition of three 2D "basic volumes" in G-Rep

---

[16] In principle, it would be sufficient to specify a partial order on the set of basic volumes, assigning a preference relation in each pair of interacting basic volumes. But this seems more cumbersome and less intuitive then the way we proposed. Especially it means to foresee all possible interactions between geometric entities in advance. In order to combine the "ease of modeling" provided by the density approach with more flexibility one could allow the geometric modeller to specify the density of a new basic volume dynamically as well as a re-adjustment of densities which preserves the "old" order.

4.4   Geometry Representation in G-Rep

This G-Rep way of declarative geometric modeling provides us with the main prerequisites for an expressive, logic-oriented approach to geometric representations.

The first question we have to answer is which are the geometric entities we (or a knowledge based system) is interested in. The next question is which information we want to assign them (and to retrieve from a geometric representation by query answering).

In G-Rep we can represent geometric entities of any kind: volumes, surfaces, lines, points, topological elements. We can represent relations between them in any way we want (as far as first order logic allows us) where the meaning of these relations

- may be left implicit ("primitive relations" [21]);
- may be expressed in (logical) terms to other relations and entities in the form of generic knowledge (i.e., definitions or consistency information); or
- may reflect the "built-in semantics" of physical space.

Whereas the first two items are KR issues "in general", the last one is characteristic for geometric knowledge representation: volumes are separated from their outside by surfaces, faces on a boundary representation are part of a corresponding surface, an edge in a boundary representation connects (at least) two faces, etc. These typical geometric relationships have to be built into a geometric knowledge representation, the geometric modeller has to reason on these relations.

In G-Rep a three-level representation has been chosen for this purposes (see [8] for details):

The main categories of geometric entities which may be relevant are volumes, surfaces and lines, and topological elements. In order to represent the special geometric relations they are involved, G-Rep provides the corresponding representational means (see fig. 3):

- surfaces are associated with basic volumes (and vice versa);
- topological entities (faces, edges) are associated to their generic entities (inheriting the corresponding information as types, parameters,etc.);
- the topology is completely represented (with all those information needed to make it unambiguous).

The distinction between generic geometry and topology is useful because different forms of knowledge can be related to them (for instance type and parameter information to generic entities, or connectedness information for topological entities), and because generic entities are not affected by geometric modeling operations (addition of new basic volumes) - in contrast to topology.

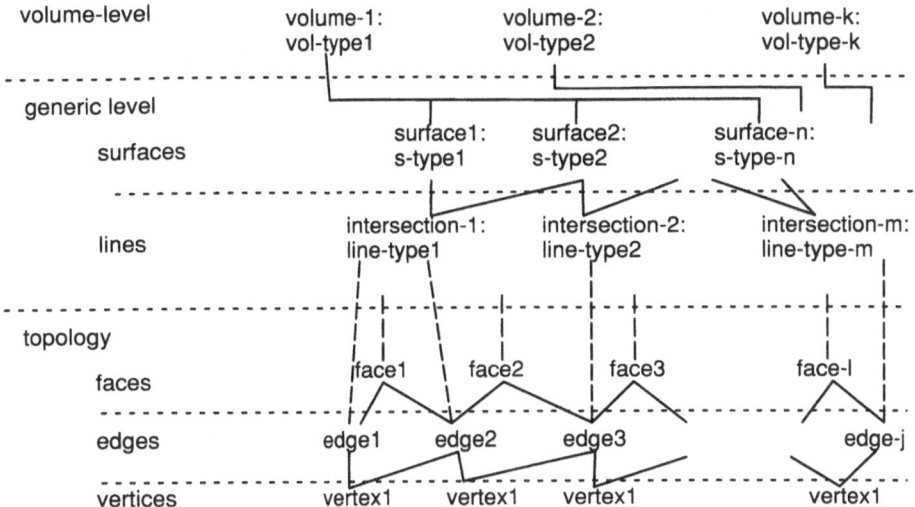

Fig. 3: Schematic view on the integrated tree-level geometry representation of G-Rep: Volumes are associated with a (complete) set of boundary surfaces, each related
to it by a relation. Each surface "knows" its intersection lines with the other surfaces, each surface and line "know" their topological elements (indicated by dashed lines) which form a complete and unique boundary representation. Each element may be directly connected to any other by definition of relations.

4.5   Geometry Generation in G-Rep

The expressive and declarative approach to geometry modeling which is provided by G-Rep can be used in geometry generation.

Take the example given in fig. 1:

First we have the definitional knowledge about 'body' and 'cylinder' objects (or features):

$$\forall X : \ body(X) \leftrightarrow \exists C \exists D :$$
$$plane(C) \wedge surface1(X, C) \wedge plane(D) \wedge surface2(X, D) \wedge \ldots$$
$$\forall X : \ cylinder(X) \leftrightarrow \exists A \exists B \exists C :$$
$$plane(A) \wedge top(X, A) \wedge plane(B) \wedge bottom(X, B) \wedge$$
$$cyl - surface(C) \wedge border(X, C)$$

Next we have, for instance, a domain specific constraint which expresses[17] that all bodies have to have a cylindrical hole (first constraint) and how this hole is

---

[17] By reasons of space limitations we only outline how it works - in reality the definitions and constraints are, of course, more complex.

related to the body (second constraint):

$$\forall X : \ body(X) \to \exists Y : cylinder(Y) \land contained(Y, X)$$
$$\forall X \forall Y \forall Z \forall V : \ body(X) \land cylinder(Y) \land contained(Y, X) \to$$
$$surface1(X, Z) \land top(Y, Z) \land surface2(X, V) \land bottom(Y, V)$$

If now the requirement is specified that there should be a body instance (where the name 'b1' may be given explicitly or may be generated by the geometry modeller):

$$G = \{body(b1)\}$$

then in order to fulfill this requirement a solid Γ will be generated:

$$\Gamma = \{body(b1)\}$$

which also implies that all the other geometric objects mentioned in the 'body' definition will be generated and related to 'b1').

The logical view on the knowledge represented in this example "tells" the geometry modeller that for each 'body' instance that should be a 'cylinder' instance as a holw in that body. Consequently the solid will be modified to contain this cylinder instance, too:

$$\Gamma' = \{body(b1), cylinder(c1)\}$$

which will be generated in a way which guarantees that the constraints outlined above are fulfilled (see also fig. 1).

In this way a geometry can be generated incrementelly without paying attention to the sequence in which the various requirements had been generated and fulfilled.

4.6   Features in G-Rep

Features emerged in recent years as an appropriate means to associate various types of information (including design intent or manufacturing plans) to geometry (for a recent review about feature technology see, for instance, [20]). Without going into the details of the discussion about what a feature is, it can be said that G-Rep with its expressive and declarative geometry representation provides essential pre-requisites for feature technolgy.

From a logical viewpoint a feature can be defined as any other class (generic) or object (instance): by logical implication or equivalence the feature can be related to a set of other geometric or non-geometric entities (including other features), and it can be constrained in every way the basic language provides. Thus the full expressiveness of the the underlying first order language can be used.

The is one advantage of this understanding of features: it is left to the user to say what s/he thinks what a feature should be in her/his application domain. Thus the problem of feature definitions is transfered to one of knowledge representation or knowledge modeling - as it is encountered in many other fields of knowledge representation, too.

5   Discussion

In the following we will discuss the relations between G-Rep and "classical" approaches to geometry modeling, relate our approach to some other recent topics in geometry modeling (persistent naming, geometric constraints), and conclude with some remarks about open problems and future research.

5.1   Comparison between CSG, B-Rep, Sweeping, and G-Rep

The basic idea behind G-Rep is to provide geometry modeling techniques which better fit the needs of knowledge representation. This has been achieved in a way which allows us to combine "pure" geometry modeling with a lot of knowledge representation and constraint processing methods. But not all what is possible in classical CSG or B-Rep modeling has an analogon in G-Rep: some restrictions had to be defined in G-Rep's geometry modeling facilities. Without being able to discuss the relations comprehensively some remakrs will be given which stress the main points:

Comparison between CSG and G-Rep The density-based geometry modeling in G-Rep mainly corresponds to (regularized) union and difference operations in CSG.

A G-Rep solid

$$\Gamma = \{V_1 . \rho_1, V_2 . \rho_2, V_3 . \rho_3, \ldots, V_n . \rho_n\}$$

(with $V_i . r_i$ indicating the involved basic volumes V_i with their associated densities ρ_i) can be transformed into an equivalent CSG tree[18]:

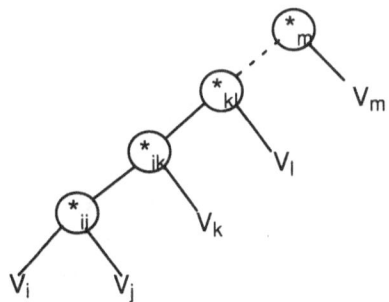

Here the $V_i \ldots V_m$ are the involved basic volumes $V_1 \ldots V_n$. Their sequence is determined by the absolute values of their densities $\rho_1, \rho_2, \ldots, \rho_n$ (in assending order). The *ij operations are the classical regularized CSG union operations, if

---

[18] CSG representations are ambiguous - i.e., for a given geometry there can be more than one geometrically equivalent CSG representations.

the densities of the involved volumes V_i and V_j have the same sign, and they are the regularized difference operations, if the signs are opposite.

The density-based geometry modeling in G-Rep allows us to modify geometry in a way which corresponds to "in-between" modifications of the equivalent CSG trees (as shown in fig. 4).

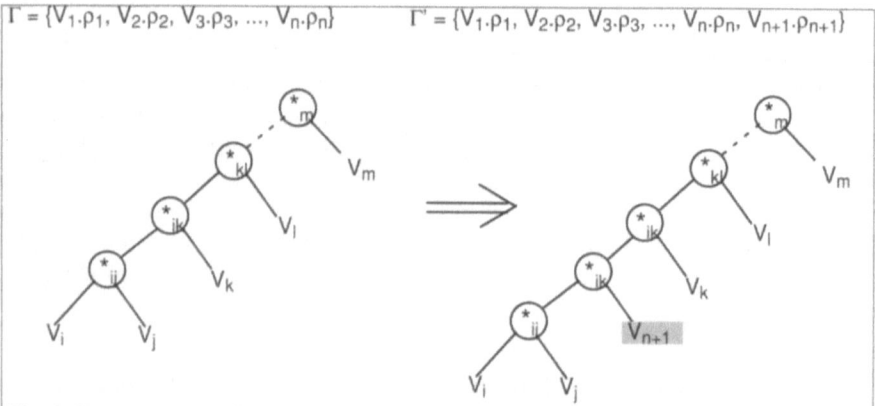

Fig. 4: Geometry modelling in G-Rep is not history based: adding a new basic volume V_{n+1} with density ρ_{n+1} to a given G-Rep solid Γ corresponds to the shown modification of the CSG tree if the absolute value $|\rho_{n+1}|$ of V_{n+1}'s density lies between the absolute values of V_j's and V_k's densities.

G-Rep's geometry modeling is equivalent to CSG trees which have a special structure: chain or list like trees (as shown in fig. 4). Geomtries which need another form of the CSG tree can only be represented in G-Rep, if the derivations from the chain like CSG tree only occur within basic volumes, but not on the level of the G-Rep solid.

The functionality of CSG intersection operations is provided in G-Rep in a restricted way: it can be used to create basic volumes, and it can be "simulated" in part by equivalent difference operatrions.

Comparison between G-Rep and Boundary Representation. One of the core ideas behind G-Rep is to store more geometry information than in classical boundary representations which then can be used for an expressive and declarative geometric modeling. Thus it will be easy to derive a boundary representation from a given G-Rep solid.

This transformation proceeds essentially in two steps (fig. 5):

1. All "constrains" topological relations will be transformed into "normal" topological relations (which in G-Rep corresponds to "defines" relations). Thereby edges related to vertices by such relations and faces related to edges will be divided into two parts.

2. The resulting topology contains the boundary representation as a subset. There are additional topological elements which do not contribute to the description of the boundary representation between material and "air", i.e., which are either completely "in air" or completely "inside material". These elements will simply be removed, resulting in a pure B-Rep.

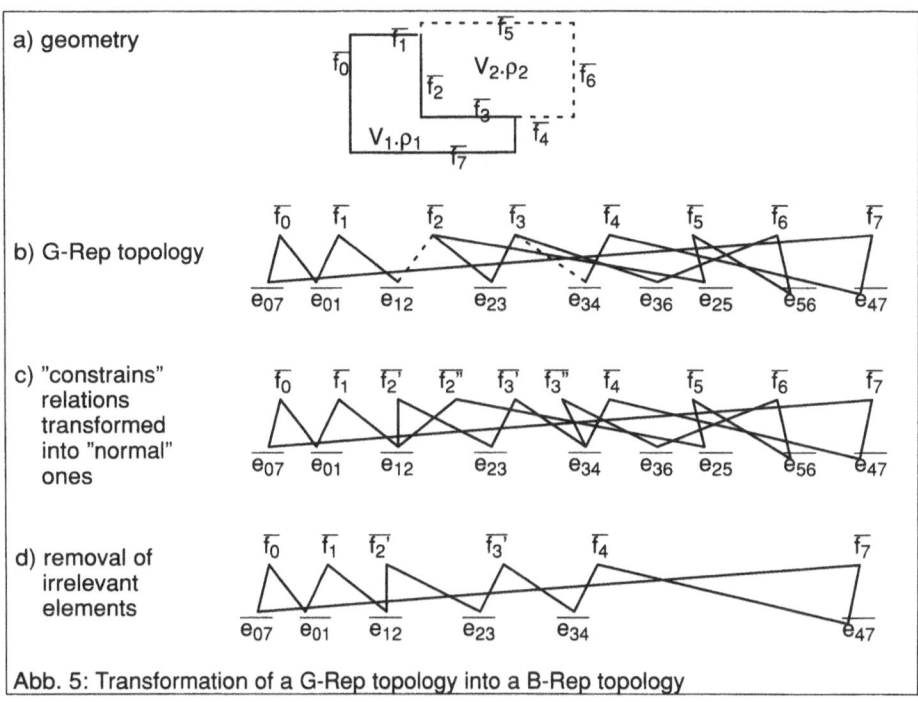

Abb. 5: Transformation of a G-Rep topology into a B-Rep topology

Sweeps in G-Rep. Also sweep modeling techniques can be integrated into G-Rep: on the level of basic volumes. In G-Rep a basic volume can be generated by a sweep representation starting with the description of the generating face. Its bounding edges which will be swept along a given curve, thus creating faces from edges and edges from vertices (in the usual way). Such a basic volume will get a density, and it can be used in G-Rep's composition in a normal way. All expressive means of G-Rep can be used to describe and constrain a basic volume generated in a sweep operation: for instance the bounding planes of the sweep operation may be left unspecified at definition time and instanciated at runtime (see also fig. 1). This can be done in the same declarative way as with basic volumes generated by the other methods.

5.2 Comparison to Related Research

There is a lot of related efforts to improve the integration of geometric modeling with other forms of information technology (including knowledge representation). In the core of these activities is the developement of feature concepts (where our understanding of the relations between features and G-Rep has been given in the previous chapter), strongly related to problems of adressing parts of a geometry ("persistent naming" [11]) and geometric constraints.

The representational means provided in G-Rep allow us to address arbitrary geometric entities in an expressive and flexible way. So the knowledge based system can address those geometric issues it is especially interested in. This also enables a persistent naming on a logical and topological basis (in constrast to a history based one as in [11]). The main problem with "persistent naming" seems to be to express the intent behind adressing a certain geometric entity: how is the entity to be adressed really characterize? The logic-based geometry representation of G-Rep (including the built-in representational) means provide us with an expressive means for this purpose.

This topic of expressiveness is also important for geometry generation: which are the real requirements a geometry to be generated should fulfill? Take the example in fig. 6: each of the three geometries can be modelled in G-Rep as in any other geometry representation. Whereas the left geometry (fig. 6a) is consistent with the requirements expressing the need for the six holes, the right geometry (fig. 6c) would be in contradiction to this requirement. This also results in a disagreement between G-Rep's knowledge level where the six holes are represented and the "real" geometry where they do not emerge. And the case in the middle? - It depends on the application if this situation is acceptable (i.e., in agreement with the requirements) or not. Appropriate expressiveness is essential here.

That's also true when *design intent* has to be captured in an adequate way: for instance, if one generates a box which is later modified by subsequent modeling operations. What was the intent behind generating this box at that position, and when is one of the modifications in contradiction with this intent? The problem with current geometry modeling approaches is that there are too many constraints left implicit. An integration of knowledge representation and geometry modeling needs more explicitness, and thus an appropriate expressiveness to represent all relevant aspects.

With this requirement we come to another important point: in knowledge representation in general, but especially in geometry, things can often be expressed in many (more or less equivalent) ways. That's true for generic as well as for case-specific knowledge. In the example given in fig. 6 , for instance, we can define a distance constraint between holes and central boss. This guarantess that only the situation in fig. 6a) is consistent, not 6b) and 6c). But equivalently, we can say that a hole (of that type) has to have a wall of at least a certain thickness around itself (which may depend on the radius of the hole). - Two problem classes result therefrom: first one needs a sophisticated modeling methodology (which allows me to avoid redundancy and inconsistencies already in the generic knowledge, and which supports the representation of design intent); and second, adequate problem solving techniques (constraint solving, model generation, etc.) are needed which allow us to solve these complicated problems.

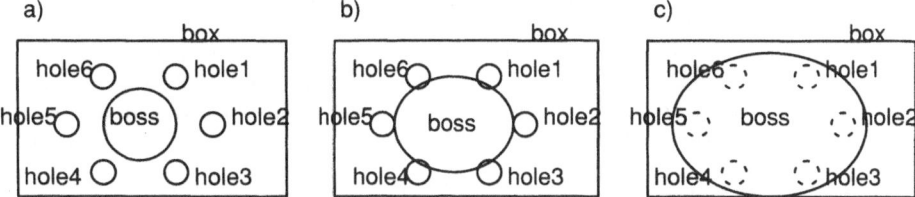

Fig. 6: Three geometries demonstrating the interactions of geometric entities:
  a) holes and central cylindrical boss do not interact;
  b) holes and boss interact which may be consistent with the requirements or not;
  c) the boss makes the six holes vanish from "real" geometry

In this sense, in G-Rep the problems of knowledge representation and constraints are "back propagated" to the level they belong to: to the knowledge level. The expressiveness really available in a modeling approach based on G-Rep strongly depends on the power of the inference engine on the knowledge level. It must provide a complete and consistent description of a G-Rep solid - which then can be processed to "real" geometry. Taking this view enables us to incorporate all the knowledge representation and constraint techniques and make them available in a knowledge based CAD environment.

Though delivering a lot of useful things, G-Rep does not mean to change recent approaches of geometry modeling in a dramatic way. Some more information has to be stored, and a second type of topological relationship has to be introduced. That's not too much [19]. The most important aspect to mention here is another one: G-Rep has demonstrated the need to modify geometry modeling in its core - in order to get a conceptual and expressive integration of knowledge representation, feature techniques, and geometry modeling. And the geometry modeling has to provide an maximum of openness in order to attach the information, constraints, etc. where it belongs to.

The modifications needed in geometry modeling seem to be straightforward. On the other hand, many of the problems encountered in recent approaches to geometry modeling will be there in G-Rep, too. So implementing G-Rep and experimenting with it will give us some more insights. On the other, theoretical side, the open questions concerning a logical view on geometry modeling should fully be worked out. What we need is a formalization of what a solid is, and how it should be generated to fulfill given requirements. This includes, for instance, what queries to a solid could be, what answers are, and how model generation has to be extended.

---

[19] Though the storage and efficiency requirements encountered in many CAD applications are such that even a factor 2 could mean a real problem (taking hardware and software limitations as they are today).

References

1. Bernardi, A., et al.: Feature Based Integration of CAD and CAPP, in: CAD-92, Hrsg. Krause, F.L. et al., Springer-Verlag, Berlin, 1992.
2. Brachman, R.J., and Levesque, H.J.: Readings in Knowledge Representation, Morgan Kaufman, 1985.
3. Cocquebert, E., et. al: State of the art and evolution of feature based modeling, Int'l. J. of CADCAM and Computer Graphics 7/2, pp. 162 - 200, 1992.
4. Hoffmann, C.M., Juan, R.: EREP - an Editable, High-Level Representation for Geometric Design and Analysis, Techn. Report, Purdue Univ., West Lafayette, 1994.
5. Gero, J. and Sudweeks, F. (eds.): Proc. of the workshop on formal design methods for computer-aided design, Tallinn, Estonia, 1995.
6. Hoffmann, C.M.: Geometric and Solid Modeling: an Introduction, Morgan Kaufman, San Mateo, California, 1898.
7. Klein, R., Buchheit, M., and Nutt, W.: Configuration as Model Construction: the Constructive Problem Solving Approach, in: G. Gero and F. Sudweeks (eds.): Proc. Int'l. Conf AI in Design, Lausanne, 1994.
8. Klein, R.: G-Rep: Geometry and Feature Representation for an Integration with Knowledge Based Systems, Proc. of the IFIP 5.2 Workshop on Geomtric Modeling in CAD, Airlie, Virginia, May 1996, Chapman Hall Publ., London, 1996 (to appear).
9. Klein, R.: Towards a logical basis of design: the GDeE approach, in: F. Brazier, T. Smithers, J. Treur (eds.): Proc. of the workshop on logic-based approaches to AI in design, San Francisco, June 1996.
10. Krause, F.L., Rieger, E., Ulbrich, A.: Feature Processing as Kernel for Integrated CAE Systems, IFIP Int'l. Conf., May 1994, Valenciennes.
11. Kripac, J.: A Mechanism for Persistently Naming Topological Entities in History Based Parametric Solid Models, in: Hoffmann, C.M., Rossignac, J. (Hrsg.): Proc. 3rd Symposium on Solid Modeling and Applictions, Salt Lake City, 1995.
12. Mäntylä, M.: An Introduction to Solid Modeling, Computer Science Press, College Park, Maryland, 1988.
13. Owen, J.C.: Algebraic Solution for Geometryfrom Dimensional Constraints, Proc. ACM Solid Modeling Conference, Austin, Texas, 1991.
14. Rappoport, A.: Geometric Modeling: a new fundamental framework and its practical implications, in: Chr. Hoffmann, J. Rossignac (eds.): Proc. of the 3rd Symposium on Solid Modeling and its Applications, Salt Lake City, 1995.
15. Requicha, A.: Representationd for rigid solids: theory, methods, and systems, ACM Computing Surveys 12:437-464, 1980.
16. Rieger, E.: Semantikorientierte Features zur kontinuierlichen Unterstützung der Produktgestaltung, Hanser Verlag, München, 1995.
17. Shah, J.J.: Assessment of feature technology, J. for Computer Aided Design 23(93), 331-343.
18. Salomons, O.W., van Houten, F.J.A.M., Kals, H.J.J.: Review of Research in Feature Based Design, Journal of Manufacturing Systems, 12/2, pp. 113 - 132, 1993.
19. Shoham, Y.: A semantical approach to non-monotonic logics, Proc. of the Int'l. Conference on Artificial Intelligence IJCAI, Milano, Italy, 1978.
20. Shah, J.J.,Mäntylä, M., Nau, D.S.: Advances in Feature Based Manufacturing, Elsevier Publ., Amsterdam, 1994.
21. Baader, F. et al.: Concept Logics, in: Loyd, J.W. (ed.): Computational Logic, Springer-Verlag, Berlin, 1990.

Projection Operation for Multidimensional Geometric Modeling with Real Functions

A. Pasko and V. Savchenko

University of Aizu, Aizu-Wakamatsu, Fukushima, 965-80 Japan

Abstract. We discuss different techniques to projecting geometric objects defined by real functions of several variables. The result of this operation is an object of the lower dimension with its own defining function. We discuss several approaches: analytical methods, approximate projections and global maximum searches. The accuracy is compared for two algorithms: the union of maximal cross-sections and the global search with the quadratic interpolation. The following applications are illustrated: 3D solid projection onto a 2D plane; 4D to 3D projection for sweeping by a moving solid; 3D reconstruction from the medial axis.

Introduction

This paper deals with the geometric projection operation in 3D modeling of surfaces/solids and in multidimensional geometric modeling. To project a surface or a solid in geometric modeling is a much more difficult problem than to project points and line segments in computer graphics. This operation has to result in a valid lower dimensional geometric model, which again can be used as an operand of further geometric transformations and analysis.

We consider the projection operation in the context of geometric modeling with real functions of several variables. The *function representation* (or *F-rep*) defines a geometric object by a single real continuous function of several variables as $F(X) \geq 0$. We do not require the defining function $F(X)$ to be polynomial or of any other specific type. The function can be defined with an analytical expression, or with a function evaluation algorithm, or with scattered data and an appropriate interpolation procedure. This representation combines many different models like classic "implicits", skeleton based "implicits", set-theoretic solids, sweeps, volumetric objects, parametric and procedural models [6],[9]. Set-theoretic operations are closed on this representation with the use of R-functions - C^k continuous definitions introduced by Rvachev [8] (see a survey in [10]). Many geometric operations are also closed on F-rep. They are blending, offsetting, Cartesian product, bijective mapping, metamorphosis and others (see [6] for details). These operations generate new real continuous defining functions and provide the closure property of the representation.

One of the main F-rep features is dimension independence. Geometric objects of various dimensions can be treated uniformly by most of operations. There are also operations that change the object dimension: the Cartesian product increases and the projection decreases it. Several applications of the projection operation are well known:

- Traditional 3D to 2D projection;
- 4D to 3D projection for sweeping by a moving solid;
- Reconstruction from the medial axis transformation;
- Construction of an envelope of a parametric family;
- Implicitization of parametric curves and surfaces;
- Calculation of the surface area by integration.

Some more specific applications are known in path planning, optimization, catastrophe theory, systems of differential equations and statistical analysis.

Problem Statement

To support real multidimensionality a geometric modeler has to provide the robust and effective projection operation. Usually the projection operation is defined by a linear transformation matrix A with $detA = 0$. This definition cannot be interpreted in terms of bijective mappings because it is not one-to-one mapping and an inverse mapping does not exist.

Let the initial object G_1 in E^n is defined by the real function $f_1(X_n) \geq 0$, where $X_n = (x_1, x_2, ..., x_i, ..., x_n)$. We call the geometric object G_2 a projection of G_1 to E^{n-1} with the defining function $f_2(X_{n-1})$, which is more or equal zero in the given point $X_{n-1} = (x_1, x_2, ..., x_{i-1}, x_{i+1}, ..., x_n)$ only if the point X_n exists where $f_1(X_n) \geq 0$. The object G_2 can be thought as a union of cross-sections of G_1 by the infinite set of hyperplanes $x_i = C$. The resulting function for the projection can also be defined as

$$f_2(X_{n-1}) = max_{x_i}(f_1(X_n)) \tag{1}$$

Let us formulate several requirements to the function $f_2(X_{n-1})$ and its evaluation algorithms:

- $f_2(X_{n-1})$ is a continuous real function;
- an evaluation algorithm should treat the initial function $f_1(X_n)$ as a "black box", i.e., a procedurally defined function with no preliminary information given about its nature or behavior;
- distance property of the function $f_2(X_{n-1})$ should be provided that is important for the further operations on the projection.

Other Works

There are several different approaches to the projection operation in geometric modeling. The conversion of parametric curves and surfaces into implicit form can be considered as a projection operation. There are several analytical conversion techniques such as resultants and Gröbner basis for the polynomial functions [2].

Kutsenko in [3] proposed the following defining function for the projection from 3D to 2D along z-axis:

$$\varphi^*(x,y) = \int_a^b F(x,y,z) + |F(x,y,z)| dz, \qquad (2)$$

where $F(x,y,z)$ defines the initial 3D object and $\varphi^*(x,y)$ correctly defines the internal and boundary points of the projection. Although this approach is quite attractive for the analytical and numerical implementation, the behavior of the resulting function restricts its application in the F-rep framework. The function $\varphi^*(x,y)$ is equal zero everywhere outside the projection. Many operations such as blending, offsetting or metamorphosis require distance-like behavior of the defining function. So, the function $\varphi^*(x,y)$ is not enough satisfactory.

Analytical methods for eliminating a variable from two equations are also used in modeling sweeping by a moving solid. The analytical solutions were found in several specific cases of sweeping by a sphere and by a cylinder [13], [4]. Sweeping by a practically arbitrary solid is approximately modeled in [5]. The moving solid is converted to the approximate ray-representation, then the swept solid is represented as a union of series of generator instances (cross-sections). Note that here the projection operation is closed on the ray-representation.

Solving the problem of collision free path planning Wise and Bowyer [14] project an interference hypersolid into the configuration space. The result of this procedure is a recursive division of the configuration space (multidimensional generalization of bin-, quad- and octrees) into hyperboxes with different properties. So, the application of the projection operation requires here the conversion from the initial CSG model to the representation with a spatial occupancy enumeration.

Reconstruction of geometric objects from their medial axis is an actual problem in geometric modeling. The medial axis transformation represents an object as its lower dimensional skeleton with the radius of the maximal inscribable disk/sphere given at every point of the skeleton. The reconstruction of the original object from the medial axis can be thought as the projection operation from $n + k$-dimensional space into n-dimensional space, where k is a dimension of the skeleton and n is a dimension of the maximal disks. Papers [12] and [1] present the reconstruction algorithms resulting in the object boundary.

Proposed Solutions

As it is mentioned in the previous section, the implementation of the projection operation often changes the representation of the object. Our goal here is to propose several approaches that provide closure of this operation on the function representation.

Analytical Approach: Case Study

Let us describe here the case study of the analytical reconstruction of the F-rep object from the medial axis transformation. A one-dimensional skeleton consist-

Fig. 1. Analytical 3D reconstruction from a skeleton

ing of line segments is given in 3D space. For every point of the skeleton the radius of the maximal inscribable sphere is defined. The problem is to obtain a real function defining the original object. Let the skeleton consist of line segments with the variable sphere radius on it. Then, for a 4D object defined by j-th segment of the skeleton with endpoints (x_{1j}, y_{1j}, z_{1j}) and (x_{2j}, y_{2j}, z_{2j}) we get the description:

$$f_{1j}(x, y, z, t) = R_j^2(t) - (x - x_{0j}(t))^2 - (y - y_{0j}(t))^2 - (z - z_{0j}(t))^2, \quad (3)$$

where $x_{0j}(t) = x_{1j} + l_{xj}t$, $y_{0j}(t) = y_{1j} + l_{yj}t$, $z_{0j}(t) = z_{1j} + l_{zj}t$, and $l_{xj} = x_{2j} - x_{1j}$, $l_{yj} = y_{2j} - y_{1j}$, $l_{zj} = z_{2j} - z_{1j}$. To project this object into 3D space following Equation (1), we have to find the maximal value of the function $f_{1j}(x, y, z, t)$ for the given 3D point (x, y, z). Assuming the linear function $R_j(t) = R_{1j}(1 - t) + R_{2j}t$, it can be shown that the following parameter value corresponds to the function f_{1j} maximum:

$$t_{0j} = \frac{l_{xj}(x - x_{1j}) + l_{yj}(y - y_{1j}) + l_{zj}(z - z_{1j}) - R_{1j}(R_{2j} - R_{1j})}{l_{xj}^2 + l_{yj}^2 + l_{zj}^2 + (R_{2j} - R_{1j})^2} \quad (4)$$

Then, for the whole reconstructed 3D object the defining function is

$$f_2(x, y, z) = max_j(f_{1j}(x, y, z, t_{0j})). \quad (5)$$

Figure 1 shows a 3D object reconstructed from its skeleton using this approach. The skeleton was modeled with "random walk" calculations with transition probabilities depending on the direction of motion. This type of skeleton can be observed in plants, blood vessels and porous media.

Numerical Algorithms

We discuss and compare here the numerical evaluation algorithms for the defining function of the projection. They are approximation and global maximum

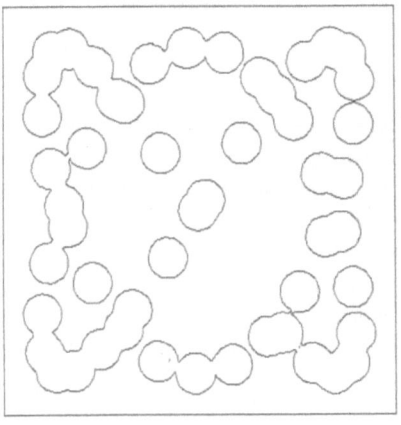

(a) Union of cross-sections

(b) Global maximum search with the quadratic interpolation

(c) Union of maximal cross-sections

Fig. 2. 3D to 2D projection of a tubular surface by three algorithms using 40 cross-sections

search algorithms. These algorithms deal with the given procedurally defined initial function $f_1(X_n)$ and evaluate the function $f_2(X_{n-1})$ of the projection in the given point.

- **Union of cross-sections.** By selecting a finite set of hyperplanes $x_i = C_j$ one can approximate the projection result applying R-union functions:

$$f_2(X_{n-1}) = (f_{11} \vee f_{12}) \dots \vee f_{1j}) \dots \vee f_{1N}), \tag{6}$$

where $f_{1j} = f_1(x_1, x_2, \ldots, x_{i-1}, C_j, x_{i-1}, \ldots, x_n)$, N is the number of cross-sections and

$$h \vee g = h + g + \sqrt{h^2 + g^2} \tag{7}$$

is one of the R-functions proposed by Rvachev [8] for the exact description of the set-theoretic union of two objects described by real functions h and g. The accuracy of the result of this projection operation highly depends on the step between adjacent cross-sections. Figure 2 shows a 3D to 2D projection of a tubular surface, which is defined as a 2D disk changing its position in xy-plane and moving in z-direction. Figure 2 (a-c) illustrates the comparison of three different algorithms (see the discussion below).

- **Blending union.** One can try to change the standard union to the blending union operation [6]:

$$h \vee_b g = h + g + \sqrt{h^2 + g^2} + \frac{a_0}{1 + (\frac{h}{a_1})^2 + (\frac{g}{a_2})^2}. \tag{8}$$

Blending union of cross-sections can produce much better result than standard union but the choice of parameters a_0, a_1, a_2 depends very much on geometry of the projected object (for example, on variations in the skeleton curvature). There is no any automatic method yet to choose these parameters and to establish their dependance on the features of the initial object.

- **Union of maximal cross-sections.** This approximate projection applies set-theoretic union to the interpolation terms between adjacent cross-sections taken with a regular step:

$$f_2(X_{n-1}) = (f_{11} \vee f_{11}^*) \vee f_{12}^* \ldots \vee f_{1j}^* \ldots \vee f_{1,N-2}^* \vee f_{1,N}, \tag{9}$$

where N is number of cross-sections,

$$f_{1j}^* = f_1(x_1, x_2, \ldots, x_{i-1}, C_j^*, x_{i-1}, \ldots, x_n). \tag{10}$$

Here the constant $C_j^* = C_j + C_0 dx_i$ defines the maximal function $f_1(X_n)$ between three cross-sections, where C_j is the value of x_i in the j-th grid node with the grid step dx_i. Parameter C_0 is calculated using the quadratic interpolation:

$$C_0 = \frac{1}{2} - \frac{f_{1,j+1} - f_{1,j}}{f_{1,j+2} - 2f_{1,j+1} + f_{1j}} \tag{11}$$

Note that if $C_0 < 0$, then $f_{1j}^* = f_{1j}$, and if $C_0 > 2$, then $f_{1j}^* = f_{1,j+2}$. Figure 2c shows a quite good result obtained with this algorithm.

- **Global maximum search.** The projection problem can be formulated as a global extremum search (see Equation (1)). We have tested an algorithm that applies the Newton quadratic interpolation with Equations (10,11) only once at the point of maximal $f_1(X_n)$ taken from the regular grid of samples [11]. This algorithm obviously requires fewer number of function $f_1(X_n)$ evaluations than the previous one but has shown more poor quality of the result for our test operations (see Figure 2b and the accuracy comparison below).

– **Combination of random and quadratic global searches**. It is well
  known that a combination of random and gradient/sequential searches is an
  effective approach to the global maximum search. We apply quasi-random
  Sobol sequences (see [7] for the description and the C-code) to generate an
  initial estimation C_j for the subsequent quadratic search with Equations
  (10,11). The search stops when the difference between two found maxima
  does not exceed the given threshold. This algorithm works more fast (about
  20%) than the previous one for the test projection (Figure 2) but its behavior
  highly depends on the threshold selection.

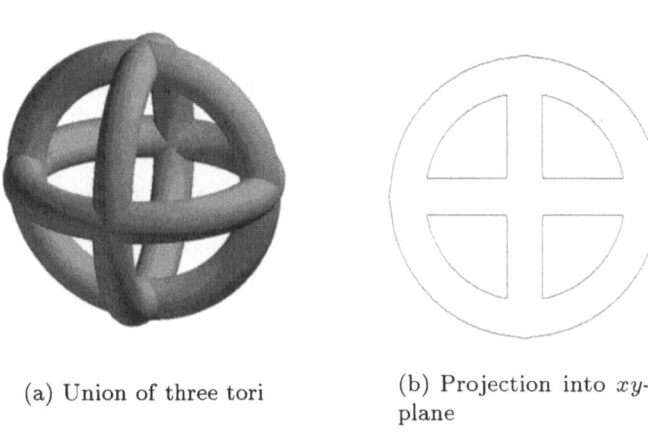

(a) Union of three tori

(b) Projection into xy-plane

Fig. 3. Projection of three tori into a 2D plane

We have selected two algorithms to compare their accuracy: the union of maximal cross-sections and the global search with the quadratic interpolation. The
test problem is shown in Figure 3: a R-union of three tori is projected into a
plane orthogonal to an axis of one of tori. We have selected 16 points in the
projection where the resulting function has to be zero. Figure 4 shows the plots
of the integral error in these points versus the number of cross-sections for both
methods. Evidently, the union of maximal cross-sections is much more stable and
converges very fast. Figure 5 illustrates a 4D to 3D projection with this method:
a solid swept by a rotating cube which moves along the triangle trajectory.

Conclusion

Projection operation is important for constructing a multidimensional geometric modeler. It is also quite useful in several applications. This operation can be

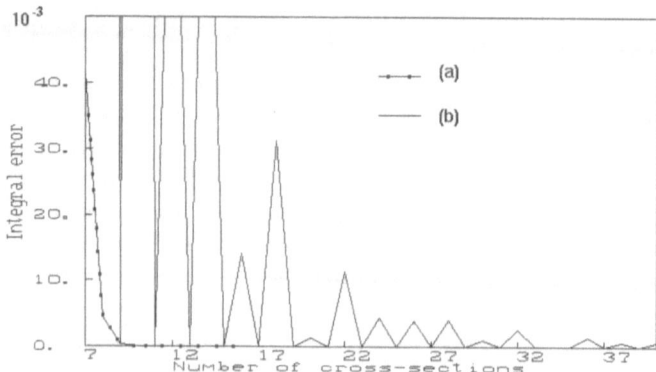

Fig. 4. Integral errors for two projection algorithms: (a) union of maximal cross-sections; (b) global search with the quadratic interpolation

Fig. 5. 4D to 3D projection: sweep by a moving cube

generally expressed in terms of real functions. Several numerical algorithms exist for the evaluation of real functions defining projections. These algorithms are time consuming because they multiply evaluate the object's defining function. This can significantly slow down the following operations on projections. To effectively utilize the general approach proposed in this paper one can incorporate it into a modeler with a voxel- or other approximate model generation for the resulting projection and with an appropriate interpolation procedure to ensure the closure property of F-rep. The central problem for our future research on the projection operation is selecting a more effective and stable global search algorithm and exploring different searches on parallel computers.

References

1. Gelston S.M., Dutta D. Boundary surface recovery from skeleton curves and surfaces, Computer Aided Geometric Design, vol.12, No.1, 1995, pp.27-51.
2. Hoffmann C.M. Geometric and Solid Modeling: An Introduction, Morgan Kaufman Publishers, 1989.
3. Kutsenko L.N. Computer graphics in the problems of projective nature, Mathematics and Cybernetics, Znanie Series, Moscow, No.8, 1990 (in Russian).
4. Martin R.R. The geometry of the helical canal surface, Computer-Aided Surface Geometry and Design, The Mathematics of Surfaces IV, A.Bowyer (Ed.), Clarendon Press, Oxford, 1994, pp.17-32.
5. Menon J., Marisa R.J., Zagajac J. More powerful solid modeling through ray representations, IEEE Computer Graphics and Applications, vol.14, No.3, 1994, pp.22-35.
6. Pasko A., Adzhiev V., Sourin A., Savchenko V. Function representation in geometric modeling: concepts, implementation and applications, The Visual Computer, vol.11, No.8, 1995, pp.429-446.
7. Press W.H., Teukolsky S.A., Vetterling W.T., Flannery B.P. Numerical recipes in C, Cambridge University Press, 1992, pp.309-315.
8. Rvachev V.L., Methods of Logic Algebra in Mathematical Physics, Naukova Dumka Publishers, Kiev, Ukraine, 1974 (in Russian).
9. Shape Modeling and Computer Graphics with Real Functions, WWW site at URL: http://www.u-aizu.ac.jp/public/www/labs/sw-sm/FrepWWW/F-rep.html
10. Shapiro V. Real functions for representation of rigid solids, Computer Aided Geometric Design, vol. 11, No.2, 1994, pp.153-175.
11. Sourin A., Pasko A. Function representation for sweeping by a moving solid, IEEE Transactions on Visualization and Computer Graphics, vol.2, No.1, March 1996, Special issue on solid modeling, pp.11-18.
12. Vermeer P.J. Two-dimensional MAT to boundary conversion, Proceedings of 2-nd Symposium on Solid Modeling and Applications, ACM, 1993, pp.493-494.
13. Wang W.P., Wang K.K. Geometric modeling for swept volume of moving solids, IEEE Computer Graphics and Applications, vol.6, No.12, 1986, pp.8-17.
14. Wise K.D., Bowyer A. Using CSG models in many dimensions to map where things can and cannot go, CSG 96 Set-theoretic Solid Modeling: Techniques and Applications, Information Geometers, Winchester UK, 1996, pp.359-376.

Breps as Displayable-Selectable Models in Interactive Design of Families of Geometric Objects

Ari Rappoport

Institute of Computer Science, The Hebrew University, Jerusalem 91904, Israel.
http://www.cs.huji.ac.il/~arir. arir@cs.huji.ac.il.

Abstract. In classical geometric modeling, the primary objects of interest are geometric pointsets, the primary representation scheme for which is the boundary representation (Brep). Modern geometric modeling focuses on parametric families of pointsets, defined using geometric operation graphs (GOGs), features and constraints. During interactive design of families of objects, users interact with an *example object* from the family. The example object is a pointset, hence is usually modeled using the Brep.

In this paper we study the issue of which modeling scheme is most appropriate for the example object. We identify two major operations which such a modeling scheme must support: *display* and *selection*. Selection can be further decomposed into *picking, invariant naming*, and *persistent naming*. We introduce the term *Displayable-Selectable Models (DS-models)* as a generic term for models providing display and selection functionality. We discuss the suitability of Breps to serve as DS-models and whether other, perhaps simpler, representations could also serve as DS-models.

1 Introduction

Geometric modeling deals with the representation and manipulation of geometric entities in a computer. The major classification of studies in geometric modeling is according to the nature of the geometric entities being studied. In classical geometric and solid modeling, the objects of interest were pointsets. Researchers have focused upon seeking unambiguous representation schemes for various types of pointsets [Requicha80].

In current geometric and solid modeling, the entities of interest are *families* of geometric objects [Hoffmann96, Shapiro95, Rappoport96a]. Such families are usually parametric (that is, members of the family are indexed using a set of external parameters), and may be specified using construction steps, constraints, and high-level application-dependent features [Hoffmann93a, Shah96]. Research on issues related to modeling families of objects gradually receives greater attention than research on the classical pointset topics.

Interactive design of geometric models is greatly aided by visualizing the current state of the designed model. Visualization of a family of geometric objects is extremely difficult. For that reason, interactive design of object families is usually done by interacting with an *example object* which belongs to the modeled family. Users express their intentions on top of the example object and use it to reconstruct a mental image of the designed family.

The example object is an important participant in interactive design systems, hence the question of which modeling scheme to use for modeling it is important. Since the example object is a pointset, in principle all modeling and representation schemes designed for modeling pointsets can be used to model it.

The boundary representation (Brep) has been a major representation scheme for pointset objects from the early history of solid modeling [Mäntylä88]. A large number of research papers have been written about Breps. Most of these papers have dealt with the design and analysis of specific Brep data structures [Woo85, Mäntylä88, Alla91, Guibas85] and with algorithms operating on Breps, such as Boolean operations [Requicha85, Hoffmann89]. The essence of a Brep is best described as explicitly representing open, dimensionally uniform, connectivity components of intersection entities generated by a set of geometric pointset carriers. This essence was formalized in the notion of a Selective Geometric Complex (SGC) [Rossignac88]. It is generally felt that Breps are well-understood.

Current interactive design systems almost exclusively use the Brep to represent the example object. This choice has important implications on the architecture and performance of the system, because an accurate Brep is not easy to compute robustly and efficiently.

In this paper we discuss the issue of modeling the example object in interactive design systems. Using a geometric modeling framework we have previously developed [Rappoport95], we analyze the requirements from a modeling scheme for the example object, concluding that the two major queries which such a modeling scheme must support are (1) displaying the object, and (2) selecting object boundary entities in a persistent manner. The latter query is further decomposed into three operations: picking, invariant naming, and persistent naming. Picking identifies boundary entities of the example object. Invariant naming translates picking results to an entity name which is identical to the name of that entity in all members of the family and which depends only on the boundary (or SGC) of members in the family. A persistent name is similar to an invariant name, but it can also depend upon additional information, e.g. the design history of the family.

We introduce the term *Displayable-Selectable Model (DS-model)* for a model supporting the above queries. We then examine the suitability of the Brep to serve as a DS-model, asking two complimentary questions: (1) does the Brep support the two queries in an efficient manner, and (2) are there Brep characteristics which are essential for any modeling scheme supporting the two queries (or equivalently, are there modeling scheme which are better suited as DS-models). We show that a complete Brep is an over-kill for display and picking, and that it may be essential for invariant naming. Invariant naming is an important open problem in geometric modeling which is not fully understood at present.

In summary, the main contributions of the paper are:

- Raising the issue of a modeling scheme for the example object in interactive design of families of geometric objects;
- Introducing the term *Displayable-Selectable Model (DS-model)*, encapsulating the requirements from such a modeling scheme;
- Examining the suitability of the Brep to serve as a DS-model.

The structure of the paper is as follows. In Section 2 we briefly review the geometric modeling framework described in [Rappoport95]. In Section 3 we sketch a particular class of representation schemes used to model families of geometric objects, which we refer to as Geometric Operation Graphs (GOGs). In Section 4 we discuss interactive design of GOGs and introduce the term displayable-selectable models.

In Section 5 we examine the essence of the Brep by briefly describing a new generalization of the selective geometric complex. In Sections 6, 7 and 8 the display, picking, and invariant naming operations are studied in detail. For each operation we discuss the amount of support given by the Brep to the operation and the characteristics of Breps which are essential to support the operation.

Although this paper describes several new concepts, its goal is not to provide a detailed report on research results but rather to sketch the larger picture and guide future research. A deeper understanding of the issues brought forth in the paper would perhaps enable us to finally design and utilize hybrid representations which combine the merits of Breps and other representation schemes while masking their drawbacks.

2 Modeling and Representation Schemes

In this section we briefly review the geometric modeling framework presented in [Rappoport95], upon which we base the analysis in this paper.

Requicha's seminal paper [Requicha80] introduced terminology and definitions for the concepts of a representation and a representation scheme, and described several representation schemes for three-dimensional solids. A representation was defined to be a structure of symbols from an alphabet, and a representation scheme to be a relation between an abstract modeling space containing the mathematical entities we want to model and the set of representations. This 'symbol structure' definition is based on the *data* contained in a representation.

Most modern approaches to system analysis and design in general and software engineering in particular emphasize not the data contained in a representation but the capabilities that the data enables and the interface to them. Abstract data types (ADTs), encapsulating data through usage of access functions, have long been advocated as an elegant and practical design and implementation paradigm. Object-oriented analysis and design take this view further by studying the inter-relationships between different object classes [Rumbaugh91].

In [Rappoport95], we presented a geometric modeling framework which enhances the symbol structure definition by combining operations and data. An abstract *modeling space* contains the entities we want to model. An entity specifies a set of operations which it supports. There are two kinds of operations: *queries* and *synthesis operations.* The queries are the operations for which the entity is modeled in the first place; they provide the functionality of the model as important to the external world. Synthesis operations are the operations using which entities are created, modified, edited and combined. Synthesis operations are not part of the external functional interface to the entity and are reflected to the outside only as user interface operations or when they are mirrored in the supported queries.

As in Requicha's definitions, the data of each entity has a representation. However, now the main concept is that of a *model.* A model is a representation which supports the required queries and synthesis operations. A *modeling scheme* is a concrete implementation of the models, including their queries and synthesis operations.

In geometric modeling, as in any other system modeling discipline, models are built and stored for the sake of *doing* something with them. It is important to analyze clearly what our models should do (queries) and how we want to specify them (synthesis operations) before choosing a specific modeling scheme.

3   Geometric Operation Graph Representations

By definition, a Brep can only represent knowledge about the pointset of a geometric object. Using a selective geometric complex (SGC), internal structures can be represented as well as the object's boundary. However, clearly there are plenty of applications in which we want to model more than the pointset of the object. We would even say that this is the case in the majority of applications.

In design applications we want to document the history of the design and explanations for design decisions. These are important when additional groups of people are being made involved in some aspect of the model, for example when a manufacturing team finds it necessary or economical to modify the design and it needs to determine whether and which modifications are allowed. Such documentations are important also in concurrent engineering. It is clear that a Brep cannot be used to store such information.

Many applications which do not need history or versioning still need to be capable of viewing the model as more than a pointset. Storage of application-specific features is essential in many cases. Some features can be represented by using attributes associated with entities of a Brep. However, it is not clear whether this is always possible, since there are features which comprise components not present in the Brep at all. For example, a slot can be created by subtracting a rotating block whose axis of rotation is not part of the Brep. It is possible that an SGC can be used to represent such features, but this has not been shown yet.

Modeling spaces containing features, history, versioning, and general modifying operations to geometric objects can be represented by a class of models which we refer to as the *Geometric Operation Graphs (GOG)* [Rappoport96d].

Other tightly related terms are *generative* [Hoffmann96] and *parametric* [Shah96] representations, as discussed below.

A GOG (see Figure 1) is a directed graph whose nodes contain arbitrary operations (or functions), which are usually geometric in nature. An arc from node A to node B denotes the fact that the corresponding output of the operation in A serves as an input to the operation in B. Usually, but not always, one of the outputs of an operation node is a geometric object. The GOG is almost always used in order to represent a *family* of geometric objects rather than a single object, and this family is usually parametric (that is, members of the family are indexed through a parameter vector; see below).

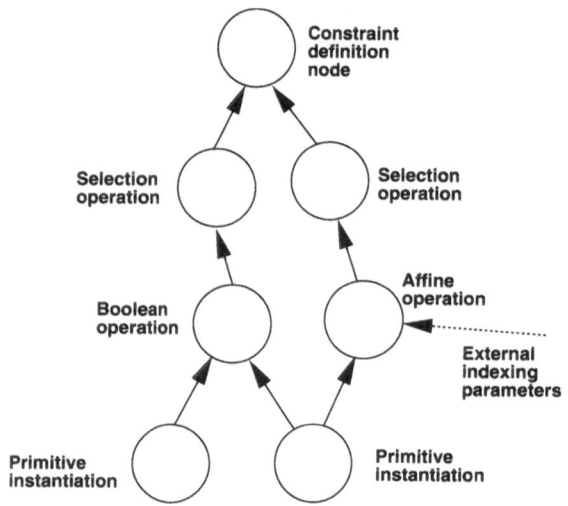

Fig. 1. A General Operation Graph (GOG).

The GOG is a generalization of several known representation schemes in geometric and solid modeling. Some procedural representations can be written as GOGs. However, a GOG can in principle support operations which establish a relation between two objects already present in the system. The semantics of such an operation is that the relation, or constraint, should be maintained from now on. Hence GOGs are more powerful than ordinary procedural representations. On the other hand, although a GOG can be enhanced to support control constructs such as loops, this would take it a bit afar from its intended spirit.

A hierarchical assembly graph is a simple GOG in which operation nodes contain affine operations. [Rappoport93] presented a scheme to directly store information belonging to a single instance of a part in an assembly in a GOG node.

The most obvious example for a GOG in geometric modeling is of course CSG, in which the operations in the nodes of the graph are primitive instantiations, affine operations, and Boolean operations. In CSG every node has a single output, which is a geometric object. The GOG point of view is that it is the operation graph which is the essence and not the specific operations used. CSG is mostly associated with the Boolean operations (indeed, the affine operations can be propagated into the leaves of the graph and factored out), while the GOG emphasizes the algebraic structure of the resulting model.

Several specific GOG systems have been studied by Rossignac: offset and blending nodes in CSG [Rossignac86a], constraints in CSG [Rossignac86b], and operations as general procedures [Rossignac89]. A GOG with constraints between coordinate systems and an underlying procedural programming language was described in [Emmerik90].

The GOG is an abstract generalization of what some people call 'parametric' representations. However, in practice, the term 'parametric' is defined today by the architecture of a specific commercial product and combines two notions which we feel are separate: (1) specifying the name of a member in a family of represented geometric objects by a parameter vector [Rappoport96a], and (2) the sequential solution of a system of constraints. We feel that the term 'parametric' is perfect for the first notion but is a poor choice for the second notion. Because of the types of modeling spaces that the GOG will be used for, in most cases it will support the first notion. However, the GOG can have operation nodes which declaratively specify constraints without necessarily specifying a solution sequence. For example, such an operation node may be implemented by a numeric constraint solver (a scheme commonly referred to as 'variational', although the word 'variational' is not appropriate since it implies a family of objects which are all variations on the same theme rather than a specific method to specify and compute such a family). Declarative constraints are actually present in some 'parametric' systems, but only at the bottom level of the graph in the form of defining a 2-D sketch to be extruded later. Hoffmann's Erep [Hoffmann93a] is such a GOG. Note that the term 'generative' is sometimes used when referring to such architectures [Hoffmann96]; if desired, the acronym GOG can be interpreted as 'Generative Operational Geometry'.

We view the GOG as a representation for design intent and for the different types of knowledge accumulated and specified during the various stages of the life of a model. At present, the GOG concept is not completely understood, especially the role of declarative constraints nodes. However, there is enough evidence to support the view that the GOG is a major class of modeling schemes in modern geometric modeling.

4   Displayable-Selectable Models

Breps are probably not suitable to serve as the main modeling scheme for entities which are more than simple geometric pointsets. The class of geometric operation graphs is an attractive modeling scheme for such entities. In this section we give a general description of interactive applications involving operation graphs and characterize the functionality required from the 'example object' used by such systems.

One of the primary applications of geometric operation graphs is to represent parametric families of objects. Thus, a major query for GOGs is the indexing query: given a parameter vector, compute the corresponding object. This object should itself be represented in some representation scheme. There are several usages for such as object, two of which are the most common: visualization and analysis. Since the Brep is considered to be an efficient representation for these operations, it is natural to use Breps as the representation in which the output of the indexing query will be represented. In this context a Brep is an 'evaluated' GOG, in the same way that the process of conversion from CSG (a specific GOG) to a Brep is called 'boundary evaluation'.

In this paper we are more interested in the specification and creation of GOGs than in the queries they support. We assume the following scenario for interactive applications used to define GOGs [Hoffmann93b] (see Figure 2). At any point in time, the user is shown a visualization of the designed GOG through an *example object* which belongs to the family represented by the GOG. The user is capable of adding or deleting a new operation node, and sees the result of applying the operation on the example object. By visualizing the result of the operation, the user can determine whether the operation achieves the desired effect. The user may also want to modify the example object by specifying different parameters. These can be specified numerically, through direct manipulation on the graphical view (shown as a dotted line in Figure 2), or using some other method.

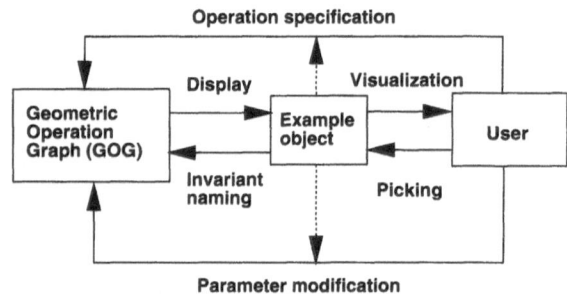

Fig. 2. The relationship between the GOG and the Display, Picking, and Invariant Naming operations.

The first operation which the system must thus support in this kind of interactive scenario is Display: given a GOG, a parameter vector, and display parameters, display the resulting object. We deliberately do not say that the system must support boundary evaluation, of course, since our goal in this paper is to specify what needs to be done (here, visualization) rather than how to do it (boundary evaluation). We discuss Display in Section 6.

The other operations which the system must support in the scenario described above are related to the way in which the user specifies new operation nodes. An operation node needs to know the type of the operation, which can be easily specified by the user, and the arguments that the operation should receive. There are two main types of arguments: numeric and symbolic data serving as additional indexing parameters, and geometric data derived from the current GOG. The former type is easy to specify. The latter arguments are specified by selecting the relevant data from the current GOG. Depending of the actual data, this selection can be done on the example object, directly on the operation graph (perhaps using some form of graph visualization), or on other views of the model [Emmerik93].

Operation nodes differ in the number and types of arguments they receive as input. Some operations operate on a whole object as a single unit. Affine and Boolean operations are such operations, which is the deep reason why CSG is so simple to define and implement. However, some operations require as input sub-parts of an object. A sub-part can be specified as the part lying in some spatial location, or more commonly, as a Brep entity. That is, vertices, edges, faces and cells of the example object can be specified as inputs to the new operation node. Systems which support the specification of Brep entities as input to operation nodes are much more powerful and allow a much greater amount of knowledge on the model to be present. It is in this type of selection that we are interested. It is convenient to think of such a selection as a GOG operation node, as shown in Figure 1.

For technical reasons we divide the selection process into three stages, which we regard as operations sufficiently different to deserve different names: picking, invariant naming, and persistent naming. We collectively refer to picking and invariant naming as *selection*, since the word 'selection' best describes how the user thinks about the functionality of these operations.

In the first stage, the user identifies the desired entity of the example object. The output of this operation is the name of this specific entity of the example object; normally, this output would be represented as a pointer to the entity. Respecting common computer graphics terminology, we call this the Picking operation. We discuss Picking in Section 7.

The output of Picking is useful only in conjunction with the example object. However, the user is allowed to arbitrarily modify the indexing parameters to produce a different example object. The arguments given to any operation node in the GOG should thus not be tied to a specific example object. Hence, we must translate the output of Picking to a representation which depends only on the properties which are present in all objects in the GOG family. Specifically, it should not depend on the unique geometry of the example object.

The problem of storing arguments to operation nodes such that they will be valid for every choice of parameter vector is usually called *persistent naming* [Kripac95, Hoffmann93b, Lequette96]. The easier problem of giving Brep entities names which are invariant under modifications to the geometry of the Brep carriers is called *invariant naming* [Rappoport96b]. Persistent naming is strongly tied to the semantics of the GOG operation nodes and to the specific application for which the GOG is designed. Due to the inherent technical differences between the two problems and to the difficulty of defining and solving them, they can be discussed separately. In the Erep project [Hoffmann93a], the general persistent naming mechanism is reported in [Chen95], while the issues related to invariant naming (in the Erep context) are reported in [Capoyleas96].

Both invariant naming and persistent naming are difficult problems, but since the scope of the former is more limited it is easier to study it mathematically. In addition, once invariant naming is available, persistent naming can be implemented as a post-process taking into account the intended semantics of GOG operations. For these reasons we do not deal with persistent naming in this paper, although in principle we should, since Breps may be strongly tied to any persistent naming mechanism. The GOG operation **Invariant Naming** is discussed in Section 8.

The functionality required from the example object is best encapsulated by the term *Displayable-Selectable Models (DS-models)*. A DS-model is a model (in the sense of Section 2) which supports the 'display' and 'selection' queries in their various manifestations. This term emphasizes the functionality required from the model and is neutral regarding specific implementations of that functionality. In the rest of this paper we examine the suitability of Breps as DS-models and whether they possess characteristics which are essential for any DS-model.

5   The Essence of the Brep

The boundary representation is usually described as 'representing the boundary of the object using entities such as faces, edges and vertices'. This vague definition was made precise by Rossignac and O'Connor, who introduced the concept of the *Selective Geometric Complex (SGC)* [Rossignac88]. In this section we give an informal overview of the SGC, including a new relatively minor practical generalization. A major generalization of the SGC, the *Generic Geometric Complex (GGC)* [Rappoport96b, Rappoport96a] is described in Section 8 in the context of the invariant naming problem.

There are three major conceptual stages in the specification of an SGC: carrier definition, intersections, and selection of active entities. Each stage is briefly described below.

A selective geometric complex in R^n is defined using a set of *carriers* embedded in R^n. A carrier is any pointset such that it is convenient for us to think about it as a single unit. The two major types of carriers are implicit and parametric hyper-surfaces. Implicitly defined surfaces, parametric curves, and parametric surfaces are all carriers in R^3. Even a well-defined fractal object may be a carrier. It is mostly in the definition of a carrier that our version of the SGC differs from the original one, in which only implicit half-spaces were allowed.

In general there are no special requirements from a carrier and every pointset is qualified to serve as a carrier. However, there are certain properties and applications for which is would make real sense to limit the generality of the carriers somewhat. To simplify the discussion, in the rest of this section we only deal with 2-D and 3-D SGCs. In practice, there are three major types of carriers: implicit surfaces (defined by a single algebraic equation) spline (piecewise parametric polynomial) curves, and spline surfaces.

The only requirement we impose upon these carrier curves and surfaces is that they do not possess self-intersections. That is, in a parametric curve or surface two different parameter values should not yield the same point, and in an implicit surface there should not be singular points. In case a carrier violates this requirement, it should be decomposed into parts each of which obeys the requirement. The decomposition of the carrier should itself be represented in an SGC.

The next stage in the definition of an SGC is the computation of all mutual intersections of the carriers and splitting of the carriers accordingly into open, dimensionally uniform connectivity components. In general, the intersection between a curve and another curve or surface is a set of curve segments and points; the intersection between two surfaces is a set of surface patches, curve segments, and points. Points which are isolated carriers or intersection results are called vertices; maximal curve segments which are isolated carriers, intersection results or split carriers are called edges; maximal surface patches of isolated or split carriers are called faces; and maximal connectivity components of space are called cells[1].

We now have a set of entities: vertices, edges, faces and cells. The importance of the intersection process is that all points in an entity possess the same characteristic function with respect to the carriers (that is, they belong to exactly the same set of carriers). In a sense, all points in an entity thus possess the same 'name'. We immediately emphasize that entities are not necessarily distinguishable according to these names. Points in different entities could end up with the same characteristic function.

As an example for an SGC, consider Figure 3. There are two carriers (the ellipses), four vertices, eight edges, and six cells. Note that the vertices cannot be distinguished by carrier identities, since they all possess the same classification with respect to the carriers. The edges are different connectivity components of split carriers and hence are indistinguishable by carrier identity as well. In Section 8 we will see that this figure demonstrates a fundamental difficulty with generic geometric complexes.

The final stage in the definition of an SGC is a selection of 'active' entities. The union of the pointsets of the active entities comprises the pointset of the object which the SGC represents. The subset of the active entities whose points lie on the boundary of that object is the Brep of the object. Note that the SGC can thus represent objects with internal structures and with mixed dimensionalities. We call the objects represented by SGCs 'decomposed pointsets'. The selection process can be guided by a graph of Boolean operations defined over the carriers, as is the case when the SGC is used to represent the result of boundary evaluation of a CSG graph, or simply be guided by a Boolean operation performed

---

[1] In [Rossignac88], the term 'cell' was used for what we call here an 'entity', because we wanted to endow the word 'cell' with its common interpretation in R^3.

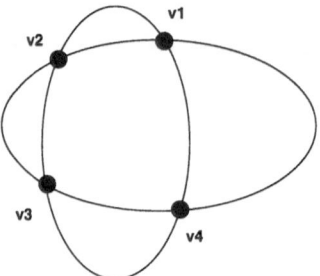

Fig. 3. A selective geometric complex (SGC) presenting fundamental invariant naming difficulties.

between two SGCs. There are many additional topics related to SGC's which we do not discuss here. For a deeper study, the reader should consult [Rossignac88].

From an operational interface point of view, a Brep supports the queries listed below. We do not enumerate the synthesis operations through which a Brep can be constructed and edited, since the reason of being of a model is defined only by the queries it supports.

- What types of carriers are present?
- What are the geometries of the carriers?
- What are the entities defined by the spatial inter-relationships of the carriers?
- What are the geometries of the entities?
- What are the adjacencies between the entities?
- What are the local relative orientations between the carriers in an entity?

A concrete model which can answer these queries is called a *complete Brep* in this paper. Naturally, implementations of a complete Brep should address efficiency issues, such as which adjacency relations are explicitly stored, should entities be sorted along each other when possible (for more efficient traversal) etc.

To summarize, the essential ingredient of the boundary representation as it is commonly understood is the explicit subdivision of space induced by a set of pointset carriers. Issues related to time and space efficiency are not considered to be essential ingredients.

6   The Display Operation

The most common operation in geometric modeling, an operation which is an inherent necessity in interactive design applications, is display of the designed object. In any interactive design application we need to see the object we design. Visualization is such an important requirement that is it useful even when the designed system or object do not possess an immediate geometric form. In this section we study the relationship between the Display operation in GOGs and the Brep.

Recall that `Display` receives three input arguments: a GOG, a parameter vector, and various parameters related to the desired visualization (camera parameters, display style etc). `Display` has three variants, depending upon which of the three arguments was modified by the user (obviously, if no argument was modified then there is no need to recompute `Display`). The variants have different implications on the performance required from the `Display` operation. In the first variant, the GOG defining the family was modified, by adding or deleting an operation node. In this case the user is willing to wait for the computation of the `Display` operation a few seconds or even more than that if the changes made to the GOG are substantial. In the second variant, the parameter vector indexing the family was modified. For this variant to be useful and to truly support rapid navigation in the modeled family, it is desired that the computation takes less than a second. That is, navigation (or 're-generation') requires `Display` to be at least an order of magnitude faster than GOG modification. Current parametric systems are very far from this goal: re-generation is usually not done interactively. In the third variant only the visualization parameters were modified, in which case it is certainly plausible to require interactive computation.

Due to practical considerations we would strongly prefer to utilize standard graphics display APIs such as OpenGL for the computation of `Display`. In such APIs the basic primitive rendered is the planar polygon. They support modeling and display transformations, rasterization (computation of the pixels covered by a polygon), hidden surface removal (usually using a z-buffer), masking operations on the image and z-buffers, texture mapping, and more. For higher-quality shading, normals to polygons and polygon vertices are needed as well. For optimization purposes, it helps if the polygons given to the graphics system are triangles, or better, triangular meshes in which the triangles are traversed such that triangles which share an edge are given to the system consecutively.

To see how the Brep relates to the `Display` operation, we will ask two questions: first, how much support is given by the Brep to the operation; and second, can the `Display` operation be supported as well by other representations (or equivalently, are there characteristics of the Brep which are not needed by the `Display` operation).

6.1 Brep's support for `Display`

The primary advantage of the Brep for the `Display` operation is that all the polygons which we want to see lie on Brep faces. That is, the boundary representation is attractive not only because it suffices to represent the boundary in order to represent the object uniquely, but also because the boundary is the only thing we are interested in for the purposes of the `Display` operation. Note that when we are interested in seeing internal structures of the object, they are explicitly represented by the Brep's SGC as well.

Given a Brep, the `Display` operation can be implemented by triangulating or polygonalizing the Brep faces and sending them to the graphics system. In addition, it is easy to compute the normals of polygons and polygon vertices. Thus, Brep's support for `Display` is strong but not immediate.

6.2 Display's need of the Brep

As noted earlier, the essence of a complete Brep is that intersections between carriers are computed and represented explicitly. Following are several indications that this may not be necessary for Display. Most of these are relevant for the relationship of Breps and visualization in general, not only in the context of general operation graphs.

Edges and vertices. When the display style is with hidden surfaces removed, Display does not need to display the intersection edges and vertices. They are implicitly visualized by the fact that faces are full. When the display style is wire-frame, the edges need to be shown, but this effect can be achieved using the z-buffer without actually computing the edges [Emmerik93]. Even if the edges are displayed after computing them explicitly, this computation is only needed to be carried to the resolution of the display device, which is normally orders of magnitude larger than the resolution of floating point computations.

Face interpenetration. Small inter-penetrations of faces into each other, which are avoided using great efforts by algorithms which compute complete Breps, do not harm visualization.

Curved face tessellations. Visualization of a curved face is done by tessellating it into polygons, and this has to be done even if the intersection of the face with other faces was computed exactly. This observation suggests an approach in which the order of operations is reversed or at least adaptively combined. Since Display needs an order of magnitude less accuracy than boundary evaluation, this approach is potentially more efficient.

Union. The result of the 'union' operation between two objects can be displayed by simply displaying the objects one after the other. The z-buffer ensures that the resulting image is correct. In this case no computation of geometric intersections is needed.

Fast display of Booleans. Algorithms for rapid visualization of CSG objects on standard platforms are beginning to appear. For example, [Rappoport96c] describes an algorithm which combines graph re-writing, hierarchical convex differences and efficient geometric algorithms (based on convex hulls of 3-D points) to visualize non-trivial CSG models interactively , using the standard graphics pipeline with a z-buffer and a stencil bit plane.

To summarize, complete Breps are not *essential* for performing the Display operation correctly; in some (perhaps even most) situations, computation of a complete Brep with all adjacency information is an over-kill. In applications providing interactive, high-level manipulation, model visualization can be achieved by computing crude linear approximations. This is especially true when the display mode is with hidden surfaces removed, but may also be true for wire-frame displays.

7   The Picking Operation

Recall that Picking refers to identification of Brep entities of the example object through graphical interaction, without persistent storage of the result.

7.1 Brep's support for Picking

There are two main methods to implement Picking in graphics systems: a graphics-based method and a geometric method. The latter is simply done by intersecting the scene with a ray from the eye to the mouse location, computing the nearest object. The former uses the graphics system in order to detect the identity of the visible polygon closest to the mouse. Which method is faster depends upon the relative performance of the graphics system and the CPU. The graphics-based method is probably easier to implement.

Since the mouse location is discrete, it cannot be expected to fall exactly on an edge or a vertex even if the user intends their selection. Therefore, in both methods, we should first identify faces and then infer selected edges or vertices from them. The alternative, computing the distance from the eye-mouse ray to every edge and vertex, is probably not practical. The Brep supports both methods since all faces are explicitly available and it can be inferred whether or not a ray intersects the face. Actually, Breps support Picking even more than they support Display since they give the information needed for Picking explicitly, while Brep faces need to be further polygonalized or triangulated.

7.2 Picking's need of the Brep

It may seem that by definition a complete Brep of the example object must be computed in order to support Picking. However, this is not the case.

Following is a sketch of a possible algorithm which does not need a complete Brep a-priori. Let us assume that the Display operation is executed correctly. As a result, a graphics-based method can be used to select the visible face(s) nearest the mouse location. It should not be too difficult to identify whether or not the user has meant to select a face, an edge or a vertex by analyzing the number of faces visible in the vicinity of the mouse. Thus, even if Brep entities are not explicitly represented, they can be computed on-demand by this 'discrete local boundary evaluation' algorithm.

Note also that the performance required from Picking is probably slower than that required from Display. Since the Picking operation identifies an argument to another operation, it is a discrete decision in the design process which does not involve numerical parameters. Usually, the numerical parameters are the ones which we want to be capable of modifying interactively. For example, we want to modify affine operations and visualize interactively the results of Boolean operations defined on the object [Rappoport96c], or we modify constraint parameters and want to see the geometric and physical behavior of the object. As a result, a slower, incremental on-demand computation of Picking is acceptable. If the user does not demand highlighting of Brep entities while moving the mouse (which is a nice aid to Picking) then the computation of Picking is allowed a few seconds.

In summary, it is likely that Picking does not need the full power of a complete Brep, although a complete Brep supports it more efficiently.

8   Invariant Naming and the Generic Geometric Complex

After obtaining the concrete names of the entities of the example object which were specified by the Picking operation, they have to be transformed to a representation in which they are invariant under parameter modifications. The general problem is called 'persistent naming'.

Technically, it is convenient to consider two types of parameter modifications, both of which preserve certain entities of the SGC representing the example object. The first type assumes only knowledge about the SGC of the example object; the second type assumes knowledge about the whole GOG. The Invariant Naming operation maps concrete entity names into generic entity names assuming parameter modifications of the first kind. For modifications of the second kind a more elaborate naming scheme is needed. Such a naming scheme typically must consider the intended semantics of GOG operations and is difficult to specify mathematically. Hence in this paper we only discuss invariant naming.

In order to precisely define the invariant naming problem, we briefly describe the *Generic Geometric Complex (GGC)* [Rappoport96b]. The GGC is a model for a family of decomposed pointsets, each modeled by an SGC, supporting the following queries:

1. *Entity-to-name:* given an SGC whose carriers possess generic names having equal status and an entity in it, return a unique generic name for the entity, guaranteed to be identical to that returned for the corresponding entity in all members of the given SGC's family.
2. *Name-to-entity:* given an SGC in the modeled family and a generic name previously returned by entity-to-name, return the entity having that name.
3. *Membership classification:* given an SGC, determine whether or not it belongs to the modeled family.
4. *Example* (optional): return a member of the modeled family.

The GGC can be viewed as the family 'spanned' by an example member. That is, the family of SGCs obtained by modifying the geometries of the carriers of a concrete SGC while preserving the existence of generically named entities designated as essential.

A distinction is made between the *nature* of a carrier and its geometry. The nature of a carrier is the representation scheme used for its pointset plus symbolic and integer parameters used in that representation. For example, an implicit surface and a parametric surface are of different natures, as are two parametric surfaces of different degrees.

The geometry of the carriers is allowed to change without changing properties which were defined to be required. Such properties may include the number of the carriers[2], the nature of each carrier, geometric properties of carriers, geometric relationships between the carriers, the existence of certain entities, geometric

---

[2] Actually, there are applications for which the number of carriers should be allowed to change as well, e.g. when the number of holes in an object is an external parameter. Discussion of such GGCs will take us too far from our present focus.

properties of entities, symbolic properties of entities, etc. The names of required entities are specified in a way which makes them independent of things which are allowed to change. Carrier geometry modifications are valid as long as they preserve the above properties, including the names of the required entities.

There are several ingredients from which names of required entities in a GGC can be composed. The major ones are:

- Carrier identities: as noted earlier, all points in an entity have the same classification with respect to the carriers. The name of an entity can utilize the names of the carriers to which all of its points belong.
- Invariant geometric properties, such as convexity, tangency etc.
- Classification with respect to carrier signs (separation): some of the carriers induce a singed function on space; all points in an entity have the same sign vector with respect to these carriers. [Shapiro93] can be viewed as studying the possibility of unique naming using this ingredient alone.
- Adjacencies: the name of an entity can recursively utilize the names of adjacent entities. It is preferable to keep the depth of the recursion as small as possible.
- Ordered adjacency and ordered usage: the name of an entity can utilize the relative local orientations of its carriers and carriers of adjacent entities.
- Local ordering: ordering along parametric entities whose parameter space can be ordered.
- Local naming: a recursive invocation of the invariant naming process on a single carrier.
- Arbitrary naming of an entity. This ingredient is needed when there is no other way to distinguish between several entities having the same invariant name. Once an entity has been arbitrarily named, names of other entities can depend upon this name through other adjacency and ordering ingredients.

Figure 4 is a simple example that shows why several ingredients are needed. On the left, the two vertices have the same carrier identities but a different local orientation of the line with respect to the circle (denoted by the dotted curves). On the right, the two marked cells have the same classification with respect to all carriers, but they can be distinguished according to the vertices they are adjacent to. The vertices can in turn be distinguished by carrier identities since they are the result of intersecting different lines.

Figure 3 is a very simple example that shows that there are situations in which all ingredients (but arbitrary naming) are not sufficient. All four vertices in the figure have the same carrier identities. Applying the orientation ingredients, they can further be divided into two sets of a pair of vertices each. However, the adjacency ingredient cannot be applied because no edge or cell can be named without naming the vertices or cells first. This SGC does not possess a unique answer to name-to-entity queries.

In [Rappoport96b] we give an algorithm based on equivalence classes and sequential introduction of name ingredients to induce invariant naming on entities in a GGC. The algorithm is very general, and always finds a unique naming if one exists.

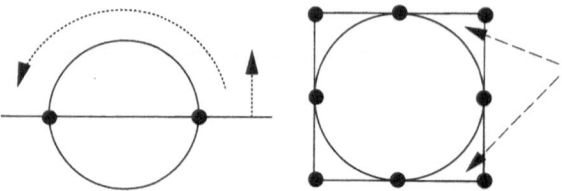

Fig. 4. Examples for the need of several ingredients in invariant entity names in generic geometric complexes.

Certain situations always admit a unique naming. For example, in [Rappoport96b] we proved that a necessary and sufficient condition for naming a connectivity component of a 2-D complex is that a name exists for a single entity in it.

The situation in 3-D is not fully understood at present. The naming scheme in [Capoyleas96] can be viewed as a scheme which utilizes some of the above ingredients in the context of a specific system, and even there it does not solve the problem completely. In some restricted (but useful) cases the problem is easy to solve. For example, if all the carriers are linear half-spaces then each pair can have at most a single intersection entity of each dimension, hence the carrier identity ingredient suffices for naming. Carrier signs are sufficient when all entities are separated by carriers, as studied in [Shapiro93].

8.1   Brep's support for Invariant Naming

In light of the discussion above, it is clear that the Brep provides substantial support for Invariant Naming. The information present in a Brep about adjacencies, and information easily derived from it about relative local orientations, are of great assistance to any invariant naming scheme. However, we must keep in mind that a Brep alone does not suffice to solve the problem in all cases.

8.2   Invariant Naming's need of the Brep

With Invariant Naming we are in a different situation from the former two operations. Display and Selection are well supported by Breps, and the only issue was whether or not they actually need the full power of a complete Brep. Here, we do not know at this stage even if a complete Brep suffices to solve the invariant naming problem.

Carrier identity, which is probably the important of the three naming ingredients given above, can be computed adaptively in an on-demand basis. The complete adjacency structure of the Brep is not essential in the case of linear half-spaces and separated entities. Perhaps these observations serve as indications that it would be possible to find naming schemes which would not require a complete Brep.

In summary, both invariant naming and persistent naming are fundamental operations in geometric and solid modeling. Both are not well understood at present. Finding a satisfactory and efficient naming scheme which does not need the full adjacency structure of an SGC is an extremely important open problem in geometric and solid modeling.

9   Discussion

Two important operations which we have not discussed in this paper are collision detection and meshing for finite element simulations. Many applications need to detect collisions between moving objects and act accordingly. Data structures supporting efficient collision detection should probably contain (hierarchical) spatial information not present in Breps. It is not clear whether complete Breps are essential for this application. For example, obviously penetration of one boundary surface into another in the same object does not harm collision detection.

Regarding meshing, we should note that mesh generators are usually interested in a *simplified* Brep and not necessarily in a complete Brep. For example, the primitive geometric operation in the paving family of meshing algorithm is the projection of an arbitrary spatial point on a surface. This primitive can perhaps be better implemented in an incremental, adaptive manner on the GOG model itself or on some partially evaluated version of it.

The reason we have not discussed collision detection and meshing is that at present they are considered as post-processes and are not well integrated with the design process. Clearly, this situation should be changed in the future to support effective optimizations and full product models.

In this paper we have also not discussed object and shape analysis applications. Breps are certainly needed by a large number of such applications.

In summary, we feel that complete Breps, that is, SGC-like Breps explicitly storing all topological entities and their (ordered) adjacency relationships, are probably an over-kill for two important queries required from the example object in interactive design of families of geometric objects. Both Display and Picking can be supported directly by a GOG model or by a partial Brep derived from it adaptively on-demand. On the other hand, valid and efficient support for Invariant Naming (and for persistent naming) is an important open problem in geometric and solid modeling for which a complete Brep may be essential.

Breps are perceived today as essential, or at least as very practical, for solid modeling. This is evidenced by the success enjoyed by the Acis Brep modeler. Modern modeling schemes supporting high-level knowledge, well-defined storage of design intent, features, parameterizations, constraints and history mostly use Breps because they need support for visualization of the model and selection of entities from it. These requirements are best encapsulated by the term Displayable-Selectable Model (DS-model), which emphasizes the functional nature of the desired representation rather than its data and implementation. The discovery of DS-models different from Breps may improve the efficiency and sophistication of solid modeling systems.

Acknowledgement: I thank Vadim Shapiro for his comments.

References

[Ala91] Ala, S.R., Design methodology of boundary data structures. *Intl. J. Comp. Geo. and Apps.* 1(4):207-226, 1991.

[Capoyleas96] Capoyleas, V., Chen, X., Hoffmann, C.M., Generic naming in generative, constraint-based design. *Computer-Aided Design*, 28(1):17-26, 1996.

[Chen95] Chen, X., Hoffmann, C.M., On editability of feature-based design. *Computer-Aided Design*, 27(12):905-914, 1995.

[Emmerik90] Emmerik, M.J.G.M. van, Interactive design of parameterized 3D models by direct manipulation. Ph.D. Thesis, Delft University Press, 1990.

[Emmerik93] Emmerik, M.J.G.M. van, Rappoport, A., Rossignac, J., Simplifying interactive design of solid models: a hypertext approach. *The Visual Computer*, 9:239-254, 1993.

[Guibas85] Guibas, L., Stolfi, J., Primitives for the manipulation of general subdivisions and the computations of Voronoi diagrams. *ACM Transactions On Graphics*, 4(2):74, 1985.

[Hoffmann89] Hoffmann, C., Geometric and Solid Modeling: an Introduction. Morgan Kaufmann, 1989.

[Hoffmann93a] Hoffmann, C.M., Juan, R., Erep, an editable, high-level representation for geometric design and analysis. In: P. Wilson, M. Wozny, and M. Pratt, (Eds), *Geometric and Product Modeling*, pp. 129-164, North Holland, 1993.

[Hoffmann93b] Hoffmann, C.M., On the semantics of generative geometry representations. *19th ASME Design Conference*, Albuquerque, New Mexico, September 1993.

[Hoffmann96] Hoffmann, C., Rossignac, J.R., A road map to solid modeling. *IEEE Transactions on Visualization and Computer Graphics*, 2(1):3-10, 1996.

[Kripac95] Jiri Kripac, A mechanism for persistently naming topological entities in history-based parametric solid models. Proceedings, *Third ACM/Siggraph Symposium on Solid Modeling and Applications (Solid Modeling '95)*, pp. 21-30, ACM Press, 1995.

[Lequette96] Lequette, R., Considerations on topological naming. Presented at the *IFIP Workshop on Geometric Modeling in CAD*, May 1996, Airlie, VA.

[Mäntylä88] Mäntylä, M., An Introduction to Solid Modeling, Computer Science Press, Maryland, 1988.

[Rappoport93] Rappoport, A., A scheme for single instance representation in hierarchical assembly graphs. *IFIP Conference on Geometric Modeling in Computer Graphics*, Genova, Italy, June 1993. Published in: Falcidieno, B., Kunii T.L. (eds), Geometric Modeling in Computer Graphics, pp. 213-224, Springer, 1993 (an updated version is available from the author).

[Rappoport95] Rappoport, A., Geometric modeling: a new fundamental framework and its practical implications. Proceedings, *Third ACM/Siggraph Symposium on Solid Modeling and Applications (Solid Modeling '95)*, May 1995, Salt Lake City, pp. 31-42.

[Rappoport96a] Rappoport, A., Parametric and declarative modeling of families of geometric objects. Presented at the *IFIP Workshop on Geometric Modeling in CAD*, May 1996, Airlie, VA.

[Rappoport96b] Rappoport, A., The Generic Geometric Complex (GGC): a model for families of decomposed pointset. Technical Report, Institute of Computer Science, The Hebrew University, 1996.

[Rappoport96c] Rappoport, A., Spitz, S., Interactive 3-D Boolean operations for conceptual geometric design. Technical Report, Institute of Computer Science, The Hebrew University, 1996.

[Rappoport96d] Rappoport, A., The Geometric Operation Graph (GOG) representation for geometric modeling. In preparation, 1996.

[Requicha80] Requicha, A.G., Representations for rigid solids: theory, methods and systems. *ACM Computing Surveys*, 12:437-464, 1980.

[Requicha85] Requicha, A.G., Voelcker, H.B., Boolean operations in solid modeling: boundary evaluation and merging algorithms. *Proc. of the IEEE* 73(1):30-44, 1985.

[Rossignac86a] Rossignac, J.R., Requicha, A.A.G., Offsetting operations in solid modelling. *Computer-Aided Geometric Design*, 3:129-148, 1986.

[Rossignac86b] Rossignac, J.R., Constraints in constructive solid geometry. *ACM Symposium on Interactive 3D Graphics*, pp. 93-110, ACM Press, 1986.

[Rossignac88] Rossignac, J.R., O'Connor, M.A., SGC: a dimension-independent model for pointsets with internal structures and incomplete boundaries. In: Wozny, M., Turner, J., Preiss, K. (eds), *Geometric Modeling for Product Engineering*, North-Holland, 1988. Proceedings of the 1988 IFIP/NSF Workshop on Geometric Modeling, Rensselaerville, NY, September 1988.

[Rossignac89] Rossignac, J.R., Borrel, P., Nackman, L.R., Interactive design with sequences of parameterized transformations. *Intelligent CAD Systems 2: Implementational Issues*, P. ten Hagen, T. Tomiyama (eds), Springer-Verlag, 1989.

[Rumbaugh91] Rumbaugh, J., Blaha, M., Premerlani, W., Eddy, F., Lorensen, W., Object-Oriented Modeling and Design. Prentice-Hall, 1991.

[Shah96] Shah, J., Mäntylä, M., Parametric and Feature-Based CAD/CAM: Concepts, Techniques, and Applications. Wiley, New-York, 1996.

[Shapiro93] Shapiro, V., Vossler, D., Separation for boundary to CSG conversion. *ACM Transactions On Graphics*, 12(1):35-55, 1993.

[Shapiro95] Shapiro, V., Vossler, D.L., What is a parametric family of solids? Proceedings, *Third ACM/Siggraph Symposium on Solid Modeling and Applications (Solid Modeling '95)*, pp. 43-54, ACM Press, 1995.

[Woo85] Woo, T.C., A combinatorial analysis of boundary data structure schemata. *IEEE Computer Graphics and Applications* 5:19-27, 1985.

Synthesis of a Unified Approach to Shape Modeling

Abel Gomes and Alan Middleditch

Centre for Geometric Modelling and Design,
Department of Computer Science and Information Systems, Brunel University,
Uxbridge Middlesex UB8 3PH, England

Abstract. Geometric solid modeling and form feature-based modeling are the two main aproaches to model the shape of polyhedra. This paper proposes a unified shape representation for polyhedra. It is a result of some efforts for the last few years to understand the shape of polyhedra, and how current shape models represent and manipulate it.

1   Introduction

Solid modelers work to a large extent with finite symbolic representations (*i.e.*, data structures) of mathematical models of physical solid objects. For early solid modelers it was not of primary importance to define precisely which class of polyhedra should be representable and to which polyhedra certain operations should apply. The objective was to find a solution for a given practical problem. According to Pratt [18], the appearance of solid modeling in the 70s was not only due to the need for 3D graphic representations on a computer, but also the engineering requirement for the computation of volume, mass, and moments of inertia of objects. Subsequent theoretic treatment of solid models provides a posterior justification of development.

Notably, a compact, regular, and semi-analytic set of points in $\mathbb{R}^3$ has been accepted as a standard model in solid modeling [19][4]. An alternative mathematical model for solids is that one based on surfaces without boundary [5]. These two mathematical models gave rise to the two most well-known computational models for solids, CSG (Constructive Solid Geometry) and BRep (Boundary Representation), respectively. Recall that semi-analytic sets are studied in algebraic geometry, whereas (combinatorial) surfaces without boundary are studied in algebraic topology. Both are thus algebraic models.

In fact, one important property of semi-analytic sets is that they are closed under (regularized) set union, intersection, and difference operations, and thus form a Boolean algebra [13]. As noted by Shapiro [23], this facilitates the representation of solids in terms of Boolean operations on simpler solids which is crucial in user interfaces and many applications. Besides, the algebraicity inherent to a solid model is not only important to support the description of the interactive construction sequence of a solid, but it is also desirable because a computer is an algebraic machine [14][6]. The algebraicity and finiteness (or finite describability) properties of a mathematical model of solids are probably the most important of the properties which guarantee its computability [10].

Analogously, a BRep model is also based on a finite algebraic model. Basically, a closed 2D manifold bounding a solid is subdivided into a finite number of n-cells

($n \leq 2$) using a finite set of Euler operators. These topological operators satisfy the well-known Euler-Poincaré formula for polyhedra. [2][5].

The work of Weiler [24] on non-manifold solid models started a new period in geometric modeling. The main novelty of his model was that the represented objects were no longer required to be regular, or homogeneously three-dimensional. Weiler's motivation was to achieve a unified representation of wireframe, surface, solid modeling forms simultaneously in the same modeling environment. Since then, similar non-manifold models have been developed, for example, those of Wu [25], Masuda *et al.* [16], and Yamaguchi and Kimura [26].

Following the development line based on the algebraic geometry, other researchers extended the representation of semi-analytic sets to dimensionally non-homogeneous, non-closed point sets with internal structures, for example, those of Rossignac and O' Connor [21], Rossignac and Requicha [22], Middleditch [17], and Djinn's researchers [4].

Despite the evolution of geometric modeling towards the unification of the shape representations, a long way has still to be covered. So far, all development has focussed on the geometry. As so many times referred in the geometric modeling literature, the geometry provides a complete description of the (geometric) shape of a solid, or polyhedron. The geometric completeness of a mathematical model for solids implies, in principle, the representational completeness (completeness, for short). Recall that complete representations define solids unambiguously [20]. With a complete representation of solid S, it should be possible to decide for any point $p \in \mathbb{R}^3$ whether p is in S, on the boundary of S, or out of S. The choice of a mathematical model is determined to a large extent by a set of restrictive computational requirements as representational finiteness, constructivity, and completeness which are related to the properties of the mathematical model, namely, finiteness, algebraicity, and geometric completeness, respectively. But a mathematical model also should take into account other requirements imposed by the downstream engineering applications; form feature-based modeling is no more than an attempt to overcome the practical deficiences of current geometric modelers in some engineering applications. To the best of our knowledge, there is no theory to support form feature-based modeling.

2   Shape and Shape Equivalence

What is shape? What is form? To say that two objects have the same shape has an intuitively obvious, but very imprecise meaning [7]. A mathematics of unified shape description and analysis is greatly needed in solid modeling, mainly for two reasons: on the one hand to clarify the notions of shape and shape equivalence, and on the other hand to synthesise the shape aspects involved in geometric solid modeling and form feature-based modeling.

In rather abstract terms, *shape* is an ordered pair $(S,\mathcal{E})$, where S is a set (of "undefined" elements) and $\mathcal{E}$ is an equivalence relation defined in the set of all subsets of S. The shape $(S,\mathcal{E})$ studies those and only those properties of a subset $X \subset S$ that X has in common with all subsets equivalent to X; these are the invariant properties.

Topologies, geometries, and morphologies are quite different shapes that may be defined over the same space. They arise from different definitions of equivalent subsets. As will become apparent throughout this paper, two subsets that are equivalent in the sense of Euclidean geometry (congruent or metrically equivalent subsets) are also morphologically and topologically equivalent, but not conversely. This justifies calling

Euclidean geometry $(S, \mathcal{C})$ a *subshape* of morphology $(S, \mathcal{M})$, and $(S, \mathcal{M})$ a subshape of topology $(S, \mathcal{H})$, where $\mathcal{C}$, $\mathcal{M}$, and $\mathcal{H}$ are their equivalence relations, respectively.

2.1   Geometric Equivalence

All we are familiar with some aspects of geometry. A very specialised geometry is 3-dimensional Euclidean geometry $(\mathbb{R}^3, \mathcal{C})$, where $\mathbb{R}^3$ is the 3-dimensional Euclidean space and $\mathcal{C}$ is the equivalence relation on the set of all subsets of $\mathbb{R}^3$ that defines two subsets to be equivalent if and only if they are *congruent*. This idea of congruence is often expressed as a principle of superposition; that is, two subsets are equivalent from the viewpoint of Euclidean geometry, or are *congruent*, if one subset can be *rigidly moved* such that it is exactly superimposed on the other subset. Thus Euclidean geometry studies properties which are left unchanged by so-called *rigid motions* or *transformations*. Intuitively, a rigid transformation of an object does not alter its shape. Examples of properties preserved by rigid transformations are lenght, area, volume, and angle.

In more formal terms, rigid transformations are known as isometries. Suppose X and Y are metric spaces, with metrics d_X and d_Y respectively. A function $i : X \to Y$ is an **isometry** iff $d_Y(i(\mathbf{p}), i(\mathbf{q})) = d_X(\mathbf{p}, \mathbf{q})$ for all $\mathbf{p}, \mathbf{q} \in X$. Two metric spaces are *isometric* iff there is an isometry of one onto the other [8]. A property is said to be *metric* iff it is preserved by isometry, that is: if X and Y are isometric, and one has a metric property, so does the other. A translation, a rotation, and a reflection of a subset (*e.g.*, a solid parallelipiped) of $\mathbb{R}^3$ are examples of isometries of the Euclidean space $\mathbb{R}^3$ onto itself.

To summarise, in the Euclidean space $\mathbb{R}^3$ (our geometric modeling space), two geometrically distinct subsets belong to different equivalence classes of congruent sets.

2.2   Topological Equivalence

In topology, the objects under study are spaces which are informally sets with some kind of structure. The need for a structure in order for a set to qualify as a space may be rooted in the feeling that a notion of "nearness" (in some sense not necessarily quantitative) is inherent in the concept of a space. Thus a space differs from the mere set of its elements by possessing a structure which in some way (however vague) gives expression to that notion [3, p.14]. There are a few ways to define such a structure. In this paper, we are interested in spaces which are topological spaces, because, as will become apparent later, polyhedra as subsets of $\mathbb{R}^3$ are topological spaces. The definition of a topological space may be achieved in various ways, using the notion of *neighbourhood* of a point, or the notion of *open set*, or the notion of *closed set*, etc [11, p.266]. Any definition of a space should contain *enough information* to define the notion of *continuity* for functions between spaces [1, p.12], in order to precisely define the notion of *topological equivalence*. Such *information* is related to the concept of neighbourhood of a point $\mathbf{p}$ in an abstract space X.

Let f be a function between two Euclidean spaces, say $f : \mathbb{R}^m \to \mathbb{R}^n$. The traditional definition of continuity for f follows: f is continuous at $\mathbf{p} \in \mathbb{R}^m$ if given $\epsilon > 0$ there is $\delta > 0$ such that $\| f(\mathbf{q}) - f(\mathbf{p}) \| < \epsilon$ whenever $\| \mathbf{q} - \mathbf{p} \| < \delta$. The function f is continuous if it satisfies this condition for each $\mathbf{p} \in \mathbb{R}^m$. Call a subset N of $\mathbb{R}^m$ a *neighbourhood* of the point $\mathbf{p} \in \mathbb{R}^m$ if, for some real number $r > 0$, the closed ball centre $\mathbf{p}$ radius r is a proper subset of N. Now we can to rephrase the above definition of continuity as follows: f is *continuous* if given any $\mathbf{p} \in \mathbb{R}^m$ and any neighbourhood N of $f(\mathbf{p})$ in $\mathbb{R}^m$, then $f^{-1}(N)$ is a neighbourhood of $\mathbf{p}$ in $\mathbb{R}^m$. In order to construct a space independently of any metric, mathematicians retain the concept of neighbourhood but relieve themselves of any dependence on a distance function as used above; topological equivalence does not preserve distance.

Inspection of the properties of point neighbourhoods in a Euclidean space leads to the definition of a topological space. A *topological space* is a set X together with a collection $\mathcal{N}$ of subsets of X, called neighbourhoods of $x \in X$, that satisfy a set of axioms (see [1, p.13] for details). This collection $\mathcal{N}$ of neighbourhoods of $x \in X$ is called **topology** or **topological structure on the set** X.

Now we can to define precisely what we mean by a continuous function and a homeomorphism. Let X and Y be topological spaces. A function $f : X \to Y$ is *continuous* if for each point $x \in X$ and each neighbourhood N of $f(x) \in Y$ the set $f^{-1}(N)$ is a neighbourhood of x. A function $h : X \to Y$ is called a *homeomorphism* if it is one-to-one, onto, continuous, and has a continuous inverse. When such a function exists, X and Y are called *homeomorphic* (or *topologically equivalent*) spaces.

Intuitively, homeomorphims behave like *elastic transformations* of subsets made of perfectly elastic rubber. However, we must be careful to ensure that distinct points remain distinct; we are not allowed to force two different points to coalesce into one point. Therefore, two subsets are topologically equivalent iff one subset can be made to coincide with the other by an elastic transformation. For example, a solid sphere can be elastically deformed into a solid cube, or even a cube with a protrusion on one of its faces; they are said to be homeomorphic.

Thus, an equivalence relation $\mathcal{H}$ is established on the set of all subsets of a set S by defining two subsets to be equivalent if and only if they are homeomorphic. If the group of transformations is taken to be all continuous one-to-one mappings of the 3-dimensional space onto itself (homeomorphisms), the associated shape is the **topology** of the 3-dimensional space. In more abstract terms, a *topology* $(S, \mathcal{H})$ is a shape that recognises only the continuity properties of subsets of a set S, $\mathcal{H}$ being a topological equivalence relation in S that establishes equivalence classes of homeomorphic subsets.

2.3   Morphological Equivalence

In this subsection we introduce the new concepts of resemblance and morphological equivalence in order to distinguish objects which are not topologically distinct but have distinct morphological structures (protrusions and depressions) as, for example, a cube with a protrusion and a cube with a depression.

Similarity. When the equivalence classes are such that similar objects are within one and the same class, *dilation* and *contraction* (also known as *uniform scalings*) are permitted transformations. For example, the dilation of a solid cube gives rise to a bigger solid cube. Thus, similarities do not preserve distances. However, as with isometries, they preserve angles. In formal terms, a funtion $s : X \to Y$ is a **similarity** iff there is a positive number r such that $d_X(s(\mathbf{p}), s(\mathbf{q})) = r.d_Y(\mathbf{p}, \mathbf{q})$ for all $\mathbf{p}, \mathbf{q} \in X$ [8, p.45]. The number r is the *ratio* of s. Therefore, two metric spaces are similar iff there is a similarity of one onto the other. Note that an isometry is a similarity with $r = 1$, and a similarity is a homeomorphism, but not conversely. That is, a similarity is a continuous function placed somewhere between the group of isometries and the group of homeomorphisms.

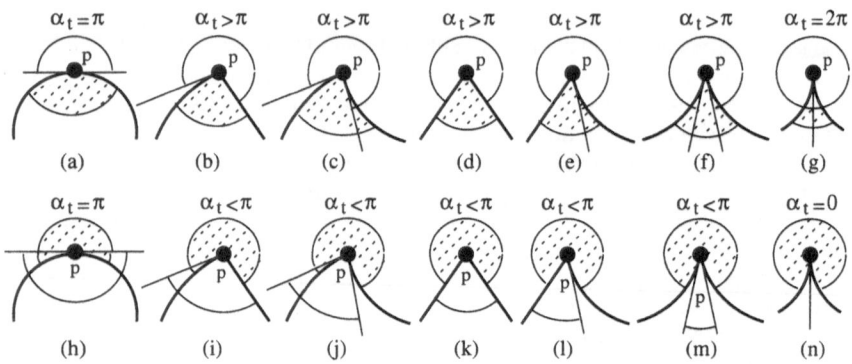

Fig. 1. Angle ranges: (a)-(g) convex range, and (h)-(n) concave range.

Strong and Ordinary Resemblance. In order to define morphological shape equivalence, we require a continuous function more general than similarities, but less general than homeomorphisms. This continuous function *preserves* neither distances nor angles, but just *vexity angles*. In the study of the morphology of polyhedra in $\mathbb{R}^3$ (polygons in $\mathbb{R}^2$), two angle ranges are used, the *concave range* $]0, \pi[$ and the *convex range* $]\pi, 2\pi[$, respectively. Under a morphology, the angle between two edges of a polygon in $\mathbb{R}^2$ incident at a vertex may range within either the concave range or the convex range. The same is true for the angle between two faces of a polyhedron in $\mathbb{R}^3$ along the points of their intersection edge. The angle α between two edges incident at a vertex p is defined as the the angle between two lines intersecting at p, each passing through one of the two points that results from the intersection between one edge incident at p and the boundary of a ball $B(p, \epsilon)$ centered at p and radius ϵ. Fig.1 shows how the angle α between two edges (two faces) varies around their intersection vertex (intersection edge), $\alpha \in]\pi, 2\pi[$ in (a)-(g) and $\alpha \in]0, \pi[$ in (h)-(n); but, its corresponding angle α_t between the lines tangent to both edges at the vertex p is either in $[\pi, 2\pi]$ or $[0, \pi]$, respectively.

The vertex p is said to be *convex* (*concave*) if α is in the convex (*concave*) range. But, Fig.1(e,f,l,m) shows that a convex (concave) vertex p does not imply that its semi-neighbourhood is convex (concave). A semi-neighbourhood $\aleph(p, \epsilon)$ of a point p, radius ϵ, is a neighbourhood of p that is homeomorphic to $\mathbb{R}^2_+$. Thus, a function that is vexity angle-preserving preserves the convexity or concavity of vertices of a polygon in $\mathbb{R}^2$ (respectively, edges of a polyhedron in $\mathbb{R}^3$), but not the convexity or concavity of their semi-neighbourhoods. Consequently, a convex edge (face) of a polygon (polyhedron) may be morphologically transformed into a concave edge (face). In this paper, we assume that each edge of a polygon (respectively, each face of a polyhedron) is either convex or concave.

In Fig.2, three collections of distinct polygons are shown. Any two polygons of the first collection (a)-(e) can be morphologically transformed into each other without varying the curvature of their edges. A function $r : X \rightarrow Y$, with $X, Y \in \mathbb{R}^2$, that preserves vexity angles and the edge curvature in $\mathbb{R}^2$ is termed a **strong resemblance**, and X and Y are said to be **strongly morphologically equivalent**. In respect to the collection of polygons (f)-(j), edge curvature-preserving is relaxed and we say that there is an **ordinary resemblance** between two polygons, X and Y; alternatively, X and

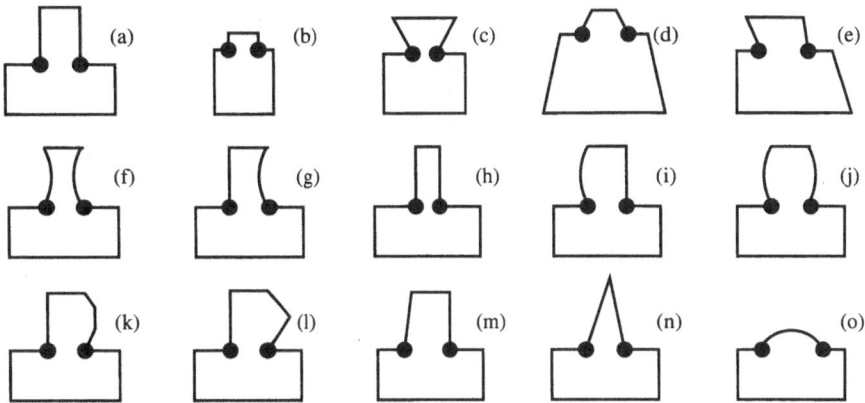

Fig. 2. Morphologically equivalent polygons.

Y are said to be **ordinarily morphologically equivalent**. Obviously, any polygon in the collection (a)-(e) is ordinarily morphologically equivalent to any polygon in the collection (f)-(j).

Weak Resemblance. In this subsection, we implicitly assume that the complexes K_1 and K_2 of two morphologically equivalent polygons (polyhedra), P_1 and P_2, respectively, are combinatorially equivalent (or isomorphic), that is, there exists a biunique, incidence-preserving correspondence between the elements (vertices, edges, and faces) of K_1 and those of K_2.

However, for the purposes of engineering, in particular for form feature-based modeling in CAD, we are interested in further weakening *local* resemblance between two subsets of a space (*e.g.*, polygons in $\mathbb{R}^2$ and polyhedra in $\mathbb{R}^3$). Two **weakly resemblanced** (**weakly morphologically equivalent** or **weakly morphomorphic**) subsets are not required to be isomorphic (*e.g.*, the collection (k)-(o), Fig.2). Thus, a weak resemblance function (or weak morphomorphism) $r_w : X \to Y$ is less restrictive (*i.e.*, more general) than an ordinary resemblance function $r_o : X \to Y$ or a strong resemblance function $r_s : X \to Y$. In this sense, all the polygons in Fig.2 are weakly resemblanced.

The major problem in comparing two weakly resemblanced polyhedra is that their complexes are not necessarily isomorphic. An isomorphism between the complexes of two polyhedra enforces their local shape properties; for example, the convexity of a vertex of the first complex is preserved by its corresponding vertex of the second complex. This means that two polyhedra possessing non-isomorphic complexes are not morphologically comparable, simply because a cell of a complex does not necessarily meet a corresponding cell in the other complex.

This suggests that a weak resemblance function does not require an isomorphism between cells (*i.e.*, in local terms) but an isomorphism between collection of cells (*i.e.*, in zonal terms). Each collection of cells of a complex results from a cell decomposition of either a protrusive or depressive zone of the boundary of a solid polyhedron. That is, two polyhedra are weakly resemblanced iff there exists an isomorphism between their "zones". Because weak resemblance preserves the one-to-one correspondence between

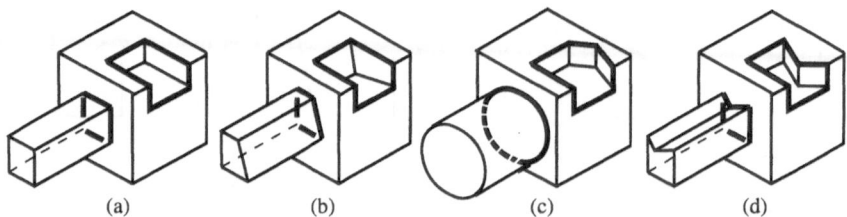

Fig. 3. Morphologically equivalent polyhedra.

protrusive zones (respectively, depressive zones) of two polyhedra, it is zonal vexity angle-preserving. For example, the four polyhedra in Fig.3 are weakly morphologically equivalent, because each includes two protrusive zones and one depressive zone, independently of local vexity of their points. Roughly speaking, two protrusive zones are separated by a path of concave edges, whereas a protrusive zone and a depressive zone are separated by a path of convex edges. Thus, the vexity is preserved along a path separating two zones. In Fig.3, these paths separating protrusive and depressive zones are highlighted by thick edges. Note that zonal vexity angle-preserving does not imply local vexity angle-preserving. For example, in weakly morphologically transforming the polyhedron (3.c) into the polyhedron (3.d), just one concave edge of the depressive zone in (3.c) became a convex edge of the depressive zone in (3.d).

A **protrusive zone** Z_P of a polyhedron $X \subset \mathbb{R}^3$ is defined as a subset of the boundary of X, $Z_P \subset \partial X$, whose cell decomposition corresponds to a 2-chain $C(Z_P)$ bounded by a 1-cycle that results from the sum of two or more minimal 1-cycles, each adding at least one concave edge to the resulting 1-cycle[1]. Note that the 1-cycle (or closed edge path) bounding a protrusive zone may contain convex edges. Nevertheless, it includes at least a 1-chain (edge path) of concave edges, each 1-chain separating it from another protrusive zone. Each concave edge separates two minimal protrusive zones. A **minimal protrusive zone** is defined as the union of the semi-neighbourhoods of the points of a concave edge. Thus, a protrusive zone includes at least one minimal protrusive zone.

On the other hand, a **depressive zone** Z_D is also a subset of the boundary of a polyhedron, whose cell decomposition corresponds to a 2-chain $C(Z_D)$ consisting of a collection of 2-chains of non-disjoint protrusive zones. It can be demonstrated that $\partial C(Z_D)$ is a collection of 1-cycles of convex edges.

In order to establish a zonal morphological equivalence relation between two polyhedra, we must decompose each polyhedron into corresponding protrusive and depressive zones. This kind of polyhedral decomposition is called **weak morphological decomposition**, and the collection of protrusive and depressive zones is termed the weak

---

[1] A *k-chain* is a sub-collection of k-cells ($k \leq 2$) of a cell complex associated with a polyhedron. [11, p.148]. The sum $C_1 + C_2$ of two k-chains C_1 and C_2 is the set of cells contained in C_1 and C_2, but not contained in both. (In set theory, this operation is known as symmetric difference.) The *boundary of a k-chain* C, ∂C, is the (k-1)-chain whose (k-1)-cells are incident an odd number of times with the cells of C; the boundary operator is additive, $\partial(C_1 + C_2) = \partial(C_1) + \partial(C_2)$. A *k-cycle* is a k-chain with null boundary. A k-cycle is said to be *minimal* if it bounds just one (k+1)-cell. (The sum operation and the set of k-chains is a group in which the identity is the empty set $\emptyset$ of k-cells.)

morphological covering. A **weak morphological covering** of a polyhedron P is a family $\{Z_i\}$ of subsets of ∂P such that $\bigcup Z_i = \partial P$, where each Z_i is a depressive or a protrusive zone.

In Fig.4, polyhedra with different weak morphological coverings are depicted.

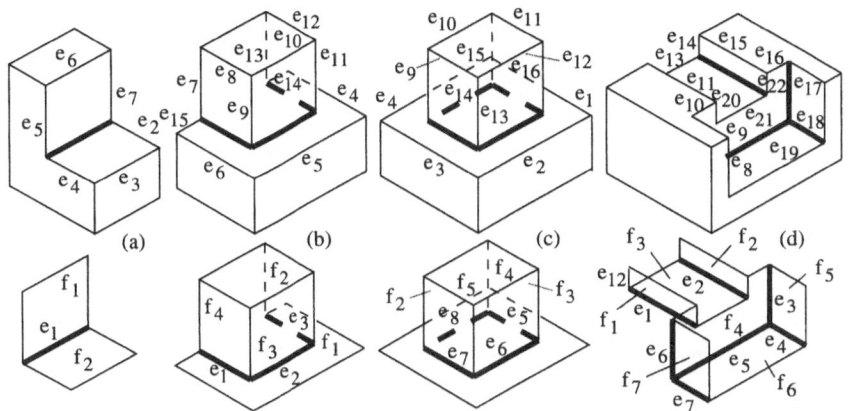

Fig. 4. Protrusive and depressive zones.

Example 1. In Fig.4a, there is just one concave edge e_1 and two minimal protrusive zones corresponding to 2-chains $C(P_{Z_1}) = \{f_1\}$ and $C(P_{Z_2}) = \{f_2\}$ bounded by two 1-cycles $Z(P_{Z_1}) = \partial f_1 = e_1 + e_5 + e_6 + e_7$ and $Z(P_{Z_2}) = \partial f_2 = e_1 + e_2 + e_3 + e_4$, respectively. The union of P_{Z_1} and P_{Z_2} is a depressive zone D_{Z_1} such that $C(D_{Z_1}) = C(P_{Z_1}) + C(P_{Z_2})$ and $Z(D_{Z_1}) = Z(P_{Z_1}) + Z(P_{Z_2}) = e_5 + e_6 + e_7 + e_2 + e_3 + e_4$. This 1-cycle separates D_{Z_1} from a third protrusive zone P_{Z_3} consisting of the remaining faces of the polyhedron.

Example 2. The second polyhedron shown in Fig.4b contains three concave edges, e_1, e_2, e_3, and, consequently, six minimal protrusive zones. But, taking into account that the three concave edges constitute just one 1-chain (or edge path), we can say that the polyhedron (4.b) has two protrusive zones, P_{Z_1} and P_{Z_2}, whose 2-chains are $C(P_{Z_1}) = \{f_1\}$ and $C(P_{Z_2}) = \{f_2, f_3, f_4\}$, respectively; their boundaries are the following 1-cycles $Z(P_{Z_1}) = \partial(f_1) = e_1 + e_2 + e_3 + e_{14} + e_4 + e_5 + e_6 + e_{15}$ and $Z(P_{Z_2}) = \partial(f_4 + f_3 + f_2) = (e_1 + e_7 + e_8 + e_9) + (e_2 + e_9 + e_{10} + e_{11}) + (e_3 + e_{11} + e_{12} + e_{13}) = e_1 + e_7 + e_8 + e_{10} + e_{12} + e_{13} + e_3 + e_2$), respectively. These two protrusive zones give rise to just one depressive zone D_{Z_1} such that $C(D_{Z_1}) = C(P_{Z_1}) + C(P_{Z_2})$ and $Z(D_{Z_1}) = Z(P_{Z_1}) + Z(P_{Z_2}) = e_7 + e_8 + e_{10} + e_{12} + e_{13} + e_{14} + e_4 + e_5 + e_6 + e_{15}$. This 1-cycle separates D_{Z_1} from a third protrusive zone P_{Z_3} containing the remaining faces of the polyhedron.

Example 3. The polyhedron depicted in (4.c) has eight minimal protrusive zones (four concave edges) distributed into two maximal protrusive zones, P_{Z_1} and P_{Z_2}, each bounded by two 1-cycles, that is, $\partial P_{Z_1} = Z_1(P_{Z_1}) + Z_2(P_{Z_1}) = (e_9 + e_{10} + e_{11} + e_{12}) + (e_5 + e_6 + e_7 + e_8)$ and $\partial P_{Z_2} = Z_1(P_{Z_2}) + Z_2(P_{Z_2}) = (e_5 + e_6 + e_7 + e_8) + (e_1 + e_2 + e_3 + e_4)$, with $Z_2(P_{Z_1}) = Z_1(P_{Z_2})$. The union of P_{Z_1} and P_{Z_2} is a depressive zone, D_{Z_1}; hence, $\partial D_{Z_1} = Z_1(P_{Z_1}) + Z_2(P_{Z_2})$. ∂D_{Z_1} separates D_{Z_1} from two aditional protrusive zones,

P_{Z_3} and P_{Z_4}; P_{Z_3} includes only the top face of the polyhedron, while P_{Z_4} consists of the four side faces and the bottom face of the polyhedron.

Example 4. The boundary of the polyhedron (4.d) contains fourteen minimal protrusive zones (seven concave edges) which form three 1-chains of concave edges, namely: $\{e_1\}$, $\{e_2\}$, and $\{e_3, e_4, e_5, e_6, e_7\}$. Yet, in general, the number of 1-chains of concave edges is equal to the number of depressive zones, the polyhedron (4.d) has actually two (and not three) depressive zones because e_1 and e_2 belong to the same 1-cycle $Z(P_{Z_1})$ bounding the protrusive zone P_{Z_1} whose 2-chain is $C(P_{Z_1}) = \{f_3\}$. The first depressive zone D_{Z_1} is the union of three protrusive zones, P_{Z_1}, P_{Z_2}, and P_{Z_3}, such that $Z(D_{Z_1}) = Z(P_{Z_1}) + Z(P_{Z_2}) + Z(P_{Z_3}) = (e_1 + e_{13} + e_2 + e_{21}) + (e_{20} + e_{11} + e_{12} + e_1) + (e_2 + e_{14} + e_{15} + e_{22}) = e_{20} + e_{11} + e_{12} + e_{13} + e_{14} + e_{15} + e_{22} + e_{21}$. The second depressive zone D_{Z_2} is the union of four protrusive zones, P_{Z_4}, P_{Z_5}, P_{Z_6}, and P_{Z_7}; hence, $Z(D_{Z_2}) = Z(P_{Z_4}) + Z(P_{Z_5}) + Z(P_{Z_6}) + Z(P_{Z_7}) = (e_4 + e_{19} + e_7 + e_5) + (e_3 + e_{17} + e_{18} + e_4) + (e_6 + e_7 + e_8 + e_9) + (e_6 + e_{10} + e_{20} + e_{21} + e_{22} + e_{16} + e_3 + e_5) = e_{19} + e_8 + e_9 + e_{10} + e_{20} + e_{21} + e_{22} + e_{16} + e_{17} + e_{18}$. Notice that edges e_{20}, e_{21}, and e_{22} belong to both depressive zones, D_{Z_1} and D_{Z_2}; as a consequence, we may consider that D_{Z_1} and D_{Z_2} are sub-zones of a third depressive zone $D_{Z_3} = D_{Z_1} \cup D_{Z_2}$, with $Z(D_{Z_3}) = Z(D_{Z_1}) + Z(D_{Z_2}) = e_{19} + e_8 + e_9 + e_{10} + e_{11} + e_{12} + e_{13} + e_{14} + e_{15} + e_{16} + e_{17} + e_{18}$. Besides, $Z(D_{Z_3})$ separates the depressive zone D_{Z_3} from the eighth protrusive zone P_{Z_8} that consists of the remaining faces of the polyhedron.

The definitions of protrusive and depressive zones suggest an algorithm to determine the weak morphological covering of a polyhedron P: (i) identify the concave edges of P; (ii) determine the protrusive zones separated by concave edge paths; (iii) determine the depressive zones from the union of protrusive zones; (iv) determine the remaining protrusive zones outside of the depressive zones.

Therefore, in principle, we are able to establish a zonal or weak morphological equivalence between polyhedra starting by relating their concave edge paths. This zonal or weak morphological equivalence preserves zonal vexity angles. Two polyhedra are said to be weakly morphomorphic iff one polyhedron can be "elastically" transformed into the other preserving zonal vexity angles. A weak morphomorphism is an homeomorphism with the restriction that zonal vexity angles are preserved. This explains why the a rectangular protrusive zone in (3.a) can be morphologically transformed into a cylindrical protrusive zone in (3.c).

To conclude, we can define a shape $(S, \mathcal{M})$, called a **morphology**, where S is a set and $\mathcal{M}$ is a morphological equivalence relation defined on the set of all subsets of S.

3   Shape Taxonomy

In the preceding section, we established a top-down shape hierarchy with reference to shape functions (homeomorphisms, morphomorphisms, and isometries) as follows: topologies, morphologies, and geometries. This section is primarily concerned with point set-based shape classification.

3.1   Geometric Shapes

One normally thinks of a geometry as being a set S associated with a structure. This structure is a set of subsets of S; in $\mathbb{R}^3$ these subsets are called *points*, *lines*, and *surfaces*.

From a geometric point of view, a solid polyhedron P is a sub-geometry of $\mathbb{R}^3$, *i.e.*, a subset of $\mathbb{R}^3$, that is a union of its finite set of (open) n-dimensional sub-geometries, $n \leq 3$, namely: vertices ($n = 0$), edges ($n = 1$), faces ($n = 2$), and solids ($n = 3$). These n-dimensional geometries that constitute a polyhedron are its *local geometric shapes*. Thus, a polyhedron as a whole is a *global geometric shape* that is a "sum" of its *local geometric shapes*. The algebraic properties of the real numbers and of $\mathbb{R}^3$ are used to define these sub-geometries and to draw conclusions about a specific geometry. This correspondence between a set of points (a geometry) and $\mathbb{R}^3$ allows us to transform geometric problems into algebraic ones. Thus, it is possible to *finitely* describe and compute properties of geometries with an uncountable number of points. For example, a point in $\mathbb{R}^3$ may be defined in absolute coordinates through a 3-tuple (x, y, z), a straight line segment or surface patch may be defined by the coefficients of a parametric equation, a plane may be described by its coordinates (A, B, C, D) from the analytic equation $Ax + By + Cz + D = 0$, a solid sphere may be represented implicitly by the coefficients of a semi-algebraic function, and so on.

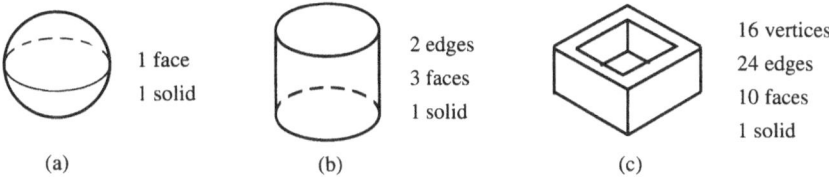

Fig. 5. Geometric shapes in polyhedra.

Every geometry or geometric shape has a well-defined dimension. If a sub-geometry is bounded, or, intuitively, has finite extent, it is a local geometric shape. For example, in Fig.5, the sphere with radius r consists of two local shapes: one spherical face $x^2 + y^2 + z^2 = r^2$ and a spherical solid $x^2 + y^2 + z^2 < r^2$ (5.a); the cylinder consists of six local shapes: two circular edges, three faces (two planar and one cylindrical), and one cylindrical solid interior (5.b); and, the solid parallelepiped with a through hole consists of a number of local shapes: sixteen vertices, twenty-four edges, ten faces, and one solid (5.c).

3.2 Topological Shapes

In the previous subsection, a geometric viewpoint was used in order to classify point sets in $\mathbb{R}^3$. In this subsection, we are interested in classifying the same point sets from a topological point of view. In particular, we are interested in which topological shapes are definable in a polyhedron.

A **polyhedron** is a Hausdorff space X in which each point has a neighbourhood homeomorphic to either $\mathbb{R}^n$ or $\mathbb{R}^n_+$ (a closed semi-hyperspace of points $\mathbf{x} = (x_1, x_2, \ldots, x_n)$, such that $x_n \geq 0$). The *interior* of X consists of the points of X which have a neighbourhood homeomorphic to $\mathbb{R}^n$. On the other hand, the *boundary* of X consists of those points $\mathbf{x} \in X$ for which there is a neighbourhood N, and a homeomorphism $f : \mathbb{R}^n_+ \to N$ such that $f(0) = \mathbf{x}$. This definition is valid for solid polyhedra, surfaces without boundary (*e.g.*, surfaces bounding a solid polyhedron), and surfaces with boundary, etc.

It is remarkable that the theory of topological polyhedra can be used to *classify* polyhedra and thus determine exactly how many different polyhedra types exist. In fact, a surface classification theorem establishes that every compact, connected surface without boundary is topologically equivalent or homeomorphic to a sphere or a connected sum of tori [11, p.122]. Recall that the connected sum (#) of two surfaces is defined as follows: remove an arbitrary disc from each surface and connect the two resulting boundary circles by an arbitrary general cylinder. This suggests a method of construction for every surface without boundary: take the 2-sphere S^2 (Fig.6a), remove two disjoint discs and then add a cylinder by sewing its two boundary circles to the boundaries of the holes in the sphere. The result is a 2-sphere S^2 with a handle, which is nothing more than (is homeomorphic to) the 1-fold torus T^2, Fig.6b. By repetition, we are able to construct a sphere with two, three, or any finite number k of handles, or, equivalently, a 2-fold torus $T^2 \# T^2$ (Fig.6c), a 3-fold torus $T^2 \# T^2 \# T^2$ (Fig.6d), or in general a k-fold torus $T^2 \# T^2 \# \ldots \# T^2$, respectively.

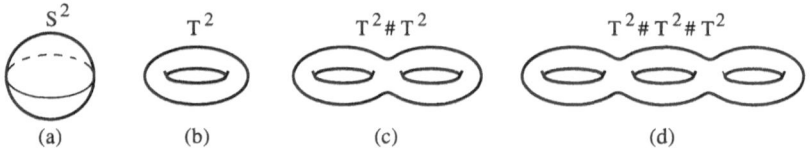

Fig. 6. Classifying topological surfaces without boundary.

We can generalise this surface construction process to solid polyhedra by treating a surface as a "degenerate" solid polyhedron, whose interior has been reduced to $\emptyset$ (*i.e.*, its interior is void). Before establishing a classification for solid polyhedra, we consider first the Euler characteristic $\chi(P) = \beta_0 - \beta_1 + \beta_2$ valid for every polyhedron P in $\mathbb{R}^3$.

The *zero-th Betti number* β_0 is the number of components C of a polyhedron. A component of a polyhedron is a maximal connected subset, that is, a connected subset that is properly contained in no other connected subset. If a polyhedron is connected, then it has only one component [12]. For example, $\beta_0 = 4$ for a polyhedron consisting of four 2-spheres $S^2 = \{\mathbf{p} = (x, y, z) : |\mathbf{p}| = 1\}$, and for a polyhedron composed of four 3-spheres $S^3 = \{\mathbf{p} = (x, y, z) : |\mathbf{p}| \leq 1\}$. Thus, β_0 is the same for every polyhedron with the same number of components.

The *first Betti number* β_1 is sometimes called the connectivity number of a polyhedron. A surface requiring k cuts or surgeries in order to render it homeomorphic to S^2 is said to be *n-tuply connected*, with $n = 2k$. Each cut traverses a handle along a non-self-intersecting closed curve such that a surface with two distinct disc holes is created. Thus, k cuts are required to transform a surface with k handles (a k-fold torus) into a surface homeomorphic to S^2. This creates a surface with $2k$ disc holes, which must be filled with $n = 2k$ discs. The first Betti number of a surface is exactly two times the number of handles (or the number C_h of holes through a surface), $\beta_1 = 2.C_h$; hence, $\beta_1(S^2) = 0$, $\beta_1(T^2) = 2$, $\beta_1(T^2 \# T^2) = 4$, $\beta_1(T^2 \# T^2 \# T^2) = 6$, and so forth. The number of handles of a surface is known as its *genus*.

Fig.7 illustrates the type of modification we have in mind for the 2-fold torus $T^2 \# T^2$. We begin with a non-self-intersecting closed curve which does not separate the surface into two pieces, but does cut completely across a handle. After surgery

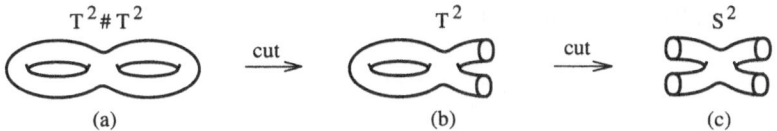

Fig. 7. Transforming a 2-fold torus $T^2 \# T^2$ into a 2-sphere S^2.

along this curve, we fill each of the two resulting holes with a disc. The result is a surface homeomorphic to a 1-fold torus T^2 (7.b). Further surgery provides the 2-sphere S^2 (7.c). For a mathematical proof of this procedure see corollary 7.12 in [1, p.163]. Obviously, for solid polyhedra, the number of cuts required to obtain a solid homeomorphic to a 3-sphere S^3 is exactly the number of holes through components and there is no need to repair any disc holes; hence, $\beta_1(S^3) = 0$, $\beta_1(T^3) = 1$, $\beta_1(T^3 \# T^3) = 2$, $\beta_1(T^3 \# T^3 \# T^3) = 3$, and so forth. That is, the first Betti number of a solid polyhedron is the number C_h of holes through it, $\beta_1 = C_h$.

The *second Betti number* is the number C_c of cavities in components, also called voids. Clearly, for any surface without boundary P, the number C_c of cavities is equal to the number C of components; hence $\beta_2 = \beta_0$, and the corresponding Euler characteristic is $\chi(P) = 2C - 2C_h$. But, for a solid polyhedron P, the number of cavities in components is not necessarily identical to the number of components; hence, $\chi(P) = C - C_h + C_c$.

The above discussion about the polyhedron classification leads to the conclusion that two polyhedra are homeomorphic (or topologically equivalent) iff their corresponding Betti numbers are the same. Equivalently, two polyhedra are of the same *topological shape* iff they have the same number for components, through holes, and interior cavities. For example, in spite of the local shape differences of the polyhedra depicted in Fig.8 —their complexes are not isomorphic—, they have the same global shape in some as yet unspecified sense. In fact, they have the same topological shape: one component ($C = 1$), one through hole ($C_h = 1$), and zero interior cavities ($C_c = 0$).

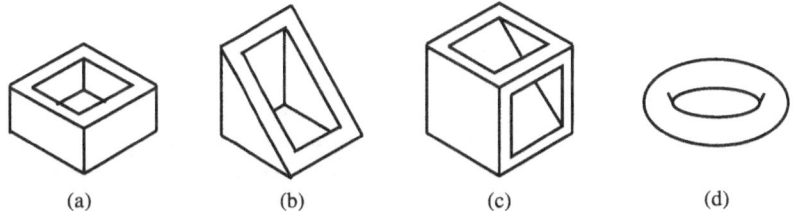

Fig. 8. Polyhedra with identical β_i are topologically equivalent.

In short, components, through holes, and interior cavities are the *global topological shapes* of a polyhedron, whereas the n-cells of its complex are its *local topological shapes*.

3.3   Morphological Shapes

As seen in section 2.3, the shape resemblance in which we are interested falls between the (too-rigid) traditional geometric shapes and the (too-relaxed) topological shapes.

This shape resemblance is zonal vexity-preserving such that two polyhedra are weakly morphologically equivalent iff they have isomorphic weak morphological coverings.

Taking into account that every protrusive zone Z_{P_i} (respectively, every depressive zone Z_{D_j}) of a polyhedron P is a *retract*[2] of a protrusion P_i (respectively, a depression D_j), a weak morphological covering $\{Z_{P_i}, Z_{D_j}\}$ may be considered as a retract of a weak morphological structure $\{P_i, D_j\}$ such that $\partial P = \bigcup \{Z_{P_i}, Z_{D_j}\}$ and $P \subseteq \bigcup \{P_i, D_j\}$, respectively. Note that the family of protrusive zones is a sub-covering of ∂P, wheras the family of protrusions is a sub-covering of P, because $\partial P = \bigcup Z_{P_i}$ and $P = \bigcup P_i$, respectively.

Fig.9 depicts the coarsest and finest weak morphological structures for the polyhedra shown in Fig.4a,b,c. Their coarsest weak morphological structures are all $P_1 \cup D_1$, whereas their finest weak morphological structures are $(P_2 \cup P_3) \cup D_1$, $(P_2 \cup P_3) \cup (D_2 \cup D_3 \cup D_4)$, and $(P_2 \cup P_3) \cup (D_2 \cup D_3 \cup D_4 \cup D_5)$, respectively. This leads to the conclusion that there is a *resolution* inherent in each morphological structure. Besides, their finest *protrusive* weak morphological structures $(P_2 \cup P_3) \cup D_1$, are also the same. This explains why these polyhedra appear us to be resemblanced in some sense.

It is worthy of note that although a protrusive morphological structure may be useful for design, it is not necessarily suited to process planning. For example, process planning for a part to be machined must take into account the material to be removed. This is a collection of delta volumes which correspond to depressions in the morphological structure of a polyhedron. Obviously, a designer may construct a part in different ways, that is, using different morphological structures. However, since a designer's morphological structure is always a coarse weak morphological structure of a polyhedron, and it is always possible to find (as outlined in section 2.3) the corresponding finest weak morphological structure, we are able to determine the delta volumes (depressions) needed for process planning. For example, from the designer's point of view, the polyhedron in Fig.9(b) is easier to model using two protrusions than a protrusion from which three depressions (delta volumes) are to be removed.

Yet protrusions and depressions are not required to be convex sets, every protrusion is either (i) decomposable into irreducible protrusions or (ii) contained in a convex protrusion. A protrusion is called *reducible* when it can be expressed as the union of proper protrusions, and is called *irreducible* otherwise. For example, in Fig.10, the protrusion P_1 of (10.a) is reducible to two protrusions, P_2 and P_3, that is, $P_1 = P_2 \cup P_3$, but neither the protrusion of (10.b) nor the protrusion of (10.c) are reducible, because no edge is concave. Nevertheless, because the protrusion P_1 of (10.c) is non-convex, it can be viewed as a proper protrusion of another protrusion P_2 such that $P_2 = P_1 \cup D_1$. It is clear that the same arguments are valid if we consider depressions instead of protrusions.

---

[2] Let X be a polyhedron. Then, $R(X) = \{x : x \subseteq \partial X\}$ is called a **retract** of X. Let X and Y be two polyhedra such that $X \cap Y \neq \emptyset$. The set of points $\partial Y \subset X$ is called the **retract of Y in X**, and is denoted by $R(Y \sqcup X)$, whereas the **retract of X in Y** $R(X \sqcup Y)$ is the set of points $\partial X \subset Y$. A **depressive zone** $Z_D = R(Y \sqcup X)$ of a polyhedron $P = X \setminus Y$ is a retract of a polyhedron (or depression) Y in a polyhedron (or protrusion) X. On the other hand, the set of points $\partial Y \not\subset X$ is called the **retract of Y outside of X**, and is denoted by $R(Y \sqcap X)$. A **protrusive zone** $Z_P = R(Y \sqcap X)$ of a polyhedron $P = X \cup Y$ is the retract of a polyhedron Y outside another polyhedron X. X and Y are to said to be **protrusions** of P, while X is a protrusion in relation to Y, and vice-versa.

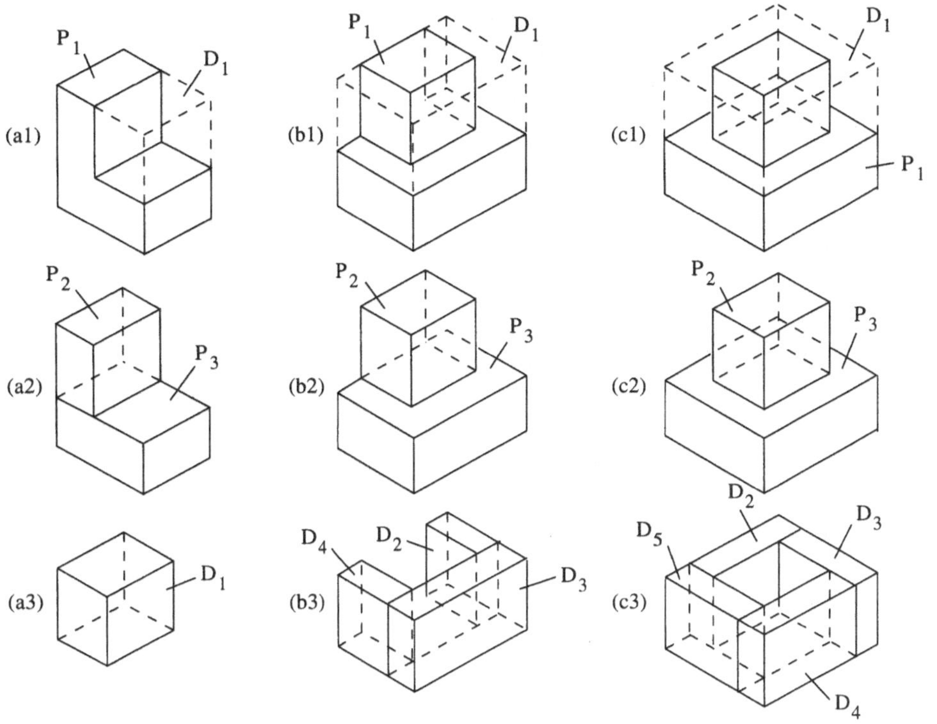

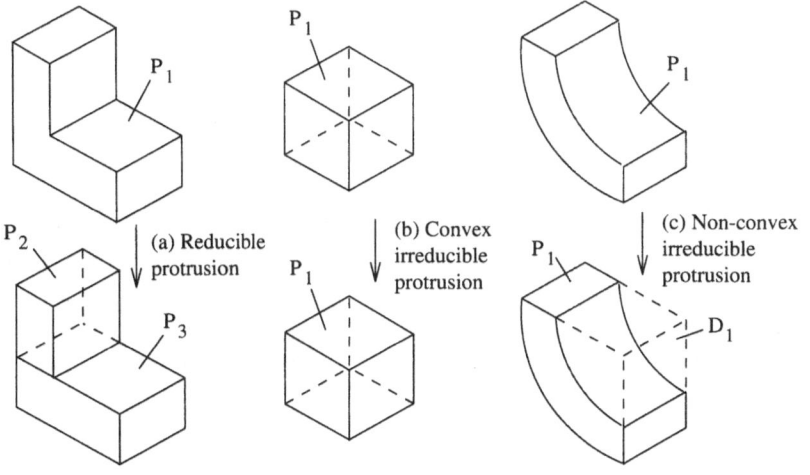

Fig. 9. (a1,b1,c1) Coarsest weak morphological structures; (a2,b2,c2) finest *protrusive* and (a3,b3,c3) *depressive* weak morphological structures.

Fig. 10. Convexity-based classes of protrusions.

As previously discussed, a polyhedron may have several different morphological structures, but its weak morphological covering is proved to be unique. Besides, a morphological structure corresponds to only one polyhedron, because it is the union of its protrusions. In a way, this is similar to the cellular decomposition of a polyhedron. Analogously, we can establish connectivity relationships between morphological shapes as they are established for the cells of a polyhedron. Thus, two morphological shapes are connected iff they are path-connected [1, p.61]. Recall that a *path* in a space X is a continuous map $f : \mathbf{I} \to X$; if $f(0) = a$ and $f(1) = b$, one says that f is a path from a to b; a space X is *path-connected* if, for every $a, b \in X$, there is a path in X from a to b. Thus, two morphological shapes X and Y are *path-connected* if, for every $x \in X$ and $y \in Y$, there exists a composite path $h = f \circ g$ from x to y, where f and g are paths in X and Y, respectively, such that $f(0) = x$, $f(1) = g(0)$, and $g(1) = y$. A protrusion (respectively, depression) connected to n protrusions (respectively, depressions) is said to have a **morphological genus** equals to $+n$ (respectively, $-n$), and it is denoted by P^{+n} (respectively, D^{-n}), with $n \in \mathbb{N}$; P^{+n} denotes a **n-protrusion**, while D^{-n} denotes a **n-depression**. Some examples of protrusions and depressions are shown in Fig.11.

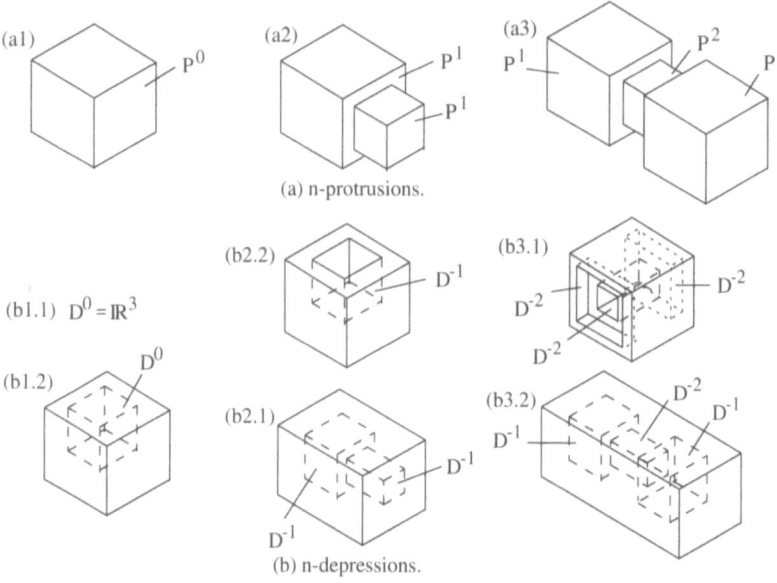

Fig. 11. Connectivity-based classes of protrusions and depressions.

Fig. 12 shows a weak morphological structure of a polyhedron (12.a), and its morphological connectivity graph (12.b). Note that, for example, the depression that results from the connectivity between P_1^1 and P_1^2 is not included in the data structure, though it can be easily determined using the algorithm that computes a depression from its depressive zone.

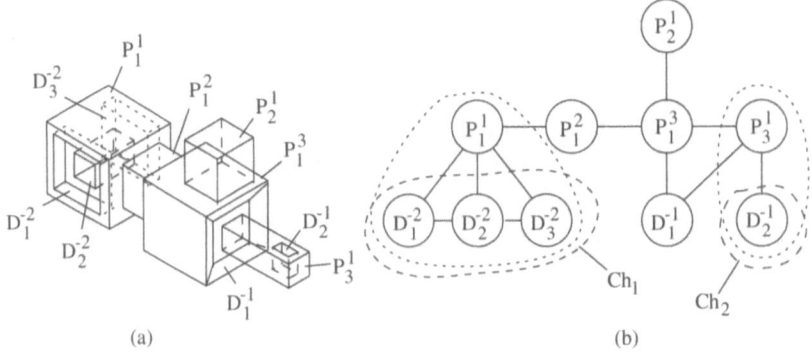

Fig. 12. (a) A weak morphological structure and (b) its connectivity graph.

4   CW Topological Structure-Based Shape Kernel Architecture

The shape kernel architecture (SKA) shown in Fig.13 mirrors the shape taxonomy proposed in the previous section, as well the axioms relating the different shapes. Although the axioms of our shape theory have not been dealt with in formal terms, some are intuitively apparent. For example, the axiom relating topological shapes to morphological shapes: a topological shape contains at least one morphological shape. Informally, one component has at least one protrusion, one through hole or interior cavity includes at least one depression.

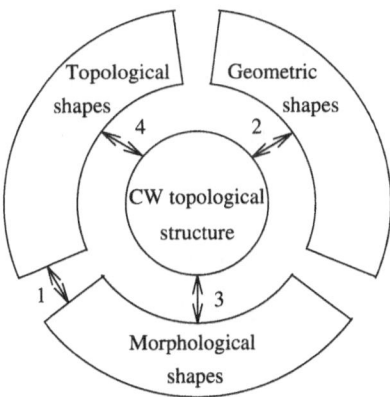

Fig. 13. Shape kernel architecture.

In Fig.13, these containment relations between topological and morphological shapes are denoted by the double-sided arrow 1. Note that, there is no direct relationship between topological/morphological shapes and geometric shapes. Such relationships are indirectly set up through local topological shapes or cells of a CW cellular data structure (see [9] for details). Thus, this architecture emphasizes the separation between cellular topology and the shape of a polyhedron, as a BRep does in relation to topology and geometry of a polyhedron.

But, unlike a (non-manifold) BRep, a CW complex is a collection of subcomplexes, each of which is a collection of n-dimensional cells. It is this cellular hierarchical structure (complex $\leftrightarrow$ subcomplex $\leftrightarrow$ cell) that makes possible not only to establish a one-to-one correspondence (double-sided arrow 2) between a local geometric shape and a local topological shape (or cell), but also between a morphological or topological shape and a subcomplex of a polyhedron (double-sided arrows 3 and 4, respectively). For example, the through holes C_{h1} and C_{h2} in Fig. 12 should point to their subcomplexes, $S(C_{h1})$ and $S(C_{h2})$, respectively; $S(C_{h1})$ contains three subcomplexes, each corresponding to a distinct depression of C_{h1}, while $S(C_{h2})$ includes only one subcomplex which is the cellular counterpart of the depression of C_{h2}.

5   Alternative Shape Kernel Architectures

Current shape modelers can be divided into two major categories: geometric (solid) modelers and form feature-based modelers. The former are basically geometry-processing machines, whereas the later are functional shape-processing machines. Geometric solid models represent the geometry of a polyhedron, that is, the *set of points* of that polyhedron. In this sense, geometric solid models are provided with *geometric completeness*, because every point of a polyhedron is represented directly or *indirectly* (i.e., it is *computable*). In fact, every polyhedron P is viewed as a set of points (each point **inside** P). As a consequence, geometric solid models are not able to represent significant regions **outside** a polyhedron; for example, through holes and depressions are not represented in general. In informal terms, this means that they suffer *shape incompleteness*.

5.1   Boundary Representation Modelers

The shape incompleteness of BRep's (Boundary Representations) is in large part because its mathematical model does not provide generalised aggregation entities for cells (subcomplexes). Exceptions are shells for components and interior cavities, and loops for face boundaries. In fact, the design of a BRep data structure is based on the topological concept of (open) cell: 0-dimensional cell (vertex), 1-dimensional cell (edge), and 2-dimensional cell (face). Therefore, a polyhedron (global geometric shape) is treated as a collection of k-cells ($k \leq 2$), each provided with suitable geometry (local geometric shape), *e.g.*, a point, a line segment, or a surface patch. It is not viewed as a collection of more complex shapes as protrusions and depressions. A possible solution for this problem would be to introduce 2-chains (as well as their boundaries which are 1-cycles) into the BRep data structure somewhere between shells and faces, each 2-chain being a cellular representation of either a protrusive or depressive zone, and extend the mathematical model underlying the BRep in order to include surfaces with boundary. In this way, it would be possible to represent retracts of topological and morphological shapes in the same data structure.

5.2   Constructive Solid Geometry Modelers

Every CSG polyhedron results from combining a collection of finite primitive polyhedra through the so-called Boolean operators. According to the arguments used in section 3.3, a CSG data structure could be the nucleous of a shape kernel architecture (SKA) in such a way that its primitive polyhedra may function as CSG representations of protrusions and depressions. Unfortunately, the geometry of a CSG polyhedron is not evaluated; its vertices, edges, and faces are not represented explicitly. As a consequence, human-machine interaction as required in design activities is only possible on a CSG modeler if a boundary evaluator is available. This means that a CSG-based SKA has to include a BRep and a boundary evaluator pipeline between both data structures. Obviously, this hybrid CSG-BRep nucleous would be greatly improved if an usual BRep was replaced by an augmented BRep. Furthermore, a (CSG-BRep)-based SKA would need a space decomposer to decompose primitive polyhedra to reflect the protrusions and depressions.

5.3   Non-Manifold Geometric Modelers

Non-manifold geometric (NMG) models are intended to be unified representations for wireframes, surfaces, and solids. These unified representations meet the dimension-independent requirements for design, and have also enlarged the geometric domain. In this category of models we include non-manifold boundary representations (NM-BRep's) [24] [15] [26] and selective geometric complex representations (SGC-Rep's) [21].

In spite of their generality, NM-BRep's keep basically the same aggregation entities (shells and loops) as conventional BReps. This means that the idea of a polyhedron as a collection of cells remains. However, because these representations include 3-cells (also called volumes or regions), it is possible to devise aggregation entities of 3-cells outside their topological structures. For example, in [15], such aggregation entities are called *primitive objects*. Thus, a (NM-BRep)-based SKA is feasible provided that its nucleous includes such primitive objects, each primitive object representing the cellular structure of either a topological or a morphological shape.

SGC representations are also extensions of boundary representations. A selective geometric complex is —like any complex— a collection of n-cells whose underlying point set is closed. However, it differs from a common complex in the ability to represent non-closed objects and control *selectively* the cellular decomposition of the represented point set. The *selectivity* of a SGC representation comes from attaching a set of attributes to each cell (or collection of cells). In particular, the *active* (FALSE, TRUE) attribute specifies whether the point set of a cell should be included in the point set defined by the SGC. The union of all active cells with the same attribute values constitutes a *region*. Note that regions are entities indirectly represented in a SGC data structure via an attribute-based aggregation mechanism. Thus, regions as aggregation entities are external to the SGC data structure, and play a role similar to primitive objects of other NM-BRep's. That is, a SGC data structure may function as a nucleous in a SKA since it is supplemented by region entities in order to establish a one-to-one correspondence between topological/morphological shapes and regions.

5.4   Non-Homogeneous Set-Theoretic Modelers

Non-homogeneous set-theoretic representations (NH-STRep's) are extensions to CSG representations. As far as we know, only two models of this category are known, namely: CNRG (Constructive Non-Regularized Geometry) [22] and CST (Cellular Set-Theoretic) [17] representations.

CNRG trees are representations of aggregates (*i.e.*, unions) of mutually disjoint regions. Each *region* **is a point set in** $\mathbb{R}^n$ and need not be connected, regular, or even dimensionally homogeneously. This means that primitive homogeneously n-dimensional point sets (volumes, surface patches, curve segments or points), boolean combinations of homogeneously n-dimensional point sets, or even point sets that result from applying topological operations (*e.g.*, interior, closure, boundary and regularisation) are regions. A CNRG region is not the same as a SGC region, because the former is a pointset and the later is a collection of point sets (cells of a selective geometric complex). Therefore, a SGC region corresponds to the cell decomposition (structure) of a CNRG region. Each CNRG region not only defines a point set, but also a structure (decomposition) of the point set. However, unless the incidence-based structure of SGC's is also used, CNRG's are merely a set theoretic-based structure.

Similarly, CST's represent the point set of a polyhedron via its set theoretic structure. Regions are here called simply cells. Each **cell** is a primitive entity (point, line segment, surface patch, or solid patch, all of which are open with respect to their embedding geometry), a regularised set combination of two cells, or a conventional set intersection of two cells. CST's provide more operators than CNRG's. These operators, e.g. aggregation and partitioning operators, are suited to many industrial applications.

Thus both CNRG-based SKA and CST-based SKA are suited to shape modeling provided that it is possible to establish a one-to-one correspondence between topological or morphological shape and CNRG regions (CST cells, respectively), and between geometric shapes and primitive homogeneously n-dimensional point sets (primitive cells, respectively).

6   Conclusions

A shape-theoretic framework for polyhedra has been proposed in this paper. Each kind of shape is associated with a *group of transformations*, or one-to-one mappings of the space onto itself, that leave intact certain properties of their subsets. And conversely, the specification of a group of one-to-one mappings of a space onto itself determines a shape (*e.g.*, isometry mappings for Euclidean geometry, morphomorphims for morphology, and homeomorphisms for topology). From these one-to-one mappings of a space onto itself we have established equivalence classes of polyhedra in geometry, morphology and topology, and, consequently, we have achieved a general shape taxonomy. In addition, containment and connectivity amongst various shapes have been established in order to derive a general shape kernel architecture (SKA) for polyhedra.

SKA is a result of an effort to develop a unified shape model for polyhedra that is independent of the functional aspects of shape in form feature-based modeling. Because of its generality, SKA could play the role of a shape kernel of a feature-based modeler as part of a CAD system. However, an efficient feature-based modeler requires a better understanding of the relationships between the functional aspects of engineering and the shape of parts in order to achieve a theoretical model for feature-based modeling. In this respect, form features appear to be the entities capable of supporting such a theory since they are carriers of both engineering and shape information. SKA also seems to be suited to both design and downstream applications (*e.g.*, process planning) because it supports the transformation between morphological structures of different resolutions.

References

1. M. Armstrong. *Basic topology.* McGraw-Hill Ltd., 1979.
2. B. Baumgart. A polyhedron representation for computer vision. In *AFIPS Conf. Proc.*, volume 44, pages 589–596, 1978.
3. L. Blumenthal and K. Menger. *Studies in geometry.* W.F. Freeman and Company, 1970.
4. A. Bowyer, S. Cameron, G. Jared, R. Martin A. Middleditch, M. Sabin, and J. Woodwark. Djinn: a geometric interface for solid modeling. Technical Report (Section 2, Draft 5), The Geometric Modeling Society & Information Geometers Ltd., 1996.
5. Ian Braid, Robin Hillyard, and Ian Stroud. Stepwise construction of polyhedra in geometric modeling. In Ken Brodlie, editor, *Mathematical Methods in Computer Graphics and Design*, pages 123–141. Academic Press, 1980.
6. D. Cohen. *Computability and logic.* Ellis Horwood Ltd., 1987.
7. J. Cordier and T. Porter. *Shape theory: categorical methods of approximation.* Ellis Horwood, Chichester, 1989.
8. G. Edgar. *Measure, topology, and fractal geometry.* Springer-Verlag, 1989.
9. A. Gomes. CW topology-based data structure for polyhedra and its applications. *(to be re-submitted)*, 1996.
10. A. Gomes and J. Teixeira. A mathematical framework for set-theoretic solid models. In A. Middleditch and A. Requicha, editors, *Set-Theoretic Solid Modeling: Techniques and Applications*, pages 19–33. Information Geometers Ltd., 1994.
11. M. Henle. *A combinatorial introduction to topology.* W.F. Freeman and Company, 1979.
12. J. Kelley. *General topology.* Van Nostrand Publ., 1955.
13. K. Kuratowski and A. Mostowski. *Set theory.* North-Holland, 1967.
14. J. Lipson. *Elements of algebra and algebraic computing.* Addison-Wesley Publishing Company, 1981.
15. H. Masuda. Topological operators and boolean operators for complex-based non-manifold geometric models. *Computer Aided Design*, 25(2), 1993.
16. H. Masuda, K. Shimada, K. Numao, and S. Kawabe. A mathematical theory and applications of non-manifold geometric modeling. In F. Krause and H. Jansen, editors, *Advanced Geometric Modeling for Engineering Applications*. International GI-IFIP Symposium 89, North-Holland, 1989.
17. A. Middleditch. Cellular models of mixed dimension. Technical Report BRU/CAE/92:3, Department of Computer Science, Brunel University, West London, 1992.
18. M. Pratt. Solid modeling—survey and current research issues. In D. Rogers and R. Earnshaw, editors, *Computer Graphics Techniques: Theory and Practice*. Springer-Verlag, 1990.
19. A. Requicha. Mathematical models of rigid solid objects. Technical Report TM 28, Production Automation Project,University of Rochester, 1977.
20. A. Requicha. Representations of rigid solids: Theory, methods, and systems. *ACM Computing Surveys*, 12(4), 1980.
21. J. Rossignac and M. O'Connor. SGC: A dimension-independent model for pointsets with internal structures and incomplete boundaries. In M. Wozny, J. Turner, and K. Preiss, editors, *Geometric Modeling for Product Engineering*. IFIP, Elsevier Science Publishers B.V. (North-Holland), 1990.

22. J. Rossignac and A. Requicha. Constructive non-regularized geometry. *Computer Aided Design*, 23(2), 1991.
23. V. Shapiro. Real functions for representation of rigid solids. *Computer Aided Geometric Design*, 11(2), 1994.
24. K. Weiler. *Topological structures for geometric modeling*. PhD thesis, Rensselaer Polytechnic Institute, Troy, New York, 1986.
25. T. Wu. Towards a unified data structure for geometrical representations. In F. Kimura and A. Rolstadas, editors, *Computer Applications in Production and Engineering*. IFIP, Elsevier Science Publishers B.V. (North-Holland), 1989.
26. Y. Yamaguchi and F. Kimura. Nonmanifold topology based on coupling entities. *IEEE Computer Graphics & Applications*, 15(1), 1995.

Virtual Topology Construction and Applications

Alla Sheffer[1], Ted Blacker[2], Jan Clements[2], Michel Bercovier[1]

[1] Institute of Computer Science, The Hebrew University, Jerusalem 91904, Israel.
e-mail: sheffa@cs.huji.ac.il
[2] Fluid Dynamics International, 500 Davis St., Suite 600, Evanston, Illinois 60201,
USA. e-mail: ted@fdi.com

Abstract. A new enhanced B-Rep for solids is introduced, using the concept of virtual topology in which the geometric and topological entity definitions as defined in the standard B-Rep are uncoupled, thus allowing topological entities not corresponding to geometric ones. The need for such enhancement is explained as well as the construction process starting with an object described as a standard B-Rep and resulting in the desired modified topology, suitable for the user needs.

1   Introduction

B-Rep has been accepted in recent years as the most common representation of solid objects with complex geometrical surfaces. It provides a complete and unambiguous definition of the solid through a concise representation of the bounding topology and a robust definition of the underlying geometry. It is the most common representation in such areas as CAD/CAM, analysis, modeling, etc.

When using the solid for various applications, the B-Rep must be edited to suit the application. For instance, when meshing a solid for use in analysis, small features and other minor details such as short edges, fillets etc. may need to be eliminated or replaced. Complex solids often need to be dissected or decomposed into simple, "meshable" pieces. Typically these changes can be applied to the topology only, since many applications, like meshing, rely more on the object topology than on the geometry.

Objects built with free-form B-Rep surfaces often have "small" inconsistencies, such as a common edge defined twice, small gaps and holes resulting from rounding errors in the manipulation of the surfaces or from erros in translation between different formats. Such inconsistencies are an obstacle to any automated analysis of the object, and have to be fixed.

In the current boundary representation scheme, the topology of the object corresponds completely to the geometry, and therefore any adjustments of the topology require changing the geometry as well. Changing the geometry directly is typically an expensive, complex and often undesirable procedure. Therefore providing methods for topology adjustment independent of geometry changes is an important problem in solid modeling.

The existing methods for adjustments to a B-Rep solid are based on two general concepts of topology analysis:

- Feature recognition - recognizing insignificant features, removing them from the B-Rep and updating the geometry accordingly. [DeFloriani89, Shah91]
- Medial object construction - building the medial object of the body, gathering from it the simplified topology for the object and again updating the geometry accordingly. [Armstrong95]

Most of the methods introduced require changes of the underlying geometry [Jones95, Butlin96], which is non-trivial and often tedious. In [Armstrong95] a distinction between logical and physical faces and edges is suggested, allowing merge operations, but no solid representation allowing such operations is presented.

We introduce here a new concept of virtual topology which allows uncoupling of the B-Rep geometry and topology, thus enabling changes of the topology without changing the underlying geometry. While adjustment of the topology is commonly a necessary step in object analysis, the adjustment of the geometry is many times unnecessary.

2   Topology Editing

When using a boundary representation of objects all the access to the object is typically done through the topological entities of the object and not directly through the geometry. For example when faceting the object envelope for display each topological face of the object is faceted separately. Many applications, like meshing, rely more on the object topology than on the geometry. For instance meshing an object that has the topology of a cube (six faces with four edges each) is almost as simple as meshing a real cube without any regard to the geometric complexity of the surfaces.

In the standard boundary representation, the topological entities correspond one-to-one with geometric entities. Thus each surface of the object is represented by a face in the B-Rep. In this approach, any adjustments of the topology require changing the geometry as well, in order to preserve the unique correspondence of topology and geometry.

Editing the topology is an essential pre-processing stage for many applications. For instance, when meshing a solid for use in analysis, the analyst may want to take out all kinds of features that complicate the meshing but add nothing to the solution, this includes eliminating or combining into one mass of small pieces such as short edges, fillets, small holes, etc. Such simplification of the geometry simplifies meshing and results in more regular meshes, thus enabling considerably faster finite-element analysis. From St. Venant's principle, a geometric detail which is small compared to the amount of material surrounding it will only cause local stress perturbation, therefore such modifications have very small effect on the analysis result.

For meshing needs,it is sometimes necessary to decompose (or compose) the object into pieces that can be meshed with customizable and predictable results, by allowing application of mapping (boxlike) algorithms or sweeping (2.5D) algorithms. Allowing the user to do those adjustments dynamically allows the modeler great flexibility in producing an adequate mesh.

Affecting these modifications by changing the geometry directly is typically an expensive, complex and often undesirable procedure.

For example, one of the most common changes required is providing an approximation of the object which disregards insignificant features (small holes, bumps, etc...).

Actually removing such features from the geometry requires modification of the surface descriptions, which is not trivial. Such modification is also irreversible, and if a more accurate approximation of the object is needed, the whole approximation process would have to be repeated with the new tolerances. Using virtual topology, the actual features are not removed, thus a finer mesh will capture these features as desired.

Using virtual topology allows changing the object topology without modifying the geometry, making the process much more efficient and simple. Since the actual geometry is not affected, the topology can change dynamically (like changing the position of a "splitting" vertex) without the need for topology reconstruction.

Another advantage is that since the actual geometric descriptions are not modified, no accuracy is lost. If a more exact analysis of the object is required, all the user needs is a denser sampling. Since the actual geometry of the object is unchanged during the topology editing process, a restoration of the original object topology is possible at any stage as opposed to geometric modifications which usually are irreversible.

3   Mesh Generation

A finite-element modeling system is designed to accept a general problem definition containing a geometric model of the component and its required physical attributes and to return results of prescribed accuracy. Adaptive, automatic mesh generation is an essential preprocessing step for generating analysis results of prescribed accuracy for the given computational domain. Most mesh generation schemes produce either triangular or quadrilateral elements in 2D and tetrahedral or hexahedral elements in 3D. Due to various analysis and design benefits quadrilateral/hexahedral meshes are prefered.

The quality of the mesh can play an essential factor in both the accuracy of the FE analysis result and the speed of the solution. Some of the factors of mesh quality are:

- Number of elements - less elements result in faster solution.
- Element size and structure - equally sized and more square-like elements produce better and faster analysis results.
- Mesh regularity - more ordered, grid-like mesh simplifies the matrices computed for the analysis and since simplifies the analysis process.

Several techniques had been suggested for automated quadrilateral/hexahedral meshing as reviewed in [Field95], most of them however impose some restrictions on the meshing domain. One of the most commonly used technics for quadrilateral meshing of a general 2D (surface) domain is paving [Blacker91] and its 3D extension plastering [Blacker93]. This technique necklaces rows of well formed, boundary sensitive elements around each region boundary (both external and internal boundary loops), until the necklaces, or fronts collide. Elements are inserted into the necklaces, or reduced from them when dictated by angle criteria to maintain element aspect ratios. When collisions or intersections are detected, angle criteria dictate where connections will be generated between the opposing fronts. This technic is used in most of the examples shown in this paper.

4   Virtual Topology Structure

Using the new concept, the B-Rep of an object is constructed of topological entities of two types:

- *Real*, that correspond to actual topological entities with their attached geometric entities (surfaces, curves, vertices and volumes);
- *Virtual*, that have no geometric definition of their own, but that are based on other entities.

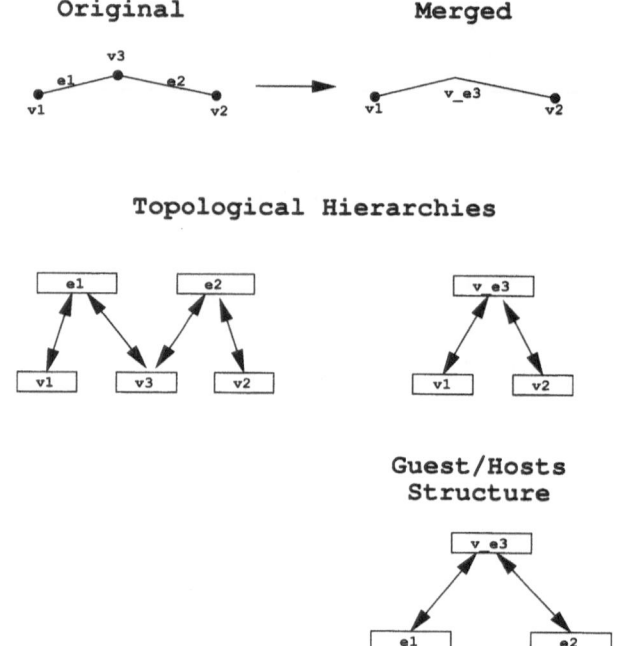

Fig. 1. An enhanced B-rep for a merged virtual edge with the resulting topological and the virtual hierarchies.

Using virtual entities in the boundary representation of an object allows us to suitably modify an object's topology, while maintaining the geometry. Virtual geometry also allows us to handle incomplete and erroneous geometric inputs (which are very common when working with "imported" geometry) without the tedious task of *fixing* the geometry like in [Butlin96].

Four types of virtual entities are introduced:

- entities that correspond to a subset of another entity (e.g. a virtual edge lying on part of another edge);
- entities that contain a number of other entities (e.g. a virtual face including a number of underlying faces that have common boundaries);
- virtual entities that lie on an entity of higher dimension, and have minimal geometric properties (e.g. a virtual edge lying on a face, with the edge only constrained to be on the face).
- entities that replace in the topological hierarchy a set of duplicate (erroneous) representations which should have been one entity, including the "stitching" of

upper topology (e.g. a single virtual vertex replacing a set of different geometric representations for a same vertex).

The entities on which the virtual entity relies are called *hosts* of the entity. A virtual entity provides all the geometric properties of a real entity, by accessing the hosts' data. An entity that relies on another entity is called it's *guest*.

The enhanced boundary representation consists of: a set of topological entities (vertices, edges, etc...); a topological description of connectivity and orientation of the entites; and for each of the topological entities it has a corresponding geometric entity which contains for *real* entyty, it's actual mathematic (geometric) definition, and for *virtual* entity it contains references to it's *host* entities and the virtual type.

This way we have two hierarchical structures, the topology conectivity structure and along with it a hosts/guests structure, which holds in it's leafs *real* entities, and through which the virtual entities derive their geometric properties.

A simple example of the enhanced B-rep structure with a simple merged edge based on two straight *host* edges is shown in Fig. 1.

The *host* topological entities in most cases are no longer part of the B-Rep of the object, but only part of the virtual hierarchy (as in Fig. 1).

This separation of topology and geometry allows common simplified interface for all types of real and virtual geometry entities.

5   Construction Process

The process suggested for adjusting the object topology starts with a standard B-Rep of an object and then applies a set of operations on the object until a topology that suits the needs of the aplication is built. The operations applied include:

- Merger of adjacent entities of similar dimension, in order to unite small features with their surrounding entities and to combine similar geometry.
- Splitting of entities into parts more suitable for the analysis.
- Construction of individual virtual entities, used mainly as "topological supports" for future operations.
- Connection of multiple representations of a single entity into one.

5.1   Merge

Merging geometric entities into a single topological entity is the main tool for combination of similar geometry and elimination of insignificant details in the topological representation (Fig. 3 Fig. 2). It eliminates the requirement during meshing to place nodes at the shared boundary between the merged entities, and thus removes significant constraints from the meshing algorithm. It is especially important when one or both of the merged entities are relatively small with respect to the desired element size.

For example in Fig. 2 the paving algorithm will consider the circle boundary edge as a mandatory starting front for the paving advancing-front rows, thus creating highly unordered mesh on the surrounding face. Creating a volume mesh using the resulting face meshes would be virtually imposible. By merging the bulges with the surrounding faces we remove this constraint, and allow construction of fully ordered mesh for both the faces and the volume.

The methods provided for merging entities of similar dimension are:

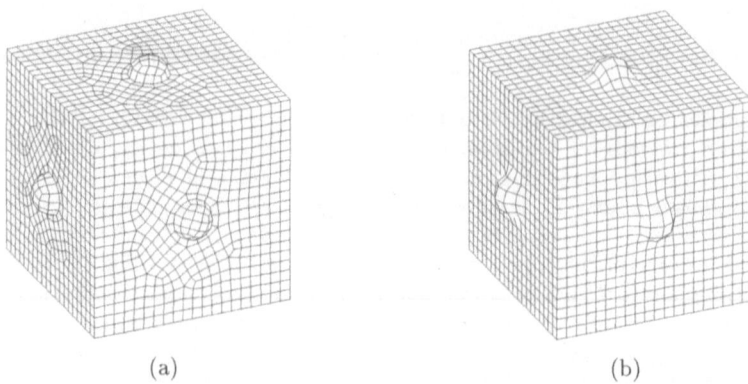

Fig. 2. (a) Mesh of the faces of a brick with spherical bulges; (b) Mesh of the faces after merging the bulges with their surrounding faces. This is a good example of the disturbance caused by insignificant features, and the way to eliminate it. (with coarser mesh element size the bulges would be ignored completely)

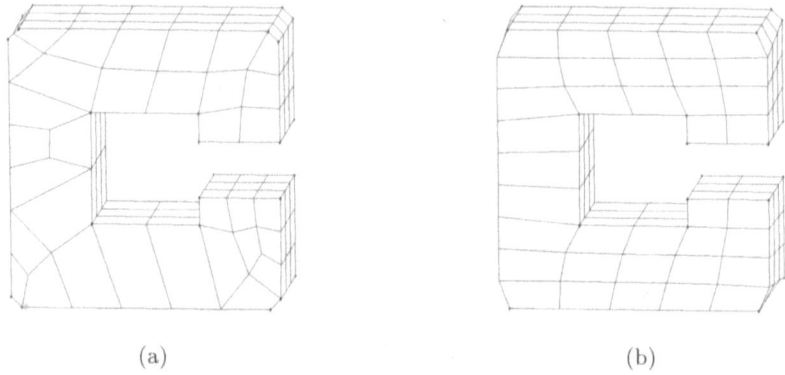

Fig. 3. (a) Mesh of original object; (b) Mesh after the corner edges were merged with left/right ones. Note that nodes are no longer placed at the deleted (merged) vertices.

- Edges: Two edges can be merged if they have a common vertex and belong to the same set of faces. In this case a new virtual edge is created with the two original edges as hosts and the unshared vertices of the host edges as end vertices.
- Faces: Two faces can be merged if they have at least one common edge and belong to the same set of volumes. The resulting face has the two faces as hosts and has as edges the unshared edges from the two faces.

The new entity resulting from a merge replaces the host entities in the upper topology definition (e.g. when merging two faces belonging to a single volume, the new face replaces the two host faces in the bounding faces list of the volume) as in Fig. 1.

5.2   Split

Splitting of edges, faces and volumes can be used in order to create a simpler topology more suitable for analysis. For example splitting a face into several faces with simpler (boxlike) topology (Fig. 4).

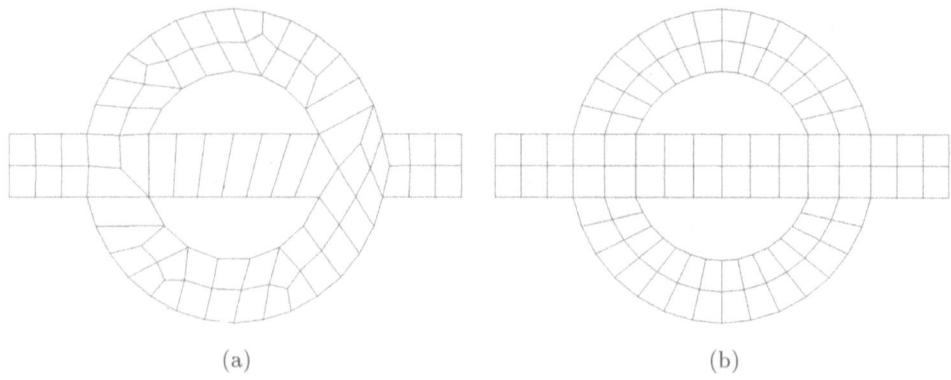

(a)   (b)

Fig. 4. (a) Mesh of the original face; (b) Mesh after the face was split into three simple faces

Another use of split is in collapsing of minor details by "dividing" them between their neighbours. Such collapse operation is shown in Fig. 5 where we first split the conic face into three triangular parts (creating splitting boundary edges from the cone's top vertex to the vertices shared by neighbour faces) then each part is merged with the adjacent neighbour face, resulting in a final simple brick-like topology.

In order to split an entity we must provide a boundary to use as the splitting boundary and for this either a real entity is created or a virtual entity on higher dimension (like edge on face) has to be introduced.

The following splitting techniques are provided:

- Splitting an edge: Create a vertex at a given parameter value on the edge (real or virtual). Create two virtual edges on the two resulting parts of the original edge.
- Splitting a face: Create an edge lying across a face (real or virtual). Create two faces on the two resulting face parts which use the original face as a host and use as bounding edges the resulting edge loops.
- Splitting a volume: Create a face from a loop of edges on the volume. Create two volumes using as boundary the two resulting groups of faces.

The entities resulting from a split replace the host entity in any upper topology definitions (e.g. when splitting an edge belonging to a number of faces, the two new edged replace the split edge in the edge lists of the faces).

5.3   Virtual Entities on Higher Dimension Hosts

A virtual entity can be created on an entity of higher dimension (e.g. a virtual vertex on an edge). Such entities are useful mostly as "topological support" in the more complex

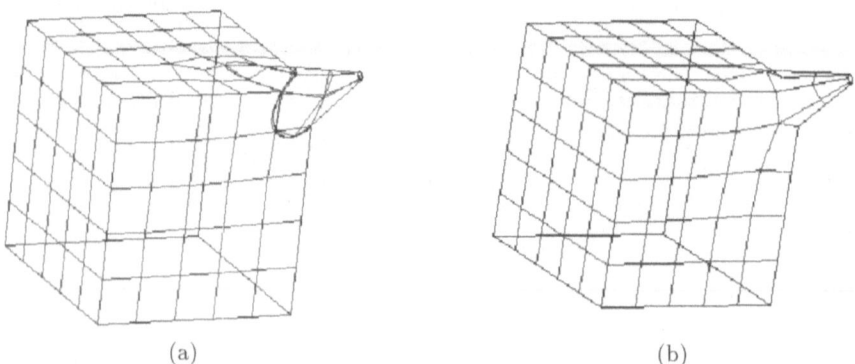

(a)   (b)

Fig. 5. (a) A brick united with a cone at a corner; (b) Mesh of the object (3 faces) after collapsing the conic face between it's neighbour faces.

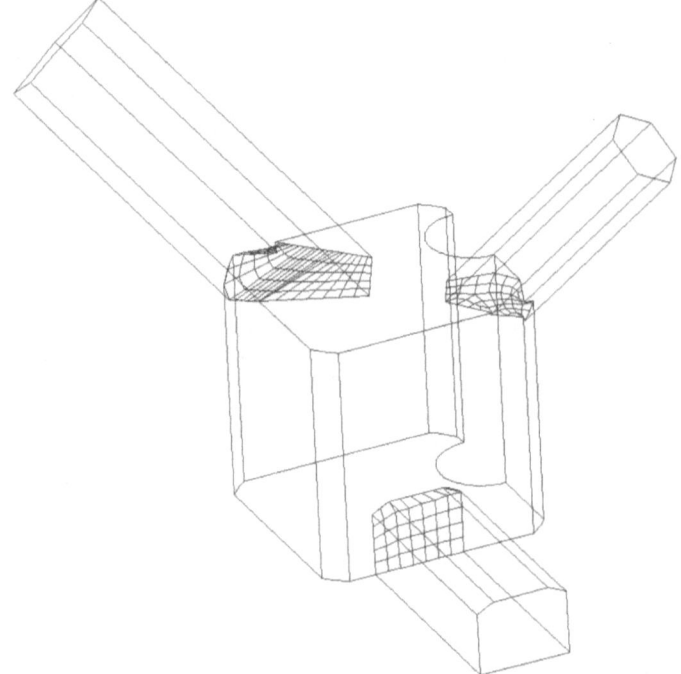

Fig. 6. Mesh on three virtual faces on the volume, splitting the volume into four parts.

virtual operations (usually in splits). For example when splitting the face in Fig. 4 (b), four virtual edges are created which have as host the original face. These edges are used as the splitting edges, when splitting the face into three virtual faces.

Since the higher dimension entity can not define the lower entity's geometry, it must have some type of geometrical definition. Therefore such virtual entity has a minimal geometry of its own and is constrained to be on its host entity.

- A virtual vertex has a location as its geometry.
- An edge on a face host, is regarded as a straight line between its end vertices projected on to the face.
- A virtual face on a volume is regarded as a least square fit plane to the edges that bound it. The definition is sufficient to answer the geometric queries about the face. During meshing nodes are usually allowed to migrate off the plane during smoothing to produce a "spider web" like effect for the mesh on the face. Virtual faces are particularly useful during decomposition when defining an actual geometric surface using the edges would be difficult or impractical (Fig. 6).

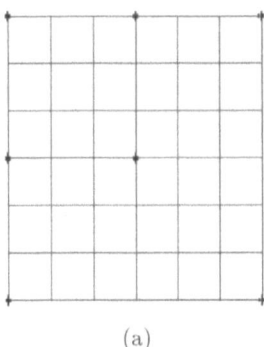
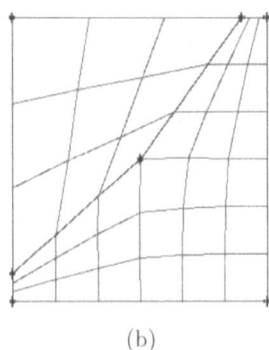

(a)   (b)

Fig. 7. (a) Mesh on a split face; (b) The mesh movement following the move of two virtual vertices

One of the advantages of using virtual entities is their mobility, i.e the geometry of a virtual entity can be changed (in the bounds of its host entity) automatically updating the geometry of upper topology entities connected to it, but without any topology change. If a virtual vertex is used to split an edge, the vertex is free to float along its host entity, and so when the two edges are created after split the mesh on them will follow the vertex movement. In a similar manner if a virtual edge is used to split a face, it is free to float across its host face or volume , moving the meshes of the split face parts along with it. This freedom of movement allows direct local control of the mesh element density without increasing the element count, as shown in Fig. 7.

5.4   Connect

Another use of virtual geometry is to provide correct topology on erroneous geometric input. The most common problem is having more then one geometric representations for the same entity, resulting from different entities of the upper topology (e.g. several vertex representations from the adjacent edges). Until these duplicate topologies are joined, a valid B-Rep cannot be constructed. The reasons for such inconsistencies are usually either too low tolerances during construction or non-perfect translations from one representation format to the other.

In such cases a new virtual entity is constructed as a connect of those entities. When the geometric data of this virtual entity is queried it returns some interpolation

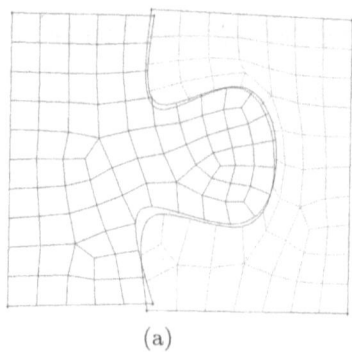
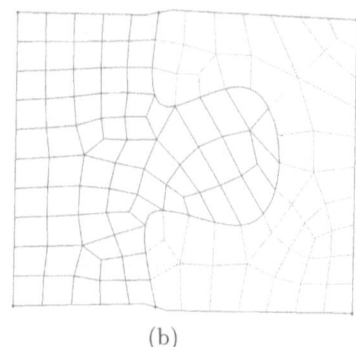

(a)   (b)

Fig. 8. (a) Two faces which were supposed to have a common edge (b) The mismatched edges are replaced by one virtual edge, which is a connect of the two. The faces are replaced by virtual faces that take into account the connected edge.

of the data of the host entities. This way we avoid the complex task of finding "real" geometric interpolation of the entities and with it the necessary "fix" of the geometry of the upper topology entities.

For example in Fig. 8 to solve the common error of two representations for the same edge (from two adjacent faces) one virtual edge is created on top of them which when interrogated, will give as geometric data some interpolation of the two host edges. The duplicate end vertices of the two edges are also replaces by connected vertices.

It is not enough to connect just the similar entities, but it is necessary to introduce virtual upper topology as well. For example when merging two edges of adjacent faces the adjacent faces have to become virtual since their geometry must also incorporate the change of borders as result of the new virtual edge. (Notice the change of the geometric description of the edges sharing the connected vertices in Fig. 8.)

6   Accessing the Geometry of a Virtual Entity

Access methods of topological entities can be divided roughly into three groups according to the query argument:

- methods that query the entity as a whole;
- methods that receive as an argument a point on or near the entity;
- methods that get as an argument a value in the parametric representation of the entity.

For the first two types of methods the enhancement to virtual topology is immediate. When accessing an entity via a point argument the algorithm simply finds the host entity nearest to the point and derives the information from it.

Parametric representation of virtual entities can be more problematic. For virtual edges the lengthwise parameterization is used and based on it the query is passed to the appropriate host.

Both in the paving algorithm that we used and other commonly used technics for general surface meshing like medial-axis based meshing [Tam91, Price95], the surface parametrisation is not used and mesh nodes are simply projected onto the face. Reparametrization of faces is complex and can be provided only for special cases.

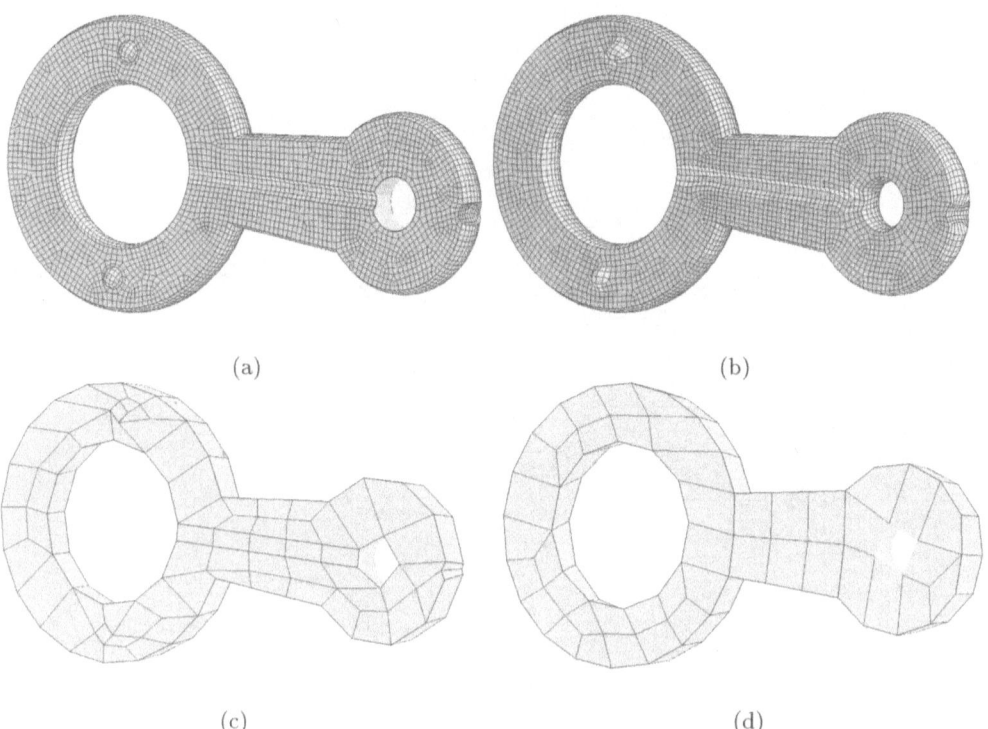

(a)   (b)

(c)   (d)

Fig. 9. Mesh of a connecting rod: (a) Mesh of the original topology with element size 0.5 (b) Mesh of re-edited (virtual) topology with size 0.5; (c) Mesh of the original topology with size 4 (d) Mesh after re-editing with element size 4; Note that with coarser element size, the bumps and the canal elements are "flattened" by both meshes but in the original topology they continue to cause disturbance to the mesh.

7 Implementation and Results

An example of a complex virtual topology is shown in Fig. 9 where the toplogy of a simple connecting rod is modified using both merge and split operations until it is suitable for meshing.

The new object topology was constructed by first merging the front face of the object with the minor features adjacent to it (the bumps and the channel), and then splitting the resulting face into three simple faces (the two circles and the rectangle in between). In order to split the face two virtual edges were constructed on top of the merged face to be used as split "aids". Several short edges were merged with adjacent edges, again to simplify the topology. A part of the virtual B-Rep structure of the connecting rod in Fig. 9 is shown in Fig. 10.

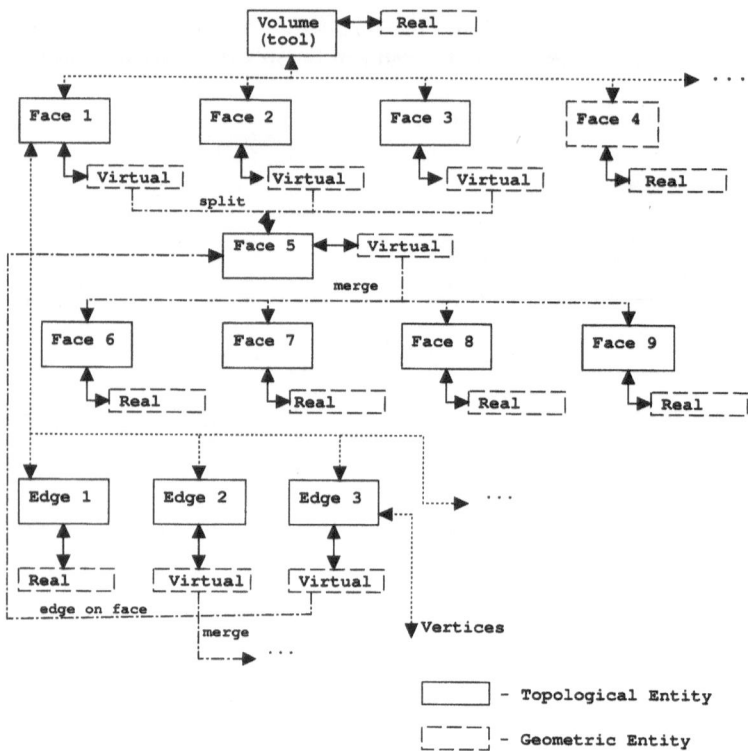

Fig. 10. Part of the virtual B-Rep graph of the connecting rod in Fig. 9

The virtual topology is implemented as part of the geometric modeler package of the FIDAP 8.0 Fluid Dynamics Analysis Package. As seen in the examples it provides great flexibility in editing of object topology, and adjusting it to meshing and analysis needs while maintaining the object geometry. The system is written in C++ (using inheritance so as to make the access to real/virtual entities completely transparent for the interface) and it runs on multiple platforms.

8   Conclusions

With the introduction of virtual topology, we significantly enhance the ability of the user to edit the geometry. By adjusting the topology independent of the geometry, many complex changes can be achieved without laboring geometric operations, while maintaining transparent simplistic access to both real and virtual entities geometry, and without the loss of accuracy caused by real geometry modification.

Insignificant features can easily be removed, through merges. Decomposition using virtual splits can be easily introduced and flexibly adjusted. Connect operations using virtual geometry simplify the clean up of "dirty" geometry into a valid B-Rep without the ominous task of geometrically equivalencing dissimilar geometry.

The virtual geometry concept as introduced so far, does not provide the automatic decision making process of which virtual entities should be created. Some sugestions on automated detail suppresion using virtual topology tools are introduced in [Sheffer97]. Other technics like feature recognition and/or skeleton based methods may be used to assist in the recognition of such details as well.

References

[Armstrong95]   C. G. Armstrong, D. J. Robinson, R. M. McKeag, T. S. Li, S. J. Bridgett, R. J. Donaghy, C.A. McGleenan, Medials for Meshing And More. Proc. 4th International Meshing Roundtable, Albuquerque, New Mexico, October 1995.

[Blacker91]   Blacker T.D., Stephenson M.R., Paving: a new approach to automated quadrilateral mesh generation, International Journal of Numerical Methods in Engineering, 32, 1991, pp. 811-847.

[Blacker93]   Blacker T.D., Meyers R.J., Seams and wedges in plastering: a 3D hexahedral mesh generation algorithm, Engineering with computers, 9, 1993, pp.83-93.

[Butlin96]   Butlin G., Stops C., CAD Data Repair, 5th International Meshing Roundtable, Pittsburgh, USA, October 1996.

[DeFloriani89]   De Floriani, L., Feature extraction from boundary models of three-dimensional objects, IEEE Transactions on Pattern Analysis and Machine Intelligence, 11(8), August 1989.

[Field95]   Field, D. A., The legacy of automatic mesh generation from solid modeling, Computer Aided Geometric Design, 12(7), November 1995.

[Jones95]   Jones M.R., Price M.A., Butlin G., Auto-Meshing, 4th International Meshing Roundtable, Albuquerque, New Mexico, October 1995.

[Price95]   Price M.A., Sabin M.A., Armstrong C.G., Fully Automatic Quad and Hex meshing, Proc. 5th International Conference on Reliability of Finite Element Methods for Engineering Applications, pp 356-367, Amsterdam, 10-12 May, 1995.

[Shah91]   J.J. Shah, "Assessment of feature technology", Computer Aided Design, Vol. 23, No. 9, pp. 58-66, June 1991.

[Sheffer97]   Sheffer A., Blacker T., Clements J., Bercovier M., Clustering: Automated Detail Suppression Using Virtual Topology, To be presented at Symposium on Trends in Unstructured Mesh Generation, 1997 Joint ASME/ASCE/SES Summer Meeting. June 1997.

[Tam91]   Tam T., Armstrong C.G., 2D Finite Element Mesh Generation by Medial Axis Subdivision, Advances in Engineering Software, Vol 13, 5/6, pp 312-324, Sep/Nov 1991.

Generalization of Modified Octrees for Geometric Modeling

N. Hitschfeld

Dpto. Ciencias de la Computación, Univ. de Chile
Blanco Encalada 2120, Santiago, CHILE
E-mail: nancy@dcc.uchile.cl

Abstract. This paper discusses several aspects of modified octrees that can be generalized in order to obtain solid representations using less primitive elements than the traditional modified octree. The aspects under study include the use of elements of different type as internal nodes, a general refinement approach and cuboids, pyramids, prisms and tetrahedra as final elements. These concepts can be applied to the generation of mixed elements meshes for different applications. In particular, the new ideas are presented here for the generation of mixed element meshes that satisfy Delaunay condition. Examples are given to compare a new implementation with previous approaches.

1 Introduction

Since the last fifteen years, modified octrees have been used very often in geometric modeling and mesh generation[1,2]. The modified octree approach works as follows: The 3-D domain is enclosed in a cube, whose octants are repeatedly refined at their edge midpoints until the boundary and internal quantities are sufficiently approximated. Elements with and without edge midpoints are partitioned into tetrahedra. In case of mesh generation, the final elements have to fulfill the requirements imposed by the numerical method.

The modified octree approach has the following drawbacks: (1) The use of only cubes as internal nodes while fitting the device geometry does not allow to stop the refinement as soon as the part of the object intersecting a cube can be represented with a pre-defined set of *well-shaped elements*. (2) The refinement into eight similar elements increases unnecessary the number of nodes for several applications. (3) In case of mesh generation, the most common is to use tetrahedra as final elements. Tetrahedra are required for the finite element method but if some problem can be solved using the *Box method*[3], several element types can be used.

This paper discusses several aspects in the generation of octrees and modified octrees that can be generalized in order to get a final domain representation that contains less basic elements than the former approaches: (1) The domain can be enclosed for a cuboid. A cuboid has rectangular faces. (2) The internal elements (nodes) can belong to a set of *well shaped elements*, such as pyramids, prisms and tetrahedra of rectangular basis, and cuboids. The set of elements that is called *well-shaped* depends on the application. This set has to be closed under the *refinement operator*, i.e, each element can be refined in such a way that all newly generated elements belong to this set. The trees that can handle different element types as internal nodes are called *mixed element trees.* (3) The refinement can be either bisection or what we have called *intersection* based approach. Using the bisection based approach the refinement is always made at the edge midpoints. Using *intersection* based approach the refinement is made at the most convenient edge point. The best point—the one whose associated refinement generates sons with the smallest aspect ratio —is chosen from the available Steiner points (points generated by the refinement of the edge neighbors) and intersection points (points generated by the intersection between the object geometry and the current element). (4) Internal elements can be divided into a different number of elements and into elements of different type. It depends on the type of the internal node and on the refinement direction. For example, if a refinement is required along one, two, or three coordinate axes, cubes, are subdivided into two halves, four quadrants, and eight octants, respectively. (5) The set of final elements are defined by the application. They can be of the same type of the ones used as internal nodes, or of other type. What we kept from the modified octree approach is that the refinement is parallel to the axes of the coordinate system.

2   Characterization of well-shaped meshes

This section introduces the concept of well-shaped mixed element meshes independent of the application.

2.1   Basic algorithm to generate a well-shaped representation

The discussion of this paper is focused on the algorithms that are extensions of modified octrees and generate solid representations refining coarse elements. Independent of the application, the following consecutive steps are used:

1. Generate a macro-mesh that fits the geometry of the modeled device exactly
2. Refine due to variations in some internal values and certain geometrical parameters
3. Generate a proper mesh for the current application
   - make the mesh 1-irregular
   - look for proper tessellations
4. Store the information required by the application

2.2   Macro-elements

A macro-mesh is composed by macro-elements. Macro-elements are used to fit the device geometry.

The following theorem characterizes the set of macro-elements used in the generation of well-shaped mixed element meshes.

Theorem 1. *Let P be a set of polyhedra. P leads to well-shaped meshes if each polyhedron $p \in P$*
(i) fulfills the restrictions imposed by the current application, and
(ii) can be refined in such a way that all newly generated polyhedra also belong to P (P is closed).

Proof: Condition (i) guarantees that the macro-elements fulfill the restrictions imposed by the current application. Condition (ii) guarantees that for each element generated through the refinement process it will be possible to fulfill condition (i).

2.3   Different element refinement approaches

The most common way of refinement is bisecting an element, i.e., each element edge is bisected. This method is easy to analyze and implement but it does not allow flexibility in choosing the most appropriate refinement point. In the following, the refinement approach that allows to chose the refinement point is called *intersection-based* approach. (The bisection-based approach is a particular case of the intersection-based approach.)

2.4   Elements with Steiner points

Irregular macro-elements are elements with edges split at least once. The point splitting an edge is called a *Steiner point*. Irregular elements appear between coarse and fine regions after the density requirements are satisfied. In order to complete the mesh using as few elements as possible, the tessellation of irregular elements is necessary. 1-irregular elements are elements with edges split at most once.

Definition 2. Let p be an irregular convex polyhedron. The tessellation t of p is well-shaped if and only if t satisfies the conditions of the current application.

2.5   Final elements

Final elements are the elements that compose the final mesh. In the current version of the algorithm, we are not considering that final elements will require further refinement. For this reason, the set of final elements can include more basic elements than the ones included in the set of macro-elements used to fit the device geometry.

   The set of final elements are defined by the application. They can be of the same type of the ones used as internal nodes, or of other type.

3   Applications

The next sections show applications of mixed elements meshes. A complete description is given for mixed elements meshes that satisfy the Delaunay condition.

3.1   Delaunay meshes

Definition 3. A tessellation T of a set of points S is a *Delaunay tessellation* if there exists point-free circumsphere for each tessellation element.

   We use the term *Delaunay tessellation* and not *Delaunay triangulation* because our meshes include other element types than tetrahedra.

Macro-elements

   The following theorem characterizes the set macro-elements used in the generation of Delaunay mixed element meshes.

Theorem 4. *Let P be a set of polyhedra. P leads to Delaunay meshes if each polyhedron*
$p \in P$
(i) has co-circular vertices. They define the point-free circumsphere, and
(ii) can be refined in such a way that all newly generated polyhedra also belong to P (P is closed).

Our set of macro-elements is composed of rectangular pyramids, rectangular prisms, bricks, rectangular tetrahedron and its complement inside a cuboid (Fig. 1). They are elements that satisfy Theorem 3 and can be properly refined as will be shown in the next section.

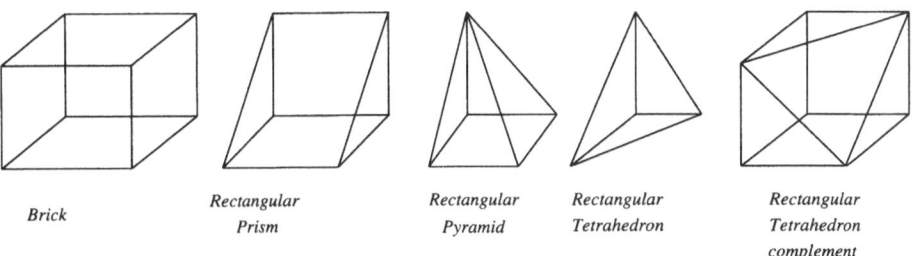

Fig. 1. Set of elements used to fit the device geometry

Element refinement

Since the bisection-based approach is a particular case of the intersection-based approach, the current section only shows the refinement for each macro-element under the general refinement approach.

Using an intersection based approach, elements are rarely split at the edge mid-points. The refinement location along the edges is determined using the current aspect ratio and the location of the current Steiner-points.

Refinement of bricks

Bricks can be split into two halves, four quarters or eight octants as before but edges are not necessary bisected. Figure 2 shows the different ways to split a brick using arbitrary refinement points. The only restriction is that parallel edges have to be split at the same relative position from their endpoints in order to generate bricks and not general polyhedra.

Refinement of prisms

Rectangular prisms can be partitioned in the same way as before, but their triangular faces impose an additional restriction: the refinement point at the diagonal edge determines the refinement points for other edges as shown in Fig. 3. The triangular face (a) shows a triangular face with two possible refinement points. In the triangular face (b), a vertical and a horizontal dotted line illustrate how this face would be refined if both points were used: four new faces would be generated and the two shaded faces

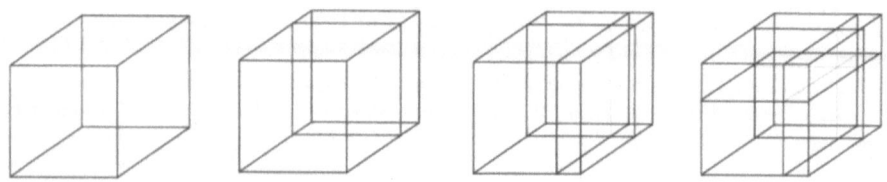

Fig. 2. Bricks refined in one, two, or three directions generate two, four, and eight bricks, respectively

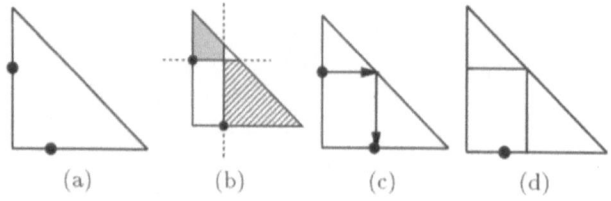

(a)   (b)   (c)   (d)

Fig. 3. Refinement of a triangular face

would not belong to any of the macro-elements. edge defines the location of the vertical line as shown in the triangular face (c). Triangular face (d) depicts the final status. Figure 4 shows the whole prism refinement depending on the direction(s) required.

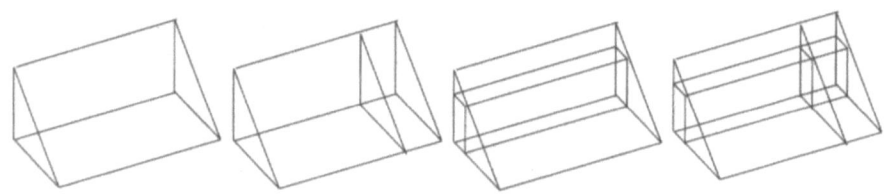

Fig. 4. Prism refinement in one, two, or three directions generates two prisms, one brick plus two prisms, and two bricks plus four prisms, respectively

Refinement of pyramids

The refinement of a rectangular pyramid in a general position is governed by the restrictions impose by its triangular faces. The intersection point that generates subelements with the smallest aspect ratio is chosen. Figure 5 shows its partition using the refinement point p. Pyramid (a) shows the main parameters: the point p, the edge e (where p is located), and the main diagonal d. Let f be the plane with normal vector in direction of e and passing through the point p. The point c corresponds to the intersection point between the plane f and the edge d. The coordinate values of c define the refinement points in the other directions (pyramid (b)). The result is shown in pyramid (c).

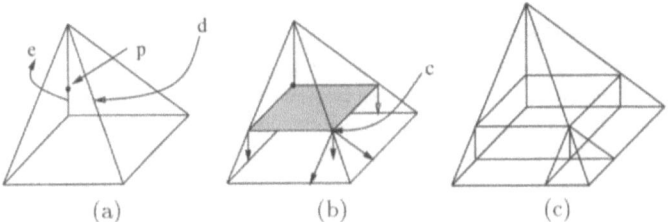

Fig. 5. Pyramid refinement in three directions generates one brick, two prisms and two pyramids

Refinement of the tetrahedron and its complement

Figure 6 shows the refinement of the rectangular tetrahedron and its complement. The rectangular tetrahedron is refined into three similar elements and a rectangular tetrahedron complement (left). The rectangular rectangular tetrahedron complement is refined into three similar elements, four bricks and one rectangular tetrahedron (right).

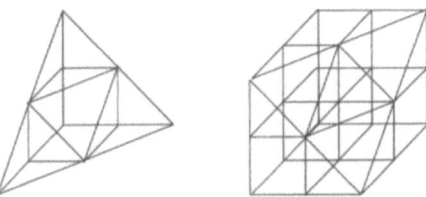

Fig. 6. Refinement of the tetrahedron and its complement

1-irregular elements

Irregular macro-elements appear after refining the grid elements according to given density requirements. Irregular macro-elements, i.e., elements with edges bisected more than once are made 1-irregular looking for well-shaped tessellations. The next definition characterizes such an element.

Definition 5. Let l be a 1-irregular macro-element. l is a well-shaped if no Voronoi point of l lies outside its convex hull (in this case, the 1-irregular macro-element itself).

The previous definition can be fulfilled by finding a Delaunay tessellation of the 1-irregular macro-element that satisfies the following theorem:

Theorem 6. *Let* $S \subset \mathbb{R}^n$, $n \leq 3$ *be a set of points, C the convex hull of S, and T a Delaunay tessellation of S. Then no Voronoi point of S lies outside C if and only if for each face f_{ijk} in 3-D of T on the surface of C, the circumsphere of f_{ijk} with the center in the middle of f_{ijk} is point-free.*

The proof of this theorem can be found in [4]

The final mesh is generated only after checking that each 1-irregular element fulfills Definition 5. Theorem 6 imposes the restrictions to the tessellation of each 1-irregular macro-element: the local tessellation must be Delaunay and there exist a point-free circumsphere for each face on the surface of a 1-irregular macro-element. The last restriction guarantees that exist a point-free circumsphere for neighboring elements.

Final elements

The current set of final elements is shown in Fig. 7. A final element is any one whose vertices are co-circular. This set of elements solves around 80% of the 1-irregular configurations for a cuboid [5].

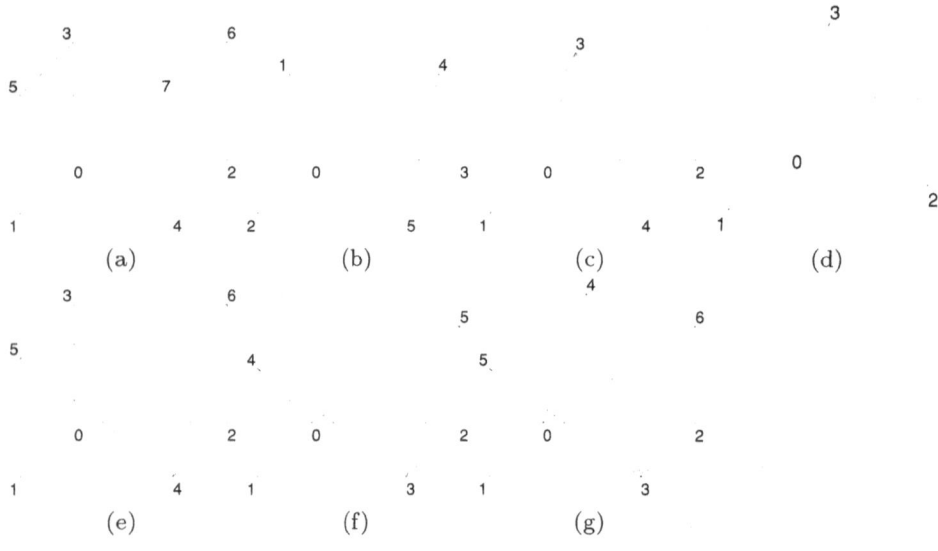

Fig. 7. Set of final elements

3.2   Meshes for Box-method

The meshes required by the control volume discretization method are a subset of Delaunay tessellations. Definition 7 characterizes these meshes:

Definition 7. *A Delaunay tessellation of a set of points S is adequate for the control volume discretization method if the corresponding Voronoi diagram fulfills the following* conditions: (a) No Voronoi point is outside the boundary of the tessellation (b) Each Voronoi face intersects its Delaunay edge

The previous Definition implies that elements lying at the boundary or material interfaces have additional restrictions. Compared to those inside the material, no Voronoi point (center of the element) must lie outside the faces at the border or at a material interface. The number of macro-elements is then reduced to rectangular prisms, rectangular pyramids and cuboids. They contain the center of the circumsphere that surrounds them. More details are given in in [5].

3.3 Solid modeling

Mixed element trees can be used as a solid representation model. It belongs to the methods based on cell decomposition. Depending on the kind of the application it is also possible to use incomplete mixed element meshes.

A complete comparison with the most used representation model would require the revision of the properties of the representation models.

4 Description of different implementations

4.1 Mixed element trees

The trees that can handled different element types as internal nodes are called *mixed element trees.*

Definition 8. Let T be a tree. T is a *mixed element tree* if
(i) each node (internal or leaf) is a polyhedron
(ii) each internal node is labeled with the axes across which the node is refined.

Theorem 9. *Let T be a mixed element tree. T leads to well-shaped meshes by construction if and only if*
(i) each internal node is one of the macro-elements described in the previous section,
(ii) each leaf is an irregular macro-element satisfying Definition 5 or a macro-element without Steiner points (regular element), and
(iii) each 1-irregular leaf can be tessellated into the final elements.

This representation allows the generation of well-shaped meshes by construction because it permits the implementation of the concepts introduced in the previous section in a natural way.

4.2 Mixed element trees using a bisection based approach: Ω_{mebi}

The main algorithm looks as follows:

1. Generate a mesh that fits the geometry of the modeled device exactly. This initial mesh consists of cuboids, rectangular prisms, rectangular pyramids, rectangular tetrahedra and its complement inside a cuboid. The mesh is handled as a forest where each element is the root of a tree.

2. Refine due to variations of some internal values and certain geometrical parameters. In order to fit physical and other geometrical parameters, an irregular mesh is generated by refining each element independently of the others.
3. Generate a proper Delaunay mesh. A finite element mesh is obtained after tessellating all the irregular elements into tetrahedra, pyramids, prisms and cuboids.

The most serious problem of Ω_{mebi} [6] originates from the algorithm that fits the original device geometry. This algorithm generates an initial (tensor product based) mesh that is a complete partition of the device (i.e. mesh elements have no Steiner points). A complete partition is required in order to be able to use a bisection based approach. During the generation of the initial mesh, small geometry features are propagated (by inserting planes) to the boundaries of the device. Therefore, the initial mesh contains a high number of unnecessary elements with a very bad aspect ratio. In addition, the repetitive generation of new points due to intersections between the inserted lines (planes in 3-D) and some boundary or material interfaces makes it impossible to fit several device geometries.

4.3   Mixed elements trees using an intersection based approach: Ω_{mein}

The same consecutive steps are used to generate a mesh using an *intersection*-based approach but each step is focused in a different way.

The geometry is fitted by refining the mesh elements at the best possible point. The best point—the one whose associated refinement generates sons with the smallest aspect ratio—is chosen from the available Steiner points and intersection points. Elements are bisected for generated sons with bad aspect ratio.

After the geometry is completely fitted, the irregular macro mesh is further refined until the density requirements are fulfilled. Elements are partitioned in the required direction according to the best located Steiner point.

The mesh is made 1-irregular before looking for proper tessellations. Subsequently, the algorithm checks the *splittable* condition, i.e., a condition that guarantees the existence of a proper tessellation. If an element is non splittable, proper points are inserted by looking inside the problematic element.

Once all elements are splittable, each local tessellation is computed using an algorithm to compute Delaunay tessellations inside 1-irregular macro-elements [5]. The set of final elements was shown in Fig. 7.

4.4   Comparison

The implementation of a general refinement approach is quite more difficult than the bisection based approach. The main difficulties are in:

- The algorithms and data structures to consistently keep the geometrical information generated at each refinement step [7].
- The tessellation of elements with edge midpoints can be solved pattern-wise. The tessellation of elements with Steiner points at arbitrary position must be solved using an algorithm that recognize the basic elements [5].
- The propagation of points among the neighbors must be controlled very carefully.

5   Comparison using examples

The following examples are used to analyze empirically the influence of the aspects that have been generalized from modified octrees in the generation of mixed element meshes.

5.1   Complete meshes

Figure 8 shows two meshes for a diode; the left one was generated using Ω_{mebi}, and the right one was generated using Ω_{mein}. Ω_{mein} fulfilled the density requirements specified by the user using one half of the elements required by Ω_{mebi}. The exact number of elements of each type used in the final mesh is given in Table 5.1.

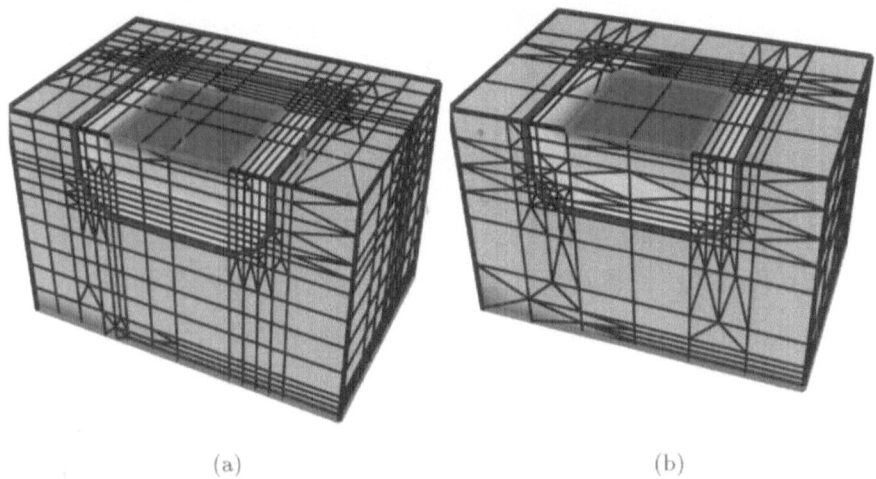

(a)   (b)

Fig. 8. A final mesh for a diode (a) using Ω_{mebi}, (b) using Ω_{mein}

The main reasons for the strong reduction in the number of elements are: (1) Ω_{mein} uses a larger set of final elements than Ω_{mebi}, (2) Ω_{mein} uses an algorithm to compute the tessellation of 1-irregular configurations while Ω_{mebi} uses a template-based solution. The template-based solution only includes the most common 1-irregular configurations that can be tessellated using cuboids, prisms, pyramids and tetrahedra.

	Cuboid	Prism	Pyramid	Tetr.	Tetr. Compl.	Deformed Prism	DBC	Total
Ω_{mebi}	2628	1682	404	153	0	0	0	4867
Ω_{mein}	692	588	575	478	93	1	0	2427

5.2   Incomplete meshes

Figure 9 shows the same two views of the silicon part of a ECL bipolar transistor. The device geometry is fitted in Fig. 9(a) using Ω_{mebi} approach and in Fig. 9(b) using

Ω_{mein}. The partial mesh shown in Fig. 9(b) strongly reduces the number of unnecessary mesh points and elements. In addition, the intersection-based approach avoids the propagation of mesh planes to the whole device allowing the fitting of more complicated geometries because small changes in the geometry can be fitted locally. The aspect ratio of the elements is controlled according to user parameters.

Empirically, the algorithm to fit the device geometry using an intersection-based approach is much faster. For example, Ω_{mein} is four times faster than Ω_{mebi} for the example shown in Fig. 9.

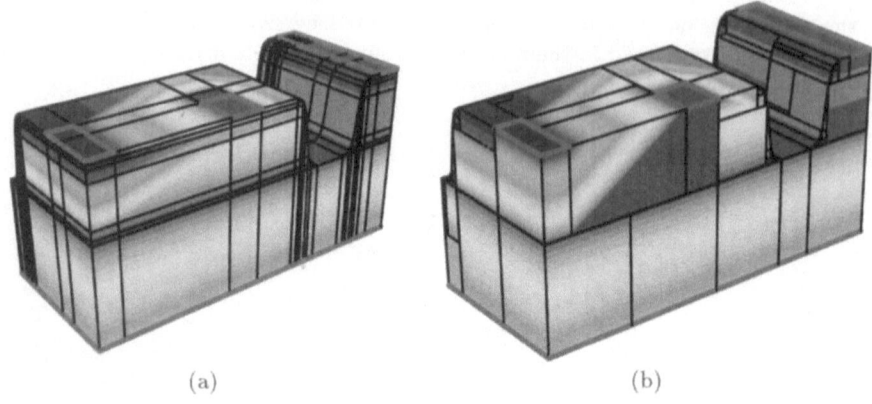

(a)   (b)

Fig. 9. Fitting the device geometry for the bipolar transistor (a) 2365 pts (b) 881 pts

Figure 10 shows the next step in the generation of a proper mesh. The density requirements specified by the user are fulfilled in both meshs. However, the partial mesh generated using the intersection-based approach (Fig. 10(b)) needs fewer mesh points and elements. The main reason is that a better macro-mesh is generated to fit the device geometry.

The intersection based approach shows clear advantages with respect to the bisection based in the first steps of the mesh generation process. Unfortunately, it is still not possible to compare the final mesh for this example. Previous tests have shown that the current implementation of the next steps will generate more elements than Ω_{mebi}. The problem is that the current strategy to stop the propagation of points among the neighbors while doing the mesh 1-irregular is too simple. It does not use the neighboring information. This is a very important aspect that is currently studied in order to reduce the number of elements in the final meshes in examples with complex geometries.

6   Conclusions

This paper presents several extensions to modified octrees that have being implemented using mixed element trees. Ω_{mebi} generates mixed elements meshes using bisection based approach. In comparison to modified octrees, it uses several elements types to represent the device geometry, elements are only refined in the required direction and

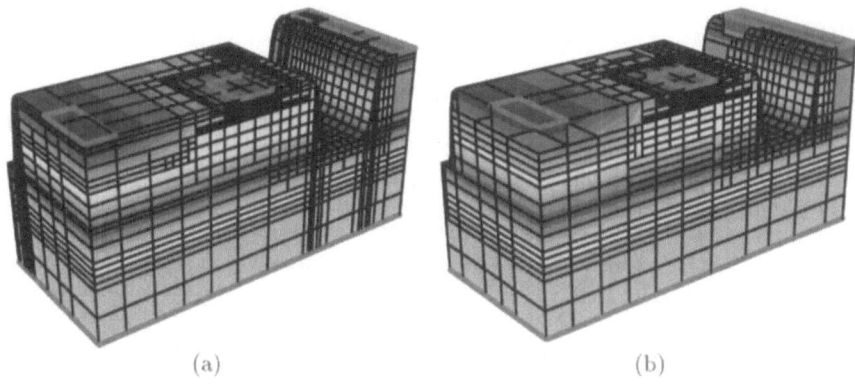

(a)   (b)

Fig. 10. Achieving the desired mesh density for the bipolar transistor (a) 10,376pts (b) 7,097pts

uses cuboids, prisms, pyramids and tetrahedra as final elements. This allows to represent exactly more complex geometries and using less final than mesh generators based on the modified octree approach. Ω_{mein} introduces the refinement at any position, the design of an algorithm to automatically compute Delaunay tessellations for 1-irregular macro-elements, and a larger set of final elements than Ω_{mebi}. The test examples have shown that the fitting of the device geometry and the fulfilling of the density requirements are done more efficiently in Ω_{mein} than in Ω_{mebi}. The last steps of the mesh generation process need still more testing to give confident conclusions. For the objects with simple geometry like the diode, Ω_{mein} shows a better performance and reduces strongly the number of elements of the final mesh. For complicated geometries, like the bipolar transistor, new strategies have to be included to stop the propagation of points among the neighboring elements.

7   Current work

The generation of mixed element meshes using an intersection based approach are still under study. The complete mesh generation process is being improved in the following parts:

- the algorithm that controls the propagation of points among the neighbors
- the algorithm that generates automatically tessellations for 1-irregular macro-elements can include more final elements than the ones presented here.
- the robustness of the complete process.

The generation of more test examples is necessary to give complete comparison.

8   Acknowledgments

This work was partially supported by FONDECYT project No. 1940323 in 1996.

References

1. M. A. Yerry and S. Shephard, "Automatic Three-dimensional Mesh Generation by the Modified-Octree Technique," *Int. J. Numer. Methods Eng.*, vol. 20, pp. 1965–1990, 1984.
2. M. S. Shephard and M. K. Georges, "Automatic Three Dimesional Generation by the Finite Octree Technique," in *International Journal for Numerical Methods in Engineering*, vol. 32, pp. 709–749, 1991.
3. R. E. Bank, D. J. Rose, and W. Fichtner, "Numerical methods for semiconductor device simulation," *IEEE Trans. on El. Dev.*, vol. ED-30, no. 9, pp. 1031–1041, 1983.
4. N. Hitschfeld and W. Fichtner, "3-D Grid Generator for Semiconductor Devices using a fully flexible Refinement Approach," in *Int. Conf. on Semiconductor Devices and Processes, pub. in Simulation of Semiconductor Devices and Processes*, vol. 5, pp. 413–416, Springer-Verlag, 1993.
5. N. Hitschfeld and R. Farías, "1-irregular element tessellation in mixed element meshes for the control volume discretization method," in *Proceedings of the 5th International Meshing Roundtable*, pp. 195–204, Pittsburgh, Pennsylvania, U.S.A., October 10-11, 1996.
6. N. Hitschfeld, S. Müller, and W. Fichtner, "Generation of 3-d Delaunay Meshes for Complex Geometries using Iterative Refinement," *Ifip Transactions. Algorithms, Software, Architecture. Information Processing 92.*, vol. I, pp. 388–394, 1992.
7. N. Hitschfeld, "Algorithms and data structures for handling a very flexible refinement approach," in *Proceedings of the 4th International Meshing Roundtable*, pp. 265–276, Sandia National laboratories. Albuquerque, October 1995.

Boundary Generation from Voxel-based Volume Representations

R. Joan-Arinyo, J. Solé

Universitat Politècnica de Catalunya
Departament de Llenguatges i Sistemes Informàtics
Av. Diagonal 647, 8a, E–08028 Barcelona

Abstract. A method to generate polygonal boundary representations from voxel-based volume representations is presented. The algorithm makes use of a face octree as an auxiliary data structure. That face octree is extracted from the voxel-based volume representation. The precision of the computed boundary is given by the length of the main diagonal of the voxels in the initial representation. Experimental results that illustrate the method are presented.

Keywords: Brep, voxel-based representation, face octree, surface extraction, deformable models, simplification.

1 Introduction

Voxel-based volume models are widely used in scientific visualization, [8,12]. This is because voxels are the simplest way of representing spatial data stored in 3D arrays, and because this kind of data is what remote sensing and scanning technology generate from nondestructive examination of the internal structure of objects. Unfortunately, voxel-based volume schemes have some disadvantages. Some of them are: The huge amount of storage required to encode even simple solids, high time-consuming rendering algorithms, and the difficulty to apply shape recognition, geometric measurements or to perform a sequence of geometric operations.

There are several reasons that justify the design of an efficient algorithm to convert huge voxel data representations into simple, closed, polygonal boundary representations. First, the conversion will yield a reduction of the amount of storage needed for the representation. Second, rendering of solids represented as polygonal boundary representations will be faster than rendering original voxel-based models, and the quality of the generated images will be improved. Third, simple, closed, polygonal boundary representations can be used for measurements and to perform some geometric operations.

In this paper we present a new algorithm to extract a polygonal boundary representation from a voxel-based volume representation. The algorithm reads as input a voxel-based representation consisting of a binary three-dimensional array that encodes in each cell either the presence or absence of solid matter. The algorithm consists of three major steps. First, a set of non planar quadrilaterals that cover the set of voxels in the boundary of the volume data set is generated. Then, a face octree is extracted from the volume data representation. Finally, a valid boundary representation is derived from the set of non planar quadrilaterals with the help of the face octree computed in the previous step. The planes associated with face octree nodes are used as embedding planes where quadrilaterals inside each octree node are projected. This has the effect of "ironing" high frequencies derived from the voxel data set. Non planar quadrilaterals spanning two or more octree nodes are split into two triangles to complete the boundary construction. The extracted representation is a correct boundary representation because it is non-self-intersection and closed. In fact, the computed boundary representation is a polyhedron that stabs all and only the boundary voxels.

The precision of the boundary representation is given by the length of the main diagonal of the voxels in the initial data set except for some subsets of voxels forming walls or crests that are one voxel thick which are pruned.

2   Related Work

Hyerarchical spatial encodings like classical octrees [13,19,18] and extended octrees [1,4,6] have been proposed to simplify models and to reduce storage requeriments. They allow the use of simple algorithms to perform boolean operations between solids. Face octrees are an intermediate scheme between classical octrees and extended octrees; they are more concise than classical octrees and are particularly well suited to approximate representations of objects with complex surface boundaries [2,3]. The most important drawback in face octrees representations is that they are not closed models.

Schmitt et al. describe in [20] an adaptive subdivision method of fitting surfaces to sampled data. The technique approximates the sample data points with bicubic Bézier surface patches. In [5], DeHaemer et al. reported several variations on the Schmitt et al. basic approach in order to deal with a number of shortcomings like edge gaps and aliasing.

Schroeder et al. proposed in [21] a method called *triangle decimation* to reduce the number of triangles needed to model an object. The algorithm takes as input a dense triangle mesh that approximates a surface and makes multiple passes using local geometry and topology to remove vertices that pass a distance or angle criterion. The vertex removal generates holes that must be patched later using a local triangulation process. Turk, [23], generates different levels of approximations to a given surface defined by a large number of polygons. The original data set is simplified by successively removing vertices and locally re-tiling the surface in a way that matches the initial surface. The original and the new data set coexist in a data structure called *mutual tesellation*.

Hoppe et al. presented in [9] a method for reducing the number of vertices in an initially dense mesh of triangles that represents a set of data points scattered in three dimensions. The approach minimizes an energy function that explicitly models competing factors such as conciseness and fidelity to the data. The method can be applyed to surface reconstruction from unorganized points. Park and Kim described in [15] an adaptive method for smooth surface approximation from scattered three-dimensional points. The approximated surface is represented by a piecewise cubic triangular Bézier surface.

Kalvin and Taylor, [11] developed an algorithm for symplifying polyhedral meshes by reducing the number of vertices, edges and faces. The algorithm, called *Superfaces*, uses a bounded approximation approach which guarantees that the simplified mesh approximates the original mesh to within a prespecified tolerance. Unfortunately the method can generate self-intersecting models.

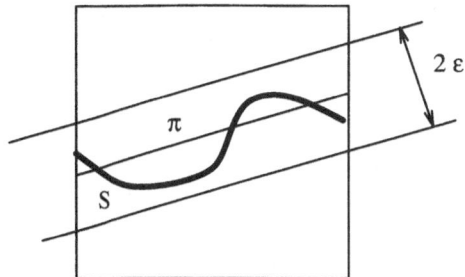

Fig. 1. Face node and associated band.

He *et al.* developed a method for object simplification for applications where gradual elimination of details is desired, [7]. The approach samples and low-pass filters voxel based object representations by appliying the marching cubes algorithm, generating a multi-resolution triangle-mesh hierachy. The method simplifies the genus of the objects.

3   Face Octrees

Just for the sake of completeness let us recall the basic concepts involved in octree representations. Octrees are one of the classical approximate decomposition schemes. They are trees that represent solids by encoding the recursive subdivision of a cubic finite universe where the solids are placed, [13,17].

The nodes allowed in classical octrees are either terminal or non-terminal nodes. Terminal nodes are labeled as black or white depending on whether they are fully inside or fully outside the solid boundary. The solid boundary is approximated by a layer of cubes of a minimum specified size; they are also labeled as black or white depending on some criterion [13,19]. Non-terminal nodes are labeled as grey nodes.

In order to avoid the verbosity and the approximate character of the classical octree scheme, several proposals of new octree encodings have appeared in the literature. Extended octrees were introduced by Ayala *et al.*, [1] and polytrees were introduced by Carlbom *et al.*, [4]. Besides the white and black nodes, these octrees also represent face, edge and vertex nodes. These new types of nodes are terminal nodes and allow the exact representation of the solid boundary of polyhedra while reducing verbosity and keeping a reasonable complexity for the boolean operations.

Face octrees were introduced by Brunet in [2]. A face octree is an octree with white, black, face and grey nodes together with a tolerance ε that controls the degree of approximation of the representation. White, black and grey nodes are defined as in classical octrees. Face nodes contain a connected part of the object boundary and each of them has an associated equation of some plane, π, that approximates the boundary S within the node with a given tolerance ε. See Figure 1. Grey nodes are those that can not be labeled as white, black or face nodes. They represent regions of the object surface that are not flat enough. When the recursive subdivision reaches the minimum predefined node size, grey nodes are terminal grey nodes. Face octrees are halfway between classical and extended octrees; they are more concise than classical octrees and are well suited to approximate representations of objects with complex surface boundaries, [3,16].

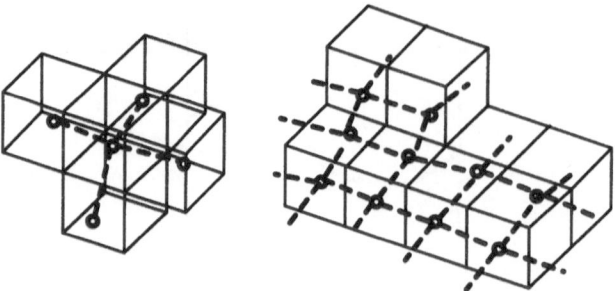

Fig. 2. Interconnecting representative points.

4   The Algorithm

The algorithm has three major steps. First, a set of non planar quadrilaterals that cover the set of boundary nodes in the volume data is generated. Then a face octree is extracted from the volume data representation. Finally, a valid boundary representation is derived from the set of non planar quadrilaterals with the help of the face octree computed in the previous step.

4.1   Definitions

It is assumed that the voxels in the representation are labeled as either black or white; they represent space regions that are, respectively, inside or outside the original solid. Two voxels in the representation are *neighbours* if they share a face, an edge or a vertex. We will refer to either a *vertex neighbour*, an *edge neighbour* or a *face neighbour* whenever we want to make explicit the neighbourhood relationship between two given voxels. A voxel that has at least one white voxel in its neighbourhood is said to be a *boundary voxel*.

4.2   Non Planar Quadrilaterals

A set of non planar quadrilaterals that cover the set of boundary voxels is computed in the form of a geometrically deformed model (GDM), [10,14]. The GDM is an elastic network shaped by a set of points as follows. We start by associating a point with each boundary voxel; we shall call these points the *representative points*. Next each representative point is placed in the boundary voxel center and a link is defined between each representative point and every point that represents a face neighbour of the considered voxel. See Figure 2. A cost function is associated with every representative point in the network. The network is relaxed by minimization of the cost function while each representative point is constrained to stay inside the limits of the voxel that it represents. This restriction guarantees that no self intersections will occur between different boundary elements. If the links between representative points are interpreted as edges, the relaxed network of representative points defines a set of non planar quadrilaterals such that covers the set of boundary voxels and all the edges are wholy inside these boundary voxels. A detailed discussion can be found in [22].

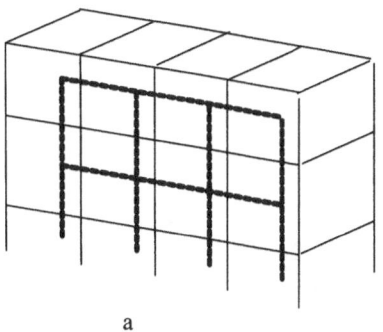 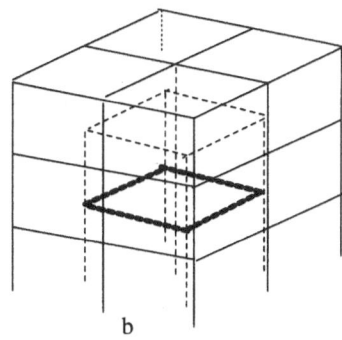

Fig. 3. Extra quadrilaterals in bold dashed lines. a) Dangling quadrilaterals. b) Internal quadrilaterals.

This set of quadrilaterals contains quadrilaterals from which a valid boundary representation can be derived plus some extra quadrilaterals that must be pruned. The undesired quadrilaterals are generated when boundary voxels are arranged forming crests with a thickness of one or two voxels, as illustrated in Figure 3. When the crest is one voxel thick, extra quadrilaterals are dangling quadrilaterals. They are characterized by the fact that the quadrilaterals on the top of the crest have at least one vertex shared by less than four quadrilaterals. These quadrilaterals are recursively pruned starting from those on the top of each crest. Note that as a result of this pruning some information present in the initial data set is lost. When the crest has a thickness of two voxels, quadrilaterals would be located inside the final boundary. In this case, the number of quadrilaterals sharing each vertex of each extra quadrilateral is greater than the number of representative points to which the vertex is connected. These extra quadrilaterals are pruned by a linear walk through the set of quadrilaterals.

It is worth to note that since all the boundary voxels are used to compute the set of quadrilaterals our method does not suffer from the branching problem, [14], that is, the algorithm naturaly handles the situation where several branches of voxels depart from a compact set of voxels.

4.3   Face Octree Generation

A face octree is computed by applying the procedure reported in [10] to the set of relaxed representative points obtained in Section 4.2.

First, a plane is associated with each boundary voxel in the volume representation as follows. For each boundary voxel, the representative points in its face neigbourhood are sorted circularly taking as center the representative point of the considered voxel. Then, for each triangle defined by the center and two consecutive sorted representative points, a normal is computed. Finally, a normal defined as the average of the normals to the triangles weighted with the respective triangle area is associated with the voxel considered. The plane defined by the average normal and the representative point is the GDM plane.

The face octree is extracted from the relaxed representative points and the GDM planes associated to the corresponding boundary voxels, by means of a compaction process. Initially a face octree induced by a space subdivision coincident with the

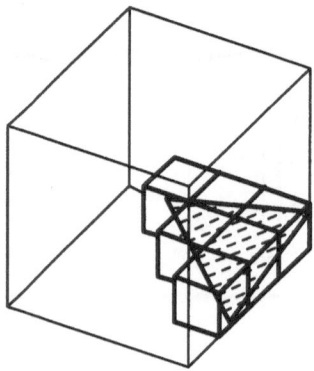

Fig. 4. Face nodes compaction.

planes that define the volume data voxels is defined. The face nodes in this face octree are the boundary voxels with the associated GDM planes. The construction of this first face octree guarantees that the space subdivision defined by face octrees generated in the compaction process will never intersect the interior of a data voxel. The extraction process compacts face nodes in the face octree as much as possible while preserving the volume data precision given by the length of the main diagonal in the data voxels. As illustrated by the simple example in Figure 4, compaction is performed by trying to fit a plane that is interior to all boundary voxels inside the region bounded by the current face octree node and does not intersect any other voxel in the node, [10]. It is worth to note that the face octree computed so far, approximates the local planarity of the boundary voxels in the volume data.

4.4   Boundary Generation

We have a set of boundary voxels each with an associated representative point that has been relaxed. We have a set of non planar quadrilaterals defined on this set of relaxed points that covers the set of boundary voxels. Furthermore, there is a face octree that induces a space subdivision compatible with the voxels in the volume data; i.e., the interior of each boundary voxel is either inside or outside a given face node in the face octree. Generating a valid boundary representation from these informations is straightforward.

For each face node in the face octree, those relaxed representative points of data voxels that are inside the octree node are projected on the plane associated with the node by projecting orthogonally the representative points. See plane π in Figure 5. As a result, projections of non planar quadrilaterals whose vertices are inside a given face node are planar quadrilaterals embedded in the plane associated with the face node, and are inside the same face node, too. These projections are collapsed on one planar polygon defined by the subset of the projections of representative points such that for each projection at least one of the quadrilateral edges incident on it is partially inside and partially outside the face node. We shall refer to these polygons as *planar polygons*. Polygon in boldline in Figure 5 shows an example.

Quadrilaterals generated by projection of non planar quadrilaterals whose vertices belong to data voxels located in different face nodes obviously have different vertices

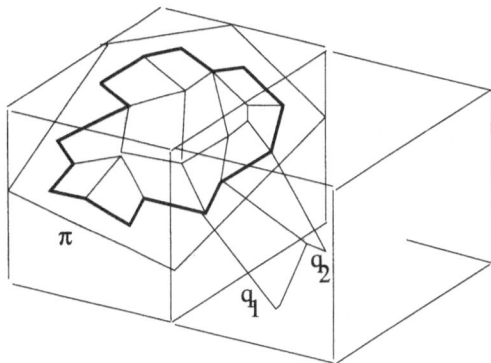

Fig. 5. Quadrilaterals generated by projection of relaxed representative points (thin line) and planar polygon defined (bold line).

inside different face nodes. We will call them *seam quadrilaterals*; in general they are non planar and are split into two *seam triangles*. These triangles complete a valid polyhedral boundary representation by seaming planar polygons that belong to adjacent nodes in the face octree.

Figure 6 shows two face nodes in the octree that are neighbours. Quadrilaterals q_1 to q_6 are projections of non planar quadrilaterals; q_1 and q_2 are inside node a, q_5 and q_6 are inside node b; all of them are planar quadrilaterals embedded in the plane associated with the respective face octree node and will be included in the corresponding planar polygon. Quadrilaterals q_3 and q_4 are seam quadrilaterals. They have vertices in different face octree nodes and, in general, they are not planar and are split into two seam triangles. For an in depth discussion see [22].

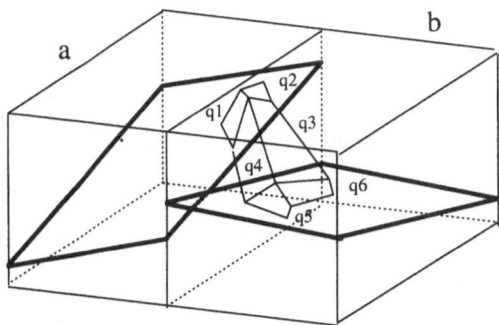

Fig. 6. Projected quadrilaterals. Quadrilaterals q_1 and q_2 are inside node a. Quadrilaterals q_5 and q_6 are inside node b. Quadrilaterals q_3 and q_4 have vertices in both nodes.

5   Geometric Accuracy

An important issue is to guarantee that the extracted representation is a good approximation of the original volume data set. This problem has been considered in a number of different ways, [11].

Since our main goal is to generate boundary representations to perform subsequent geometric operations, we focus in preserving accuracy when classifying a point with respect to the extracted representation. We shall show that, except for those subsets of voxels forming crests one voxel thick that have been lost in the pruning process, the resulting boundary representation approximates the data with a precision given by the length of the main diagonal of the voxel in the initial voxel-based volume representation,

When classifying a point with respect the polyhedral boundary representation, two different cases can arise depending on whether the boundary element is a planar polygon or a seam triangle. Let us consider first the case of the planar polygon. By construction, the plane that supports the planar polygon associated to the face node stabs every boundary voxel in the face node and only these voxels. Thus, the worst situation will happen when the planar polygon and the surface of the initial object respectively share with the boundary voxel the opposite vertices of the main diagonal of the voxel. In this case the error is bounded by the length of the main diagonal of the voxel. Figure 7 illustrates the situation in two dimensions where point p will be incorrectly classified.

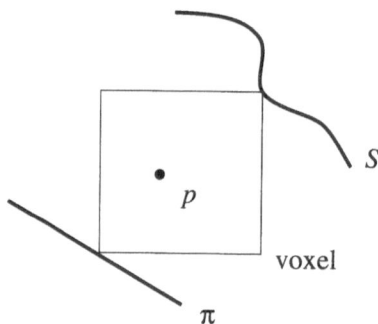

Fig. 7. Worst case when classifying a point with respect to a planar polygon.

Now consider the case where a point must be classified with respect to a seam triangle. First note that as has been said in Section 4.2, the edges of the seam quadrilaterals connect boundary voxels that are face neighbours. Since the edges are inside the set of boundary voxels so are the seam quadrilaterals, and the seam triangles. Figure 8 depicts in two dimensions the situation. Now point p is classified according to plane π that supports the seam triangle and, clearly, the worst case corresponds to the situation where the distance from point p to plane π is equal to the length of the main diagonal of the voxel.

Therefore, the boundary representation generated with our approach classifies a point with an error bounded by the length of the main diagonal of the original voxels.

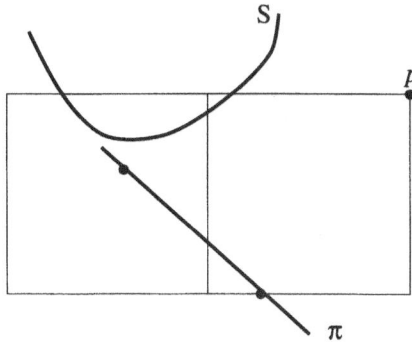

Fig. 8. Worst case when classifying a point with respect to a seam.

6   Results

The method has been implemented and applied to voxel data sets generated from solids defined with a geometric solid modeler. We present three case studies. For each case study we have sampled the geometry of the initial solid models with three different precision levels. The lowest level was generated by dividing into 32 voxels the edges of the cubic space region where the solids were embedded, for the intermediate level each edge was divided into 64 voxels, and for the highest precision the edges were divided into 128 voxels. Numerical values are shown in Table 1. The second column gives the precision, the third gives the number of voxels in the data set, the fourth is the number of nodes in the face octree, the fifth column gives the number of polygons in the boundary representation, and the last gives the ratio of the number of polygons in the boundary representation with respect to the number of voxels in the volume data set. From the numerical results it is clear that, except for the lowest precision level, the

Table 1. Case studies.

Solid	precision	# voxels	# fo nodes	# polygons	# polygons / # voxels
	32	1320	653	1718	1.30
A	64	7599	1373	5185	0.68
	128	53645	2886	16768	0.31
	32	4502	1351	4285	0.95
B	64	19773	2596	11470	0.58
	128	229593	16662	51753	0.23
	32	4926	1510	4965	1.00
C	64	33373	3175	14694	0.44
	128	246365	8582	48272	0.19

polygonal boundaries generated are much more compact than the initial volume data sets. As one could expect, the larger the precision the larger the resulting compaction.

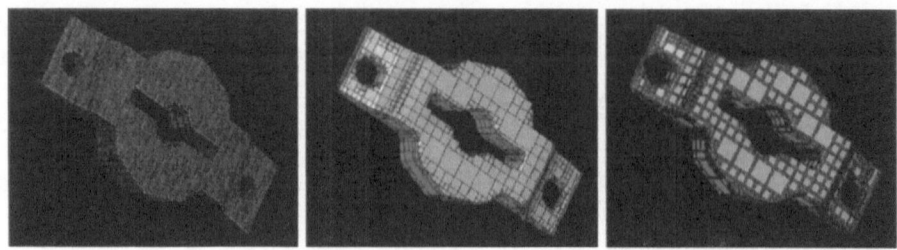

Fig. 9. Case study solid A.

Figures 9, 10 and 11 illustrate the quality of the boundary representations for solids in Table 1. The pictures were computed for the 128 by 128 by 128 voxel data sets. In each of the three sets of pictures, the picture on the left side shows the set of data voxels, the picture in the middle shows the associated face octree, and the picture on the right side shows the boundary representation.

In the Figures on the right side, the planar polygons resulting from collapsing projections of non planar quadrilaterals on the octree face nodes planes can be easily distinguished as clean areas with uniform colour. The dark lines are the chains of seam triangles resulting from splitting the non planar seam quadrilaterals which sew the gaps between planar polygons.

We applied the algorithm to a voxel data set corresponding to a human skull that contains a total of 28837 voxels. The skull was embedded in a 128x128x128 voxels[3]. Figure 12 shows from left to right the original voxel data set, the face octree associated and the boundary representation. The number of face nodes in the face octree is 17079 and the number of polygons in the boundary representation is 49412.

In this case the number of faces in the final boundary representation is larger than the number of voxels in the original data set. The rational behind this fact is that the skull is a thin solid and while the number of voxels needed to capture the presence or absence of solid is rather small, a valid, polygonal boundary representation is still verbose.

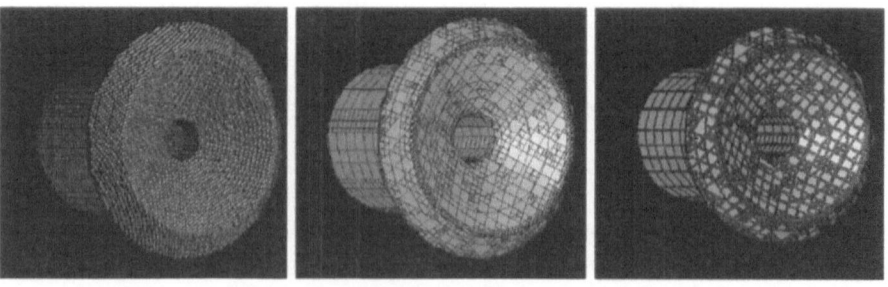

Fig. 10. Case study solid B.

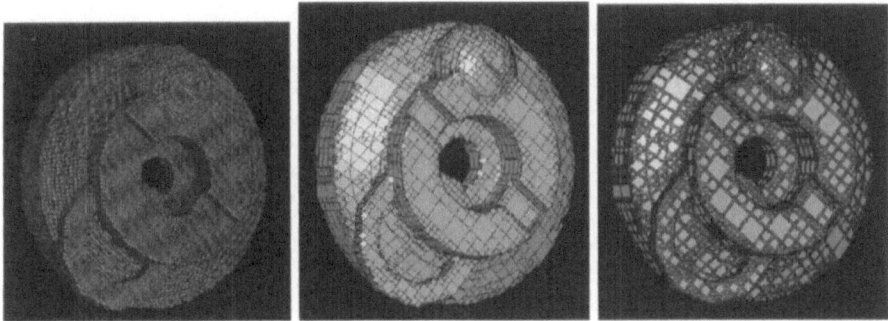

Fig. 11. Case study solid C.

7   Summary

Given a voxel-based model, the algorithm presented generates a closed, non self-intersecting, polygonal boundary representation, with a high degree of compaction. The degree of approximation of the initial data set is preserved except for those parts where there are crests of voxels one voxel thick. We have demonstrated the use of this algorithm in several practical examples.

The algorithm is based on GDM techniques. It has been used in computing both a collection of non planar quadrilaterals that approximate the data set boundary, and in computing a face octree which works as an auxiliar data structure. As a result, the algorithm inherits several interesting properties exhibited by GDM models. One interesting property is that the cost of extracting the boundary representation depends on the complexity of the object, not on the size of the original voxel data set. Furthermore, since all the boundary voxels are involved in the computation of the GDM, the algorithm explicitly handles the branching problem.

The extraction of valid boundary representations from large sets of volume data is always a time-consuming problem. To reduce computation time, a parallel version of the algorithm presented here is now being programmed on a CM-2 SIMD machine. This is part of a larger project comprising parallelization of a number of geometric operations on solids representations.

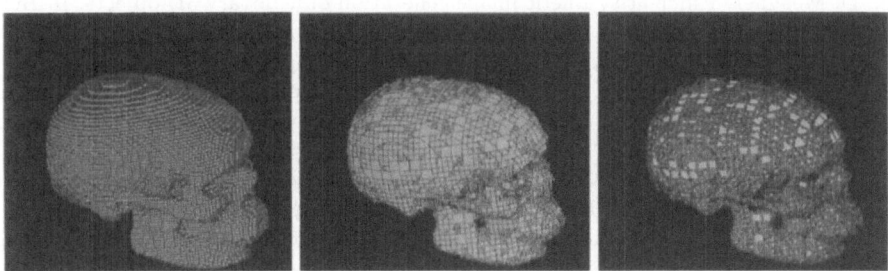

Fig. 12. Case study human skull.

Acknowledgement

The first author has been partially supported by CICYT under grant TIC95-0630-C05-04. We also thank Prof L. Pérez-Vidal for his help.

References

1. D. Ayala, P. Brunet, R. Juan, and I. Navazo. Object representation by means of nonminimal division of quadtrees and octrees. *ACM Transactions on Computer Graphics*, 4:41–59, 1985.
2. P. Brunet. Face octrees: Involved algorithms and applications. Technical Report LSI-90-14, Universitat Politècnica de Catalunya, Department of LiSI, 1990.
3. P. Brunet, I. Navazo, and A. Vinacua. A modelling scheme for the approximate representation of closed surfaces. *Computing*, 8:75–90, 1993.
4. I. Carlbom, I. Chakravarty, and D. Vanderschel. A hierarchical data structure for representing the spatial decomposition of 3-D objects. *IEEE Computer Graphics and Applications*, pages 24–31, April 1985.
5. M.J. DeHaemer and M.J. Zyda. Simplification of objects rendered by polygonal approximations. *Computers and Graphics*, 15(2):175 – 184, 1991.
6. M.J. Dürst and T.L. Kunii. Integrated polytrees: A generalized model for integrating spatial decomposition and boundary representation. In *Theory and Practice of Solid Modeling*, Springer Verlag, 1989.
7. T. He, L. Hong, A. Kaufman, A. Varshney, and S. Wang. Voxel based object simplification. In G.M. Nielson and D. Silver, editors, *Visualization'95*, pages 296–303, Atlanta, GA, October 29 – November 3 1995.
8. G.T. Hermann and H.K. Liu. Three-dimensional display of human organs from computer tomograms. *Computer Graphics and Image Processing*, 9, (1):1–21, 1979.
9. H. Hoppe, T. DeRose, T. Duchamp, J. McDonald, and W. Stuetzle. Mesh optimization. In *SIGGRAPH'93, Computer Graphics Proceedings, Annual Conference Series*, pages 19 – 26, 1993.
10. R. Juan-Arinyo and J. Solé. Constructing face octrees from voxel-based volume representations. *Computer-Aided Design*, 27(10):783–791, 1995.
11. A.D. Kalvin and R.H. Taylor. Superfaces: Polyhedral approximation with bounded error. *IEEE Computer Graphics and Applications*, 16(3):64–77, May 1996.
12. B.H. McCormick, T.A. De Fanti, and M.D. Brown (Eds). Visualization in scientific computing. *Computer Graphics*, 21:1–21, 1987.
13. D. Meagher. Efficient synthetic image generation of arbitrary 3D objects. In *IEEE Computer Society Conf. on Pattern Rec. and Image processing*, pages 473–478, 1982.
14. J.V. Miller, D.E. Breen, W.E. Lorensen, R.M. O'Bara, and M.J. Wozny. Geometrically deformed models: A method for extracting closed geometric models from volume data. *Computer Graphics*, 25, 4:217–226, 1991.
15. H. Park and K. Kim. An adaptive method for smooth surface approximation to scattered 3D points. *Computer-Aided Design*, 27(12):929–939, 1995.
16. N. Pla-Garcia. Recovering a smooth boundary representation from an edge quadtree and from a face octree. *Computer Graphics Forum*, 13(4):189–198, 1994.
17. A. Requicha. Representations for rigid solids: Theory, methods, and systems. *Computing Surveys of the ACM*, 12:437–464, 1980.

18. H. Samet. *Applications of Spatial Data Structures.* Addison Wesley Publ., Reading, MA, 1989.

19. H. Samet. *The Design and Analysis of Spatial Data Structures.* Addison Wesley Publ., Reading, MA, 1989.

20. F. Schmitt, B. Barsky, and W. Du. An adaptive subdivision method for surface-fitting from sample data. *Computer Graphics,* 20(4):179–188, 1986.

21. W. J. Schroeder, J. A. Zarge, and W. E. Lorensen. Decimation of triangle meshes. *Computer Graphics,* 26(2):65 – 70, 1992.

22. J. Solé. *Parallel Operations on Octree Representations Schemes.* PhD thesis, Department LiSI, Universitat Politècnica de Catalunya, 1996.

23. G. Turk. Re-tiling polygonal surfaces. *Computer Graphics,* 26(2):55 – 64, 1992.

Conversion of Binary Space Partitioning Trees to Boundary Representation

João Comba[1] and Bruce Naylor[2]

[1] Computer Science Department, Stanford University, USA *
[2] Spatial Labs, Inc. **

Abstract. Binary Space Partitioning Trees (BSP-Trees) have been proposed as an alternative way to represent polytopes based on the spatial subdivision paradigm. Algorithms that convert from Boundary Representation (BRep) to BSP-Trees have been proposed, but none is known to perform the opposite conversion. In this paper we present such an algorithm, that takes as input a BSP-Tree representation for a polytope and produces a BRep as output. The difficulty in designing such algorithm comes from the fact that the information about the boundary is not explicitly represented in the BSP-Tree. The solution we present involves a recursive traversal of the tree to compute lower dimensional information, along with a gluing algorithm that combine the convex regions defined by the BSP-Tree, removing internal features. A new data structure is proposed (a Topological BSP-Tree), that augments the traditional BSP-tree with topological pointers and is used to store intermediate results used in the reconstruction of the BRep.

1 Introduction

Boundary representation (BRep) is a widely used representation of solid geometry, based on the description of an object by its boundary, as a collection of faces, edges and vertices[3]. On the other hand, Binary Space Partitioning Trees (BSP-Trees) consist of convex hierarchical decompositions of the space, where the object is represented by the union of convex regions. The basic operation for the construction of this decomposition consists of a partition of the underlying space by a given hyperplane. This partition is represented as a binary tree, where each node is identified by a hyperplane, and the left and right subtrees of the node represent the two halfspaces obtained in the partition. The recursive application of this operation creates convex hierarchical decompositions of the space. In Solid Modeling, BSP-Trees have been used along with BReps, and many algorithms have been proposed, like the conversion from BRep to BSP-Tree [11], or the one that computes boolean operation with BSP-Trees [7].

In this paper we consider the problem of converting a BSP-Tree representation of a polytope into a BRep. This problem is similar to the conversion of CSG to BRep (also called the *Boundary Evaluation*), because the BSP-Tree can be first converted into a CSG by computing the union of all the convex regions that the BSP-Tree defines, and

---

* e-mail: comba@cs.stanford.edu, web: http://www-graphics.stanford.edu/
** e-mail: naylor@spatial-labs.com

applying over the result any of the Boundary Evaluation algorithms proposed in the literature([9], [10]).

However, this approach does not exploit the spatial subdivision information that the BSP-Tree encodes, which may lead to more efficient algorithms. In order to use this information, we propose an algorithm that converts the BSP-Tree to a BRep by working directly in the structure of the BSP-Tree. In Fig. 1 we illustrate the problem

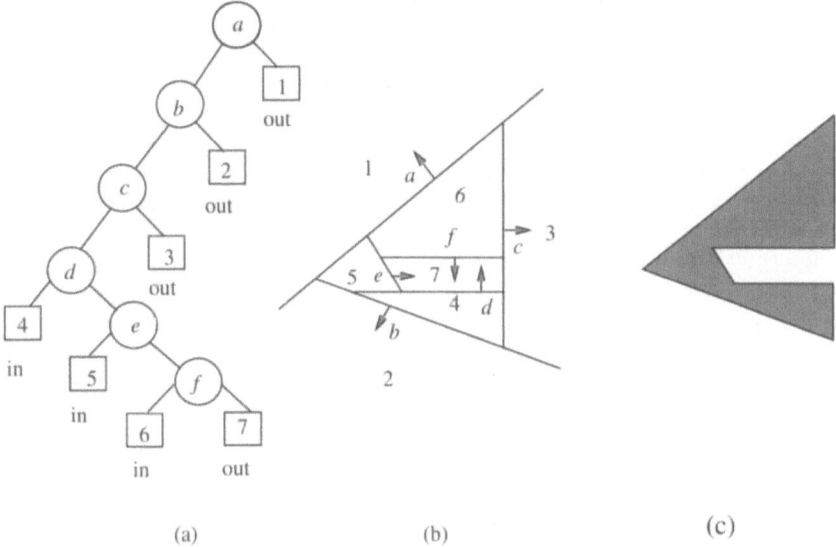

(a)   (b)   (c)

Fig. 1. (a) BSP-Tree, (b) Partition induced by the BSP-Tree. The resulting object is composed of 3 different regions (4, 5, 6), corresponding to the IN nodes (c) Computing the boundary requires gluing different regions and removing internal elements

of reconstructing the boundary of an object obtained by a partition induced by the BSP-Tree. In this example the union of regions 4, 5 and 6 (Fig. 1c) form the object. Looking at the example we identify some difficulties in the design of the conversion algorithm:

- The BSP-Tree does not have lower dimensional information: Intersections need to be computed to obtain the faces, edges and vertices on the boundary.
- The BSP-Tree does not have adjacent-topological information.
- The BSP-Tree represents a convex decomposition of solids as convex regions, and the boundary may be partitioned in multiple independent components that are not minimal.

In the following sections we present how we solved each of the above problems, and give the algorithm to compute a BRep from a BSP-Tree. Initially we review some basic concepts about BSP-Trees. The algorithm that computes the BRep from a BSP-Tree is discussed next, where we describe the Topological BSP-Tree (TBSP-Tree), an auxiliary data structure that stores intermediate results and topological relations among them.

2   Review of BSP-Trees

2.1   Basic Concepts

Binary Space Partitioning trees (BSP-Trees) are spatial search structures used in many different aspects of Computer Graphics and Geometric Modeling. Applications in Solid Modeling [6] [7] [11] [13] , Visibility Orderings [1] [2] [12] and Image Representations [8] , among others, can be found in the literature. In order to describe the concepts of BSP-Trees, it is always nice to first explain the relation it has with Binary Search Trees.

Binary Search Trees have been used in Computer Science in many ways, but mostly as a data structure to accelerate search queries based in symbolic values. A geometric interpretation of this data structure is a hierarchy of binary partitions of the real line, where the partitioner is a point and each partition obtained represents an interval. The problem with this interpretation is that it does not directly generalize to higher dimensions, as points do not partition such spaces. The misleading information with this interpretation is that the partitioner that was supposed to be a point is in fact a hyperplane. In general, for a D-dimensional space the partitioner corresponds to the hyperplane to that space (a (D-1)-D element), and the partition has the same dimension of the underlying space. BSP-Trees and Partition Trees [4] use this analogy to extend the concepts of Binary Search Trees to higher dimensional spaces, but we choose to work here with BSP-Trees as they have a more natural correspondence with the representation of solids than Partition Trees.

One advantage of BSP-Trees is the ability to combine a search structure with a representation scheme in one unique data structure. The use of BSP-Trees as a solid representation scheme reveals one application where this combination is well exploited. Solids are represented using BSP-Trees by a union of convex regions, which are identified by associating attributes to the leaves of the tree. These regions may be either inside or outside the solid, and are defined by the hyperplanes that are in the path from the root to the leaf node. In order to precisely define the region, the leaves have associated an attribute that indicates that the region is inside or outside the solid.

2.2   Formal Definitions

Many algorithms in BSP-Trees are better explained if we represent the main concepts formally. In this section we present some definitions and properties that are going to be used in the paper.

Definition 1. Hyperplanes and Halfspaces: The hyperplane is the basic element to recursively partition the space, and it is described by the following equation:

$$h = \{(x_1, ..., x_d) \mid a_1 x_1 + ... + a_d x_d + a_{d+1} = 0\}$$

The hyperplane separates the space in two halfspaces, the positive and the negative halfspace. Each of them is expressed as:

$$h^+ = \{(x_1, ..., x_d) \mid a_1 x_1 + ... + a_d x_d + a_{d+1} > 0\}$$
$$h^- = \{(x_1, ..., x_d) \mid a_1 x_1 + ... + a_d x_d + a_{d+1} < 0\}$$

The normal of the hyperplane is defined by the vector $(a_1, a_2, ..., a_d)$. The positive halfspace corresponds to the one that lies in the direction of the normal.

Definition 2. BSP-Tree Nodes and Leaves: A BSP-Tree node represents the information of the binary partition being performed on the space. It consists of a partitioning hyperplane, and left and right children that point to BSP-Tree representations of the positive and negatives halfspaces. The hyperplane that defines the node n is denoted with a $H(n)$.

A BSP-Tree leaf contains attributes associated to a given region. It contains the labels IN (if the region is inside the solid), or OUT (if the region is outside the solid), but may also contain additional attributes, like color or density.

Definition 3. Region Path: A Region Path corresponds to the path that leads from the root of the tree to another node of the BSP-Tree. This path represents one given partition of the space, and is represented by an ordered list of nodes L.

$$RP(n) = L = \{root, ..., parent(parent(n)), parent(n), n\}$$

Definition 4. Region: A region of a given node represents the geometric interpretation of the partition defined by the region path $RP(n)$. It corresponds to the intersection of all positive halfspaces in the region path, and can be formulated as:

$$R(n) = \{\cap H^\diamond(\nu) \mid H^\diamond(\nu) \in RP(n), \diamond = +, -\}$$

Definition 5. Sub-hyperplane: A sub-hyperplane s of a hyperplane h at a given node n is defined by the intersection of the hyperplane h with the region $R(n)$ defined in the node.

$$s = SUB(h, n) = \{h \cap R(n)\}$$

Definition 6. Projected Hyperplane: A projected hyperplane p is a one dimension lower hyperplane that is obtained by projecting the intersection of two hyperplanes $h1$ and $h2$ orthogonally to one of the coordinate axis.

Definition 7. Path Partial Ordering $\prec_p$: Let h_1 and h_2 be two hyperplanes in a common region path p from the root of the tree. We define $h_1 \prec_p h_2$ if h_1 is in the left subtree of h_2. Otherwise, $h_2 \prec_p h_1$.

3   Computing the BRep from a BSP-Tree

The BSP-Tree represents solids as the union of convex regions, and the boundary associated with each of these regions is defined by the intersection of halfspaces in the tree. Each convex region is identified by two attributes: a leaf node l labeled as IN (i.e. not empty), and a region path $RP(l)$ associated with l. The boundary of this region is obtained by the intersection of the hyperplanes in its corresponding region path.

In the simple case where the BSP-Tree consists of a single convex region we need to consider only one region path, and to obtain the boundary we compute the intersection of all hyperplanes in the region path (Fig. 2).

In a more complicated BSP-Tree, with more than one convex region, it is likely that some of the region paths associated with these regions will share nodes in the tree. In Fig. 3, region paths of the regions R1, R2, R3 and R4 share the nodes a and b, and the intersection of $H(a)$ and $H(b)$ is on the boundary of all these regions. In fact, any

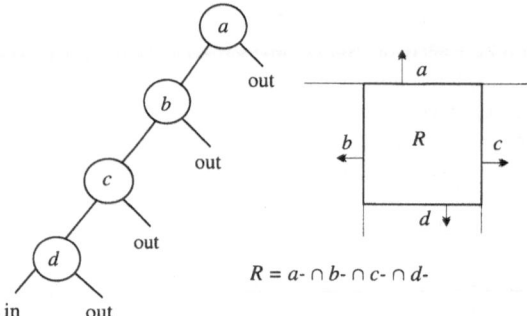

Fig. 2. Simple BSP-Tree generating only one convex region

node in the tree may contribute to the boundary of all regions that contains the node in its region path. A related consequence of this fact is that when computing a given region we may not need to compute the intersections of all hyperplanes in the region path, as some may not contribute to the region (redundants to the region),

The sharing of nodes among different regions implies that the computation of intersections to find the boundary of regions may require repeated computations. This fact reveals a structure common in Dynamic Programming (DP) problems. In each node of the tree we can partition the problem of computing the boundary into smaller problems, corresponding to the boundary in the positive and negative halfspaces of the node. In order to obtain such boundaries, we need to compute intersections among hyperplanes in each of the subtrees (separated subproblems) and ancestor hyperplanes (shared subproblems), which identifies the DP structure.

One way to avoid re-computation of common intersections shared by subproblems is to store them in intermediary data structures. The way of representing these inter-

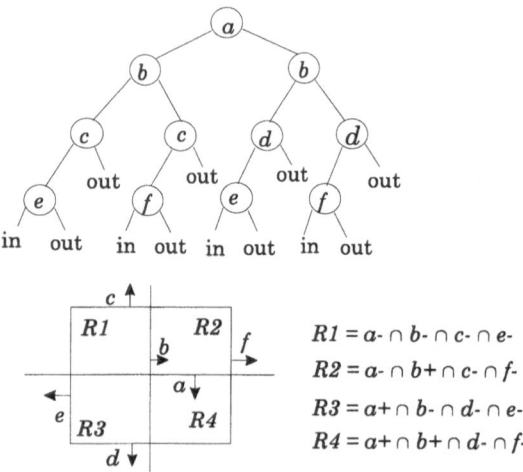

Fig. 3. Complex BSP-Tree generating more than one convex region

sections, here called lower dimensional information, is described in details in the next section.

3.1   Storing Lower Dimensional Elements

One method of representing lower dimensional elements in a BSP-Tree relies on using BSP-Trees of lower dimension, which results in a structure called a multi dimensional BSP-Tree. In 1990, Naylor [5] proposed but did not elaborate the use of a pure BSP-Tree model to represent solids. The proposal was based in the extension of the standard BSP-Tree model to represent explicitly the multi-dimensional information defined by the structure of the tree. In 1991, Vanecek [13] used similar ideas and proposed the BRep-Index, which consisted of a multi-dimensional BSP-Tree (called MSP) attached to a BRep, with the goal of providing efficient access to the BRep structures. The BRep-Index is constructed in such a way that there is a correspondence between (0,1,2)-d nodes of the MSP with the vertices, edges and faces of the BRep.

In this paper, we represent the lower dimensional information obtained in a different kind of multi-dimensional BSP-Tree. In the BRep Index of Vanecek the topological information is stored in the BRep structure and not in the MSP. In the data structure proposed in this paper, the topological information is stored in a multi dimensional BSP-Tree, augmented with additional topological pointers connecting elements topologically adjacent. We call this data structure a *Topological BSP-Tree* (TBSP-Tree).

The motivation for creating such a structure comes from the fact that in order to obtain lower dimensional information we must compute the intersection of hyperplanes, which gives certain information about the way that elements are topologically related. In the example of Fig. 4, when computing the intersection of the lines a and b we obtain a point ab that we need to insert into the lower dimensional BSP-Trees associated with a and b. In other words, a is topologically adjacent to b by ab. In order to preserve this information, we connect the nodes we obtain in this intersection with topological pointers.

Besides the addition of these topological pointers, the TBSP-tree is different from the MSP of Vanecek because we keep not only one, but two pointers to lower dimensional dimensional BSP-Trees, corresponding to the subdivisions formed over both sides of the hyperplane. One reason for this choice includes the simplification of the incremental algorithm to build the topological BSP-Tree, which requires a partial ordering among the hyperplanes, that can be simplified if we process the subdivisions in both sides separately.

3.2   Navigating the TBSP-Tree

The TBSP-Tree stores topological information about the intersections computed in a pre-order traversal of the tree. This is achieved by creating copies of the same intersection and connecting them with topological pointers. In general, for each intersection of two hyperplanes h_1 and h_2 we keep three copies in the TBSP-Tree. The node higher located in the tree receives one copy, stored in the lower dimensional tree being used in the current region path. The other two copies are stored in the two lower dimensional trees associated with the other node. These last two copies are connected to the first copy created by a topological pointer. In the case where $h_1 \prec_p h_2$, the first copy is stored in the negative lower dimensional tree of h_2, because h_1 is in the left subtree

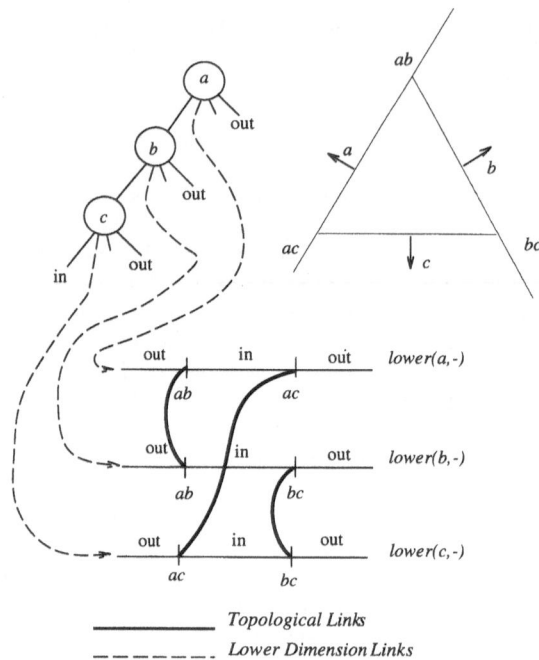

Fig. 4. Topological BSP-Tree (TBSP-Tree) example in 2D

of h_2. The other two copies are stored in the positive and negative lower dimensional trees of $h1$.

Using the information stored in the TBSP-Tree we are able to recover the basic elements of the Boundary Representation. This can be illustrated by looking at the partial TBSP-Tree representation for a cube in Fig. 5. The faces of the cube are defined by the hyperplanes $H(a)$, $H(b)$ and $H(c)$, which are represented in a 3D BSP-Tree by the nodes a, b and c. The intersection $H(a) \cap H(b)$ is represented in the lower dimensional trees associated with a and b as ab and ba respectively. The edges that belong to the face $H(a)$ are defined in the lower dimensional trees associated with node a, which in this case has one empty tree, as the partition occurs in only one of the sides of a.

In order to traverse the faces of the cube directly from the TBSP-Tree we make use of the following operators:

- **TreeParent(node) and TreeSon(node,side):** Traditional pointers in trees for parents and sons.

- **DimensionParent(node) and DimensionSon(node,side):** Pointers that reflect an incidence relation in the dimension of the space. DimensionSon(node,side) returns a pointer to the lower dimensional BSP-Tree associated with the node in the specific side, and DimensionParent(Node) returns an upper-dimensional BSP-Tree associated with the node.

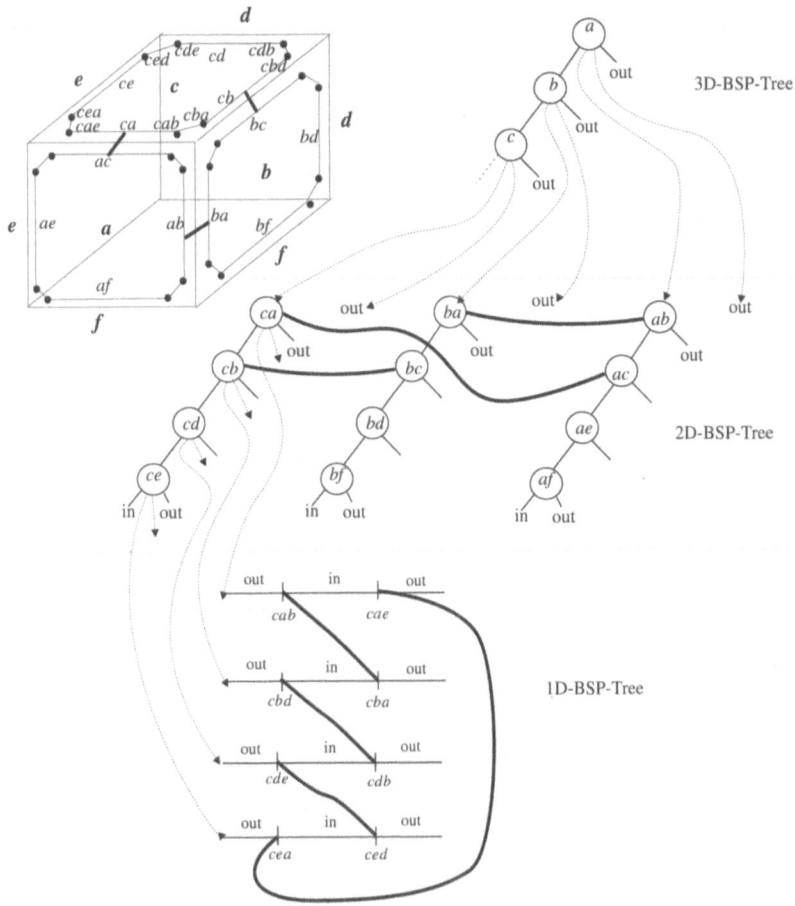

Fig. 5. Topological BSP-Tree (TBSP-Tree) example in 3D

— **TopologicalCopy(node,side):** The application of this operators returns the topological copy of a node at a given side. In Fig. 5, TopologicalCopy $(ab,-)$ returns
  the copy ba^-, which corresponds to the same intersection ab, but stored in the
  negative lower dimensional BSP-Tree of b.

— **TopologicalNeighbor(node):** In 1D it is possible to define an adjacency relation.
  The Topological neighbor of a node corresponds to the next node in ascending order
  in the real line. This order is defined in terms of the projected hyperplane normal,
  and reflects our convention for the orientation of loops in the representation of the
  boundary. In Fig. 5, the 1D-BSP-Trees are oriented by the hyperplane normal to
  each c edges. The direction of the hyperplane normal is defined by the projected
  hyperplane and the current side of the lower dimensional tree being used.

These operators allow us to access the boundary elements for the cube. In Fig. 6
we show the procedure to visit the elements of a face (**VisitFace**). The application of

this procedure to the leaf node *IN* (next to node *cea*) will visit edges (*cea*, *ced*), (*cde*, *cdb*), (*cbd*, *cba*) and (*cab*, *cae*), which correspond to the edges of the face *c*.

```
procedure VisitFace(1D-BSPTree node1D)
  // The starting node corresponds to a leaf containing
  // an IN attribute in the 1D-BSP-Tree
  currentNode1D = startNode1D = DimensionParent(node1D);
  do
    nextNode1D = TopologicalNeighbor (currentNode1D);
    VisitEdge (currentNode1D, nextNode1D);
    currentNode1D = TopologicalCopy (currentNode1D);
  while (currentNode1D != startNode1D)
end VisitFace;
```

Fig. 6. Procedure to visit a face using the TBSP-Tree

3.3   Incrementally Computing the TBSP-Tree

For every node visited in the traversal of the tree, we discover more information about how the BSP-Tree partitions the space. When a new node is reached, we may compute the intersection of the hyperplane that defines the node against all hyperplanes in the region path of the node. These intersections are used to update the lower dimensional BSP-Trees of all the nodes in this region path.

Every intersection obtained is projected onto one of the coordinate hyperplanes before insertion into the lower dimensional trees, corresponding to the concept defined before of a projected hyperplane. One way to understand this operation is to remember the way that Gaussian Elimination (GE) solves linear systems. In GE, the solution of a linear system involves first a triangulation step, where each of the columns under the pivot (the diagonal element) is eliminated (replaced by zeros). In fact, at every elimination step the columns of the matrix, which can be interpreted as the coefficients that define a hyperplane, are projected into one of its dimensions. This is exactly the same operation we are performing here to obtain the projected hyperplanes. In order to compute all lower dimensional information induced by the BSP-Tree on a specific hyperplane we compute the intersections with the other hyperplanes, and project these intersections in a direction orthogonal to one of the coordinate axis. This has the same effect of sweeping with zeros a column in GE. For simplicity, from now on when we refer to intersections we are in fact referring to the projected hyperplanes of the intersections.

The update step of the lower dimensional trees is performed after an intersection is found, consisting of the insertion of projected hyperplanes in lower dimensional trees. This insertion operation is guided by a partitioning operation, which involves the classification of a hyperplane against a partitioning hyperplane. Depending on the result of this classification, the intersection between the hyperplanes is computed and the hyperplane being inserted is partitioned into two sub-hyperplanes. The sub-hyperplanes are recursively inserted into the left and right subtrees of the partitioner node. However, we exploit one unrecognized property of the BSP-Trees that guarantees that we need to perform this insertion into only one subtree of the node.

Suppose that in the BSP-Tree described in Fig. 7 we want to compute the lower dimensional information associated with node a. Performing a pre-order traversal of

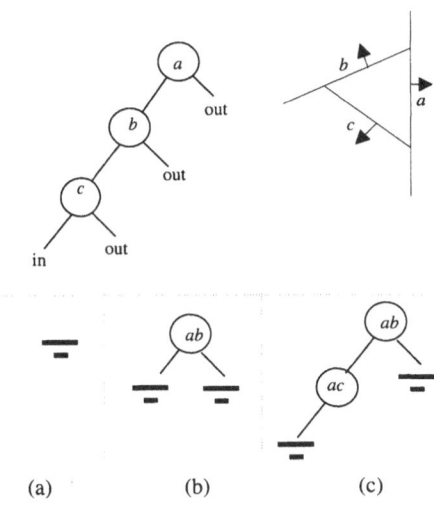

(a)   (b)   (c)

Fig. 7. Simple case in the incremental construction of the TBSP-Tree. The status of the lower dimensional BSP-Tree associated with a after the visit of nodes a, b and c is illustrated in (a), (b) and (c)

the tree we visit in this order nodes a and b in the tree. In this case b intersects a, therefore we compute the intersection ab and insert as a node into the negative lower dimensional BSP-Tree of a, as $b \prec_p a$. The next node visited is c, and the insertion of ac into the lower dimensional tree of a involves a partition against ab, to decide in which side of ab we need to insert ac. Using the the path partial ordering defined by the tree, we observe that $c \prec_p b$, and therefore we must have that $ac \prec_p ab$, which means that ac needs to be inserted into the left subtree of ab.

But it is easy to formulate a case where the insertion operation can not be guided by such partial ordering. In Fig. 8, b does not intersect a. This is only possible because b is parallel to a, as b is a child of a. Clearly, one of the subtrees of b (left or right) does not need to be tested against a (in this case the right subtree). Therefore, even if nodes c and d intersect a we do not need to partition one against the other, as b is *shielding* one of them from intersecting a.

A more complicated example involves the case where some node (not a child) does not intersect a. Consider the case in Fig. 9 where b and e intersect a, but c does not intersect a. The insertion of af into the lower dimensional BSP-Tree associated with a requires first a partition against ab. Using as above the partial order $f \prec_p b$ we propagate the result to the left subtree of ab. Now we need to partition af against ae, but the path partial ordering is not defined between e and f. Note that the fact that c does not intersect a tells us that one of its subtrees does not need to be tested for intersection against a, because as before c and its ancestors shield one of its subtrees from intersecting a. In this case, c and b shield f from intersecting a

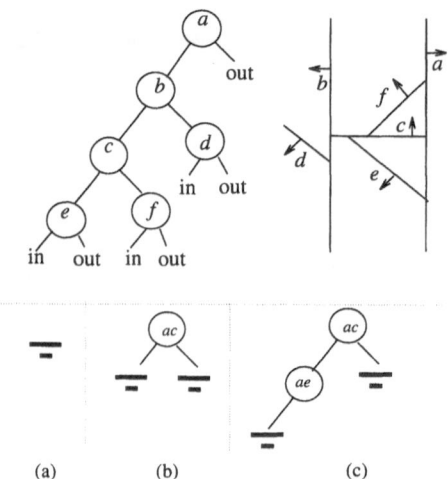

Fig. 8. First case of Partial Ordering not defined. The status of the lower dimensional BSP-Tree associated with a after the visit of nodes a, c and e is illustrated in (a), (b) and (c)

Using such results, we may decide in which side to insert a given hyperplane when we perform the partitioning operation. We partition the hyperplane by computing the intersection with the partitioner, and propagate the result to the side consistent with the partial ordering defined by the current path. When a leaf node is reached, we evaluate the sub-hyperplane partitioned, and if it is not empty we create a new node with undefined left and right attributes, and update all the 1D-BSP-Trees of the most recent partitioner hyperplanes found along the way.

One difficulty in this incremental construction is that we only discover the attributes (IN or OUT) when we visit a leaf node. At this point, we need to propagate this attribute to all lower dimensional BSP-Trees in the current region path and create leaf nodes with the attribute just discovered. In fact this difficulty is a particular case of the insertion that we are performing in the other cases, only that in this case we are not inserting a hyperplane but an attribute, that can be localized in the tree using the partial ordering as above.

Another important operation that needs to be performed when computing intersections and inserting them into lower dimensional BSP-Trees is to keep track of the place where we insert the different copies of an intersection. For instance, the same intersection h_{12} of h_1 and h_2 needs to be inserted into the lower dimensional BSP-Tree associated with h_1 and with h_2. Topological pointers are created to connect the copies of this intersection, which will allow the reconstruction of the BRep in a next step. The procedure to compute the lower dimensional information is described in Fig. 10.

3.4   Using the TBSP-Tree to Reconstruct the Boundary

The incremental construction of the TBSP-tree built information about topological relations among the elements in all dimensions. A node in the tree has all the information necessary to reconstruct the boundary when both lower dimensional trees have been

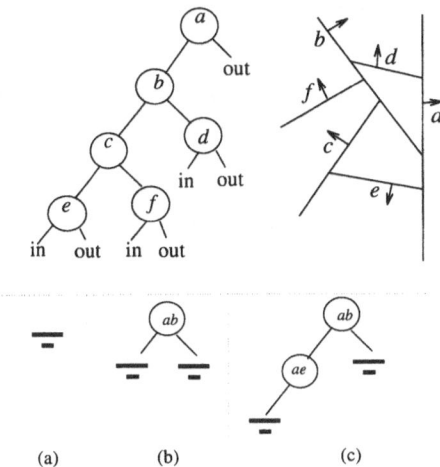

(a)   (b)   (c)

Fig. 9. Second case of Partial Ordering not defined. The status of the lower dimensional BSP-Tree associated with a after the visit of nodes a, b and e is illustrated in (a), (b) and (c)

completely computed. In order to extract the boundary we perform a gluing operation in the dimension of the embedding space, which remove edges internal to faces and faces internal to solids.

In Fig. 11 we have an example where the partitions in both sides of the hyperplane h need to be glued together to remove internal elements. The gluing operation is responsible for combining the results from the two subtrees in such way that the result is a valid representation for the boundary defined in the subtrees of the node. In order to glue the information in the positive and negative halfspaces of a node, we perform the symmetric difference of the lower dimensional BSP-Trees in both sides of the hyperplane, which is a boolean operation that can be executed by a tree merging algorithm for BSP-Trees[7]). The result of the symmetric difference operation is a tree whose *IN* regions are elements of the boundary of the object.

Note that the gluing process to remove internal features entails a recursion on dimension that first glues lower dimensional features. In other words, in order to glue two faces we first glue the elements internal to each of these faces separately, which will correspond to the removal of internal edges of a face. The algorithm initially performs two gluing operations in the positive and negative lower dimensional trees, and a symmetric difference operation that combines the results obtained. The algorithm for gluing TBSP-Trees is dimension-independent, and the pseudo-code is described in Fig. 12.

The gluing process can be implemented in such a way that the boundary relations of the different dimensioned elements is preversed. In Fig. 13 we illustrate the gluing operation for an example in 2D. The complicated cases arise in nodes f and c, where we have partitions induced by the BSP-Tree in both sides of the hyperplanes. In Fig. 13a we show the partition induced by the tree, and the information stored in the TBSP-Tree is illustrated by the cycles of edges and vertices in each cell. When performing the symmetric difference we not only remove internal features, but we also join the cycles

```
procedure ComputeLowerDimension(BSPTree *current, BSPTree *path)
   for each hyperplane h_p in the current path
      if current hyperplane h_c is a leaf
         // We update the lower dimensional trees with the attribute
         Insert(DimensionSon(h_p, side(h_p)), current.attribute, path);
      else
         Compute intersection h_pc between h_p and current h_c
         if h_pc is not empty
            // We insert the intersection into the lower BSP-Trees
            // of hc and hp and link them with topological links
            Insert(DimensionSon(h_c, +), h_pc, path);
            Insert(DimensionSon(h_c, -), h_pc, path);
            Insert(DimensionSon(h_p, side(h_p)), h_pc, path);
         endif
      endif
   endfor
   ComputeLowerDimension(current.left, path ∪ current^+)
   ComputeLowerDimension(current.left, path ∪ current^-);
end ComputeLowerDimension;
```

Fig. 10. Procedure to Compute Lower Dimensional Information

in both sides of the hyperplane. In Fig. 13b we show the result of the application of the gluing operation to the node f. The corresponding features are identified by performing the symmetric difference operation for both lower dimensional trees in one dimension lower, and the cycles are joined together when we have mutually incident relations. In this case, the edge generated by the hyperplane f is removed, and the cycle of edges of both copies of f are joined together. In a similar way Fig. 13c shows the result of the gluing operation for node c.

It is important to demonstrate that the boundary information obtained is in fact minimal. This is achieved due to the fact that all internal features are removed by the recursion in dimension performed by the gluing algorithm. This is an important consequence, because the convex decomposition created by the BSP-Tree decomposition may generate a fragmentation of the boundary when non-convex objects are represented. The gluing operation, as proposed, provides an elegant solution to the problem of reconstructing the minimal BRep from a BSP-Tree representation of a polytope.

4   Conclusions

In this paper, we have presented an algorithm to convert a BSP-Tree representing a polytope to a BRep representation. The storage of lower dimensional information in a TBSP-tree allowed us to reconstruct the information necessary to recover the boundary of the object. The TBSP-Tree is incrementally computed during a pre-order visit of the tree, which computes all lower dimension information in both sides of each node. A gluing step is performed when all information about a node is discovered. which involves a recursion in the dimension of the space to remove internal features and glue

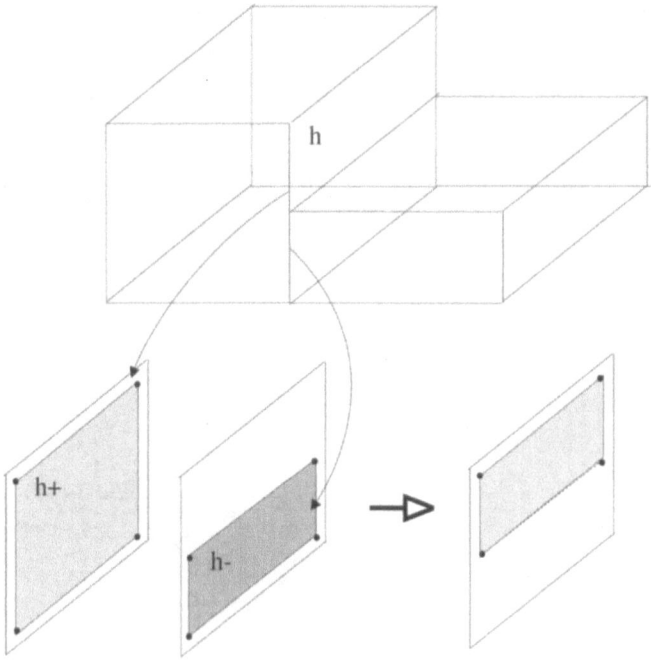

Fig. 11. Gluing opposite faces to obtain the boundary

the boundary representation in each side of the node. As a consequence of this process, the boundary representation obtained at the end corresponds to a minimal BRep.

We believe that the importance of the BSP-Tree representation can be extended by having such algorithm available. BSP-Trees and BReps are both representation of polytopes used in Solid Modeling, each one with its own advantages. For example, BSP-Trees are more efficient where visibility orderings or boolean operations need to be computed, whereas in some other applications, like topological deformations, we would prefer to use the BRep. By having both conversion algorithms available we can exploit the advantages of each model more efficiently.

Another application where the conversion algorithm proposed in this paper can be used refers to the problem of finding a near-optimal BSP-Tree representation of a

```
procedure GlueBSPTree(BSPTree node, Dimension dim)
   if dimension > 1
      tree1 = GlueBSPTree(DimensionSon(node,+), dim-1);
      tree2 = GlueBSPTree(DimensionSon(node,-), dim-1);
   endif
   return SymmetricDifference(tree1, tree2);
end GlueBSPTree;
```

Fig. 12. Procedure to glue TBSP-Trees

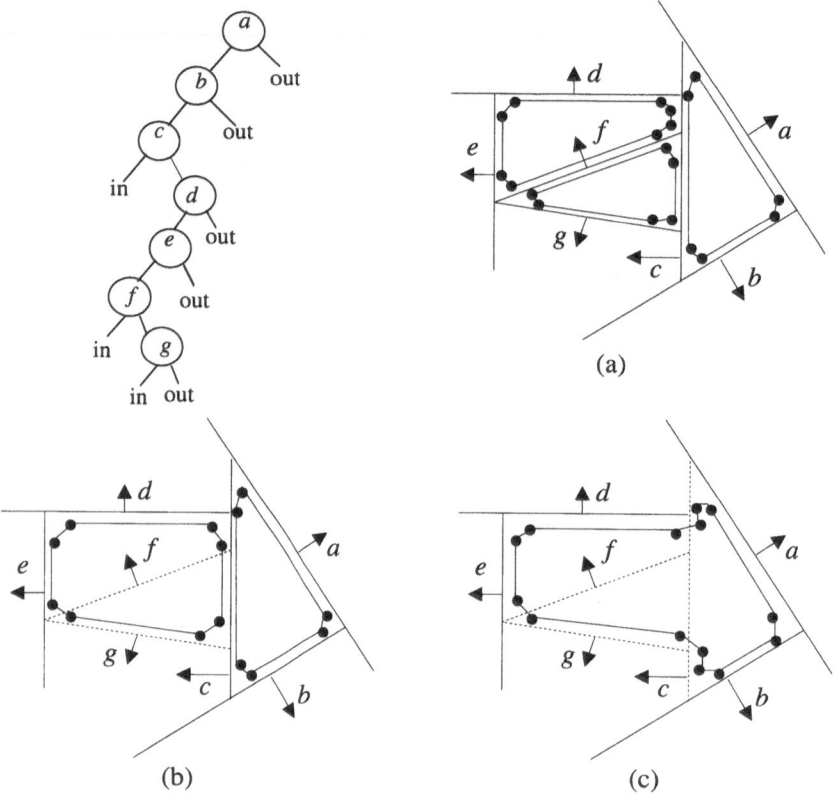

Fig. 13. Example of the gluing Procedure in 2D

polytope. The BSP-Tree representation is not unique, and many different trees may represent a given polytope. The problem of finding a minimal tree is even more important when we perform successive boolean operations by applying tree merging algorithms, which may generate trees that need to be re-structured. Unlike Binary Search Trees, where we have algorithms to keep a tree balanced, the balancing of a tree in higher dimensions is a much more difficult problem. By applying the conversion described here we obtain a minimal BRep, which aggregates the geometric information expressed by the tree. The further conversion from BRep to BSP-Tree may generate a much more balanced tree, as the topological information of the BRep gives better heuristics in the construction of the tree.

Finally, the combination of topological information in the BSP-Tree, which resulted in the TBSP-Tree, shows promising applications in Solid Modeling. In future work we plan to extend BRep algorithms, like topological deformations, to work directly with TBSP-Trees.

5   Acknowledgments

The first author would like to thank Charles Loop for the proposal of the problem of converting BSP-Trees to BRep, as well as orientation in the first attempts to solve the problem; Leonidas Guibas for helpful discussions and a grant from Brazilian Agency CNPq under process number 200789/92.9.

References

1. H. Fuchs, Z. M. Kedem, and B. F. Naylor. On visible surface generation by a priori tree structures. *Computer Graphics (SIGGRAPH '80 Proceedings)*, 14(3):124–133, July 1980.
2. Dan Gordon and Shuhong Chen. Front-to-back display of BSP trees. *IEEE Computer Graphics and Applications*, 11(5):79–85, September 1991.
3. M. Mantyla. *An Introduction to Solid Modeling*. Computer Science Press, Rockville, Md, 1988.
4. J. Matoušek. Efficient partition trees. *Discrete Comput. Geom.*, 8:315–334, 1992.
5. Bruce Naylor. Binary space partitioning trees as an alternative representation of polytopes. *Computer-Aided Design*, 22(4):250–252, May 1990.
6. Bruce Naylor. SCULPT an interactive solid modeling tool. In *Proceedings of Graphics Interface '90*, pages 138–148, May 1990.
7. Bruce Naylor, John Amanatides, and William Thibault. Merging BSP trees yields polyhedral set operations. In Forest Baskett, editor, *Computer Graphics (SIGGRAPH '90 Proceedings)*, volume 24, pages 115–124, August 1990.
8. Bruce F. Naylor. Partitioning tree image representation and generation from 3D geometric models. In *Proceedings of Graphics Interface '92*, pages 201–212, May 1992.
9. A. A. G. Requicha and H. B. Voelcker. Boolean operations in solid modeling: Boundary evaluation and merging algorithms. *Proc. IEEE*, 73(1):30–44, January 1985.
10. Jaroslaw R. Rossignac and Herbert B. Voelcker. Active zones in CSG for accelerating boundary evaluation, redundancy elimination, interference detection, and shading algorithms. *ACM Transactions on Graphics*, 8(1):51–87, 1989.
11. William C. Thibault and Bruce F. Naylor. Set operations on polyhedra using binary space partitioning trees. In Maureen C. Stone, editor, *Computer Graphics (SIGGRAPH '87 Proceedings)*, volume 21, pages 153–162, July 1987.
12. Enric Torres. Optimization of the binary space partition algorithm (BSP) for the visualization of dynamic scenes. In C. E. Vandoni and D. A. Duce, editors, *Eurographics '90*, pages 507–518. North-Holland, September 1990.
13. G. Vanecek, Jr. Brep-index: a multidimensional space partitioning tree. *Internat. J. Comput. Geom. Appl.*, 1(3):243–261, 1991.

A Formal Approach to Multiresolution Hypersurface Modeling

Leila De Floriani[1], Enrico Puppo[2], and Paola Magillo[1]

[1] Dipartimento di Informatica e Scienze dell'Informazione
Università di Genova
Via Dodecaneso, 35 – 16146 Genova, ITALY
Email: {deflo,magillo}@disi.unige.it

[2] Istituto per la Matematica Applicata
Consiglio Nazionale delle Ricerche
Via de Marini, 6 (Torre di Francia) – 16149 Genova, ITALY
Email: puppo@ima.ge.cnr.it

Abstract. Multiresolution geometric models support the representation and processing of geometric entities at different levels of detail, and are useful in several application fields, such as geographic information systems, CAD systems and scientific visualization. The aim of this paper is to provide a framework for multiresolution geometric modeling, independent both of the dimension of spatial objects under consideration, and of the specific application. This paper introduces a formal model, called the Multiresolution Simplicial Model (MSM), capable of capturing the characteristics of most multiresolution models proposed in the literature. The paper provides an analysis of the relationships between the intrinsic structures of different multiresolution models, as well as a definition of relevant application-independent operations on them. Major data structures used to encode multiresolution models are reviewed, as well as algorithms which implement the operations on each data structure.

1 Introduction

Multiresolution geometric models support the representation and processing of geometric entities at different levels of resolution. Such models are useful in several application fields to handle geometric data at different levels of detail, depending on specific application needs. The main advantage of a multiresolution model is in speeding up data processing because of data reduction, wherever a representation at low resolution is adequate.

Multiresolution models were proposed in the literature for topographic surface representation, for generic surfaces embedded in 3D space, for 3D volumes in CAD and object recognition, and for volume data in scientific visualization. A major approach to multiresolution modeling is to stick to a decomposition modeling scheme, and to control resolution through the level of refinement of the domain decomposition: a key concept is that resolution is somehow proportional to the number of cells decomposing the domain.

In spite of the number of models proposed in the literature, there is a substantial lack of a systematic description, independent both of the dimension of spatial objects under consideration, and of the specific application. This paper is an effort towards the development of a formal model that captures the characteristics of most models based on mutiresolution decompositions known in the literature. Key issues in our analysis are the relationships between the intrinsic structures of different multiresolution models, the relevant operations on them, and the data structures to encode them.

The assumption on which we rely is that a d-dimensional geometric object embedded in $\mathbb{E}^n$ can be represented (at a given approximation) through a d-dimensional cell complex whose domain coincides with (or approximates) the object. The number and size of the cells of such complex determine the resolution of the description; representations of the same object at lower/higher resolutions are obtained by decreasing/increasing the density of cells.

For the sake of simplicity, the theory is developed in this paper for representations based on simplicial complexes, although it is not difficult to extend it to general cell complexes. The model presented here is based on a multi-resolution decomposition of a domain, called a *Multiresolution Simplicial Model* (MSM). The two-dimensional version of the model, called a Multi-Triangulation, is introduced in a companion paper, and both its properties and its application to the representation of terrain surfaces are investigated in detail [14]. The MSM is thought as a generalization over a broad class of multiresolution models. The basic idea underlying the model is that a large number of different decompositions of a domain can be obtained on the basis of a relatively small set of atomic components, which can be combined in different ways to cover the domain. Components can partially overlap, and they are arranged into a partially ordered set, where the order relation depends on spatial interferences of components, and on the possibility to combine them to obtain simplicial complexes describing the given object at different resolutions, and possibly at different levels of resolution through the domain.

Multiresolution models proposed in the literature are usually classified into *hierarchical* and *pyramidal* models on the basis of the characteristics of their domain subdivision [8]. Hierarchical models are characterized by a nested subdivision of the domain which can be effectively described by a tree. In pyramidal models arbitrary collections of cells are replaced by other arbitrary collections among different components subdividing the same portion of the domain. Alternatively, a pyramidal model can be viewed as a historical sequence of changes made to an initial coarse subdivision while progressively refining it [3,13,12]. In this paper, we show that all such models can be seen as special cases of MSM, which thus defines a unifying approach to multiresolution modeling based on domain decomposition.

We also define a basic set of general, application-independent operations on MSMs, such as the extraction of a description of the object at a user-defined level of resolution, navigation of the model within a given resolution, and the solution of spatial queries at an assigned level of resolution. In the final part of the paper, we analyze within the above framework different data structures that have been proposed in the literature for multiresolution models [1,3,6,12–15,17].

The paper is organized as follows. In Section 2, we introduce the multiresolution simplicial model, related definitions and properties, and show how models proposed in the literature can be interpreted as MSMs. In Section 3, we define three basic operations on MSMs, namely the extraction of a representation at a given resolution, navigation, and interference queries. In Section 4, we analyze the various data structures for multiresolution models proposed in the literature. Section 5 briefly summarizes techniques for performing the previously defined operations on multiresolution models. Finally, Section 6 contains some concluding remarks.

2   Multiresolution Simplicial Model

In this Section, we first introduce auxiliary definitions and notations related to partial orders and simplicial complexes [4]. Then, we provide the definition of the Multiresolution Simplicial Model, and properties of such model. The model described in this section directly extends a model proposed, in two dimensions, in [14]. Results are stated without proof, since they immediately extend similar results given in [14] for the two-dimensional case.

2.1   Partially Ordered Sets

Let C be a finite set. A *partial order* on C is an antisymmetric and transitive relation $<$ on its elements. A pair $(C, <)$ is called a *partially ordered set* (*poset*). For every $c, c' \in C$, notation $c \prec c'$ means $c < c'$ and $\not\exists \, c''$ such that $c < c'' < c'$. An element $c \in C$, such that for all $c' \in C$, $c \le c'$, is called a *minimal* element in C. A *maximal* element is defined in a symmetric way. If there exists a unique minimal element $c \in C$, then c is called the *minimum* of C.

A subset $C' \subseteq C$ is called a *lower set* if $\forall c' \in C'$, $\forall c < c'$ then $c \in C'$. Intuitively, a subset C' of C is a lower set if it contains all elements that precede each of its elements. For any $c \in C$, the set $C_c = \{ c' \in C \mid c' \le c \}$ is the smallest lower set containing c, and it is called the *down-closure* of c. We also define the *sub-closure* of c as $C_c^- = \{ c' \in C \mid c' < c \} = C_c \setminus \{c\}$. Given a lower set $C' \subseteq C$, a *consistent order* on C' is any total order that extends the partial order $<$ locally to C': a relation $<_{C'}$ on the elements of C' is a consistent order if $<_{C'}$ is a total order on C' and $\forall c, c' \in C'$, $c < c' \Rightarrow c <_{C'} c'$.

The algebraic structure of a poset $(C, <)$ can be described as a DAG, where nodes represent elements of S and arcs encode relation $\prec$: there is a directed arc (c, c') for every pair of elements in C such that $c \prec c'$. A lower set of a poset $(C, <)$ defines a cut on such DAG.

2.2   Simplicial Complexes and Approximated Models

A k-*simplex* (or a *simplex of order* k) σ, with $0 \le k \le n$ is a subset of $\mathbb{E}^n$, defined as the locus of points which can be expressed as convex combinations of $k + 1$ affinely independent points; such points are called the *vertices* of σ. Examples of simplexes are a point (0-simplex), a straight-line segment (1-simplex), a triangle (2-simplex), a tetrahedron (3-simplex).

The *interior* of a simplex σ (denoted by $i(\sigma)$) is the locus of points which can be expressed as convex combinations of the vertices of σ, with coefficients restricted in $(0, 1)$.

We say that a q-simplex σ is a *(proper) face* of a k-simplex τ (with $q < k$) if the set of vertices of σ is a subset of the set of vertices of τ; if σ is a q-simplex, we say that σ is a *q-face* of τ.

Definition 1. A finite set Σ of simplexes in $\mathbb{E}^n$ is a *simplicial complex* when the following conditions hold:

1. for each simplex $\sigma \in \Sigma$, all faces of σ belong to Σ;
2. for each pair of simplexes $\sigma, \tau \in \Sigma$, either $\sigma \cap \tau = \emptyset$ or $\sigma \cap \tau$ is a simplex of Σ;

A simplicial complex Σ is called a *d-simplicial-complex* if d is the maximum among the orders of simplexes belonging to Σ (d is called the *order* of Σ). In this paper, a d-simplicial-complex is also called, for brevity, a d-complex.

The union of all simplexes of Σ, regarded as point sets, is the *domain* of Σ and it is denoted by $\Delta(\Sigma)$. A d-simplicial complex whose domain is a (polyhedral) set Ω is called a *simplicial decomposition* of Ω.

A d-simplicial-complex Σ is *regular* if every k-simplex in it is a face of at least one d-simplex of Σ. The domain of a regular d-simplicial complex is a homogeneously d-dimensional subset of $\mathbb{E}^n$. Here, we consider regular d-complexes.

We call any finite set of d-simplexes in $\mathbb{E}^n$ a *d-set*. A regular d-simplicial complex is completely characterized by the collection of its d-simplexes, i.e., by the *d-set* associated with Σ. In the following we will use a (regular) simplicial complex and its d-set interchangeably.

Within an application, a simplicial decomposition Σ of a domain Ω provides an approximate representation of some information distributed over Ω: an element of information is associated with every simplex $\sigma \in \Sigma$, which approximates in synthetic form the information contained in the portion of domain covered by σ. A finer decomposition of Ω leads to a more accurate approximation of the original information over Ω. A measure of the *accuracy* of the approximation within each simplex is provided by a function $\mu : \Sigma \to [0, 1]$, where a higher value of μ corresponds to a higher level of accuracy: $\mu(\sigma) = 1$ means exact representation. The accuracy of the approximation provided by a complex Σ is defined as $\min\{\mu(\sigma) | \sigma \in \Sigma\}$.

2.3   Operators on d-Complexes and d-Sets

We define two operators on d-sets: the *interference* operator $\otimes$ and the *combination* operator $\oplus$. Both operators take two d-sets as arguments and produce a d-set.

Definition 2. The *interference* of two d-sets Σ_i and Σ_j is defined as
$\Sigma_i \otimes \Sigma_j = \{\sigma \in \Sigma_i \mid \exists \sigma' \in \Sigma_j, i(\sigma) \cap \sigma' \neq \emptyset\}$.

In other words, the *interference* of Σ_i and Σ_j is the set of simplexes of Σ_i which have a proper intersection (i.e., "interfere") with some simplex of Σ_j. If $\Sigma_i \otimes \Sigma_j \neq \emptyset$, then we say that Σ_i and Σ_j are *interfering*.

Definition 3. The *combination* of two d-sets Σ_i and Σ_j is defined as
$\Sigma_i \oplus \Sigma_j = \Sigma_i \setminus (\Sigma_i \otimes \Sigma_j) \cup \Sigma_j$

In other words, the *combination* of Σ_i and Σ_j is the d-set obtained from Σ_j by adding all the d-simplexes of Σ_i, which do not interfere with Σ_j. Note that, in general, $\Sigma_i \oplus \Sigma_j \neq \Sigma_j \oplus \Sigma_i$. Some examples of intereference and combination of two d-sets are illustrated in Fig. 1.

If $\Sigma_i \oplus \Sigma_j$ is a d-simplicial complex, and $\Delta(\Sigma_i \oplus \Sigma_j) = \Delta(\Sigma_i) \cup \Delta(\Sigma_j)$, then Σ_j is said to be *compatible over* Σ_i.

We define the *combination* of a sequence of d-complexes $[\Sigma_0, \ldots, \Sigma_k]$, as the successive combination of its elements.

Definition 4. Let $[\Sigma_0, \ldots, \Sigma_k]$ be a sequence of d-complexes. The *combination* $\oplus_{i=0}^{k} \Sigma_i$ of $[\Sigma_0, \ldots, \Sigma_k]$ is defined as follows:

- if $k = 0$, then $\oplus_{i=0}^{0} \Sigma_i = \Sigma_0$;

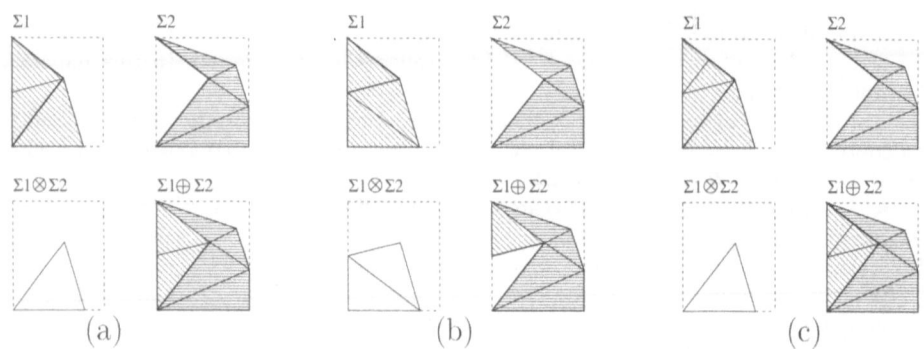

Fig. 1. Examples of interferences and combinations in 2D: (a) $\Sigma_1 \oplus \Sigma_2$ is a simplicial decomposition of $\Delta(\Sigma_1) \cup \Delta(\Sigma_2)$; (b) $\Sigma_1 \oplus \Sigma_2$ is a 2-complex, but does not cover $\Delta(\Sigma_1) \cup \Delta(\Sigma_2)$ because $\Sigma_1 \otimes \Sigma_2$ is not contained into $\Delta(\Sigma_2)$; (c) $\Sigma_1 \oplus \Sigma_2$ is not a 2-complex, because condition 2 in definition 1 is violated.

- if $k > 1$, then $\oplus_{i=0}^{k} \Sigma_i = (\oplus_{i=0}^{k-1} \Sigma_i) \oplus \Sigma_k$.

A *compatible sequence* is a sequence of complexes where every complex is compatible over the combination of complexes preceding it in the sequence.

Definition 5. A sequence of d-complexes $[\Sigma_0, \dots, \Sigma_k]$ is a *compatible sequence* if $\forall j = 1, \dots, k$, Σ_j is compatible over $\oplus_{i=0}^{j-1} \Sigma_i$;

It is easy to see that the combination of a compatible sequence $[\Sigma_0, \dots, \Sigma_k]$ is a simplicial decomposition of $\cup_{i=1}^{k} \Delta(\Sigma_i)$.

2.4 Multiresolution Simplicial Model

Let Ω be a d-dimensional polyhedral domain in $\mathbb{E}^n$.

Definition 6. A *Multiresolution Simplicial Model* (MSM) on Ω is a poset $(\mathcal{S}, <)$ where $\mathcal{S} = \{\Sigma_0, \dots, \Sigma_h\}$ is a set of d-complexes, and $<$ is a partial order on $\mathcal{S}$ satisfying the following conditions:

1. $\Delta(\Sigma_0) = \Omega$ and $\forall i = 0, \dots, h$, $\Delta(\Sigma_i) \subseteq \Omega$;
2. $\forall i, j = 0, \dots, h$, $i \neq j$,
   a. $\Sigma_i \prec \Sigma_j \Rightarrow \Sigma_i \otimes \Sigma_j \neq \emptyset$;
   b. $\Sigma_i \otimes \Sigma_j \neq \emptyset \Rightarrow \Sigma_i$ is in relation with Σ_j (i.e., either $\Sigma_i < \Sigma_j$ or $\Sigma_j < \Sigma_i$).
3. the sequence $[\Sigma_0, \dots, \Sigma_h]$ of all complexes of $\mathcal{S}$ defines a consistent order with respect to relation $\prec$ and $[\Sigma_0, \dots, \Sigma_h]$ is a compatible sequence.

The meaning of condition 2 in the above definition becomes clear if we consider the graph encoding relation $\prec$ for set $\mathcal{S}$ (see Fig. 2):

- every two complexes Σ_i and Σ_j, that are connected by an arc (Σ_i, Σ_j), are interfering;

– if two complexes Σ_i and Σ_j interfere, then they are connected through a path (of lenght ≥ 1).

Condition 3 warrants that the elements of S can be combined, and that the order relation provides a way to obtain compatible sequences.

The elements of S are called *components*. The d-set $\Sigma_S = \cup_{i=0}^h \Sigma_i$, i.e., the set of all d-simplexes of the MSM, is called the *d-set associated with S*.

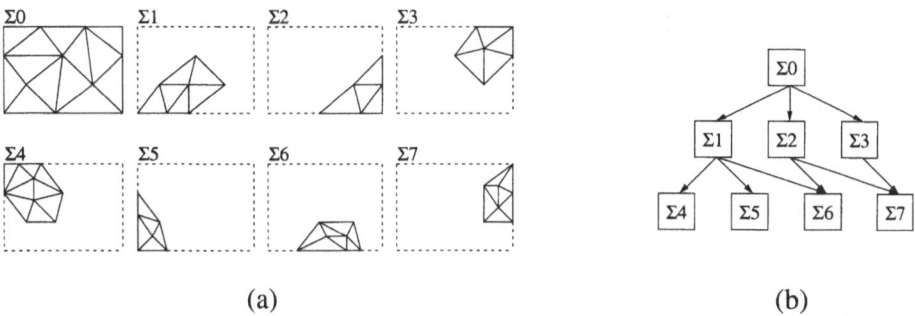

(a)   (b)

Fig. 2. An example of an MSM: (a) the components, and (b) the graph representing relation $\prec$.

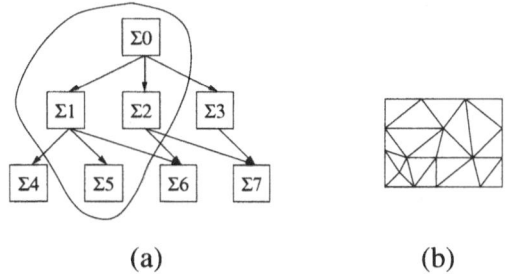

(a)   (b)

Fig. 3. (a) A lower set of the MSM of Fig. 2, and (b) its combination; the default compatible sequences is $[\Sigma_0 \Sigma_1 \Sigma_2 \Sigma_5]$; other compatible sequences are $[\Sigma_0 \Sigma_2 \Sigma_1 \Sigma_5]$ and $[\Sigma_0 \Sigma_1 \Sigma_5 \Sigma_2]$, all leading to the same combination.

Lemma 7. Σ_0 *is the unique minimum element of* $(S, <)$.

Given any subset $S' \subset S$, the total order of its element consistent with the sequence $[\Sigma_0, \ldots, \Sigma_h]$ will be called the *default order* of S'.

Lemma 8. *The default order of any lower set* $S' \subset S$ *is a consistent order.*

Lemma 9. *In an MSM* $(S, <)$, *the combination of a lower set* $S' \subseteq S$ *is independent of the specific consistent order.*

Since the combination obtained from any consistent order is unique, it will simply be called the combination of S', and denoted by $\oplus S'$. Fig. 3 shows a lower set and its combination.

Intuitively, every component of an MSM describes a portion of Ω at a certain resolution; complete descriptions of Ω at different resolutions can be obtained from an MSM as the combinations of different lower sets of S.

Definition 10. Let Σ_i be a component of $(S, <)$. The combination of the sub-closure $S_{\Sigma_i}^-$ of Σ_i is called the *support* of Σ_i. The set of triangles of the support that are interfering with Σ_i, i.e., the set $(\oplus S_{\Sigma_i}^-) \otimes \Sigma_i$ is called the *floor* of Σ_i.

In practice, the support of Σ_i is the combination of the minimal lower set of S to which Σ_i can be combined. The floor of Σ_i is composed of those triangles of the support which are replaced by the triangles of Σ_i when Σ_i is combined to its support.

Within an MSM, an accuracy function μ can be defined on the d-simplexes of its associated d-set. If we have that $\mu(\sigma) < \mu(\tau)$ for every d-simplex τ properly intersecting σ, and belonging to any component Σ_j such that $\Sigma_i < \Sigma_j$, then the multiresolution is said *strict*. A strict multiresolution corresponds to the idea that, if $\Sigma_i < \Sigma_j$, then $\oplus S_{\Sigma_i}$ and $\oplus S_{\Sigma_j}$ provide a worse and better description of Ω, respectively.

In models constructed according to geometry driven criteria non-strict multiresolution can occur (see Section 2.5). However, a finer description usually relies on a finer domain partition. For the above reasons, we are especially interested in MSMs whose ordering reflects the degree of refinement of the corresponding combinations. The following definitions provides formal characterizations of such models.

Definition 11. A multiresolution simplicial model $(S, <)$ is *increasing* if $\forall\, S', S''$ lower sets, if $(S' \subset S'')$ then $|\oplus S'| < |\oplus S''|$ (where $|\cdot|$ denotes the number of d-simplexes of a complex).

According to this definition, adding more components to a lower set increases the size of its combination. Decreasing MSMs are also of interest, and can be defined analogously. A further interesting property of some MSMs is the following.

Definition 12. An increasing MSM $(S, <)$ has *linear growth* if and only if, for each lower set $S' \subseteq S$, the size of S' is linear in the size of its combination.

Linear growth is a desirable property since it implies that the MSM does not introduce a serious overhead with respect to a single d-complex describing the domain Ω at the maximum feasible resolution, i.e., with respect to the d-complex obtained from the combination of S itself as a lower set.

In the following we introduce some special subclasses of MSM which have linear growth.

Definition 13. An MSM $(S, <)$ is called *nested* if every component Σ_i can be partitioned into a collection of d-sets, such that each d-set covers a single d-simplex of the combination of its sub-closure $\oplus S_{\Sigma_i}^-$.

Definition 14. An MSM $(S, <)$ is called *clustered* if every component Σ_i, consists of d-simplexes that are incident into a single vertex.

Nested MSM have linear growth; a clustered MSM has linear growth if we impose the additional constraint that the degree of the central vertex of each component Σ_i is bounded by a constant. In Section 2.5 we will see that nested MSMs formalize hierarchical models proposed in the literature, and clustered MSMs formalize multiresolution models built through iterative deletion of independent vertices.

2.5   Existing Multiresolution Models as MSMs

In the literature, multiresolution models are usually classified as *pyramidal models*, in which a sequence of complexes representing the whole domain at different resolutions is provided, and *hierarchical models*, which are based on a nested subdivision of the domain, where each simplex of an initial complex is recursively refined into a "local" complex [8].

There are two basic techniques for constructuing such models: *simplification* methods start from the full resolution and progressively reduce the number of vertices on which the model is based, in order to coarsen resolution; *refinement* methods start from a coarse approximation, and progressively refine it by inserting new vertices, in order to improve resolution. For pyramidal models a new representation is built for the whole domain at each iteration by either simplifying or refining an existing representation; for hierarchical models a new node in the tree is built at each iteration, by either subdividing an existing region (top-down refinement), or collapsing a group of regions into a single one (bottom-up simplification).

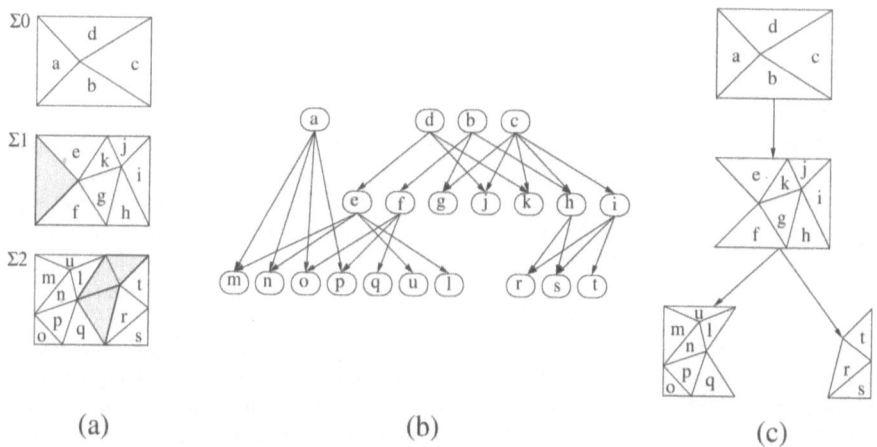

(a)   (b)   (c)

Fig. 4. An example of a pyramidal model in 2D: (a) the steps of the construction process (shaded triangles survive across consecutive stages); (b) the DAG encoding interferences among simplexes; (c) the same model interpreted as an MSM.

General pyramidal models (see Fig. 4 (a) and (b)) represent a sequence of complexes, each defining a simplicial decomposition of Ω. Interference links are defined between pairs of d-simplexes belonging to consecutive complexes in the sequence, which have proper intersection. Some models [1] avoid considering trivial interferences (i.e.,

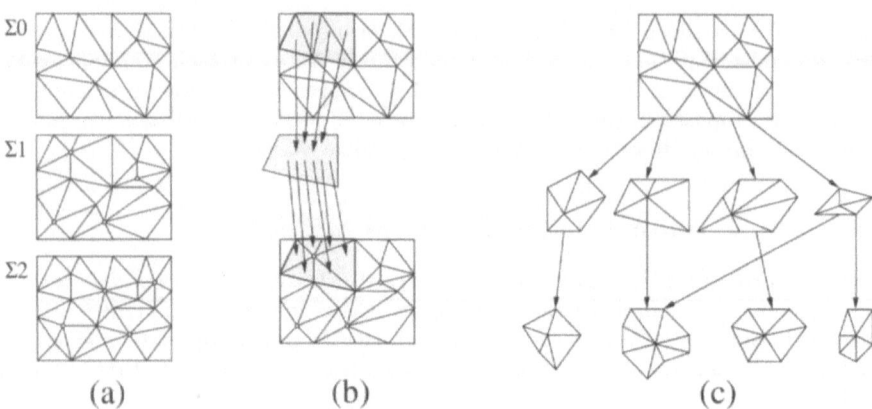

Fig. 5. An example of a clustered pyramidal model in 2D: (a) the steps of the bottom-up construction process (the vertices removed at each step are marked with circles); (b) representation of interference links within a cluster; (c) the same model interpreted as an MSM.

interferences between simplexes which "survive" across a refinement step) by representing duplicate d-simplexes only in the first complex where they appear. The first proposal of a pyramidal model is the *Delaunay pyramid* [6], which encodes a collection of *Delaunay[1] simplicial complexes* describing a domain Ω at a sequence of increasing resolutions. The Delaunay pyramid does not rely on a special construction technique, and can be built either bottom-up or top-down.

A pyramidal model can be formalized as an MSM where the set of components S is obtained by decomposing each d-complex of the model (which is produced by a step of the construction algorithm) into $(d-1)$-connected components of d-simplexes that belong to the complex at the current step but not to the complex corresponding to the immediately coarser step (i.e., the previous/next step in the case of a top-down/bottom-up construction). Let Σ_i be a component, subset of the d-complex at step k; then, $\Sigma_j \prec \Sigma_i$ holds for every component containing a d-simplex from step $k-1$ or $k+1$ (in the case of a top-down or bottom-up construction, respectively) that interferes with Σ_i (see Fig. 4 (c)).

The model proposed by de Berg and Dobrindt [5] represents a special case of a two-dimensional pyramid (see Fig. 5). Such model is built through iterative simplification of a complex representing a two-dimensional domain Ω at full resolution: at each step, a set of independent[2] vertices of small degree is removed, and the "holes" left by those vertices are re-triangulated. de Berg and Dobrindt produce a Delaunay triangulation at each step, but other criteria can be applied as well. Such model is an example of what we call a *clustered model* since modifications between two consecutive levels occur independently at each removed vertex. Interference links are maintained between the d-simplexes incident in a removed vertex P and the d-simplexes created to fill the hole

---

[1] A d-simplicial-complex Σ is a Delaunay complex if and only if, for each d-simplex $\sigma \in \Sigma$, the hypersphere curcumscribed to σ does not contain vertices of Σ in its interior.

[2] Two vertices are *independent* if they are not endpoints of the same edge.

left after the elimination of P. Clearly, a clustered pyramidal model leads to clustered MSM.

In recent alternative approaches, a pyramidal model is regarded as a historical sequence of local modifications made to an initial subdivision by progressive refinement or simplification [3,13,12]. Cignoni et al. consider historical changes determined by updating an existing triangulation through iterative insertion/deletion of vertices into/from an initial model at coarse/high resolution. Klein and Straßer adopt the same historical view based on refinement through vertex insertion. Hoppe consider historical changes determined on a triangulation at the highest resolution, by iteratively collapsing edges to their midpoints. Though such models have been proposed in the two-dimensional case, it is not difficult to extend them to higher dimensions.

The interpretation of such historical models as MSMs is straightforward: the initial mesh form the minimum element of the MSM, while each group of simplices that update the model because of a local modification form a component; the partial order is obtained by making a component Σ_j follow Σ_i in the order if some d-simplex of Σ_i is eliminated when the update corresponding to Σ_j occurs.

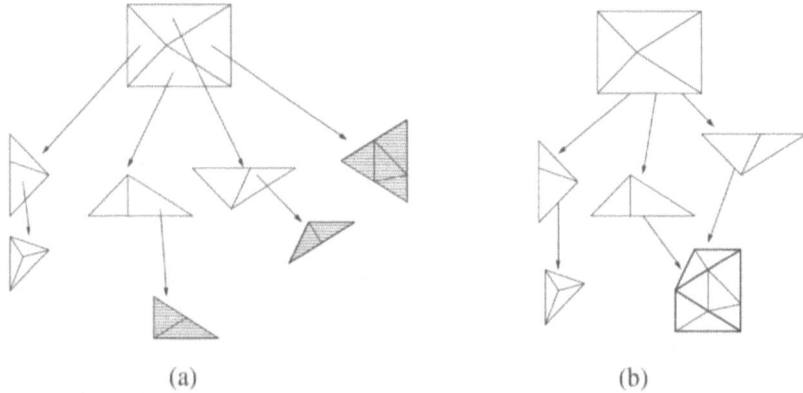

(a)   (b)

Fig. 6. An example of a hierarchical model in 2D (a), and the same model interpreted as an MSM (b); the dashed nodes in (a) are merged into a single component in (b).

Hierarchical models have been studied for a long time. A *hierarchical model* (see Fig. 6 (a)) encodes a nested partition of the domain Ω, which is obtained through a recursive refinement process: starting from a coarse representation, at each step a d-simplex σ is refined independently into a "local" d-complex whose domain is σ. The refinement of σ is performed by inserting new vertices in its interior or on its faces. A hierarchical model is represented as a tree where each node is a local complex, and each arc corresponds to containment relation between a simplex and the complex refining it.

Quadtrees[3] [2,15] and *quaternary triangulations* [11] are built based on the replication of a fixed geometric pattern (a square and a triangle, respectively) at different

---

[3] In fact, quadtrees are not simplicial models, but the theory presented in this paper can be easily generalized to complexes using different kinds of cells.

resolutions; quadtrees can be extended in generic d dimensions. Other hierarchical models are based on a top-down refinement process, which is accuracy driven rather than geometry driven [16,7]. At a generic step, every simplex is expanded into a complex by iteratively inserting points in its interior or on its faces, until the a given accuracy is reached. The refinement of the common face of two adjacent d-simplexes is done consistently, by creating congruent sub-faces.

A hierarchical model cannot be interpreted directly as an MSM whose components coincide with the nodes of the tree representing the hierarchical model. Indeed, two interfering nodes of the tree may not be compatible because their common border is subdivided differently. This problem is overcome by clustering nodes of the tree to form compatible components. An MSM $(\mathcal{S}, <)$ representing a hierarchical model is obtaineded in the following way (see Fig, 6):

- the components of $\mathcal{S}$ are obtained from the nodes of the tree by iteratively joining pairs of nodes corresponding to complexes in which two adjacent d-simplexes have been refined by decomposing their common $(d-1)$-face;
- relation $<$ is defined by imposing that, for every arc (N_1, N_2) of the tree, if node N_1 was included into a component $\Sigma_i \in \mathcal{S}$, and node N_2 was included into a component $\Sigma_2 \in \mathcal{S}$, then $\Sigma_i < \Sigma_j$ must hold.

The MSM obtained from a hierarchical model is obviously nested.

3   Operations on Multiresolution Simplicial Models

The design of a data structure to encode a multiresolution model depends on the operations that must be performed on the model. We can define a basic set of general-purpose operations which are independent of any specific application. We classify general-purpose operations into three categories:

- *Extraction of a variable-resolution representation over the whole domain*: a basic issue in multiresolution modeling is to be able to extract a single domain decomposition from the multiple representations stored in the model, in which a predefined resolution is achieved in each simplex.
- *Navigation*: in order to navigate the model we need to extract adjacency and incidence information at any resolution; for instance, finding all d-simplexes sharing a $(d-1)$-face with a given d-simplex and satisfying a predefined accuracy. The navigation of an MSM must be possible without the extraction of a representation satisfying the given resolution requirements.
- *Spatial searching*: we need to be able to position any spatial entity with respect a domain decomposition at variable resolution; this task requires solving spatial interference queries on an MSM, such as point location, line or polygon interference, without explicitly extracting a single representation.

Within a specific application, a criterion must be available to define the required resolution in each area of the domain. In order to be generic about the application, we assume that a boolean condition c is defined over each simplex of an MSC, such that, for a given simplex σ, $c(\sigma)$ is true if and only if the resolution of σ is acceptable. For example, a condition requiring a uniform accuracy over the domain Ω could be defined as: $c(\sigma) \equiv \mu(\sigma) \geq \varepsilon$. In general, the accuracy may be variable over the domain. For example, a condition requiring an accuracy that decreases according to the distance from a fixed point V (e.g., a viewpoint) has this form: $c(\sigma) \equiv \mu(\sigma) \geq g(d(\sigma, V))$, where d denotes the Euclidean distance and g is a monotonically non-increasing function. Such condition has been used for instance in [3] for terrain visualization in flight simulation. Condition c needs not necessarily depend on the accuracy of a simplex: for example, $c(\sigma)$ may be defined as true if σ is small enough, independently of $\mu(\sigma)$. We extend condition c to complexes by saying that $c(\Sigma)$ is true on a complex Σ if c is true on every cell of Σ.

3.1   Extracting a Representation at Variable Resolution

A representation of the domain Ω of an MSM at a given resolution, specified by an arbitrary condition c, corresponds to a simplicial decomposition of Ω, formed by simplexes of the MSM which satisfy condition c. Moreover, such representation must obey to a criterion of "minimal refinement", i.e., it must be the coarsest simplicial decomposition of Ω that can be built from d-simplexes of the MSM satisfying c.

More formally, given an MSM and a boolean condition c, the problem is finding the less refined complex Σ, formed of d-simplexes in $\mathcal{S}$, such that $c(\Sigma)$ is true. For an increasing MSM, such complex Σ corresponds the combination of the smallest lower set of the MSM, whose combination satisfies c (see Fig. 7).

Definition 15. Let $(\mathcal{S}, <)$ be an MSM and c be a condition defined on the simplexes of $\mathcal{S}$. The simplicial complex extracted from $(\mathcal{S}, <)$ under condition c is defined as $\oplus \mathcal{S}_c$, where $\mathcal{S}_c$ is a lower set such that

- $c(\oplus \mathcal{S}_c)$ is true, and
- $\forall\, \mathcal{S}' \subset \mathcal{S}_c$, $c(\oplus \mathcal{S}')$ is false.

3.2   Navigation

Navigation on a simplicial complex means moving from one simplex to another simplex, by following topological relations. *Topological relations* are defined between pairs of simplexes of the complex. Since in a d-complex Σ we have simplexes of $d + 1$ different orders, this leads to $(d + 1)^2$ topological relations, where a generic topological relation is defined between the p-simplexes and the q-simplexes of Σ, for fixed p and q in the range $0, \ldots, d$. Topological relations in a simplicial complex Σ can be classified into:

- *adjacency relations*, that link pairs of k-simplexes sharing a $(k-1)$-face;
- *boundary relations*, that link a k-simplex σ to lower order simplexes that are the faces of σ;
- *coboundary relations*, that are symmetric of boundary relations and link a k-simplex σ to higher order simplexes having σ as one of their faces.

Note that the boundary of a simplex σ is defined independently of the complex containing σ, while the coboundary of σ can be defined only within a simplicial complex.

A navigation query within a complex Σ is defined by providing a reference simplex $\sigma \in \Sigma$ and a topological relation $\mathcal{R}$: the answer to the query is given by the collection of simplexes of Σ which are in relation $\mathcal{R}$ with σ. The answer can be restricted to report fewer simplexes (e.g., one) by providing some constraints: for example, for adjacency relations, a constraint is defined by selecting the $(k-1)$-face that σ and the reported simplex must share.

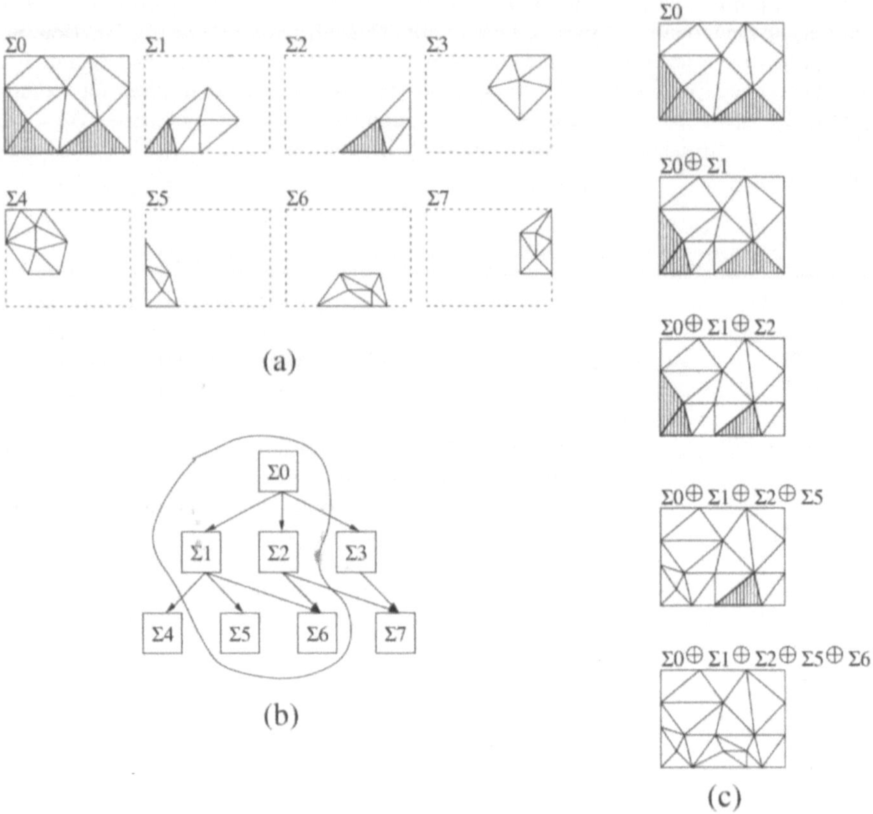

(a)

(b)

(c)

Fig. 7. Extraction of a representation related to a certain condition c on the MSM of Fig. 2: (a) c = false on dashed simplexes, and c = true on the remaining simplexes; (b) the lowerset defining the solution; (c) the extracted domain representation obtained as a combination of such lower set.

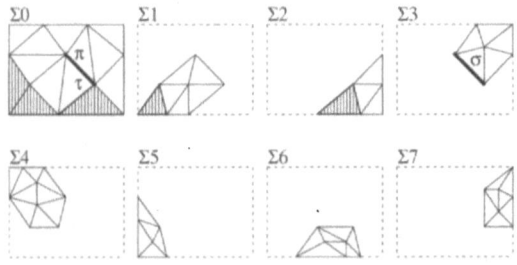

Fig. 8. Navigation related to a certain condition c on the MSM of Fig. 2: c = false on shaded simplexes, and c = true on the remaining simplexes; simplex π is the answer to a query asking for the simplex adjacent to τ along the thick edge; simplex τ is the answer to a query asking for the simplex adjacent to σ along the thick edge.

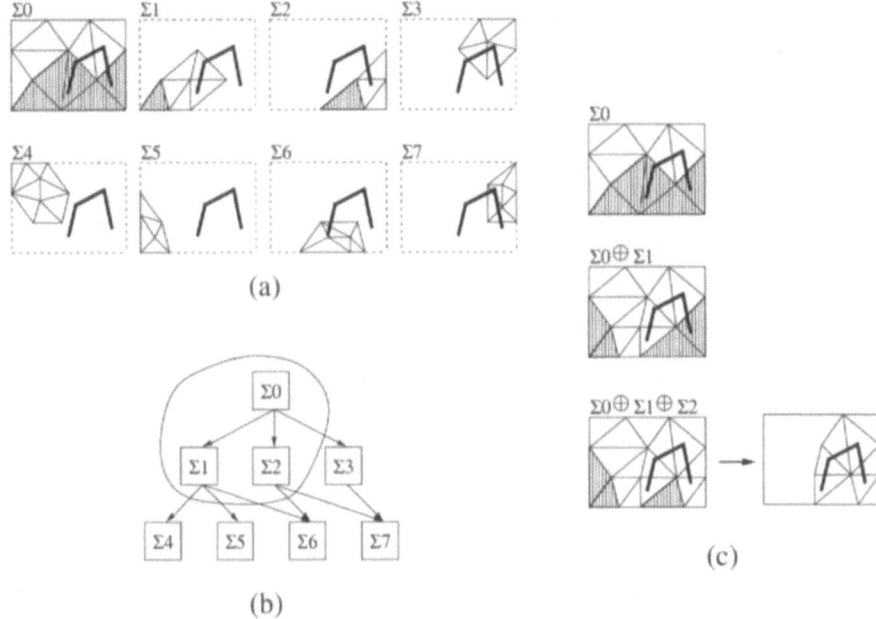

(a)

(b)

(c)

Fig. 9. An intesection query related to a certain condition c on the MSM of Fig. 2: (a) $c = $ false on shaded simplexes, and $c = $ true on the remaining simplexes, the query line is shown; (b) the lower set that contributes to the answer; (c) the answer to the query consists of the simplexes intersecting the line in the combination of such lower set.

In an MSM, topological relations are also defined between simplexes that belong to different components.

Boundary relations of a k-simplex σ in an MSM are "closed" with respect to the component containing σ: if τ on its boundary of σ, then τ necessarily belongs to the same component as σ (τ may belong to other components as well). On the contrary, *adjacency relations* and *coboundary relations* are "open": a k-simplex σ may be adjacent to, or be on the boundary of simplexes in the MSM that do not belong to the component containing σ.

A navigation query in an MSM, at a given resolution, is defined by providing a reference simplex σ, a topological relation $\mathcal{R}$, and a boolean condition c defining the desired resolution. Such a query is answered by reporting simplexes of S that are in relation $\mathcal{R}$ with σ, satisfy c, and obey a criterion of "minimal refinement": intuitively, a simplex τ is reported only if there exists no simplex that is in the given relation with σ, satisfies c, and is "less refined" than τ (see Fig. 8).

Definition 16. A *navigation query* on an MSM $(S, <)$ is defined by a reference simplex $\sigma \in \Sigma_S$, a topological relation $\mathcal{R}$, and a boolean condition c defined on the simplexes of Σ_S. The *answer* to such a query is the set of simplexes $\tau \in \Sigma_S$ such that:

1. $\sigma \, \mathcal{R} \, \tau$;
2. $c(\tau) = $ true;

3. there does not exist any d-simplex $\tau \in \Sigma_S$ satisfying both 1. and 2., and such that $i(\tau) \cap i(\tau') \neq \emptyset$ and $\tau \in \Sigma_j$, $\tau' \in \Sigma_i$, with $\Sigma_j < \Sigma_i$.

3.3   Spatial Searching

Searching problems are concerned to retrieving simplexes of the model that "interfere" with a given query entity. *Interference relations* are based on set-theoretic concepts like coincidence, element-of, inclusion, and intersection. While topological relations can only be defined between entities within a complex, interference relations are naturally defined in general between entities in $\mathbb{E}^n$:

- *Coincidence relations* are defined between entities of the same dimension. Two entities are coincident if they are formed by the same subset of points in $\mathbb{E}^n$.
- *Element-of relations* are defined between points and higher dimensional entities, and, symmetrically, between k-dimensional entities (with $k > 0$) and points. In total, we have n direct element-of-relations, that link a point to a k-dimensional entity containing it (for $k = 1, \ldots, n$) and n corresponding symmetric relations.
- *Containment relations* correspond to the containment of an entity into another entity (or viceversa), when both are considered as subsets of $\mathbb{E}^n$. Direct containment relations link an entity to other entities of greater or equal dimension; the corresponding symmetric relations link an entity to entities of the same or lower dimension. Containment relations between pairs of points can also be defined, and reduce to coincidence between points. Note that element-of relations are special cases of containment relations.
- *Intersection relations* are defined between pairs of entities of any dimension: two entities are related if they have a non-empty intersection, when considered as subsets of $\mathbb{E}^2$.

We restrict our attention to intersection relations, since coincidence and containment relations can be reduced to special cases of intersection.

Let us consider a single d-simplicial complex Σ first. Given a query entity in $\mathbb{E}^n$ (e.g., a k-simplex, with $0 \leq k \leq n$), the aim is finding all the simplexes of Σ which are intersected by the query entity. If desired, the answer can be restricted to simplexes of a certain order (e.g., all the d-simplexes).

On an MSM, the answer to an intersection query, at a resolution specified by a condition c, can be characterized as the less refined complex formed of d-simplexes from the MSM which satisfy c, whose domain completely contains the query entity (see Fig. 9).

Definition 17. An *intersection query* on an MSM $(\mathcal{S}, <)$ is defined by a query entity $q \subset \mathbb{E}^n$, and a condition c defined on the simplexes of Σ_S.
The *answer* to such query is formed by a d-set $\Sigma \subseteq \Sigma_S$ such that

1. $\forall \sigma \in \Sigma$, $\sigma \cap q \neq \emptyset$;
2. $\forall \sigma \in \Sigma$, $c(\sigma) =$ true;
3. $\forall \sigma \in \Sigma$, σ belongs to a component Σ_i of the MSM, such that, $\forall \Sigma_j \in \mathcal{S}$ with $\Sigma_j < \Sigma_i$, if $\Delta(\Sigma_j) \cap q \neq \emptyset$ then $\exists \tau \in \Sigma_j$ suct that $\tau \cap q \neq \emptyset$ and $c(\tau) =$ false.

Condition 3 in the above definition formalizes the requirement that, if a simplex σ has been included in the answer, then no simplex interfering with σ and belonging to a component $\Sigma_j < \Sigma_i$ could be included in the answer, because Σ_j, restricted to the simplexes intersecting q, does not satisfy c.

4   Data Structures for Multiresolution Simplicial Models

In this Section, we classify basic information that describe an MSM, and examine existing data structures for MSMs illustrating which information are stored within each data structure and how. Relevant information about a multiresolution model include:

- *Geometric information* about the simplexes composing the model: the *geometry* of a k-simplex is completely defined by the set of its vertices, which also uniquely determines the boundary of the simplex; thus, it is sufficient to represent the d-simplexes of an MSM. This information is necessary and sufficient to characterize the d-set associated with the MSM.
- *Topological information*, which encode adjacency, boundary and coboundary relations among the simplexes of an MSM. We distinguish between *local topology*, which is concerned with topological relations between simplexes within the same component, and *global topology*, which considers relations between simplexes not necessarily belonging to the same component (global topology includes local topology). Here, we restrict our attention to $(d-1)$-adjacencies between d-simplexes.
- *Spatial interference information*, which relate components and simplexes that overlap, providing different approximations of the information contained in their domain. Interference information can be encoded either between pairs of interfering components, or between all pairs of interfering d-simplexes belonging to such components.

Interference links *between components* are a subset of the order relation $<$: they can be represented as a DAG, called the *Multiresolution DAG* (MDAG). The MDAG (see Fig. 10 (a)) is a supergraph of the DAG encoding relation $\prec$ for the poset $(\mathcal{S}, <)$, and a subgraph of its transitive closure (encoding relation $<$):

- $\mathcal{S}$ is the set of nodes of the MDAG;
- there is an arc (Σ_i, Σ_j) in the MDAG if and only if $\Sigma_i < \Sigma_j$, and $\exists \sigma \in \Sigma_i$ such that $(\sigma \in \oplus \mathcal{S}_{\Sigma_j}^-) \otimes \Sigma_j$ (i.e., if some d-simplex of Σ_i belongs to the floor of Σ_j).

In other words, the MDAG encodes links of relation $<$ which are relevant with respect to the combination of components: Σ_j has an incoming arc from every node Σ_i containing a d-simplex that is covered by Σ_j when Σ_j is combined to its support.

Alternatively, interference information can be represented between d-simplexes instead of between components. This approach also leads to a DAG, called the *Simplex Interference DAG* (SIDAG), where nodes are the d-simplexes of an MSM and arcs correspond to spatial interferences between pairs of interfering d-simplexes belonging to adjacent components in the MDAG:

- the nodes of the SIDAG are the simplexes of the d-set $\Sigma_{\mathcal{S}}$;
- there is an arc (σ, τ) in the SIDAG if $i(\sigma) \cap i(\tau) \neq \emptyset$ and $\sigma \in \Sigma_i$, $\tau \in \Sigma_j$, where (Σ_i, Σ_j) is an arc of the MDAG.

The SIDAG can be seen as an "explosion" of the MDAG obtained by replacing every node Σ_i with as many nodes $\sigma_1 \ldots \sigma_p$ as the d-simplexes in Σ_i, and by transferring to each σ_j $(j = 1, \ldots, p)$ the arcs of Σ_i which are relevant to σ_j (see Fig. 11).

Data structures for pyramidal models [6,1] are based on the representation of interference links between pairs of simplexes; in other words, they encode the SIDAG associated with an MSM. Information about how simplexes are grouped to form the

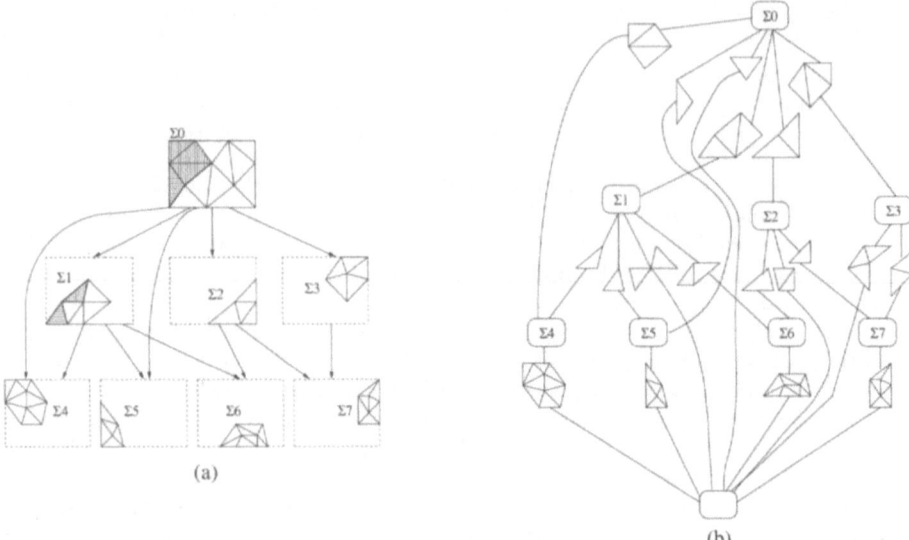

(a)

(b)

Fig. 10. (a) The MDAG associated with the MSM of Fig. 2; the MDAG contains additional arcs (Σ_0, Σ_4) and (Σ_0, Σ_4) with respect to the graph representing relation $\prec$; the floor of components Σ_4 and Σ_5 is shown. (b) The data structure implementing it (b), for simplicity, when a group of triangles shares the links to components, only one link is shown.

components of the model are also stored. In [1], each component is represented by simply listing all its d-simplexes. The structure can be extended easily to represent global topology, by linking a simplex to the list of all its adjacent simplexes.

Data structures which represent interference links between components [5,14] are based on an indirect encoding of the MDAG, which explicitly maintains relations between triangles and components, and implicitly relations between components. Each component stores links to the simplexes composing it and to the simplexes composing

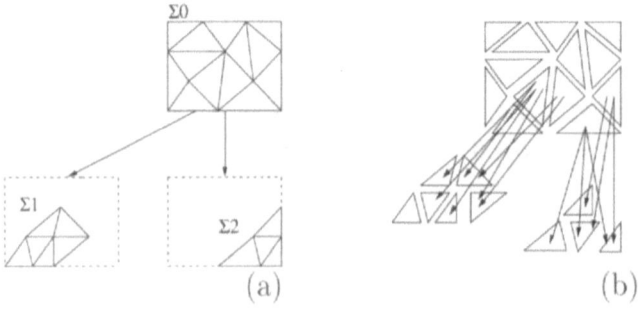

Fig. 11. Relation between the MDAG and the SIDAG: a portion of the MDAG of Fig. 10 (a); the corresponding portion of the SIDAG (b).

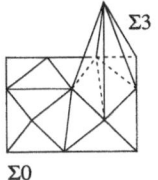

Fig. 12. Embedding of components Σ_0 and Σ_3 of the MSM of Fig. 2 in three dimensions: Σ_3 is glued over its floor.

its floor (see Fig. 10 (b)). Each simplex stores the inverse links, i.e., links to the component containing it and to the component containing it in its floor. The data structure of de Berg and Dobrindt [5] stores topology locally to each component. In the data structure of Puppo [14], topology is not represented.

In the data structure proposed by Cignoni et al. [3], the connection between a component and its floor is represented indirectly, through the storage of global topology. Adjacency relations exist between the boundary simplexes of a component Σ_i and the boundary simplexes of the floor of Σ_i, which share a common $(d-1)$-face. No information about spatial interference is represented. The d-dimensional multilevel structure is embedded into $\mathbb{E}^{d+1}$, where a a component and its floor are "glued" at their boundaries (see Fig. 12): this data structure maintains adjacencies by defining an embedding of the d-dimensional MSM as a complex into the $(d+1)$-dimensional Euclidean space.

Data structures proposed by Hoppe [12] and by Klein and Straßer [13] for their historical models are implicit, i.e., only the initial mesh, plus minimal information about the sequence of changes are maintained. In [12], a sequence of vertex split operations (which reverse edge collapse) is encoded by giving for each operation an existing vertex, plus the endpoints of an edge resulting from splitting such a vertex. In [13], a sequence of vertex insertions in a Delaunay triangulation is considered: since the Delaunay triangulation is uniquely defined by its construction algorithm, only the sequence of vertices is maintained, sorted according to the insertion order. Note that neither the components, nor the partial order of the corresponding MSM are maintained by such data structures.

Data structures developed to represent hierarchical models are based on the representation of the tree that describes the recursive domain refinement. Every node of the tree is represented separately, through an individual data structure. For every arc (N_i, N_j) of the tree, which represents containment of a d-complex N_j into a d-simplex $\sigma \in N_i$, a link between σ and N_j is stored in the data structure. Thus, interference information is represented between d-simplexes and d-complexes. These links can be thought as encoding a special situation, where the floor of a component consists of a single simplex.

The data structure used in [7] was developed for representing two-dimensional hierarchical models. Such structure encodes interference between nodes, as explained above, local topology, and partial information about global topology, based on the following remark. If σ is a boundary triangle of the node N_i where it belongs to, σ has a "free" adjacency link along the edge e lying on the boundary of N_i. The data structure uses such link to connect σ to the triangle adjacent to σ along e in the less refined (i.e., nearest to the root) node; this connection is called a "rope" since it allows to move up in the hierarchy. All the triangles adjacent to a given triangle τ can be found by following the "rope" link, which connects τ to the coarsest adjacent triangle τ', and then collecting all other simplexes adjacent to τ while descending into the subtree rooted at the node containing τ'. The technique of "ropes" extends to higher dimensions.

We conclude this Section with a comparison among existing approaches for encoding multiresolution models from the point of view of space requirements. Assuming that all structures must represent the same geometric information (i.e., the vertices of the d-simplexes of an MSM), we compare the amount of memory required for storing topological and interference information in the various approaches illustrated above.

The space required for local adjacencies is linear in the number of d-simplexes of an MSM since every such simplex has only one adjacent d-simplex for every $(d-1)$-face. The space required for global adjacencies can be quadratic in the number of d-simplexes of Σ_S. As far as interference information is concerned, we observe that the number of arcs in the MDAG is linear in the number of d-simplexes of an MSM in the worst case. In a worst-case SIDAG, every simplex σ, belonging to a component Σ_i, has a link to every simplex τ such that $\tau \in \Sigma_j$ and (Σ_i, Σ_j) is an arc of the MDAG. Thus, the size of the SIDAG can be evaluated as $\sum_{(\Sigma_i, \Sigma_j) \in \text{MDAG}} (|\Sigma_i| \times |\Sigma_j|)$, which can be roughly approximated by $k^2 l$, where k is the average size of a component and l is the number of arcs of the MDAG. The structure encoding the MDAG requires only a linear space in the number of d-simplexes; thus, it results in a very efficent way of encoding a multiresolution simplicial model; moreover, it supports general extraction and searching algorithms.

5   Performing Spatial Operations

In this Section, we consider the spatial operations defined in Section 3, and analyze which information, stored in a data structure for encoding an MSM, support each operation efficiently.

First, we focus on the extraction of a simplicial decomposition of the domain at a given resolution, defined by a condition c. Algorithms for extracting a representation of the domain at a *uniform resolution* have been developed for data structures which implement the SIDAG, and reduce the problem to finding connected components on such graph [1].

An algorithm for extracting a simplicial decomposition of the domain at a *variable resolution* is supported by data structures which represent interferences through the MDAG. This algorithm has been proposed in [14] for a two-dimensional MSM, but it directly extends to d dimensions, and is based on a traversal of the DAG in a breadth-first order.

The model of de Berg and Dobrindt comes together with a simple algorithm, which extracts a representation at variable resolution. The algorithm is based on a top-down traversal of the pyramid, and on a greedy construction of the result. Unfortunately, the greedy approach, which accepts a triangle in the solution as soon as possible, does not warrant that the desired accuracy is fulfilled everywhere. The algorithm of Puppo [14], which is valid on the model of de Berg and Dobrindt as well, overcomes this problem.

A generic extraction algorithm, working on hierarchical models, represented through the data structure of [7], is described in [9]; this algorithm performs a top-down traversal of the tree and uses adjacency information in order to detect nodes of the tree that belong to the same component of the MSM, and thus must be visited together.

Data structures representing only global topology support an extraction algorithm for the special case when condition c requires a *resolution decreasing with the distance from a predefined point V* [3]. Such algorithm, developed by Cignoni et al. [3], starts from the triangle containing V (on which the maximum resolution is required) and progressively adds adjacent triangles. The correctness of the approach is guaranteed by the special condition c considered: such an algorithm is not suitable for a generic condition c, because it is not possible to decide, on a local basis, whether the current set of d-simplexes is consistent (i.e., it is a subset of a combination of some lower set of the model). The algorithm directly extends to arbitrary dimensions. A similar approach for hierarchical models is described in [7].

Implicit data structures proposed for historical models are very compact in terms of storage requirement, but offer small support to operations. Hoppe [12] proposes an algorithm for variable resolution extraction, which incurs in drawbacks similar to that of de Berg and Dobrindt (i.e., the solution might not fulfill the desired accuracy); moreover, the computational cost might be increased by the need of explicitly performing all updates, which are maintained only implicitly in the structure. Klein and Straßer propose an algorithm that performs variable resolution extraction correctly, but has a high computational overhead due to repeated search and modifications of the model, based on properties of the Delaunay triangulation that need numerical computation; again, the computational overhead comes from the need of makeing the structure explicit.

Now, we consider the solution of interference queries on an MSM, at a resolution specified by a condition c. The representation of interference information is a basic requirement for a data structure for efficiently processing of interference queries.

As a special case of interference queries, *point location* queries can be efficiently solved on any data structure representing intereference information: either between simplexes, or between components, including tree-based data structures for hierarchical models. Algorithms for this task traverse a path in the DAG used to represent interferences until a d-simplex containing the query point and satisfying c is reached.

Solving an intersection query related to a *k-dimensional entity*, with $k > 0$, is more difficult, since the given entity may intersect more than one simplex, and an algorithm must guarantee that the set of reported simplexes defines a proper d-complex.

In hierarchical models, the nested structure of the domain decomposition guarantees that simplexes belonging to different branches of the tree do not overlap. Thus, a valid approach consists of following multiple paths in such a tree. On general MSMs, intersection queries can be solved by applying an algorithm which extracts a simplicial decomposition of the domain satisfying the given condition, restricted to simplexes which are intersected by the query entity. The same limitations outlined before for extraction algorithms apply: in data structures based on the MDAG [5,14], intersection queries related to any condition c can be processed, while data structures representing the SIDAG [1] can only process queries at a uniform resolution.

Efficient *navigation* of the model within an arbitrary resolution, specified by a generic condition c, is possible in data structures which encode global topology [3], or in data structures, like the one described in [7], which contain enough auxiliary information to reconstruct global topology efficiently.

Implicit structures do not offer support to either the solution of iterferences or navigation.

6   Concluding Remarks

The multiresolution simplicial model presented in this paper is a general model for multiresolution decomposition of d-dimensional domains, which include different other existing models as special cases. Recently, we have proposed an extension of the model to represent surfaces in space as multiresolution triangle meshes; in [10] methods have been proposed for building an MSM of the surface bounding a solid object starting from acquired scattered data or from a a boundary representation of the object, where faces are described as parametric surfaces.

We are currently extending the data structure proposed in [14] to allow efficient navigation and spatial searching. Based on such a structure, we are implementing the multiresolution simplicial model in the two-dimensional case, within a prototype system for the representation, visualization and analysis of surfaces. The system will include:

- algorithms for building the structure (through either refinement or simplification strategies);
- the algorithm for the extraction of a representation at variable resolution through the domain, described in [14];
- algorithms for answering spatial interference queries related to k-dimensional entities, for any k;
- navigation algorithms.

Acknowledgments

This work has been partially supported by the coordinated project 96.01964.CT12 "Multiresolution Models for the Visualization of Multidimensional Fields" of the Italian National Research Council, and by the BRITE European project BRPRCT96.0150 "VENICE - Virtual ENvironment interface by sensory integration for Inspection and manupulation Control in multifunctional underwater vehicles".

References

1. M. Bertolotto, L. De Floriani, P. Marzano, 1995, Pyramidal simplicial complexes, *Proceedings third ACM Symposium on Solid Modeling and Applications*, Salt Lake City, Utah.
2. Z.T. Chen, W.R. Tobler, 1986, Quadtree representation of digital terrains, *Proceedings Autocarto*, London, 475–484.
3. P. Cignoni, E. Puppo, R. Scopigno, 1995, Representation and visualization of terrain surfaces at variable resolution, *Proceedings International Symposium on Scientific Visualization*, R. Scateni, Ed., World Scientific, Cagliari (Italy).
4. F. H. Croom, 1978, *Basic Concepts of Algebraic Topology*, Undergraduate Texts in Mathematics, Springer-Verlag.
5. M. de Berg, M., K.T.G. Dobrindt, 1995, On the levels of detail in terrains, *11th ACM Symposium on Computational Geometry*, Vancouver, BC (Canada), c26–c27. Also published in extended version as *Technical Report* UU-CS-1995-12, Utrecht University, Dept. of Computer Science, 1995. Available from `ftp://ftp.cs.ruu.nl/pub/RUU/CS/techreps/CS-1995/1995-12.ps.gz`
6. L. De Floriani, 1989, A pyramidal data structure for triangle-based surface description, *IEEE Computer Graphics and Applications*, 67–78.

7. L. De Floriani, E. Puppo, 1995, Hierarchical Triangulation for Multiresolution Surface Description, *ACM Transactions on Graphics*, 14, 4, 363–411.
8. L. De Floriani, P. Marzano, E. Puppo, Multiresolution models for topographic surface description, *The Visual Computer*, 12, 7, pp.317-345.
9. L. De Floriani, P. Magillo, 1996, A Comprehensive Framework for Spatial Operations on Hierarchical Terrain Models, *Technical Report* DISI-TR-96-15, Dipartimento di Informatica e Scienze dell'Informazione, Università di Genova, Italy. Available from `http://www.disi.unige.it/person/MagilloP/` Computer and Information Science Deptartment (DISI),
10. L. De Floriani, P. Magillo, E. Puppo, Multiresolution Representation and Reconstruction of Triangulated Surfaces, *Technical Report* PDISI-96-26, Computer and Information Science Deptartment (DISI), University of Genova, Italy, 1996. Available from `http://www.disi.unige.it/person/MagilloP/`
11. D. Gomez, A. Guzman, 1979, Digital model for three-dimensional surface representation, *Geo-processing*, 1, 53–70.
12. H. Hoppe, 1996, Progressive meshes, *Proceedings SIGGRAPH 96*, in *Computer Graphics - Annual Conference Series*, ACM SIGGRAPH, New Orleans, LA, USA, August 4-9, 1996, pp.99-108.
13. R. Klein, W. Straßer, 1996, Generation of multiresolution models from CAD data for real time rendering, in *Theory and Practice of Geometric Modeling*, R. Klein, W. Straßer, R. Rau, (Editors), Springer-Velrag.
14. E. Puppo, 1996, Variable resolution terrain surfaces, *Proceedings Canadian Conference on Computational Geometry*, Ottawa (Canada), 12-15 August, 1996, to appear. Also published in longer version as "Variable resolution triangulations", *Technical Report* N.12/96, Istituto per la Matematica Applicata, C.N.R., Genova, Italy, 1996. Available from `http://www.disi.unige.it/person/PuppoE/publications.html`
15. Samet, H., 1990, *Applications of Spatial Data Structures*, Addison Wesley, Reading, MA.
16. L.L. Scarlatos, T. Pavlidis, 1990, Hierarchical triangulation using terrain features, *Proceedings IEEE Conference on Visualization*, San Francisco, CA, 168–175.
17. Von Herzen, B., Barr, A.H., 1987, Accurate triangulations of deformed, intersecting surfaces, *Computer Graphics*, 21, 4, 103–110.

Generation of Multiresolution Models from CAD-Data for Real Time Rendering

Reinhard Klein and W. Straßer

Wilhelm-Schickard-Institut, GRIS,
Universität Tübingen, Germany

Abstract. A mesh refinement and a mesh simplification algorithm are presented. Both algorithms guarantee a user-defined error tolerance and deliver a multiresolution model. After the computation of the multiresolution model triangulation of the surface patches at variable resolutions can be incrementally generated on-the-fly at rendering time. The resulting triangulations form hierarchical Delaunay triangulations in parameter space.

1 Introduction and Previous Work

The visualization of large CAD-models, like cars, trains, aero planes, etc. becomes a major challenge in the context of virtual reality. In most experiments a number of such models have to be visualized and animated simultaneously. Examples are the optimization of the cabin of a train or the optimization of a driver's position in a car. To examine the panorama in such a place, not only the train or car themselves have to be visualized but also other cars, pedestrians, buildings, etc. Despite the performance of modern graphics hardware this cannot be realized in real time without special simplifications of the objects in the scene.

The most common way in currently available CAD-systems to represent the boundary surfaces of objects is the use of trimmed NURBS. For visualization purposes the surface patches are approximated by triangle meshes, since triangle meshes are the most widely used representation of models in computer graphics. Planar polygons and in particular triangles are standard rendering primitives of common graphics workstations that can rapidly render polygons.

There is a number of algorithms concentrating on real time rendering of trimmed NURBS surfaces, see for example [17,24,21,1]. The main idea of these algorithms is to preprocess the surface patches in an appropriate way, normally generating simpler subpatches. In the second step the subpatches are uniformly tessellated into a grid of rectangles connected by triangles to points evaluated along the trimming curves of the patches. The polygons defined in the (u, v)-parameter space are transformed into facets in object space by evaluating their vertices with the surface functions, resulting in polygonal meshes in three dimensional space. These methods allow fast rendering of CAD models but have disadvantages.

The computations in the preprocessing step and the algorithms to generate triangles near the boundary of the trimmed patches are complicated. Furthermore, each surface patch is sampled with a constant sampling rate depending on the viewing parameters. In many cases this results in oversampling and many unneccessary triangles. Clipping away parts of a patch that do not belong to the view frustrum is not possible either.

Therefore, in this paper we propose a different strategy:

- To compute in a preprocessing step a complete multi resolution representation of the CAD-model.
- To generate a view dependent triangulation of the model on-the-fly during rendering time combining triangles form different levels of detail into one single mesh.

The use of a dynamic triangulation is not a new idea. There is a number of other authors working on that subject [25,5,9,2,3,8]. But the main problem of all the resulting algorithms is the additional storage space for additional data structures needed to combine different levels of detail. While this additionally needed storage space does not lead to problems for small models, it becomes a major problem for large models such as a whole car. A sufficient approximation of such a model which corresponds to the finest level of detail in the multiresolution model consists of millions of triangles. For example an approximation of the boundary surface of a car door model consisting of 446 trimmed NURBS Surface patches up to one millimeter of accuracy needs about 350.000 triangles. In most real-life applications such models cannot be loaded entirely into memory. Moreover, the access time to and from the memory may become a bottleneck.

We show in the following how we can trade storage requirements for computation power in order to generate a multilevel triangulation on-the-fly by using an incremental delaunay triangulation algorithm. The on-the-fly multilevel triangulation reduces storage requirements and the explicit determination of the number of levels.

In the next section we give a brief description of the models we are dealing with. In section 3 we describe how to generate multilevel triangulations. Section 4 deals with the on-the-fly computation of view dependent triangulations. In section 4 we show some results and an overview of our current and future work.

2   Parametric Models

We assume a given boundary representation (BREP) of the model. In general, such a boundary consists of several trimmed parametric patches $f : \Omega \longrightarrow \mathbb{R}^3$, where $\Omega \subset \mathbb{R}^2$ is a planar domain. Most of the currently available CAD-systems support as representation of f NURBS-tensorproduct surfaces.

In the case of a trimmed patch Ω is defined by a set $B = \{b_0, \cdots, b_N\}$ of *boundary polygons*. The set B includes an exterior boundary b_0 and interior boundaries (holes) $b_1, \cdots b_N$. Each boundary b_i consists of a finite number (≥ 3) of oriented curved *domain edges* $e_{i0}, \cdots, e_{in_i}$ and is defined by a set of *boundary nodes* $V \supset V_i = \{v_{i0}, \cdots, v_{in_i}\}$. Every boundary polygon b_i defines a region Ω_i. The interior of these regions $\overset{\circ}{\Omega}_i$ is located on the left side of the oriented boundary polygons . In order for Ω to be a well-defined finite domain, the following conditions must hold:

$$\complement \ \overset{\circ}{\Omega}_i \subset \overset{\circ}{\Omega}_0, \ 0 \leq i \leq N,$$
$$\overset{\circ}{\Omega}_i \cap \overset{\circ}{\Omega}_j = \emptyset, \ 1 \leq i, j \leq N, \ i \neq j,$$
$$b_i \cap b_j \subset V, \ 0 \leq i, j \leq N, \ i \neq j.$$

In most applications the edges are also given by a NURBS representation.

Surface modelers used in the car industry for the design of car hulls often do not provide topological information. Therefore, there is no a-priori knowledge about neighbourhood relations between different surface patches. For visualization purposes this is not a serious problems, since it is sufficient to approximate and visualize each patch independently from the others. The only problem one has to take care of, is that the approximation error along the boundary curves of the patches is always smaller than one pixel in screen space.

In the following we restrict ourselves to cases in which no neighbourhood relationships are known, processing each patch by its own. How to deal with the more complicated case of boundary-conforming approximations is described in [12].

3   Generation of a Multilevel Model

Following, we shortly review a refinement method for the adaptive approximation of trimmed NURBS surfaces [16,11,12], that can be used to approximate an initial NURBS surface up to a given error tolerance using a parametric distance between the approximating mesh and the original surface. After that, we show how a simplification method can be used for the same purpose. In this simplification method we use the Hausdorff distance between the approximating mesh and the original surface to control the approximation error. Through this distance measurement higher reduction rates can be achieved. Both methods deliver a multilevel triangulation which can be used to generate adaptive triangulations of CAD-models on-the-fly with controlled approximation errors.

3.1   Pre-Sampling

The first step of the refinement and the simplification method is to pre-sample the given patch in such a way that the Delaunay Triangulation of the pre-sampled points $(u_j, v_j) \in \Omega \; j = 0, \cdots, p$ in the parameter-domain delivers a piecewise linear approximation f_S of the surface with a maximum parametric approximation error $d_a(f, f_S) = sup_{(u,v) \in \Omega} \|f(u, v) - f_S(u, v)\| < \eta$. In order to build the multilevel model the piecewise linear approximation f_S is used instead of f. f_S has a corresponding Constrained Delaunay triangulation Σ that defines the finest resolution of the multiresolution model. In order to get a unique Delaunay triangulation we assume that no four sample points lie on a circle. Otherwise, the sample points are perturbed slightly.

The main objective of the pre-sampling step is to guarantee a maximum parametric approximation error between f and f_S. If bounds on the second derivative of f are known a regular pre-sampling can be done [7]. In other cases, e.g. for Bézier-Tensorproduct-Surfaces or B-spline-Tensorproduct-Surfaces, the distance between corresponding points of the linear approximation and the control-points can be used, to indicate if the subdivision of a patch is neccessary. This leeds to a quadtree-data structure in the parameter domain.

After the pre-sampling step a Constrained Delaunay triangulation of the sampled points in parameter space is computed. This triangulation is then used to build up a multiresolution model of the surface patch. We denote the triangulation Σ by Σ_M indicating that it contains M vertices.

3.2   Refinement Method

The first step of the refinement method is the generation of a coarse piecewise linear approximation Ω_S of the boundary curves of the domain Ω and the computation of an initial Constrained Delaunay triangulation Σ_m of Ω_S containing m vertices on the boundary of Ω_S. This initial static triangulation is refined by an iterative insertion of new points, one at a time. The insertion is based on an incremental insertion of points on the domain Ω. At each iteration, the point p with the maximum approximation error is inserted as a new point and the triangulation is updated accordingly. The refinement process continues until the parametric approximation error between the actual intermediate triangulation Σ_n and the finest triangulation Σ_M is zero.

For the refinement method a parametric distance is used. This simplifies the computation of the point p with maximum approximation error.

The insertion of a single point during the insertion process does not necessarily cause a decrease in the approximation error. However, the convergence of the method guarantees that the approximation will improve after some additional vertices have been inserted. This scheme for adaptive approximation was proposed by several authors (see e.g. Floriani, Falcidieno, and Pienovi [4], Lee and Schachter [18], and Rippa [20]) in other contexts.

The inserted points are stored in a list L ordered by the insertion time. In addition, for each point in the list the maximum approximation error of the reduced triangulation is recorded at the insertion time. Therefore, for each point we can assign an approximation error. Since this approximation error is independent from the view direction, it is called, hereafter, the *geometric error*. We denote by $G_\epsilon(p)$ the function which maps a given point p to its associated global geometric error.

The refinement algorithm can be outlined as follows:

1. Compute the boundary of the surface including trimming curves in parameter space.
2. Generate an initial Constrained Delaunay triangulation Σ_M of the parameter domain of the faces containing the vertices on the boundary-curves and on the trimming-curves so that the corresponding function f_S approximates f up to a predefined error tolerance η.
3. Compute a coarse approximation Ω_S of Ω and a Constrained Delaunay triangulation Σ_m of Ω_S.
4. Start with Σ_m and insert, one at a time, the point causing the maximum geometric error. The point and its associated error are stored in the list L. This continues until the maximum geometric error of the reduced triangulation is zero, or up to another prescribed user-defined error.

A detailed description of this algorithm can be found in [16,12]. Implementation aspects of the algorithm are described in [10].

3.3   The Simplification Method

In the following we briefly review our simplification algorithm for arbitrary triangle meshes in 3D [15] and show how this algorithm can be modified to also produce a multiresolution model.

The algorithm One of the main ideas of the simplification method is the use of the Hausdorff distance:

The Euclidean distance between a point x and a set $Y \subset \mathbb{R}^n$ is defined by

$$d(x, Y) = \inf_{y \in Y} d(x, y),$$

where $d(.,.)$ is the Euclidean distance between two points in $\mathbb{R}^n$. Using this definition we can define the distance $d_E(X,Y)$ *from* a set X *to* a set Y by

$$d_E(X,Y) = \sup_{x \in X} d(x,Y). \tag{1}$$

We call this distance *one-sided Hausdorff distance* between the set X and the set Y. It doesn't define a distance function on the set of all sets of $\mathbb{R}^n$, because it is not symmetric. That means that in general $d_E(X,Y) \neq d_E(Y,X)$. The *Hausdorff distance* is defined by

$$d_H(X,Y) = \max(d_E(X,Y), d_E(Y,X)). \tag{2}$$

In contrast to the one-sided Hausdorff distance it is symmetric and we have

$$d_H(X,Y) = 0 \iff X = Y.$$

If the Hausdorff distance between the original triangulation T and the simplified triangulation S is less than a predefined error tolerance ϵ, then

$$\forall x \in T \text{ there is a } y \in S \text{ with } d(x,y) < \epsilon$$

and

$$\forall y \in S \text{ there is a } x \in T \text{ with } d(x,y) < \epsilon.$$

Therefore, the Hausdorff distance between the original and the simplified triangulation is the one a user would intuitively think of.

It is worthwhile to mention that for *any* parameterized surface $f : \mathbb{R}^2 \supset \Omega :\longrightarrow \mathbb{R}^3$ that is approximated by a piecewise linear surface $f_\Sigma :: \mathbb{R}^2 \supset \Omega :\longrightarrow \mathbb{R}^3$ we always have

$$d_H(f, f_\Sigma) \leq ||f - f_\Sigma||_\infty = \sup_{(u,v) \in \Omega} ||f(u,v) - f_\Sigma(u,v)||.$$

For this reason, using the Hausdorff distance for error measurements results in higher reduction rates for the same error tolerance.

The proposed algorithm is a typical mesh simplification algorithm, e.g., it starts with the original triangulation Σ_M and successively simplifies it: It removes vertices and retriangulates the resulting holes until no further vertices can be removed from the simplified triangulation Σ without exceeding a predefined Hausdorff distance between the original triangulation and the simplified one.

A main idea of the new algorithm is to compute and update an error value for every single vertex of the simplified mesh. This value describes the *potential error*, that is the Hausdorff distance that would occur if the vertex would be removed. In each step we actually eliminate one of the vertices with the smallest potential error. At the beginning of the algorithm the original and simplified triangulation coincide. For every single vertex the potential error is computed and all vertices are stored into a list L in ascending order according to their potential errors. If a vertex is actually removed from the current simplified triangulation this list is updated. Because of the ordering of the list, the vertex that should be removed next is placed at its beginning. There are two cases where the removal of a vertex would not make sense: First so-called complex vertices, see [22], and second vertices for which the retriangulation of the resulting hole may lead to topological problems. These situations are detected by topological consistency checks, see [23]. In both cases the potential error is set to infinity.

If we remove a vertex v from the triangulation its adjacent triangles are removed and the remaining hole is retriangulated. In addition, for all neighbouring vertices $v_1, \cdots, v_n$, $n \in \mathbb{N}$ of v the potential errors need to be updated. The vertices have to be removed from the list L and reinserted into L according to their new potential error. Note that this can be done in $O(\log r)$ time, where r is the number of remaining vertices in the reduced mesh.

Modificaton of the algorithm In the original algorithm [15] for the retriangulation of the remaining hole the adjacent vertices are projected into a plane similar to the algorithm of Turk [23]. If the corresponding polygon in the plane does not self-intersect, the polygon is triangulated using a Constrained Delaunay triangulation.

For parametric surface approximation this complicated procedure is not neccessary. Each triangulation in space has a unique corresponding Constrained Delaunay triangulation in the parameter domain Ω. If a vertex of a triangulation in 3D-space has to be removed, its corresponding point in the domain Ω is removed from the Constrained Delaunay triangulation in parameter space and the 3D-triangulation is updated accordingly instead of retriangulating the remaining hole in 3D-space. In such a way it is guaranteed for each step of the algorithm that the triangulation of the domain is a Constrained Delaunay triangulation. Like in the case of the refinement algorithm the removed points in parameter space are stored in a list L in a descending order by the removal time. In addition, for each point in the list the maximum approximation error of the reduced triangulation is recorded at the removal time.

3.4   The Multiresolution Model

From the coarsest Constrained Delaunay triangulation Σ_m and the sequence of inserted or removed points $(p_{m+1}, p_{m+2}, \ldots, p_M)$ transforming Σ_m into the triangulation Σ_M at full resolution all intermediate unique Delaunay triangulations $\Sigma_m, \Sigma_{m+1}, \ldots, \Sigma_M$ can be recomputed through simple point insertion. Therefore, the multiresolution representation of a surface patch only requires a coarse Delaunay triangulation Σ_m of the domain in the XY- plane and the sequence of points transforming Σ_m into the triangulation Σ_M at full resolution. The topology of the triangulation is implicitly given by the use of the Delaunay triangulation and does not have to be stored explicitly. This leads to a massive reduction of the storage costs for the multiresolution model. Since for each point p_n of the sequence the approximation error ϵ_n between the original triangulation Σ_M and the corresponding triangulation Σ_n is additionally stored in the multiresolution model, we also have knowledge about the corresponding approximation error which is neccessary to guarantee the correctness of the visualized surface object.

The storage cost for the initial triangulation can be considered constant. Thus, the total storage cost for the multiresolution model is $6 * (M - m) + B$, where $6 * (M - m)$ is the number of floating point values of the model (2 cartesian coordinates in parameter space, 3 cartesian coordinates in 3D-space and one error-value) and $B = B(m)$ is the total storage cost for the initial triangulation.

4   Extracting triangulations from different levels of detail

In this section we show how triangulations from different levels of detail can be computed using the multiresolution model. We further show how the simplest triangulation can be computed from triangles of the multiresolution model that satisfies a user-defined resolution function.

4.1   Extracting a Triangulation with Constant Error Tolerance

To extract a triangulation that guarantees a user defined error tolerance ϵ we start with the coarsest triangulation Σ_m and stepwise insert points into the corresponding Constrained Delaunay triangulation in parameter space until the geometric error $G(p)$ of the inserted point p is less than ϵ. If we have already extracted a triangulation that guarantees a smaller error tolerance than ϵ it is not neccessary to start again with the coarsest triangulation. Instead vertices are removed from the triangulation, beginning with the vertex with the smallest index (the one which was inserted last).

Since the size of one patch is small compared to the visible part of a complex CAD-model, for many practical visualization purposes it is sufficient to extract each patch with its own constant approximation error. This approximation error depends on the viewing parameters and the bounding box of the patch in 3D-space and can easily be updated after a change of the camera position. We found that for real-world applications, due to the smooth changes of the camera, only a few vertices have to be inserted or removed from the triangulation. Using fast algorithms for inserting and removing points into and from a Constrained Delaunay Triangulation real time performance can be achieved [13,10].

4.2   Extracting a Triangulation at Variable Resolution

The following notation is adopted from Puppo [19]. We assume that for each triangle in the multiresolution model a Boolean condition $c()$ is defined. $c(t)$ is true if and only if the resolution of t is acceptable. Similarly, the notation $c(T)$ means that all triangles of a triangulation satisfy $c()$.

We consider the following problem:

> Given a multiresolution model as described above, extract the smallest triangulation T consisting of triangles from possibly different levels of detail, so that $c(T)$ is true.

Based on the ordered list L such a triangulation can also be rapidly computed without storing additional information, taking advantage of the following observations:

1. If all points p_i of the list L with corresponding maximum geometric error $\epsilon_i > \epsilon$ are inserted into a triangulation, the global geometric error is guaranteed to be less than ϵ.
2. Suppose that a triangle $\Delta(p_i, p_j, p_k)$ of a reduced Delaunay triangulation T at variable resolution in the domain is given. If the circumcircle of $\Delta(p_i, p_j, p_k)$ does not contain any points with corresponding insertion index less than $l = \max(i, j, k)$, the triangle $\Delta(p_i, p_j, p_k)$ belongs to the intermediate triangulation Σ_l and approximates the surface up to an approximation error $G(p_l)$, see Fig. 1. A proof of this observation can be found in [13]. Since such a triangle belongs to an intermediate triangulation it is called *valid*. An intermediate triangulation is called *valid* if all its triangles are valid.
3. A valid triangle $\Delta(p_i, p_j, p_k)$ of a reduced Delaunay triangulation remains in the triangulation until the point with lowest index r greater than $l = \max(i, j, k)$ contained in the circumcircle of $\Delta(p_i, p_j, p_k)$ is inserted into the triangulation.

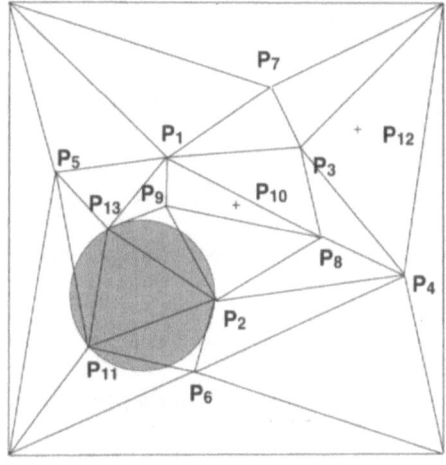

Fig. 1. Since the circumcircle of the triangle $\Delta(p_{13}, p_{11}, p_2)$ does not contain any other point with corresponding insertion index less than $13 = \max(i, j, k)$ the insertion of the points p_{10} and p_{12} will not change the triangle. Therefore, the elevation surface is approximated over the triangle $\Delta(p_{13}, p_{11}, p_2)$ up to a maximum approximation error ϵ_{13}.

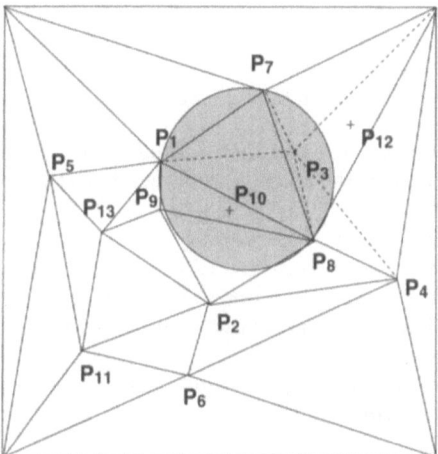

Fig. 2. Since the circumcircle of the triangle $\Delta(p_1, p_8, p_7)$ does contain point p_3 this triangle does not belong to an intermediate triangulation of the multiresolution model. Therefore, in the correction step vertex p_3 is inserted into the triangulation. After the insertion of p_3 the whole triangulation is valid.

This leeds to the following algorithm to compute a triangulation at variable resolution:

1. Start with the coarsest Constrained Delaunay triangulation Σ_m.
2. Put each triangle $\Delta \in \Sigma_m$ onto a Stack U.
3. While U is not empty
   (a) Fetch the first unmarked triangle Δ from the top of U.
   (b) If $c(\Delta)$ is not true
       determine $r := \max(p_i, p_j, p_k)$, where p_i, p_j, p_k are the vertices of Δ and find the point p_r with smallest index s, so that $s > r$ and p_s is contained in the circumcircle of Δ.
       Insert p_s into the Constrained Delaunay triangulation and correct the resulting triangulation.
       Sort all triangles that are generated during the correction step into a List V in descending order. The sorting criteria is the maximum index number of the vertices of the triangle. Mark all triangles that are deleted during the insertion step.

Note that in the average case the insertion of a point into a Constrained Delaunay triangulation can be done in constant time.

The correction step The crucial part of the above algorithm is the correction step. In general, the triangles in the list V created during the insertion of the point p_r into the Constrained Delaunay triangulation do not belong to an intermediate Constrained Delaunay triangulation $\Sigma_m, \Sigma_{m+1}, \ldots, \Sigma_M$. Therefore, no error bounds can be guaranteed for these triangles. Further points have to be inserted until the whole triangulation only consists of triangles belonging to intermediate triangulations $\Sigma_{r_1}, \ldots, \Sigma_{r_n}$ contained in the multiresolution model.

The following algorithm is based on the observations above:

1. While V is not empty fetch the first triangle not marked from the list V and determine the vertex p of the triangle with the highest index. Let us call this index l.
2. For each triangle incident to p containing vertices in its circumcircle with indices less than l, insert that vertex into the triangulation that would be a neighbour of p in a triangulation containing all points with indices less than the index of p, see Fig. 2. As shown in [6,14] for practical applications this step can be performed in constant time using a uniform grid.
3. Insert all triangles created during the inserting step into list V and mark all triangles deleted during the insertion step.
4. Go to step 1.

Note that due to the ordering of the list V triangles that are generated during the correction step and are incident to the vertex with highest index will remain in the triangulation during the whole correction step, since these triangles do not contain less vertices than the highest index in their circumcircle. Together with observation 2 this proves the validity of the resulting triangulation.

Since we can assume that the insertion of points into the Constrained Delaunay triangulation can be done in linear time, the time complexity of the algorithm is linear in the size of inserted points, and therefore in the size of resulting triangles.

5   Results and Future Work

We have presented two algorithms for building multiresoltution models for CAD-data. We showed that in addition to the sample points only a set of approximation errors is necessary to compute the camera-dependent approximations in real time using the multiresolution model. We showed furthermore, that for huge data sets, e.g. the model of a whole car the trade of storage costs and storage access time for computing power is worth to be considered.

Our future work focuses onto two parts: First, we think that an easy parallelization of the proposed algorithm can easily be done. Second, we consider how perceptional errors rather than geometric errors can be incorporated into our extraction algorithm.

Acknowledgement

We would like to thank A. Schilling and T. Hüttner for many fruitful discussions. Furthermore, I would like to thank J. Krämer for the implementation of the fast incremental Delaunay triangulation algorithm, which encouraged us in developing the presented algorithm.

References

1. S. S. Abi-Ezzi and S. Subramaniam. Fast dynamic tessellation of trimmed NURBS surfaces. In *Computer Graphics Forum*, volume 13, pages 107–126. Eurographics, Basil Blackwell Ltd, 1994. Eurographics '94 Conference issue.
2. P. Cignoni, E. Puppo, and R. Scopigno. Representation and visualization of terrain surfaces at variable resolution. In R. Scatenied, editor, *Scientific Visualization 95 (Int. Symp. Proc.)*, pages 50–68. World Scientific, 1995.
3. Leila de Floriani and Enrico Puppo. Hierarchical triangulation for multiresolution surface description. *ACM Transactions on Graphics*, 14(4):363–411, October 1995.
4. L. DeFloriani, B. Falcidieno, and C. Pienovi. Delaunay-based representation of surfaces defined over arbitrarily shaped domains. *Computer Vision, Graphics and Image Processing*, 32:127–140, 1985.
5. M. Eck, T. DeRose, T. D., H. Hoppe, M. Lounsbery, and W. Stuetzle. Multiresolution analysis of arbitrary meshes. In Robert Cook, editor, *SIGGRAPH 95 Conference Proceedings*, Annual Conference Series, pages 173–182. ACM SIGGRAPH, Addison Wesley, August 1995. held in Los Angeles, California, 06-11 August 1995.
6. Tsung-Pao Fang and Les A. Piegl. Delaunay triangulation using a uniform grid. *IEEE Computer Graphics & Applications*, pages 36–47, may 1993.
7. D. Filip, R. Magedson, and R. Markot. Surface algorithms using bounds on derivatives. *Computer Aided Geometric Design*, 3(4):295–311, 1986.
8. L. De Floriani, E. Puppo, and P. Magillo. A formal approach to multiresolution modeling. In W. Straßer, R. Klein, and R. Rau, editors, *Theory and Practice of Geometric Modeling*. Springer-Verlag, 1996.
9. H. Hoppe. Progressive meshes. In *Computer Graphics Proceedings, Annual Conference Series, 1996 (ACM SIGGRAPH '96 Proceedings)*, pages 99–108, 1996.
10. R. Klein, , T. Hüttner, and J. Krämer. Viewing parameter dependent approximation of nurbs-models for fast visualization and animation using a discrete multiresolution representation. In B. Girod, editor, *Herbsttagung '96 3D Bildanalyse und -synthese*, 1996.

11. R. Klein. *Netzgenerierung impliziter und parametrisierter Flächen in einem objektorientierten System*. PhD thesis, WSI/GRIS, 1995.

12. R. Klein and W. Straßer. Mesh generation from boundary models. In C. Hoffmann and J. Rossignac, editors, *Third Symposium on Solid Modeling and Applications*, pages 431–440. ACM Press, May 1995.

13. R. Klein and T. Hüttner. Simple camera dependent approximation of terrain surfaces for fast visualization and animation. In R. Yagel, editor, *Visualization 96*. ACM, November 1996.

14. R. Klein and J. Krämer. Fast algorithms for constructing the 2D-Delaunay-triangulation. Technical Report WSI-1994-16, Wilhelm-Schickard-Institut, Graphisch Interaktive Systeme, Universität Tübingen, 1994.

15. R. Klein, G. Liebich, and W. Straßer. Mesh reduction with error control. In R. Yagel, editor, *Visualization 96*. ACM, November 1996.

16. Reinhard Klein. Linear approximation of trimmed surfaces. In R.R. Martin, editor, *The Mathematics Of Surfaces VI*, 1994.

17. J. Lane and R. Riesenfeld. A theoretical development for the computer generation and display of piecewise polynomial surfaces. *IEEE Trans. Pattern Analysis Machine Intell.*, 2(1):35–46, 1980.

18. D. T. Lee and B. J. Schachter. Two algorithms for constructing a delaunay triangulation. *International Journal Computer and Information Sciences*, 9(3):219, 1980.

19. E. Puppo. Variable reolution of terrain surfaces. In *Proceedings Eight Canadian Conference on Computational Geometry*, August 1996.

20. S. Rippa. Adaptive approximation by piecewise linear polynomials on triangulations of subsets of scatterd data. *SIAM Journal Sci. Stat. Comput.*, 13(5):1123–1141, September 1992.

21. Alyn Rockwood, Kurt Heaton, and Tom Davis. Real-time rendering of trimmed surfaces. In Jeffrey Lane, editor, *Computer Graphics (SIGGRAPH '89 Proceedings)*, volume 23, pages 107–116, July 1989.

22. William J. Schroeder, Jonathan A. Zarge, and William E. Lorensen. Decimation of triangle meshes. In Edwin E. Catmull, editor, *Computer Graphics (SIGGRAPH '92 Proceedings)*, volume 26, pages 65–70, July 1992.

23. Greg Turk. Re-tiling polygonal surfaces. In Edwin E. Catmull, editor, *Computer Graphics (SIGGRAPH '92 Proceedings)*, volume 26, pages 55–64, July 1992.

24. Brian Von Herzen and Alan H. Barr. Accurate triangulations of deformed, intersecting surfaces. In Maureen C. Stone, editor, *Computer Graphics (SIGGRAPH '87 Proceedings)*, volume 21, pages 103–110, July 1987.

25. J. C. Xia and A. Varshney. Dynamic view-dependent simplification for polygonal models. In Holly Rushmeier, editor, *Computer Graphics (SIGGRAPH '96 Proceedings)*, volume 30(4), August 1996.

Piecewise Linear Approximation for Scientific Data

René T. Rau*

WSI/GRIS, University of Tübingen,
Auf der Morgenstelle 10, 72076 Tübingen, Germany
email: rrau@gris.uni-tuebingen.de

Abstract. The visualization of scientific data allows for a faster and better insight in measurements and numerical computations. In order to generate reliable image results, the rendering has to be based on an error control. Since many visualization techniques use linear approximation schemes, we give estimates of the approximation error in arbitrary dimensions. Our results can be considered as generalizations and improvements of already existing estimates for curves and surfaces.

1 Introduction

The faceting of surfaces is one of the necessary tasks in CAD/CAM applications. The problem consists in finding a piecewise linear approximation of a given surface and a given tolerance.

Estimates for the quality of the approximation are used in many algorithms (see, e.g. [5,14]), in particular for adaptive approximations (cf. [16,6–8]). In addition mesh reduction algorithms and multiresolution models are based on initial triangulations for which these estimates are necessary (cf. [13,4,3]). For a good performance of these algorithms it is necessary to have sharp estimates, since this allows for a significant reduction of the necessary triangles for a given tolerance.

In scientific computing the type of functions, which are considered, can be interpreted as a generalization of functions used in CAD applications. Therefore, the necessary visualization algorithms are no longer restricted to the display of curves and surfaces. Rather general fields given by

$$f : \Omega \subset \mathbb{R}^n \to R \subset \mathbb{R}^m$$

are of interest (cf. [1]) and the field of scienific visualization is concerned with the rendering of these functions.

Important examples are given by scalar fields in 3D considered in computational fluid dynamics and elasticity computations. In addition, there exists large amount of data coming from digital acquisition technology like CT-data.

For the display of these scalar fields special algorithms have been developed. Besides algorithms for the generation of slices and isosurfaces, like the Marching Cubes algorithm [11] and similar techniques, there exist various techniques for the direct rendering of these functions. Important techniques are ray casting [9,10], splatting [17],

---

* Supported by the Deutsche Forschungsgemeinschaft, SFB 382

and projected tetrahedra [15], which are based on linear and trilinear interpolation schemes.

In general, the techniques used in scientific visualization are based on the reconstruction of the field from discrete data and the subsequent resampling ([1]). Often the reconstruction is performed by piecewise linear interpolation. For a reliable visualization the quality of the interpolation needs to be controlled.

We are interested in general approximation estimates in arbitrary dimensions. Besides the norms considered in variational problems we like to have bounds for the parametric distance, i.e. bounds for the finer sup-norm. This allows for the error control of numerical integrations, which are used in ray casting as well as the control of the Hausdorff distance.

In this paper we will restrict ourself to the linear case and consider only n-simplices. The advantage of using n-simplices lies in the fact, that the linear interpolant can be defined uniquely. Additionally, n-simplices can be easily obtained from regular grids and allow for a good approximation of a given boundary. Therefore they are also considered in numerical analysis like Finite Elements.

We proceed as follows. In Section 2 we present a general estimate for Lipschitz continuous functions. In Section 3 we assume C^2 continuity and in Section 4 we discuss applications.

2   Approximation of Lipschitz Continuous Functions

Often the fields given or reconstructed do not have good regularity properties. Fields coming from Finite-Element analysis as well as fields on regular grids with trilinear interpolants are not necessarily differentiable on the domain. But it is still clear that the error for a piecewise linear approximation can be controlled for a given Lipschitz continuous function. Proposition 1 gives an estimate in this situation.

Let us first recall some definitions and notations.

A d-simplex is the convex hull of any $d+1$ affinely independent points in some $\mathbb{R}^n$ ($n \geq d$) (cf. [18]).

Let Ω denote a bounded open set in $\mathbb{R}^n$. By $C(\bar{\Omega}, \mathbb{R}^m)$ we denote the Banach space of all continuous functions from $\bar{\Omega}$ with values in $\mathbb{R}^m$ endowed with the norm $\|f\|_\infty := \sup_{x \in \bar{\Omega}} \|f(x)\|$.

A function $f : \bar{\Omega} \subset \mathbb{R}^n \to \mathbb{R}^m$ is called *Lipschitz continuous* if there exists a constant $L > 0$ such that $\|f(x) - f(y)\| \leq L\|x - y\|$ for all $x, y \in \bar{\Omega}$.

Proposition 1. *Let* $T = [P_0, \ldots, P_n] \subset \mathbb{R}^n$ *denote the n-simplex of the points* $P_0, \ldots, P_n$. *Let* $f : T \to \mathbb{R}^n$ *be Lipschitz continuous with constant L. Let* $l : T \to \mathbb{R}^m$ *be the linearly parameterized n-simplex with* $l(P_i) = f_i := f(P_i)$, $i = 0, \ldots, n$. *Then*

$$\|f - l\|_\infty \leq \frac{2n}{n+1} rL,$$

where $r := \max_{i \neq j} \|P_i - P_j\|$ *denotes the maximum edge length of T.*

Proof. Let $P \in T$ be arbitrary. Then $P = \sum_{i=0}^n \alpha_i P_i$ with barycentric coordinates $\alpha_i \geq 0$, $\sum_{i=0}^n a_i = 1$. Then there exists an j such that $\alpha_j \geq 1/(n+1)$. Without loss of generality we may assume that $j = n$. Then

$$\|P - P_n\| = \|\sum_{i=0}^{n} \alpha_i P_i - P_n\|$$

$$= \|\sum_{i=0}^{n-1} \alpha_i (P_i - P_n)\| \leq \sum_{i=0}^{n-1} \alpha_i \|P_i - P_n\|$$

$$\leq \sum_{i=0}^{n-1} \alpha_i r = (1 - \alpha_n) r \leq \frac{n}{n+1} r.$$

Therefore

$$\|f(P) - l(P)\| \leq \|f(P) - f(P_n)\| + \|f(P_n) - l(P)\|$$
$$= \|f(P) - f(P_n)\| + \|l(P_n) - l(P)\|$$
$$\leq 2L\|P - P_n\| \leq 2L\frac{n}{n+1} r.$$

$\square$

The result is not optimal in the sense that, whenever f is already a linear parameterization, the estimate in Proposition 1 does not yield zero for the distance. In order to obtain this kind of estimates one has to assume more smoothness.

3   Approximation of C^2 Continuous Functions

For more regular functions we are able to get better estimates. The idea to use at least C^2 functions lies in the fact, that the second partial derivatives of the linear approximation vanish. Therefore the right hand side of the estimate depends only on the second derivatives of the function itself.

Let us first recall some notations. We define $C^k(\bar{\Omega}, \mathbb{R}^m) := \{f : \bar{\Omega} \to \mathbb{R}^m : D^\alpha f \in C(\bar{\Omega}, \mathbb{R}^m) \ \forall \ \alpha$ multi-index, $|\alpha| \leq k\}$ with norm $\|f\|_{C^k} := \sum_{|\alpha| \leq k} \|D^\alpha f\|$.

We are now able to formulate our approximation result for arbitrary dimension.

Proposition 2. *Let* $T = [P_0, \ldots, P_n] \subset \mathbb{R}^n$ *denote the n-simplex of the points* $P_0, \ldots, P_n$. *Let* $f \in C^2(T, \mathbb{R}^m)$ *and let* $l : T \to \mathbb{R}^m$ *be the linearly parameterized n-simplex with* $l(P_i) = f_i := f(P_i)$, $i = 0, \ldots, n$. *Then*

$$\|f - l\|_\infty \leq \frac{1}{2} \frac{n^2}{(n+1)^2} r^2 \sum_{|\alpha|=2} \|D^\alpha f\|,$$

where $r := \max_{i \neq j} \|P_i - P_j\|$ *denotes the maximum edge length of* T.

Proof. Let $P \in T$. Then $P = \sum_{i=0}^{n} \alpha_i P_i$ with barycentric coordinates $\alpha_i \geq 0$, $\sum_{i=0}^{n} a_i = 1$. Then there exists an j such that $\alpha_j \geq 1/(n+1)$. As in the proof of Proposition 1 we obtain

$$\|P - P_j\| \leq \frac{n}{n+1} r.$$

We proceed similar to the proof given by Filip et al. [5, Thm. 4] for the case $n = 2$.

Let $e(x) := f(x) - l(x)$. Since T is compact there exists $P = (x_1, \ldots, x_n) \in T$ such that $\|e(P)\| = \|f - l\|$. Furthermore there exists $j \in \{0, \ldots, n\}$ such that

$$d := \|P - P_j\| \le \frac{n}{n+1} r.$$

We may assume that $j = n$. Let $v := P_n - P$ and let $(e_i)_{i=1,\ldots,n}$ denote the canonical basis in $\mathbb{R}^n$. Then $v = \sum < v, e_i > e_i$ with $| < v, e_i > | \le d$, where $< \cdot, \cdot >$ denotes the scalar product in $\mathbb{R}^n$. Since $\|e(P)\|$ is maximum we conclude $D_v e(P) = 0$, where D_v denotes the derivative in the direction v. We define $g : [0, 1] \to \mathbb{R} : g(t) := e(P - tv)$. Then

$$g'(\xi) = (e(P - \xi v))' = \sum_{i=1}^{n} \frac{\partial e}{\partial x_i}(P - \xi v) < v, e_i >,$$

$$g''(\xi) = \sum \sum \frac{\partial e}{\partial x_i \partial x_j}(P - \xi v) < v, e_i >< v, e_j > .$$

Therefore

$$|g''(\xi)| \le \frac{n^2}{(n+1)^2} r^2 \sum_{|\alpha|=2} \|D^\alpha e\| = \frac{n^2}{(n+1)^2} r^2 \sum_{|\alpha|=2} \|D^\alpha f\|,$$

since $\sum_{|\alpha|=2} \|D^\alpha l\| = 0$. Writing g in Taylor expansion yields

$$g(1) = g(0) + g'(0) + \int_0^1 g''(\xi)(1 - \xi) \, d\xi$$

and we obtain

$$0 = e(P_n) = e(P) + \int_0^1 g''(\xi)(1 - \xi) \, d\xi.$$

Therefore

$$\|e(P)\| \le \int_0^1 \frac{n^2}{(n+1)^2} r^2 \sum_{|\alpha|=2} \|D^\alpha f\|(1 - \xi) \, d\xi$$

$$= \frac{1}{2} \frac{n^2}{(n+1)^2} r^2 \sum_{|\alpha|=2} \|D^\alpha f\|.$$

$\square$

This proposition is a generalization of a theorem of Sheng and Hirsch [14] to arbitrary dimensions. For $n = 1$ we obtain the following well-known result (cf. [2]).

Corollary 3. *Let $f : [a, b] \to \mathbb{R}$ be any C^2 function and let l be the line segment with $l(a) = f(a)$ and $l(b) = f(b)$. Then*

$$\sup_{x \in [a,b]} \|f(x) - l(x)\| \le \frac{1}{8}(b - a)^2 \sup_{x \in [a,b]} |f''(x)|.$$

The estimates given by Proposition 2 for arbitrary dimension can be at least improved for $n = 2$. We are able to proof the following result. Note that by the bounding box of a set $\{P_0, \ldots, P_k\} \in \mathbb{R}^2$, $P_i = (x_i, y_i)$, we denote the axis aligned box given by

$$B := \{(x, y) \in \mathbb{R}^2 : \min\{x_i\} \le x \le \max\{x_i\} \text{ and } \min\{y_i\} \le y \le \max\{y_i\}\}.$$

Proposition 4. *Let $T = [P_0, P_1, P_2] \in \mathbb{R}^2$ be an arbitrary triangle with $P_i = (x_i, y_i)$. Let B denote the bounding box of $\{P_0, P_1, P_2\}$ and let $l_x := \max\{x_i\} - \min\{x_i\}$, $l_y := \max\{y_i\} - \min\{y_i\}$. Let $f \in C^2(T, \mathbb{R}^m)$ and let $l : T \to \mathbb{R}^m$ be the linearly parameterized triangle with $l(P_i) = f_i := f(P_i)$, $i = 0, 1, 2$. Then*

$$\|f - l\|_\infty \leq \frac{1}{8}(l_x^2 \|\frac{\partial^2 f}{\partial x^2}\|_\infty + 2 l_x l_y \|\frac{\partial^2 f}{\partial x \partial y}\|_\infty + l_y^2 \|\frac{\partial^2 f}{\partial y^2}\|_\infty).$$

The proposition improves the result obtained by Sheng and Hirsch in [14]. The constant is $1/8$ instead of $2/9$ obtained from Proposition 2. Additionally, the edge lenghts of the bounding box are considered instead of the maximum edge length. This allows for a separate control of the two directions. Furthermore, the proposition extends the results obtained by Filip et al. to the situation of arbitrary oriented triangles.

For the proof we need the following lemma.

Lemma 5. *Let $T = [P_0, P_1, P_2] \in \mathbb{R}^2$ be an arbitrary triangle with $P_i = (x_i, y_i)$. Let B denote the bounding box of T and let B_{ij} denote the bounding box of the points P_i and P_j, $0 \leq i < j \leq 2$. Then*

$$B = B_{01} \cup B_{02} \cup B_{12}.$$

Proof. It suffices to show that $B \subset B_{01} \cup B_{02} \cup B_{12}$. Without loss of generality we assume $x_0 \leq x_1 \leq x_2$.

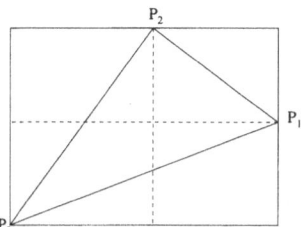

Fig. 1. Subdivision of the bounding box into 4 areas

By x_{min}, x_{max}, y_{min}, and y_{max} we define the minimum and maximum coordinate values of $\{P_0, P_1, P_2\}$. Since we consider 3 points in $\mathbb{R}^2$ and we defined 4 scalar values, we conclude that there is at least one point, which coincides with a vertex of B.

If $x_{min} = x_{max}$ or $y_{min} = y_{max}$ the triangle is degenerated and we are done. Therefore we assume $x_{min} < x_{max}$ and $y_{min} < y_{max}$. Then P_0 or P_2 is a vertex point of B. If this is true for P_0, the points P_1 and P_2 lie on the opposite edges of the bounding box. We then subdivide the bounding box into 4 rectangular areas according to Figure 1.

Thus each area is contained in a bounding box of a triangle edge. If P_0 is not a vertex of the bounding box the same arguments can be used for P_2 and we obtain the assertion. $\square$

We are now able to prove Proposition 4.

Proof. (Proposition 4) We subdivide the triangle T into similar triangles T_0, T_1, T_2, and T_3 by bisection of the edges (see Figure 2).

For the triangles T_0, T_1, and T_2 we consider the corresponding bounding boxes which we denote by B_0, B_1, and B_2. From our construction it is clear that the edge lengths of these boxes are $l_x/2$ and $l_y/2$.

We claim that $T \subset B_0 \cup B_1 \cup B_2$. To see this we only have to investigate T_3. Since the edges of T_3 are contained in the bounding boxes B_0, B_1, and B_2 we conclude from Lemma 5 that the bounding boxes of the edges are contained in B_0, B_1, and B_2. Then the bounding box of T_3 is contained in $B_0 \cup B_1 \cup B_2$ and therefore $T \subset B_0 \cup B_1 \cup B_2$.

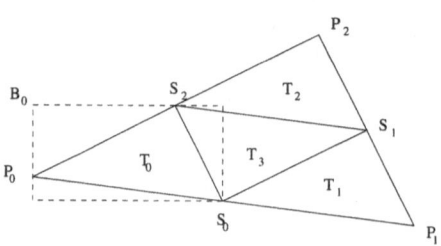

Fig. 2. Subdivision of the triangle into 4 similar triangles

Let now $P = (x, y) \in T$ be arbitrary. Then there exists $k \in \{0, 1, 2\}$ such that $P \in B_k$. Since $P_k = (x_k, y_k) \in B_k$ and $P \in B_k$ we conclude

$$|x - x_k| \le l_x/2 \quad \text{and} \quad |y - y_k| \le l_y/2.$$

Following the proof of [5, Thm. 4] for right triangle, we obtain the assertion.   □

Remark 6. A similar proof for $n = 3$ is not possible since a bisection of the edges of a tetrahedron results in 4 tetrahedra and an octahedron.

4   Conclusion and Applications

We obtained general estimates useful for a broad class of functions considered in scientific computation. The results allow for the use of visualization algorithms with a control of the absolute approximation error and therefore a control of the image quality.

In the situation of surfaces we improved existing estimates. The applications are algorithms, which are based on resampling like the one considered in [5,14,8]). Our new bounds can be used for arbitrary triangles and they reduce the amount of necessary samples compared with the estimates given in [14] by at least the factor 9/16.

In the situation with scalar fields our estimates can be used for the error control of integration techniques as well as the adaptive generation of isosurfaces (cf. [6]) with a control of the distance of the approximating mesh and the real isosurface.

In [12] an object-oriented design for the visualization of scientific data was presented based on reconstruction and resampling of given data. Here our estimates allow for a good control of the resampling rate even for functions, which are not differentiable and thus not band-limited.

References

1. K. Brodlie. A classification scheme for scientific data. In R.A. Earnshaw and D. Watson, editors, *Animation and Scientific Visualisation, Tools and Applications*, pages 125–140. Academic Press, London, 1993.
2. C. de Boor. *A Practical Guide to Splines*. Springer, 1978.
3. L. De Floriani, E. Puppo, and P. Magillo. A formal approach to multiresolution modeling. In *Proceedings Blaubeuren II)*, 1996.
4. Matthias Eck, Tony DeRose, Tom Duchamp, Hugues Hoppe, Michael Lounsbery, and Werner Stuetzle. Multiresolution analysis of arbitrary meshes. In Robert Cook, editor, *SIGGRAPH 95 Conference Proceedings*, Annual Conference Series, pages 173–182. ACM SIGGRAPH, Addison Wesley, August 1995. held in Los Angeles, California, 06-11 August 1995.
5. D. Filip, R. Magedson, and R. Markot. Surface algorithms using bounds on derivatives. *Computer Aided Geometric Design*, 3(4):295–311, 1986.
6. Mark Hall and Joe Warren. Adaptive polygonalization of implicitly defined surfaces. *IEEE Computer Graphics and Applications*, 10(6):33–42, November 1990.
7. R. Klein. Polygonalization of algebraic surfaces. In P.J. Laurent, A. Le Méhauté, and L. L Schumaker, editors, *Curves and Surfaces II*. AKPeters, Boston, 1993.
8. R. Klein. Linear approximation of trimmed surfaces. In R.R. Martin, editor, *The Mathematics Of Surfaces VI*, pages 209–212, 1994.
9. Marc Levoy. Display of surfaces from volume data. *IEEE Computer Graphics and Applications*, 8(3):29–37, May 1988.
10. Marc Levoy. Efficient ray tracing of volume data. *ACM Transactions on Graphics*, 9(3):245–261, July 1990.
11. William E. Lorensen and Harvey E. Cline. Marching cubes: A high resolution 3D surface construction algorithm. In Maureen C. Stone, editor, *Computer Graphics (SIGGRAPH '87 Proceedings)*, volume 21, pages 163–169, July 1987.
12. R. Rau. An object-oriented framework for the visualization of scientific data. *Submitted for publication*, 1996.
13. William J. Schroeder, Jonathan A. Zarge, and William E. Lorensen. Decimation of triangle meshes. In Edwin E. Catmull, editor, *Computer Graphics (SIGGRAPH '92 Proceedings)*, volume 26, pages 65–70, July 1992.
14. X. Sheng and B. E. Hirsch. Triangulation of trimmed surfaces in parametric space. *Computer Aided Design*, 24(8):437–444, August 1992.
15. Peter Shirley and Allan Tuchman. A polygonal approximation to direct scalar volume rendering. In *Computer Graphics (San Diego Workshop on Volume Visualization)*, volume 24, pages 63–70, November 1990.
16. Brian Von Herzen and Alan H. Barr. Accurate triangulations of deformed, intersecting surfaces. In Maureen C. Stone, editor, *Computer Graphics (SIGGRAPH '87 Proceedings)*, volume 21, pages 103–110, July 1987.
17. Lee Westover. Footprint evaluation for volume rendering. In Forest Baskett, editor, *Computer Graphics (SIGGRAPH '90 Proceedings)*, volume 24, pages 367–376, August 1990.
18. G. M. Ziegler. *Lectures on Polytopes*. Springer, 1995.

Part III
Systems

Provision of an Explicit Constraints Schema in the STEP Standard

Michael J. Pratt

National Institute of Standards and Technology[*]
Manufacturing Systems Integration Division
Building 220, Room A127
Gaithersburg, MD 20899, USA.
E-mail pratt@cme.nist.gov

Abstract. The International Standard STEP (ISO 10303) is intended to facilitate the exchange of data between CAD systems. The first (1994) release of the standard allows the transfer of geometric product models in terms of geometry and topology plus configuration data (version numbers, etc). Since most CAD systems now allow the creation of variational models with parametrization, constraints and features, it is necessary to extend STEP to take into account these newer capabilities. This task is being addressed by the ISO Parametrics Group, working within ISO TC184/SC4. The paper discusses some of the considerations arising in the early stages of this work; in particular, it presents a requirements analysis for a STEP schema for the representation of explicit constraints.

1 Introduction

The development of the International Standard ISO 10303 (see Appendix 1), informally known as STEP (STandard for the Exchange of Product model data), started in 1984. The first version of the standard was issued ten years later, in 1994. The time taken reflects the fact that a major infrastructure was also created for what will be an ongoing development of STEP, possibly for many years to come. However, the ten-year gap and the fact that the nature of the standardization process requires the technical content of standards to be frozen well before their eventual release has led to a situation where STEP does not reflect the full capability of modern CAD systems. In particular, these systems can now generate parametrized, constraint-based and feature-based models[5], whereas STEP currently makes no provision for the representation of parametrized entities, constraints or features.

Three major components are therefore needed for the provision of the missing variational capability in STEP. These will need to be embedded in a general framework, whose details are still being determined. In addition to (1) the constraints schema which is the main topic of this paper, the other components are (2) a means for the representation of parametrized entities, and (3) a general method for representing form features. In this contribution, a brief overview of the contrasting roles of parametrization and constraints in modeling will first be given. These topics will then be examined

---

[*] On assignment at NIST from Center for Advanced Technology, Rensselaer Polytechnic Institute, Troy, NY 12180, USA.

in the context of STEP, and a few words devoted to the related topic of form feature modeling. The remainder of the paper then concentrates solely on the topic of constraints. The nature of the overall framework for variational modeling is not discussed, though it is clearly an important topic. A contribution in this area has recently been made by Pierra et al.[11].

2   Design Freedom and Restraint

The following subsections discuss the contrasting roles of parametrization and constraints in a geometric modeling context.

2.1   'Pure' Parametrics and Design Freedom

Parametric methods are used in CAD to express the freedom which is available for redesign, variant design, design optimization or the representation of part families. This freedom is usually made available as an ability to change dimensions (either linear or angular) in the part model.

Integer parametrization is also sometimes used to specify the number of features in a pattern, e.g. the number of holes in a bolt-hole circle. This type of facility presents more problems for the implementer since it gives rise to topological changes in the model, which is not necessarily the case when parametrized dimensions are used.

It is difficult to envisage a CAD system providing a parametric capability which does not have an associated constraint capability of some kind. For example, it may be appropriate to have limitations on permissible parameter ranges, so that negative linear dimensions cannot occur in the model. It may also be convenient to restrict the values of a particular parameter to a set of discrete values. This might be so, for instance, when a company uses only a limited number of standard bolt sizes, in which case the diameters of bolt-holes in their products could be restricted to have only the appropriate values for those sizes.

A further example of parametric design is furnished by constructive solid geometry (CSG), and this method is certainly not free of constraints. Implicit parallelism and perpendicularity constraints between faces are inherent in the very notion of a rectangular block primitive. Thus, although the constraints are not imposed explicitly by the designer they arise implicitly through the semantics associated with the choice of primitive. The same is true, of course, for any other primitive type available in such a system.

2.2   Constraints and Design Restraint

In variational modeling, parametric and constraint-based capabilities are two sides of the same coin. Parametrization expresses the availability of design freedom, while constraints impose limitations on that freedom in the interests of ensuring the continued functionality of the design following a design change.

Before deciding on a representation for parametrized entities in STEP it is necessary to examine a grey area where clarification is needed. Three kinds of possibilities arise, as follows:

1. An entity is subject to a dimensional constraint, so that no subsequent change is possible.

2. An entity is defined as being parametric, in which case its parametrized dimensions may be freely changed within any limits imposed by associated parameter constraints.

3. An entity has an associated numerical dimension, but no constraint is applied to prevent this dimension from being changed in a design variation. The question here is, can this situation be considered equivalent to Case 2 above? If so, it makes sense to transmit the entity in instantiated form (i.e. with the numerical dimension) but to associate with it constraints specifying a permissible range of variation in the dimension.

The third of these possibilities may be summarized as 'If an entity is not constrained from varying, then it may be changed if required'. This may be contrasted with the second possibility, which effectively says that 'If an entity may be varied, then it should be explicitly represented as a parametrized entity'.

As will be mentioned later, the intention in STEP is always to transmit a 'current instance', i.e. an explicitly dimensioned model, together with any parametric description of a model. This somewhat lessens the contrast, since in Case 3 above the dimensioned entity itself can act as the current instance. In Case 2, it will be necessary to transmit the parametrized entity *together with* an explicitly dimensioned entity (or at least a set of numerical values for the parameters that specify a 'current instance' of the model). Since the two cases are so similar, it may be questioned whether any distinction needs to be made between them in the context of STEP.

From the implementational point of view, Case 3 appears, at first sight, to be the simpler alternative. It requires the representation of less information, but more significantly *it does not require a means for the explicit representation of parametrized entities*. Adoption of this approach could therefore significantly simplify the representation of design freedom in the STEP standard. From the CAD system point of view the specified parameter range may be regarded as an inequality constraint, and it is probably sufficient to use this merely as a check on the validity of a redesign situation, applied after a recalculation of geometry subject to constraints. The solution of the inequality simultaneously with the other constraints is thereby avoided, which simplifies the task of the constraint solver.

3   Components of a Variational Capability for STEP

3.1   Constraints

The only previous work on constraints in the context of ISO TC184/SC4 has been by a group developing the Parts Library standard (ISO 13584). This is separate from STEP, but is intended to be compatible with it[8]. The Parts Library modeling methodology is procedural, part models being represented by sequences of constructional operations. This method is entirely suitable for the definition of the part families required in the Parts Library standard, since the use of parametrized constructional entities leads to a model such that different choices for the parameters give different members of the part family. The constraints arising in this approach are *implicit* — they are inherent in the constructional procedures or modeling primitives used. An *explicit* constraint, by contrast, applies directly to geometric or topological entities occurring in an evaluated model. Two examples will be given to illustrate implicit constraints:

1. In 3D, a rectangular block primitive used in a CSG model carries implicit perpendicularity constraints between its faces where they meet at edges.
2. In 2D, a function may be defined that generates *line2* as a (directed) line parallel to *line1* and at a distance *d* from it. The relationship between the function and its inputs and output may be expressed as *line2 = parallel(line1,d)*. The new line will appear on one side or the other of the original, depending on the sign of *d*. Note that the generating function not only embodies an implicit constraint, but also implies an order of creation of the two lines concerned.

It may be noted that, in both the above cases, evaluation of the models in 3D or 2D boundary representation form may lead to loss of the constraints. Lines will indeed be parallel, or faces mutually perpendicular in the evaluated model, but this model will not necessarily contain any information to say that they have to remain so if further modeling operations are subsequently performed on it. The generation of such information would require any constraints implicit in the constructional primitives to be made explicit in the evaluated model.

An alternative approach to constraint-based modeling appears to be closer to the practice in most modern CAD systems. This uses explicit constraints, applied during the modeling process in the form of relationships between the entities composing the model. In this case the presence of a constraint does not necessarily say anything about how the model was constructed, or about the order in which the entities composing it were created.

The contrast is therefore between an implicit modeling approach that prescribes a set of instructions for building a model, and an explicit approach that gives a fully detailed 'snapshot' of the model as it exists at some stage of its history of creation.

One intention of STEP product model transfer is that modeling should be possible in the receiving system just as though the transferred model had originally been created there. This appears to require the transfer of model constructional history, in some sense, and it remains to be determined to what extent this is possible using the different available modeling approaches.

From the mathematical point of view, design constraints are often nonlinear. Some CAD systems allow the definition of groups of constraints which are required to hold simultaneously, and in this case the system has to be capable of solving systems of simultaneous nonlinear equations. Such systems in general have non-unique solutions, and following a model exchange the receiving CAD system needs to be able to determine the particular solution to choose when the constraints are solved.

A simple example of a constrained problem having multiple solutions is the determination of a line tangent to two given circles. Here there there are four solutions. If the line is taken to be directed and the positive sense of the circles is counterclockwise, the solutions correspond to the cases in which the sense of the line is the same as, or is opposed to, the sense of each of the circles at the point of tangency. It may be noted, however, that specification of the sense relationships at the points of contact removes the ambiguity, a fact exploited for many years in the creation of cutter paths using high-level NC programming languages such as APT[6].

The transmission, together with the constrained model, of a 'current instance', is seen as the answer to the problem of multiple constraint solutions in both the Parts Library and the explicit constraint approaches. In a broad sense this will convey the 'designer's intent', by indicating which particular solution is the one intended. In a narrower mathematical sense it will provide an initial approximation for the constraint solver in the receiving system to work from. This will enable it to adjust parameter values to make the model self-consistent in what may be a different computational environment from that of its original creation.

In the remainder of this paper no further consideration will be given to methods for the representation of the implicit constraints defined earlier in this section. This topic is currently being examined by the ISO Parts Library Group, and it is hoped that a U.S. project known as OCAI (Open CAD Architectures for Interoperability) will also be able to provide input in the near future. The results can then be combined with those of the present study, which is mainly devoted to methods for representing explicit constraints.

3.2   Parametrized Entities

The STEP standard currently provides no direct means for the representation of parametrized entities, dimensioned in terms of variables or in terms of expressions involving variables. All dimensions must currently be real numbers.

There exist several possibilities for overcoming this problem. Firstly, there is the possibility of modifying the existing STEP Integrated Resources. Part 42 of STEP (see Appendix 1) provides definitions of all the geometrical and topological entities used in building product model representations. At present these require all defining dimensions to be of type **length_measure** or **plane_angle_measure**, both being defined as being of base type **real**. It has been determined that a comparatively small change to allow **length_measure** and **plane_angle_measure** to take either of the types **real**, **variable** or **expression**, will allow the Part 42 geometric entities as currently defined to be used in a parametric manner. This provides a simple solution, but at the cost of changing an existing part of the standard and introducing new **variable** and **expression** data types that are not currently available. This is one example of the type of tradeoff which has to be made in revising STEP.

Secondly, it is possible to use a schema for the definition of parametrized models that already exists, in a Parts Library document[10]. This defines a new subtype for each of the dimensioned entities in Part 42 of STEP, in which the dimension or dimensions are represented by variables or expressions. A Parts Library schema also exists for the representation of variables and expressions. Both schemas are fairly complex, and the corresponding solution therefore lacks the elegant simplicity of the previous one. However, it can be implemented without making any changes to the STEP Integrated Resources.

The final possibility considered here is that, as suggested previously, each numerically dimensioned Part 42 entity, unless subject to a dumensional constraint, is regarded as a 'current instance' of a parametrized entity. Then in a redesign situation the dimension may be varied, possibly within specified limits. This method appears to avoid the necessity for defining parametric versions of Part 42 entities by (in essence) taking the view that all of them may be parametric, and that any specific dimension may be subject to change *unless specifically forbidden by an explicit constraint*.

There may, of course, be other ways of dealing with the problem of parametrization, and it is hoped that some of them will emerge during further work by the Parametrics Group.

Changes of the kind envisaged will affect the format and content of the STEP Physical File format, specified by Part 21 of the standard, and also the STEP Data Access Interface (Part 22), a standardized means for accessing STEP-formated information in databases. Upwards compatibility must be achieved as far as possible.

It will be seen that the decisions to be taken are of a delicate nature. What is desired is a compromise between representational power, conciseness, and degree of compliance with the existing released version of the STEP standard.

3.3 Features

The representation of features is within scope of the intended work, but will not be covered in the present paper. The eventual intention of the ISO Parametrics Group is to provide a resource for the general modeling of features in STEP, based on the parametric and constraint modeling capabilities currently under development. This will make available a flexible capability, useful for the modeling of a wide variety of feature classes appropriate for many application areas. This topic will be tackled when the necessary tools are in place.

4 Explicit Constraints

In this section the scope of the study of explicit constraints is further refined, and various requirements for information capture outlined.

4.1 Enumeration of Explicit Constraint Types in Modern CAD Systems

Major inputs under this heading have been the following sources:

1. A document compiled by the Japanese JSTEP Parametrics Committee, circulated at the Crystal City meeting of ISO TC184/SC4 in June 1995[9],
2. A Request for Change (RFC)[2] which seeks the addition of constraint capabilities to the U.S. data exchange standard IGES[13],
3. Information from from D-Cubed Ltd.[3,4], suppliers of the DCM constraint management software to many leading CAD system vendors,
4. A constraints schema developed by Chia-Hui Shih for the U.S. ENGEN (Enabling technologies for Next GENeration design) Project[12].

All four sources reflect the capabilities of existing CAD systems. They also agree fairly closely on what is needed. One requirement is for dimensional constraints; another is for geometric constraints such as tangency and symmetry. ENGEN does not distinguish between dimensional and other types of constraints, but D-cubed uses the distinction that a dimension is specified in terms of a numerical value, while a geometric constraint is logical (i.e. is either true or false). The distinction is not always clear-cut; for example, a logical or geometric perpendicularity constraint could be considered equivalent to an angular dimension constraint with value 90°, but certain constraints of this type occur so frequently as to merit special treatment. The D-Cubed software permits overconstrained situations (in terms of geometric constraints, provided they are not mutually inconsistent) but not overdimensioned ones. Dimensions may have variable (undefined) values in underconstrained geometric situations.

Both the IGES RFQ and the ENGEN project limit their attention to 2D constraints. The JSTEP document distinguishes an additional class of 3D assembly constraints, but gives few details. D-Cubed actually provides a 3D constraint package for part positioning in assemblies, and its documentation has been a source of useful information on this class of constraints.

There appears to be general agreement on the requirement for two primary forms of constraints. Firstly, there is a set of commonly used dimensional and geometric constraints whose semantics are very easily captured. STEP should represent these in such a way that their semantics are quite explicit. Then, regardless of how the constraint is actually represented in the sending system, its *meaning* will be clear from the exchange file, and the postprocessor will be able to generate the appropriate form in the receiving system. Such constraints will be referred to as *predefined constraints* in what follows.

A second form of representation is needed for constraints imposed for some special purpose in a particular design situation. It is impossible to catalog all possibilities for these, and we must therefore provide a general means for building constraints to meet any requirement that may arise. An algebraic formulation of what will be called *user-defined constraints* is used below to achieve this purpose. The geometric meaning of an algebraically expressed relationship is often not easy to determine by observation, and may well be impossible for an automatic translator to deduce. Accordingly, it will be desirable for any user-defined constraint to be transferred together with a textual description explaining its nature and purpose.

The intention expressed in all four documents is that constraints should be specified as relations between constructional elements of a model. In what follows it will be assumed that these are geometric elements (points, curves and surfaces) rather than topological elements (vertices, edges, and faces). This is in fact consistent with what is done in the D-Cubed software and the current version of the ENGEN model, though IGES deals in the main with bounded line segments, bounded circular arcs etc. In this paper, the simple geometric elements will be taken to be unbounded lines, circles, planes, cylinders etc.

In a typical boundary representation modeling context, a topological element will have a pointer to the geometry underlying that element. Since we propose to constrain the underlying geometry, then in a case where multiple faces of an object lie on the same surface, all those faces may be constrained by applying a single constraint to their shared surface. In what follows, it will be assumed that constraints are applied to the STEP Part 42 entities that are subtypes of **point**, **curve** and **surface**. There should not be much difficulty in extending the schema to represent relations between topological elements later if this proves desirable.

The parametric and constraint-based capabilities of current CAD systems can be broadly classified as below:

1. 2D capabilities; construction of open or closed profiles using geometric constraints and algebraic relations.
2. 3D capabilities within single part models; use of constrained 2D profiles and related 3D constructional operations including sweeps and Boolean operations. Sweep distances or rotational angles may themselves be constrained, and this results in a genuine 3D capability.
3. 3D capabilities in the modeling of assemblies. These are now emerging as CAD modelers begin to handle assemblies in a more sophisticated manner than they have in the past.

The list below is a synthesis, containing all the explicit constraint types suggested in the JSTEP, IGES, D-Cubed and ENGEN documents. Some of those documents deal only with 2D cases, and in certain cases in the list below I have suggested extensions into 3D where these may be appropriate. Algebraic constraints may be either 2D or 3D. The selection of assembly constraints given is the set implemented by D-Cubed. With the exception of the algebraic types, each constraint type is for present purposes represented declaratively in the form of a PROLOG clause. To give just one example, the interpretation of the form **parallel**(*line1,line2*) should be fairly obvious. The formulation used here will not necessarily relate closely to the one eventually chosen for STEP purposes, however.

Dimensional constraints: These are as follows

1. **distance**(e_1, e_2, d) — the distance between entities e_1, e_2 is constrained to have the value d. Here there are certain combinations of entity types for which the interpretation of the constraint is obvious. In 2D these include point/point, point/line and line/line. More generally, D-Cubed also allows a constraint in 2D on the distance between two circles, and they regard this distance as having four possible values. A line through the centers of the circles gives rise to four intersections, and the distance in question may be taken as the distance between any pair of them, chosen one from each circle. The choice in any particular case depends on the context. The question of 'legal' interpretations of distance constraints needs careful consideration in the context of STEP; it is quite possible that different systems will have different interpretations in some cases. The same overall concepts may clearly be extended into three dimensions.

2. **angle**(e_1, e_2, a) — the angle between entities e_1, e_2 is constrained to have the value a. In 2D the entities will be lines, and in 3D lines and/or planes.

3. **radius**(e_1, r) — the radius of entity e_1 is constrained to have the value r, where the entity will be a circle in 2D or a sphere in 3D. This is proposed in the JSTEP and D-cubed documents. ENGEN allows e_1 to be a *list* of circles having the same specified radius. The radius constraint appears to imply the need more generally for type-specific constraints on other individual types of curves and surfaces, e.g. on the major and minor axes of an ellipse, or in 3D the major and minor radii of a torus.

4. None of the four documents examined mentioned any requirement for *inequality* constraints on dimensions. However, it seems very likely that these will be required, for example in specifying limited parameter ranges to ensure that unwanted topological changes cannot occur as a model is modified in a receiving system. To give an example, consider a cube with sides of length 2 units, and a through hole whose axis passes through the centers of two opposing faces. If the radius of the hole is parametrized by r, then clearly we require $0 < r < 1$; if $r > 1$ then the hole no longer exists and the cube falls apart into four separate pieces.

Geometric constraints: These constraints are expressed in declarative form in the following list.

1. **on**(e_1, e_2) — entity e_1 lies on entity e_2. In the simplest case e_1 may be a point and e_2 a 2D curve, though 3D interpretations are also possible, in which points or curves are constrained to lie on surfaces.

2. **parallel**(l_1, l_2) — lines l_1 and l_2 are parallel. The IGES RFC allows an arbitrary length *list* of lines to be declared as parallel, though all the D-Cubed constraints are binary, relating just two entities. This type of constraint could be extended to cover parallelism of planes in 3D. We need to consider whether,

since STEP lines have a direction, to distinguish between 'parallel' and 'antiparallel', according as to whether the lines have the same or opposite directions. The same type of question arises for planes in 3D.

3. **concentric**$(c_1, c_2, \ldots, c_n, p)$ – the circles in the list $c_1, c_2, \ldots, c_n$ are concentric, all being centered at the point p. This case occurred only in the IGES document. In 3D, spherical surfaces could be constrained in a similar manner.

4. **parallel_offset**$(e_1, e_2, \ldots, e_n)$ — the entities in the list are parallel offsets of each other. This constraint was not specified in any of the documents consulted, but is a potentially useful generalization of the **parallel** and **concentric** constraints. It could apply to curves in 2D or surfaces in 3D.

5. **perpendicular**(l_1, l_2) — lines l_1 and l_2 are perpendicular, in the 2D case. In 3D the constrained entities could include planes.

6. **symmetric**(e_1, e_2, e_3) — entities e_1, e_2 are symmetric with respect to entity e_3. The JSTEP, IGES and D-Cubed documents only define symmetry in 2D with respect to a line. In general, e_1 and e_2 may be points, lines, circles, curves or surfaces. Symmetry may in fact be defined with respect to a point, line or plane. However, symmetry about a point seems to have little use in CAD, and therefore e_3 will be restricted to being a line in 2D and a plane in 3D. One question arising here is whether symmetry applies only to individual entities in a list, or whether it can be applied to higher-level groupings of entities, for example features.

7. **tangent**(e_1, e_2) — entity e_1 is tangent to entity e_2. The entities will be lines or curves in 2D, planes and other surfaces in 3D. The IGES document, dealing with 2D profiles, requires information about the point of tangency as it relates to each of the entities, e.g. whether it is an end point; this is because IGES generally deals with bounded rather than unbounded entities. Some discussion may be necessary before we can settle on a canonical format that handles all requirements.

   The suggestion has been made that tangency can be handled in terms of a point coincidence constraint and a coincident direction constraint. The primary problem with this is that it does not generalize well to the capture of higher-order derivative constraints.

8. **midpoint**(p, e) — point p is the center point of entity e. In 2D e will be a circle, but in 3D it could be a sphere. In 3D there are also many other similar possibilities for specialized constraints of this type, e.g. a constraint requiring that a point is the vertex of a cone.

9. **equal_length**$(e_1, e_2, \ldots, e_n)$ — the entities in the list are required to have equal arc lengths. This occurs in the IGES and ENGEN documents, and its use is unclear to me at the time of writing. However, D-Cubed say that their customers have expressed a requirement for this type of constraint.

10. **equal_radius**$(c_1, c_2, \ldots, c_n)$ — this type of constraint is implemented by D-Cubed and specified for ENGEN. It is presumably used mainly for fillet radii in 2D profiles. The effect could be achieved indirectly, by parametrizing the fillet radii, setting them all equal and then assigning a value to the parameter, but since this situation arises frequently in practice it may be simpler to provide a specialized constraint for the purpose.

11. **coincident_location**$(e_1, e_2, \ldots, e_n)$ — the specified entities must lie at the same point (this occurs only in the IGES document, with the restriction that the entities may be end-points of lines or arcs, or center-points of circles, in 2D. Again the IGES orientation towards the use of bounded elements is evident.

Constraints of this type are clearly also applicable in 3D. It has been questioned whether the topological connectivity available in Part 42 does not render this type of constraint unnecessary. However, the topology merely *reports* on the existence of a particular coincidence; it does not enforce its continuance in a redesign situation, and connectivity information alone therefore does not answer the purpose.

12. **point_at_intersection**(p, e_1, e_2) — the point p is constrained to lie at the intersection of entities e_1, e_2 in a 2D sketch. This type of constraint is suggested in the IGES document, but it is not quite clear what is constrained. Is the intention that specification of any two of the three constraint arguments constrains the third? D-Cubed handle this situation using a combination of simpler point-on-line constraints.

13. **fixed**(e_1) — the entity e_1 has its geometry, position and orientation fixed. This type was suggested in the JSTEP document, but it may also be used to handle the 'fixed_point' constraint in that document, corresponding to the IGES RFC **fixed_location** constraint, in the case where e_1 is a point. The entity e_1 may also be a direction. Directional constraints were suggested in the JSTEP proposal, but the only corresponding types in the IGES RFC were **horizontal_constraint** and **vertical_constraint**, which are presumably intended to apply in the sketch plane of a 2D profile. D-Cubed say that their customers strongly require the horizontal and vertical direction constraints, and so there may be some virtue in making special provision for these cases.

Algebraic constraints: The IGES RFC makes provision for naming real or integer variables, and for assigning values to them (presumably in accordance with the 'current instance' philosophy of always sending an explicit version of the model.) Algebraic constraint formulations are defined in terms of these variables using embedded code segments written in a subset of ANSI C. The ENGEN proposal defines a mechanism for bypassing the fact that STEP cannot currently transfer instances of variables or expressions. However, both approaches will require an additional level of file processing by a receiving system.

Assembly constraints: Suggestions under this heading arise primarily from a study of one of the D-Cubed Technical Overview documents[4]. As implemented by D-Cubed, assembly constraints differ significantly from 2D constraints. Whereas the 2D constraints apply between individual geometric elements involved in an individual part representation, the 3D constraints apply between *rigid sets* of part representation entities. Such a rigid set may be an entire part representation, in the case where that part is being positioned and oriented with respect to other parts in an assembly. Alternatively, the rigid set may be a representation of a part feature in terms of lower-level entities, in which case the 3D constraints may be used to position and orient the feature with respect to other elements of the part model. However, the actual constraints are expressed in terms of geometric elements as in the 2D case; the practical difference is that it is not a single element which has to move in order to satisfy a constraint, but the entire rigid set containing that element. In the future, when the facilities are available in STEP for defining features, this facility may need to be extended to relate pairs of entire features rather than pairs of geometric elements.

A significant aspect of the D-Cubed 3D constraints is that they often make use of auxiliary or constructional geometry rather than part geometry. Although the range of geometric elements directly supported is restricted to points, unbounded lines, planes and cylinders, the use of auxiliary geometry effectively makes this range much wider. Two examples will be given to illustrate this point:

1. A cone may be represented by its center-line for many positioning purposes, especially those involving coaxiality with some other rotationally symmetric element.
2. A toroidal duct may be smoothly joined to a cylindrical duct of the same diameter through the use of an auxiliary cylinder tangent to the torus around the circle where the join occurs.

The availability in a STEP file of auxiliary geometric elements therefore seems to be an important requirement in the data exchange context.

The D-Cubed documentation states that, in addition to dimensional constraints as discussed earlier, the following geometric constraint types are implemented in 3D: **coaxiality, tangency, parallelism, perpendicularity, coincidence**.

Some of these may be interpreted quite broadly — for example a line may be constrained to be parallel to a cylinder, in which case the line is made parallel to the *axis* of the cylinder.

Notes

1. The above list contains certain inconsistencies and is not optimally structured. In particular, some constraints are expressed in binary form, relating just two entities, while others constrain lists of entities. The list will be refined and made consistent as work progresses.
2. All of these constraints express relationships between geometric entities. For many applications STEP will also need means for relating geometric to nongeometric quantities. For example, a dimension of a part may be constrained in terms of an applied force, or conversely a relation may be defined between a key dimension and a predicted cost of manufacture. It appears that the user-defined class of constraints should be suitable not only for purely geometric constraints but also for these more general situations. This must be borne in mind in the development of a STEP constraints schema. It may be that the representations developed will also prove suitable for the modeling of constraints between physical quantities in general, but this is something to be determined at a later stage; for the moment, the aims are to handle geometry-based cases.
3. The ENGEN document proposes an additional constraint type, which constrains a point to lie at the centroid of a 2D closed profile. This was not suggested in any of the other documents, but may be added to the list.

4.2   The Mathematical Nature of Constraints

It has already been noted that, if the STEP standard is to keep abreast of current technical developments, both 2D and 3D constraint modeling capabilities will be needed. The 3D representations will be particularly valuable for enhancing the currently limited capability of STEP for the modeling of assemblies.

Other aspects of constraints concern their linearity or nonlinearity, and the question of whether they correspond in mathematical terms to equations or inequalities. These aspects will not be further discussed here. Constraint management and solution techniques lie outside the scope of the STEP work; they are assumed to be the responsibility of the receiving system. The present intention is simply to provide a means for *representing* constraints, and from that point of view little distinction need be made between equations and inequalities, linearity and nonlinearity. However, other relevant aspect of constraints in general are briefly discussed below.

4.3   Directed and Undirected Constraints

Consider an explicit constraint specifying that two lines l_1, l_2 are parallel. This may arise in two different ways:

1. Line l_1 may have been created first, and line l_2 then created parallel to it using l_1 as a *reference entity*. In this case the constraint embodies an element of constructional history, and it is important that this is captured in the constraints schema.
   Note that situations such as this do not arise only in procedural modeling systems; in many non-procedural systems the user can create a line by 'rubber-banding' or some similar technique, and move it around until the system indicates in some way that it is parallel to an existing line. If at this point the user leaves it where it is, the system automatically logs the constraint.
2. Alternatively, the parallel lines may have both existed previously, but not been constrained. In this case the constraint can be applied to them so that they will remain parallel regardless of subsequent modeling operations. Here the constraint has no reference entity, and no constructional history is implied.

These examples show the need for both *directed* and *undirected* constraints. The need for a dual treatment is even clearer in the case of algebraic constraints. For example, the height of a cylinder may be related to its radius by the relation

$$cyl.height := 2 * cyl.radius.$$

The implication here is that the radius is known beforehand, and that the height can then be determined in terms of it. On the other hand, if the relation is expressed as

$$cyl.height - 2 * cyl.radius = 0,$$

then there is no implied precedence of creation, and either quantity may be determined in terms of the other. Once again, we find a need for both directed and undirected constraints, and conclude that the STEP constraints schema must provide a general means for capturing both types.

4.4   Under- and Over-constrained models

There is practical value in being able to transmit under-constrained models, for example of products whose design is not yet finalized and whose geometry is not yet rigidly constrained. Also, as mentioned in the D-Cubed documentation, over-constrained models frequently arise in practice, usually with compatible constraint sets. The D-Cubed DCM software handles these situations by attributing a *status* to the affected elements in the model. For example, dimensional constraints may be **new** (not yet solved for), **solved** (satisfied by the current instance), **over-defined** (belonging to an overdefined but consistent set) or **inconsistent** (belonging to an inconsistent set). Geometric elements may be **unknown** (not yet solved for), **well-defined** (completely constrained), **under-defined** or **inconsistent**.

The STEP constraints schema will need some mechanism for capturing information of this type. It is primarily of interest to the constraint manager in the receiving system, and will help to resolve issues arising in the reconstruction of the received model.

4.5   Constraints and Engineering Tolerances

Another aspect of constraints which needs to be considered is their close connection with certain types of engineering tolerances[1]. Specifically, tolerances of location specify permissible variations in linear or angular dimensions, referred to some datum in modern practice. Tolerances on parallelism and perpendicularity are special cases of tolerances on angular dimensions. Thus it will be appropriate to relate this type of tolerance specification with dimensional constraints. This can also be done in the case of tolerances of size, which do not make reference to datums, e.g. a tolerance on the diameter of a cylindrical hole. Other types of tolerances, including surface finish and form tolerances such as planarity and cylindricity, are more appropriately handled as attributes of geometric or topological model entities than as attributes of constraints.

Whatever provision is made in the linking of engineering tolerance specifications to constraints, account must be taken of the existing tolerance representation capability of STEP. The relevant Integrated Resource is Part 47, at the time of writing approaching Draft International Standard status. Several other parts of STEP in the final stages of the standardization process make reference to Part 47, and it may be necessary to review their particular requirements as well.

4.6   The Numerical Accuracy of Constraints

Any CAD system can only represent geometry to a certain level of accuracy. Thus two lines related by a perpendicularity constraint may not be *exactly* perpendicular as explicitly represented in the system. The geometry created by the system only satisfies the constraint approximately, to within some error which will vary from system to system. Each system must have some numerical tolerance specifying acceptable deviations from logical constraints. The danger is that if this configuration is transmitted to another system, working to a different numerical accuracy, the receiving system might judge the deviation to be unacceptable. This would create a logical conflict between the geometry and its associated constraint.

Similar problems are already arising in the STEP transfer of boundary representation models due to conflicts between transferred geometry and transferred topology. Essentially, entities that are judged to be connected in the sending system may be found to be disconnected in a receiving system using tighter numerical tolerances. This is proving to be a major obstacle to practical data transfer, and it is desirable that measures be taken in the specification of the constraints schema to prevent, or at least minimize, such difficulties in respect of constraints.

Note that this issue is quite different from the *engineering tolerance* issue. The fact that a CAD system has internal numerical errors should have no effect on design intent as transmitted in a STEP file, provided the receiving system can satisfactorily regenerate a model from that file. However, part of that design intent may be expressed in the form of engineering tolerances. Subsequent inspection to verify whether a realized artefact meets engineering tolerances is concerned with deviations from nominal dimensions resulting from imperfections in the manufacturing process. It should be unaffected by computational errors occurring in the CAD system.

5   Issues Currently Being Resolved

Several major and minor issues are still unresolved at the time of writing.

5.1   Major Issues

One major issue remains to be resolved before progress can be made on a broad front, and that is the relation between the procedural and the explicit aspects of part models. Resolution of this issue will determine the nature of the framework provided for the representation of enhanced CAD models in STEP. The two following questions are fundamental:

1. To what extent is the capture of model construction history necessary? And how can it best be captured in an explicit snapshot-type exchange model? Is it even possible to combine, in a single standard, the means for representing both implicit and explicit models? It is worth noting that the widely used D-Cubed constraint management software takes no account of model history. However, some of the major systems not using the D-Cubed software may rely more on this type of information.

2. A related question — will it serve the purposes of STEP data exchange to represent parametrized entities as numerically dimensioned entities plus a range constraint, as suggested in Section 2.2? This might significantly simplify the provision of parametric capabilities.

Most CAD modelers do make use of procedural constructions to some extent, though this does not necessarily mean that the resulting models are stored in a format reflecting their constructional history. Indeed, most CAD modelers also provide the (possibly optional) means for capturing the sequence of commands used in constructing a model, though again this is not necessarily how the model is customarily stored or transmitted. The early years of solid modeling saw a change in emphasis from CSG-related techniques towards pure boundary representation techniques in the way models are stored. More recently, the pendulum has swung back some way, though the procedural approaches now used are much more general than the former strict CSG methodology, whose demise was primarily due to its over-restrictiveness. STEP is currently oriented towards the 'pure' boundary representations in vogue several years ago; it may be that significant changes in STEP modeling philosophy will be needed to accommodate newer modeling methods.

A recent paper by Pierra et al.[11] outlines one possible method for handling models having both procedural and non-procedural aspects. The provision of the type of constraints schema envisaged here appears to be compatible with the proposed method. This being so, the development of the schema as part of the larger framework is a worthwhile effort. Accordingly, the next section summarizes the primary requirements for the constraints schema as an initial basis for the explicit constraints subtask.

Another major issue is also emerging. It concerns the capture of constructional geometry, referenced by elements of the model but not actually part of it. A simple example is the mid-plane of a slot feature, which might be referenced by an engineering tolerance attached to the slot. More complex examples include geometric or topological entities which are created during the modeling process, constrained in some way and then deleted from the 'realizeable' part of the model by a subsequent modeling operation. It is important that such entities, which may still play some significant role in the constraint structure of the defined product, are retained in 'ghost' form in the background. Successful model exchange will require this background information to be exchanged with the explicit model. An example (due to Guy Pierra) is a rectangle, the length of whose diagonal is constrained. One of the constrained corner vertices is now removed, by filleting the corner concerned. It is often important that the effect of such a constraint persists despite the apparent absence of one of the constrained entities.

5.2 Minor Issues

Less fundamental issues that have been identified include the following questions:

- Should constraints be binary, or should list of constrained entities be allowed? (current thinking is that a list of entities subject to the same constraint captures the semantics better than a list of binary constraints of the same type),
- What is the mechanism for relating surfaces from two different part models when defining an assembly constraint (it is believed that the STEP **context** concept allows this to be done),
- Do constrained geometric entities need to be provided with back-pointers to the constraints controlling them? (this would entail changes in the existing Part 42 of STEP, but more efficient translation would then be possible).

6 Summary of Primary Requirements Identified for a Constraints Schema

Below the top (generic) level of constraint entity, there are four types of decomposition, each apparently at the same level:

1. Subdivision by dimensionality (2D, 3D). This can probably be handled implicitly, the interpretation of the STEP constraint depending on whether the constrained entities are defined in a 2D or 3D context.
2. Subdivision by supertype (dimensions, geometric or logical constraints — either of these may in principle be further subdivided into predefined and user-defined types, though whether there is a real requirement for user-defined logical constraints is not clear at present).
3. Subdivision according to whether the constraint applies between two low-level geometric entities used in a part representation or between two rigid bodies used in an assembly representation.

It seems undesirable to accept the implied subdivision of the generic constraint type into eight or more subtypes at this level. Some convenient structure will be sought that avoids this..

At the next level of decomposition we have the individual constraint types as listed in Section 4.1. This list will probably be refined as the construction of the schema proceeds. Individual constraint entities will need to have some or all of the following attributes:

- Type of constraint
- Model elements constrained (we need to decide whether constraints will be limited to binary relations or whether we can, for example, constrain a *list* of lines to be parallel to a target line)
- (Optional) senses of constrained geometric entities (could be used to cut down on possibilities for multiple solutions of constraint systems)
- Target entity for a directed constraint

- Flag for directed/undirected constraint (though the presence of a target entity may suffice for this to be unnecessary)
- Under/overconstraint indicator (for use when a constraint belongs to an under- or overdetermined set)

Additionally, it is likely that provision will be required for the representation of auxiliary constructional entities.

7   Acknowledgements

This paper has benefited greatly from my discussions with the members of the ISO Parametrics Group, from comments by others outside the Group, and from what I have learned through reading papers by various authors. The following people have provided information and had a major influence on my thinking: Bill Anderson (SCRA), Mike Atkins (D-Cubed Ltd.), Noel Christensen (Allied Signal), Edward Clapp (Autodesk Inc.), Ray Goult (LMR Systems), Akihiko Ohtaka (Toyota Soft Engineering), John Owen, (D-Cubed Ltd.), Guy Pierra (LISI/ENSMA), Chia-Hui Shih (Pacific STEP) and Peter Wilson (NIST). I would also like to acknowledge useful input from many other people too numerous to mention individually.

References

1. American Society of Mechanical Engineers, *Dimensioning and Tolerancing.* American National Standard ASME Y14.5M – 1994 (1994)
2. N. C. Christensen, 2D Geometric Constraints. IGES Request for Change #365, IGES/PDES Organization, Fairfax, VA, USA, September 1995
3. D-Cubed Ltd, Cambridge, England, *The 2 Dimensional DCM Technical Overview* (1995)
4. D-Cubed Ltd, Cambridge, England, *The 3D DCM Part Positioning Product Technical Overview* (1995)
5. J. Hoschek and W. Dankwort, *Parametric and Variational Design.* Teubner-Verlag, Stuttgart (1994)
6. International Organisation for Standardisation, *Numerical Control of Machines — NC Processor Input — Basic Part Program Reference Language* (International Standard ISO 4342, the APT language, 1983)
7. International Organisation for Standardisation, *Industrial Automation Systems and Integration – Product Data Representation and Exchange* (International Standard ISO 10303, informally known as STEP, 1994)
8. International Organisation for Standardisation, *Industrial Automation Systems and Integration – Parts Library, Part 20: General Resources* (Draft International Standard ISO DIS 13584-20, 1996)
9. A. Ohtaka, Some Considerations on Parametrics, Working Document ISO TC184/SC4 N352, International Organisation for Standardisation (October 1995)
10. G. Pierra, Parametric Product Modelling for STEP and Parts Library (V0.3). Working Document ISO TC184/SC4/WG2 N183, International Organisation for Standardisation (July 1994)
11. G. Pierra, Y. Ait-Ameur, F. Besnard, P. Girard and J.-C. Potier, A General Framework for Parametric Product Models within STEP and Parts Library. Proc. PDTAG Product Data Technology Days '96, Shell Centre, London, UK, 18 – 19 April 1996

12. C.-H. Shih, *Phase 4 Data Model for ENGEN*, Version 4.2 (draft). South Carolina Research Authority, Charleston, SC, USA, November 1996
13. US Product Data Association, Gaithersburg, MD, USA, *Computer Aided Processing of Engineering Drawings and Related Documentation (IGES)*. (IGES Version 5.2: American National Standard USPRO/IPO-100-1993, 1993

Appendix 1: ISO TC184/SC4 and STEP

The International Organisation for Standardisation (ISO) administers a range of committees and subcommittees. ISO Technical Committee 184, Subcommittee 4 (TC184/SC4) is concerned with the development of international standards for the digital representation of product data and manufacturing management data. This is the forum in which the STEP standard (ISO 10303) is being developed. The first release of STEP[7] occurred in 1994. Earlier related standards (for example IGES[13]) were intended primarily for the exchange of pure geometric data between design systems, but STEP is intended to handle a much wider range of information covering the entire life-cycle of a product.

ISO TC184/SC4 is also responsible for the development of ISO 13584 (Parts Library), a future standard for making information in libraries of standard parts accessible to CAD system users.

The STEP standard is being released in parts. The initial release contained twelve of these, but many more are in preparation, dealing with specific product ranges (e.g. automotive, AEC, shipbuilding, electrical, ...) and different aspects of the product life-cycle (design, finite element analysis, process planning, ...).

The structure of the standard is fairly complex. The lower Part numbers (100-series and below) define the infrastructure and a set of integrated resources. The actual data exchange standards are specified by Application Protocols in the 200-series, and these are defined in terms of the lower-level resources. Part 11 specifies the EXPRESS information modeling language, which is used for the formal definition of constructs in the exchange files.

The initial parts of STEP dealing with geometry transfer are two Application Protocols, AP201 (Explicit Draughting) and AP203 (Configuration Controlled Design). The first is concerned purely with 2D drawing information, while the second covers wireframe, surface and boundary representation solid models. The content of AP203 models is restricted to geometric and topological data, together with 'configuration' information relating to such matters as version control and release status.

STEP is designed to operate in the first instance as a 'neutral file' transfer mechanism. Each CAD system must be provided with a *preprocessor* and a *postprocessor*. Their functions are, respectively, to translate native data from the sending CAD system into the neutral STEP format, and to translate from the neutral format into the native format of the receiving system. This philosophy only requires the provision of $2n$ translators for exchange between any pair chosen from n systems, rather than $n(n-1)$ if 'direct' translators have to be written. As an alternative to file transmission, STEP information may be stored in a database, and a STEP Data Access Interface is being developed as part of the standard to allow the use of shared data access.

Many CAD vendors have developed or are developing STEP AP203 translators; some are already commercially available, while others are under test. Some third-party software vendors are also marketing STEP AP203 translators.

The initially released parts of the standard are

Part 1   Overview
Part 11   EXPRESS language (used in writing the standard)
Part 21   Physical file format
Part 31   Methodology and framework for conformance tools
Part 41   Fundamentals of product description and support
Part 42   Geometric and topological representations
Part 43   Representation specialisation
Part 44   Product structure configuration
Part 46   Visual presentation
Part 101   Application resources: draughting
AP (Application protocol) 201   Explicit draughting
AP (Application protocol) 203   Configuration-controlled design

Two further parts have been recently added to this list: Part 105 (Kinematics) and
AP202 (Associative Drafting). Parts of the STEP standard that have not yet reached
the Draft International Standard (DIS) stage are freely available from the Solis information server at NIST (http://www.nist.gov/sc4/).

MicroStation Modeler: The Design and Implementation of an Extensible Solid Modeling System

Brian F. Peters

Bentley Systems, Inc., 690 Pennsylvania Drive, Exton, PA 19341 U.S.A.
email: Brian.Peters@Bentley.com

Abstract. The advantages of a parametric, feature-based modeling system have been demonstrated both in the research literature and in the market place. The challenges faced by the developers of such a system are numerous. In addition to the obvious technological considerations are the many concerns that derive from analyzing the business aspects of such a product. By describing the development of Bentley System's MicroStation Modeler™ we hope to demonstrate the various design constraints involved and explain how we satisfied them.

1 Introduction

The popularity of parametric, feature-based modeling systems testifies to their advantages as perceived in the market place. A feature-based system lets users design using vocabulary and objects familiar from their application area [1]. Defining such features parametrically enables quick and easy variational studies as changes to the model are simple to make [2].

The problem for us as developers of such a system is that we do not have expertise in all specific modeling application areas. We may understand solid modeling, manifold topology, and computer science, but we are not experts in sheet metal work, mold design, or structural steel work. Our challenge then is to produce a system that allows new feature sets to be created and incorporated with existing modeling features efficiently and elegantly.

Working as a commercial software developer brings in other design criteria. We must create a product that integrates well with the company's previous offerings. Design files should be compatible across product families and releases. Because customers typically use software from many companies any new package should take advantage of industry standards to ease data transfer to systems offering complementary functionality.

The result of these considerations at Bentley Systems led to the development and release of MicroStation Modeler in January 1995. MicroStation Modeler uses an extended CSG-like tree to capture the design intent and history of a model. The nodes of this tree are not limited to the traditional Boolean union, intersection, and difference of the standard CSG approach, but rather are general operations performed on the solid. Each node stores parameters that define its current state. Sending appropriate messages while traversing this feature tree, MicroStation Modeler can create, evaluate, and edit a solid without detailed knowledge of the actions performed by the individual nodes. Such an approach allows new feature sets to be added to our product by external programmers.

Section two introduces the programming environment of MicroStation®. Section three discusses the design criteria followed in developing MicroStation Modeler. Some of the questions raised are relevant to the software industry as a whole, others are specific to CAGD. Sections four and five outline the resulting design and implementation of the product. Section six briefly discusses where this work is leading at Bentley Systems before the final section draws conclusions.

2 The MicroStation Development Environment

Before analyzing the design and implementation of our product we need to describe the environment in which it was developed. A brief introduction to the MicroStation base product, its design file format, and its customization capabilities follows.

2.1 MicroStation

The MicroStation base product is a widely used 2d/3d engineering design package. It is supported across a wide range of hardware including all popular operating systems on workstations and personal computers. Complete product information is available from the Bentley System's web site [3]. Programmatically, MicroStation can be considered an event-driven state machine responding to user actions such as mouse clicks and text key-ins.

The file format used by MicroStation is essentially a linked list of elements. To facilitate scanning, all elements share a common header that includes the offset to the next element in the file. Specific element types store the data of line strings, arcs, text nodes, NURBS curves and surface, etc. in a varying section after the header. Any element may have additional application specific user attribute data appended to it. These data files, referred to hereafter as design files, are binary compatible across all the platforms on which MicroStation is supported.

While this format is good at storing geometry it does not excel at storing associations and relations of the type used in solid modeling. The primary construct for grouping elements, the cell, simply indicates that a following specified number of elements constitute the cell; i.e. they have a single transformation matrix and behave as a single graphical unit. Cells are named by short character strings and can be nested, forming a tree-like hierarchy of complex elements.

2.2 MicroStation Development Language (MDL)

While there are many engineering design products on the market, what differentiates MicroStation is the sophistication of its development environment. MicroStation was originally developed by a small team of programmers who knew they would not have the resources to develop applications on top of it. In order to ensure a successful product, therefore, MicroStation had to be extensible. This philosophy has been maintained throughout the entire suite of MicroStation products, even as development resources at Bentley Systems have grown. The primary means to achieve this extensibility is the MicroStation Development Language (MDL™).

MDL is a complete C programming environment. Shipped with every copy of MicroStation are a source code compiler, linker, and librarian; a resource compiler, linker, and librarian; and other development utilities including a runtime source-level debugger and a platform-independent make utility. MDL application developers write in the C language making use of a library of built-in functions. The MDL compiler converts these programs into a pseudo-code that is interpreted by MicroStation at runtime.

Of particular advantage is the fact that MDL is source-level compatible across all platforms on which MicroStation ships. The make utility included with MDL assures that a single make file will work on all hardware. This frees the application developer from having to deal with porting issues arising between different operating systems running on different machines.

MDL programs run in the same address space as the MicroStation executable. This enables us to expose the internal state machine of our product to applications. Third parties can write event-driven commands that receive the same events as the commands developed by us. This, combined with the fact that multiple MDL applications can be loaded simultaneously, leads to a system where the command set of MicroStation can be seamlessly extended by external developers.

The shared address space also means that the built-in functions called by MDL programs are in fact the same internal routines used by the MicroStation core. The original version of this library contained over 1200 routines; today's version contains over three times that many. There are MDL functions to do everything a design package needs to do including:

1. Manage design files
2. Create and edit geometry in such files
3. Define a graphical user interface (GUI)
4. Query and manipulate the internal state of MicroStation

The library contains sections including:

1. Resource management utilities
2. NURBS curve and surface library
3. Routines for communicating with external programs
4. Asynchronous call-back routines for capturing events that change MicroStation's state

We assume functionality developed inside MicroStation that is of use to us will be of use to others as well and, therefore, make it available through MDL.

The resource manager, for instance, supports the storage of complex data structures in platform-independent binary resource files. Developers can define their own resource formats as well as use predefined resource types. Predefined types support all the components of a complete graphical user interface,from dialog boxes to icon palettes. Internally, we use this system to create the user interface of MicroStation and to support internationalization of the dialog boxes and items they contain.

In fact, all the visible portions of MicroStation, from the user interface to the standard work spaces and tool sets, are implemented in MDL. This explains the breadth and scope of the MicroStation Development Language. The functionality we use to create MicroStation can be used to customize and extend it; if we need some functionality, chances are some third-party developer will need it, so we make it available.

This also explains the work that has gone into MDL to make it function consistently across such a wide range of hardware platforms and operating systems; why such care has been taken to achieve binary compatibility of design and resource files. There is only a small group at Bentley Systems that needs to worry about platform issues, the rest of us are free to develop functionality independent of the targeted machine.

2.3   Dynamic Link Modules

Many operating systems now support dynamically loading compiled code at runtime. This technique allows an executing program to be extended by reading in one or more shared libraries of separately compiled, position-independent object code. Published functions from the library become available to the executable; the program behaves as if it were a single large object as opposed to multiple smaller components. Portions of the code need not be loaded until they are needed, easing system overheads.

MicroStation Development Language takes advantage of this in the form of Dynamic Link Modules (DLM). DLMs provide a platform-independent interface in the MDL environment to use stand-alone code libraries, created using a machine's native compiler and linker, to extend the library of built-in functions available to MDL applications. The supplied library of built-in functions provide access into the core of the MicroStation executable; DLMs can provide access into the core of any other software library.

Taken in its entirety, the MicroStation development environment provides services typical of many operating systems. It provides runtime execution environments for multiple applications running simultaneously. It provides memory management, input/output file manipulation, and other low level system utilities. In short, MicroStation behaves as a "CAD operating system."

3   Design Criteria

We turn now to the issues that guided the development of MicroStation Modeler. These are questions that arose early in the design of the new product. Their answers depended on many things including the resources available to the development team and the capabilities of MDL. The answers we came up with served to define what MicroStation Modeler is and how it can be used.

Limited programming resources coupled with the availability of component libraries have altered development practices in the software industry over the last few years. This phenomenon is evident in the CAGD community. Before, expert knowledge was required to develop kernel solid modeling libraries. Today such code may be purchased. Companies can now decide where development resources are to be spent; should we build a core modeler or license one and concentrate on developing a parametric, feature-based system on top of it?

A decision to license leads to many further questions. Which is the best library to choose? Is it supported on all platforms of interest? Is it provided as source code or just as linkable objects? From a business point of view it is important to consider what sort of a deal can be arranged with the providing company. What does a license cost? What protocols are in place for support and maintenance of the code? Who else is using the same library?

The architecture of the resulting product must also be carefully considered. Will the library be used as an external tool kit or adopted as core technology? To keep a library at arms length requires more discipline in its use, but leaves open the option of switching to different code at a later time. Whose data structures will be predominant? How will they interact? Where will the solid modeling data reside? This last questions was particularly relevant for us.

We wanted the base MicroStation product to be able to use the files MicroStation Modeler would create. This would provide access to the rendering, walk-through, and mark-up capabilities of other products in the MicroStation family. It would allow a coherent work flow to develop at installation sites; we knew a product in isolation would not do well. Specifically it meant that we could not extend the design file formats to include solid modeling constructs. Some other solution would need to be found.

We also knew that we would not be able to create feature sets for all the application areas in which we wanted to see the new product used. While fillet, hole, and boss may be the first to come to mind, many other design elements are amenable to implementation as parametric features. Examples we envisioned include a parametric staircase for architectural design, fold lines in a sheet metal application, and various types of loads in finite element analysis. The concept of parametric, feature-based modeling is applicable to a much broader spectrum than just mechanical manufacture.

The development team for MicroStation Modeler did not include architects, sheet metal workers, and FEA experts, though. We needed, therefore, to create a system that would allow extensions by third-party developers. If each type of feature was managed by a separate executable application, we knew that the set of features would be easily enriched after our initial product release. A system architecture that specified how parametric features would be represented and how they would interact, without specifying the details of their affects or the information they would contain had to be designed. We needed to define appropriate protocols that would capture the behavior of all the broad set of features we envisioned for MicroStation Modeler, while not restricting their specific capabilities. We needed to define features as objects that could respond to a predefined set of messages.

By taking such an object-oriented approach to the definition of features we would have an extensible feature-based modeling system. As separate executable modules would be responsible for responding to the messages sent to each feature, extension of the feature set would be independent of the core product. Other developers would be able to capture the expert knowledge specific to various application areas and write feature sets that MicroStation Modeler could easily incorporate.

We will see below that MDL not only provided the philosophical motivation, but also the means to achieve this goal. Just as MDL extends MicroStation, we enhanced MDL so that it would be able to extend MicroStation Modeler, as well.

4   Resulting Design

We did not have the resources to develop our own solid modeling kernel. Assuming the availability of a such a tool kit we decided to concentrate our efforts on developing a parametric, feature-based design system on top of a B-rep solid modeling kernel. I do not need to review the differences between CSG and B-rep solid modeling systems here; reviews and comparisons already exist [4]. Both approaches have aspects that are desirable in a feature-based modeling system.

Re-evaluating a CSG tree, for instance, hints at the reconstruction of a parameterized solid after those parameters have been changed. The ease of redefining the radius of a cylinder primitive in a CSG tree is exactly the sort of functionality users want when modifying high level features such as countersunk through holes in feature-based solids. The traditional Boolean operations of a CSG tree are a little too restrictive, however, when one wants to think at a feature level.

A B-rep scheme is needed to guarantee capturing whatever geometry may result. Complex configurations can result from the interaction of even simply manufacturing features such as slots and protrusions [5]. Furthermore, the existence of topological entities of vertex, edge, and face is explicit in a B-rep system. Users think about a solid in just such terms when applying blends to a particular set of edges of a model, or adding a protrusion to a particular face.

We needed to take advantage of both approaches to create a parametric, feature-based design package, but how could we combine such representation schemes in a single product? Prior work has investigated combining CSG and B-rep schemes in a single data model [6]. Other work has considered features as operations on solids (see [7] and [8]). We combined these ideas in our design of MicroStation Modeler.

Extending the notion of a CSG tree to include more general operations can capture the ideas of feature-based modeling easily. Instead of simply performing Boolean operations on their sub-trees, nodes in a feature tree are considered operators that take a solid(s) from their child node(s), perform modifications, and pass the resulting solid to their parents for further operation.

A union feature, for instance, still performs a union of the solids of all the nodes below it. A hole feature, however, is now seen to operate on its single child node's solid producing a modified version that contains the specified hole. Leaf nodes have no children. They are defined solely in terms of their stored parameters. Examples include the geometric primitives such as sphere, slab, cylinder, cone, and torus.

To evaluate such a model, the top node of the tree is simply asked for its solid representation. When making this request, however, MicroStation Modeler must provide the solids of child nodes as arguments for the topmost node to modify. To get these solids, the child nodes must be evaluated. This recursive technique thus descends to the leaf nodes of the tree. These primitive nodes have no children and create their solids based entirely on their stored parameters. The recursion unwinds by propagating these resulting solids back up to the topmost node of the tree.

To each node in the tree there corresponds a solid that represents that stage in the evaluation of the complete model (Fig. 1). Following [6], each node stores a pointer to a B-rep data structure representing this solid. It is exactly these solids that propagate up the tree during evaluation. The B-rep pointed to by the topmost node is the solid that represents the combination of all the features comprising the model. The B-rep pointers at each node of the tree serve to avoid a traditional drawback associated with CSG modeling, namely that of re-evaluating the entire tree for every operation. If a valid B-rep pointer exists for a node, then that solid is the one returned when the node is queried. No solid modeling calculations need to be performed; no deeper recursion is necessary.

If any node changes, it is only necessary to re-evaluate nodes above that one to the top of the tree. When the parameters defining a node are modified, MicroStation Modeler simply traverses up the tree from that node invalidating the B-rep pointers of all the nodes visited. A re-evaluation of the entire model regenerates only those features that lack valid pointers. Essentially, there is only ever one branch of the tree evaluated at a time as there is only one path from the modified node to the top of the tree.

MicroStation Modeler is not involved in the detailed generation of any particular feature node's solid. It simply sends a message, along with the B-rep solids of any underlying nodes, that the particular feature needs regeneration. The actual parametric data stored by a feature node, as well as the action it performs on the model, are

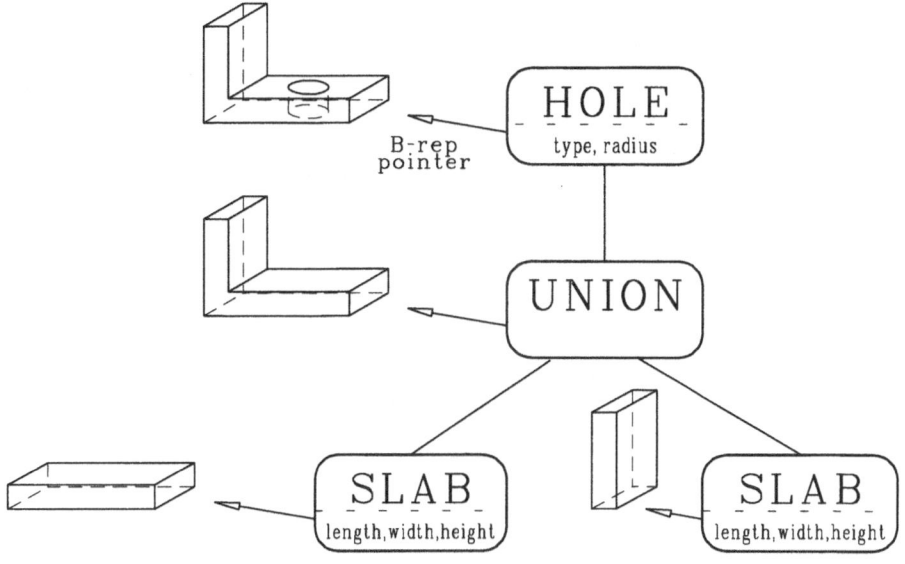

Fig. 1. Extended CSG-like feature tree.

unknown to MicroStation Modeler. The feature node itself is responsible for its own calculations. MicroStation Modeler is simply asking it for a B-rep solid. How exactly that solid is generated is not considered.

Notice the parametric dependencies implicit in such an architecture. A node may depend on anything below it in its branch of the feature tree. While this is a one-way dependency it suffices to meet the goals of our product definition. A node is guaranteed to receive a message to regenerate itself whenever any node below it in the tree is changed. If a through hole is positioned in a slab, and that slab is thickened, the hole will be instructed to recalculate itself and will be given the B-rep of the modified slab. By examining the thickened slab the through hole can assure that it remains a through hole.

The design history is implicit in such a model. The feature tree grows upwards with new nodes continually being added to the top; deeper nodes predate nodes at a higher level. Features may only depend on prior geometry. Cyclical dependencies are thus prevented. This greatly simplifies the process of evaluating such a tree, as it may be considered a directed acyclic graph. Further work at Bentley Systems has been done to address more general dependency graphs and will be mentioned below.

We see here the abstract, object-oriented definition of what a parametric feature is. A feature has internal data (its parameters) and responds to various messages (to generate its solid, for instance). What it stores and how it calculates its response to messages is irrelevant. The definition needs to be this general in order to capture what is meant by a feature in all the application areas that we want to address using parametric, feature-based modeling. As a product, MicroStation Modeler must define the methods it will invoke of feature nodes in terms broad enough to enable it to create, evaluate, and edit a feature-based solid without knowing details of the particular nodes' operations. We now need to discuss the individual feature methods in more detail.

4.1 Feature Methods

The series of required and optional feature methods took shape as we developed Micro-Station Modeler. Our challenge was to discover and codify a set of methods rich enough to capture the behavior of all the features we wanted while not imposing unnecessary limitations. We needed to consider the features we were providing in the initial release as well as those that were to come later. Certain of these methods derive from the representation of the feature tree as a directed acyclic graph. Others are more specific to the notion of feature as interpreted in CAGD. Not all are required of all feature types.

These methods are critical for MicroStation Modeler to be able to manipulate all the features in a model in a consistent manner. While a countersunk blind hole differs greatly from a varying-radius fillet or a FEA beam element, MicroStation Modeler must treat them identically while creating, evaluating, and editing feature trees. By appropriately calling the following routines Modeler accesses the behavior that is common to all feature types. It is exactly this behavior that we adopt as the definition of a feature for our product.

In order to discuss in more detail the purpose of each of the methods, explaining when and why MicroStation Modeler makes the requests, we will consider a Hole feature. This feature takes a single solid as argument and adds a hole of the appropriate type and dimension. The data that defines a hole is represented in Table 1 and will be relevant to the discussion that follows.

Table 1. Parameteric data that defines a hole feature.

Parameter	Possible Value(s)
Drill type	Through, Blind
Hole type	Simple, Countersink, Counterbore
Diameter	dimension
Depth	dimension (if Drill type is Blind)
Countersink angle	dimension (if Hole type is Countersink)
Countersink depth	dimension (if Hole type is Countersink)
Counterbore diameter	dimension (if Hole type is Counterbore)
Counterbore depth	dimension (if Hole type is Counterbore)
Tapped	True or False
Thread diameter	dimension (if Tapped is True)

generateBody. This is the primary method that all features must support. To register a feature is to support this method. It is called whenever the B-rep solid of a node is required and there is no valid pointer present. It is supplied the B-rep solids of child nodes and the particular node in the tree which triggered the call. It must return the B-rep solid that results from the action of the particular feature instance on the model.

The Hole node removes material based on its stored parameters. These define the type of hole (simple, counterbore, countersink; through hole or blind), its dimensions (radius, angle, etc.), and its location and orientation relative the solid supplied from the lower level node. Using the solid modeling tool kit the Hole node's *generateBody* function creates the volume as a union of cylinders and cones. The material removal volume is then positioned and subtracted from the supplied solid. The resulting B-rep is returned. MicroStation Modeler assigns the pointer to this B-rep structure in the feature node, preventing further recalculations until the underlying solid changes.

generateElements. This method is used to add visualization geometry directly to the model. The rest of the visualization geometry comes from a conversion of the model's fully evaluated B-rep solid. The details of model visualization are discussed below in the Sect. 6 on Implementation.

This method is useful for adding MicroStation design file elements to a model that are not part of the true B-rep solid. For example, if a hole is tapped, elements are added representing the threads directly to the visualization of the model. This alleviates the need of the B-rep solid to capture such detail, greatly easing the load on the underlying modeling tool kit.

edit. An *edit* message is sent to the feature when MicroStation Modeler's Edit Feature command is invoked for a particular feature node. This method typically responds by creating a modal dialog box that contains the value of all the settings relevant to the selected feature instance. For our Hole node, this consists of option buttons selecting the type of hole – simple, counterbore, countersink – and whether it is a through hole or blind. Based on the selection of type, various dimensional values are presented for modification: hole radius, counterbore radius, countersink angle, hole depth, etc.

Upon dismissal of the modal dialog box, the feature application informs Micro-Station Modeler if the feature has indeed been changed. If so, MicroStation Modeler invalidates the B-rep solid pointers for all nodes from the one edited to the top of the tree and re-evaluates the model. This re-evaluation invokes the *generateBody* and *generateElements* messages explained above.

preChildRemoved, postChildRemoved. If the feature that generates the face in which a hole is positioned is deleted, the Hole feature needs to be removed; Micro-Station Modeler detects this and removes the hole automatically. If all but one node below a Boolean operation node is removed then it is no longer necessary to include the Boolean operation in the feature tree. Taking the union of a single solid is a redundant operation. MicroStation Modeler does not know however, to automatically remove the Boolean node, as there is still one child below it.

To fix this behavior the Boolean operation feature nodes implement these two methods. By sending messages before and after deleting nodes from the feature tree, MicroStation Modeler gives parent nodes the opportunity to respond. Parent nodes can prevent the removal of their children if such behavior is appropriate, or delete themselves, as Boolean nodes do when there is only one child left.

generateDynamic. In order to improve performance a complex feature may implement this method to provide a simplified geometry to be used by the various commands in MicroStation Modeler that show dynamics on the screen. When moving a Hole feature, for instance, performance is accelerated if the hole is represented as a simple cylinder of appropriate radius, regardless of whether or not the hole is counterbored or countersunk, has threads or not, etc. When the user accepts the new position of the hole, the full detail does, of course, reappear.

transform, postNodeTransformed. All nodes can receive indication that they are about to be, or have been, transformed if they implement these methods. Such transformations occur when features are moved, mirrored, rotated, or arrayed by various MicroStation Modeler commands. If MicroStation Modeler's default behavior is not appropriate, the correct action can be taken by the individual feature application. These methods are optional.

showParameters. MicroStation Modeler can present a graphical representation of the feature tree of any particular model. This is an invaluable aid for understanding what features constitute a model. It is necessary for understanding which features are dependent on which others. It provides a means for direct manipulation of the feature tree, moving nodes up or down the tree, or deleting them entirely. By clicking on a node in the tree, the geometry defined by that node can be highlighted in the fully visualized feature model.

Any node pictured in the tree can be queried for a textual description of the data defining it. This method implements the response to that request by formatting multiple character strings with the data of the specific feature instance. The Hole node, for example, returns the hole's type as well as the relevant dimensional values. A Boolean union feature does not need to implement this method as there is no additional data necessary to specify the action of a particular Boolean operation.

settings. The *settings* method informs MicroStation Modeler that a feature has additional information to supply regarding annotation and/or suppression. If a feature needs to query the user about these operations this methods provides a means to do so. MicroStation Modeler will call it when appropriate actions are taking place on a feature of the relevant type. This is an optional method.

suppress. Feature suppression is an important, often over-looked capability of solid modelers. Many times it is desirable to ignore features that meet certain criteria. For instance, before meshing a solid for finite element analysis it may be convenient to ignore all holes smaller than a specified diameter. In order to support such behavior, MicroStation Modeler allows features to be tested for suppression by type. For example, the user can specify that all Hole features are to be tested before evaluation.

The *suppress* function of a feature node gets called during the evaluation of a model when that particular feature's type has been indicated for testing. It gives individual instances of the feature the opportunity to test whether or not they meet the criteria for suppression. A Hole node examines its radius and excludes itself from evaluation if it is of an insignificant size, less than a stored value. The *settings* method above is used by the Hole node to let the user set this threshhold radius value. This is an optional method and need not be implemented.

annotation. When laying out drawings of models, designers often want notes associated with particular features. The content of these notes depends upon the type of feature being annotated as well as upon the data of that particular instance. How a hole is to be machined has no relevance to a Boolean operation, for example. The *annotation* method enables a feature to specify what sort of information will be added to it by MicroStation Modeler's Annotate Feature command. Such annotations are commonly found on drawings of feature-based solids. This method automates their creation and thus speeds the production of engineering drawings. This is an optional method and need not be implemented.

generateAssocElems, preNodeRemoved. The base MicroStation product has the capability to associate elements in a design file to each other using user attribute data. (This is in addition to the correlation implied by the use of cells described in Sect. 2, but still falls far short of the richness of association in B-rep solid modeling data structures.) This capability is used, for example, to associate dimensions to geometric elements. When the geometry changes, so does the dimension.

Features may support these two methods to enable the creation of geometry in the design file that is associated to the visualization of a model. While being evaluated, a node is given the opportunity to add associated elements to the design file through the first method. The elements added differ from those added by the *generateElements* method in that they are separate from the feature tree of the model in the design file. The second method enables such associated geometry to be removed from the file when the supporting feature is removed from the tree.

5   Implementation

With few exceptions, the above discussion has not depended specifically on the architecture of the MicroStation base product. We have discussed parametric, feature-based modeling in terms that could be realized through many different programming scenarios. Now we wish to map the previous section's resulting design into MicroStation's specific programming environment, showing how we used the capabilities of MDL to arrive at our goal of an extensible, parametric, feature-based modeling system.

The decision was made to use the ACIS solid modeling library from Spatial Technologies as a solid modeling tool kit in MicroStation Modeler. In order to incorporate the library we created a dynamic link module (DLM) out of it. While it was challenging to integrate the C++ code of ACIS with the C code of MicroStation, when we were finished we had access to the functionality of both MicroStation and ACIS in a unified, platform-independent, MDL programming environment.

There are a number of advantages realized by creating a DLM out of ACIS. It gives us access to the functionality of ACIS without having to integrate it into the core of MicroStation. This is good from both business and technical standpoints. The core MicroStation product remains entirely our own, simplifying support, documentation, and internal accounting issues. It also enables us to use ACIS as a tool kit without committing irrevocably to it. By keeping the library at arms length we are keeping our options for the future open, embracing a philosophy much expounded by component software vendors.

By creating an ACIS DLM we can also offer the same development environment to our third-party developers that we have for ourselves. Viewed simply as a development platform, MicroStation Modeler provides a machine independent, C programming environment that includes the full GUI, command parsing, event-driven state machine of MicroStation, as well as all the application procedural interface (API) capabilities of ACIS. Access to this functionality is necessary in order for third-party developers to be able to create their own feature sets without having to license a separate copy of ACIS.

Individual MDL applications register responsibility for various feature types and make use of the ACIS API in the methods described above. The B-rep structure that the *generateBody* method returns is actually an opaque pointer to an ACIS BODY. Since multiple MDL applications can run simultaneously, the feature set of MicroStation Modeler may be easily extended without requiring modifications to MicroStation Modeler itself. To facilitate this we have defined a configuration variable in MicroStation that specifies the name of a directory. When MicroStation Modeler starts, it scans this directory for MDL applications and loads them all. The *main* function in each application registers responsibility for a feature type using a built-in MDL function. MicroStation Modeler stores the name of the routine supplied, as well as the application, for subsequent use while evaluating feature trees. Feature applications also add to the MicroStation command hierarchy and may open icon palettes and other dialog boxes as relevant.

Extending the feature set becomes as easy as adding new MDL feature applications to the specified directory. It is possible to have MicroStation Modeler work with different sets of features simply by redefining this configuration variable. An architectural work space could access a parametric staircase feature, as well as features defining structural steel work and roof trusses. A mechanical workspace could access an entirely different set including features representing manufacturing design objects and finite element loads.

The actual feature tree is represented as a hierarchy of nested cells in the MicroStation design file. User attribute data attached to the cells differentiates them from normal cells and stores the pointer to the B-rep solid. Further feature-specific attribute data stores the information relevant for the particular feature type. MicroStation Modeler uses the cell's name to determine the feature's type and can look up which MDL application has registered routines to manage that feature.

Cell headers are not actually visualized in MicroStation. The displayable geometry contained in the cell is the visualization geometry mentioned in the previous section. This consists of MicroStation elements that represent the wireframe and/or surface data of the B-rep ACIS BODY pointed to by the topmost node of the feature tree. The last step of the evaluation of a feature tree traverses this B-rep data structure and converts the geometry contained therein to corresponding MicroStation elements. These are distributed throughout the nested cell hierarchy so that their immediate header is the feature node that caused their creation. The circular edges and cylindrical surface of a simple through hole, for example, would be nested below the Hole node they represent. The Hole node itself may be deep within a larger feature tree, but will be nested within its parent node's cell, as seen in Table 2 and Fig. 2. By distributing the elements in such a manner the existing MicroStation element location logic can be used to identify a particular feature. For example, after the user activates the Edit Feature command geometry is selected from the screen. The base MicroStation product needs to be able to identify geometry in complex cell hierarchies and find cell headers. In MicroStation Modeler this process achieves identification of the particular feature node that is to be edited. An *edit* message is then sent to the appropriate feature application as detailed above.

Table 2. A visualized model represented as a nested cell hierarchy.

Cell name	Contains
HOLE	UNION
	Circle s, circle t, line u, line v
UNION	SLAB
	Line e, line f, line i, line k, line n
SLAB	Line j, line l, line m, line o, line p, line q, line r
SLAB	Line a, line b, line c, line d, line g, line h

Notice that we have represented parametric, feature-based solids in MicroStation design files without adding any new element types to the file format. The model may be displayed without MicroStation Modeler being loaded; visualization does not depend on the ACIS library being available. The model could not be parametrically edited, however, without access to MicroStation Modeler and the ACIS DLM it contains. This achieves exactly the goals we set out to realize regarding file formats and compatibility within the MicroStation product suite. We still need to discuss where exactly the ACIS B-rep information resides, however.

Each node in the feature tree has a pointer to a B-rep solid, an ACIS BODY data structure, representing its stage in the evaluation of the model. The resource management system of MDL provided the means to keep track of such information.

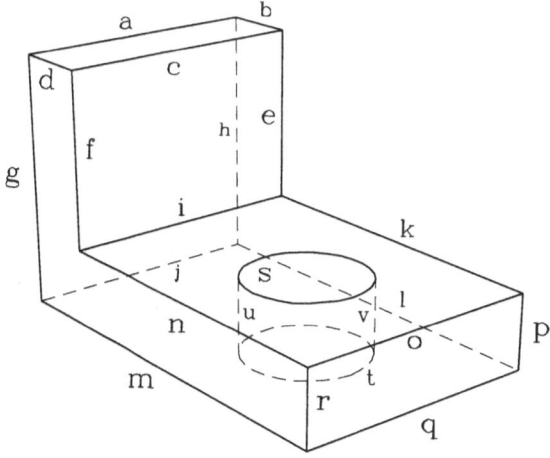

Fig. 2. The visualized model of Fig. 1.

We defined a resource type that contains an ACIS BODY. By embedding the ACIS data in a resource file associated to the design file we achieved a platform-independent means to store and communicate this data. The ACIS data structures are stored sequentially in the associated resource file. The pointer to the B-rep in the feature node is simply the index of the B-rep resource structure in the associated file.

Since the high level feature representation of a model is stored in the design file, this resource file can be recreated if it is lost. Since the feature tree captures the design history, completely re-evaluating the tree will repopulate the B-rep resource file. When a MicroStation Modeler user wants to pass a design to another MicroStation Modeler user then both the design file and the resource file are sent. Only the design file is needed to view or detail the solid in any other product in the MicroStation family.

This approach has allowed MicroStation Modeler to work well within the family of Bentley Systems products. It uses the existing capabilities of MicroStation to advantage. A solid can be created in MicroStation Modeler and then photo-realistically rendered in MicroStation MasterPiece™ , the visualization package. While a core MicroStation user can not edit the solid in terms of its feature's parameters, drawings may be laid out and dimensions added. We also see that MicroStation Modeler is a true add on product; we did not need to make modifications to the base MicroStation deliverable to get it to work.

6 Future Work

6.1 MicroStation Modeler

The first version of MicroStation Modeler was shipped with a feature set composed primarily of manufacturing features. These include primitive geometry types as well as holes, bosses, fillets, slots, protrusions, and ribs. Boolean operation features, and project and revolve operators were also present. Assembly management features were included to enable parts to be connected by joints containing varying degrees of freedom. Since its release, third-party developers have added feature sets to define the loads and elements necessary for finite element analysis of feature-based solids. Further work has been done to capture information necessary for dynamic simulations of assemblies.

Internal work is still necessary on MicroStation Modeler. We must further improve the persistent naming scheme used to identify topological entities in our models. This is an active research area in the field to which we are paying close attention ([9], [10]). There are also a few instances where we did not keep the underlying modeling tool box at arms length; some dependencies crept in. We are being careful to eliminate these in order to keep the architecture of MicroStation Modeler clean.

New feature applications are continuing to be developed. We have added capabilities to MicroStation Modeler to enable it to deal with collections of surfaces in terms of solid modeling. A new Sweep feature defines solids in terms familiar to surface modelers, as one B-spline curve swept along another. Other features are being added to enable sewing various surface patches together into a solid.

6.2 Objective MicroStation

The feature tree of MicroStation Modeler is a directed acyclic graph; dependencies are one-way and each node has a single parent. Furthermore, features fall short of being true objects as their code is located apart from their data. Ongoing work at Bentley

Systems has addressed the problems of managing more general dependency graphs and has extended the object-oriented ideas nascent in MicroStation Modeler.

While MDL has enabled expanding the tool set of MicroStation, as discussed above, extending the file format has always been restricted. Objective MicroStation™, the foundation of future Bentley Systems products, redefines the idea of the design file in an object-oriented way that enables its extension as well. Design files simply contain the data of users' geometry; code to manipulate that data exists either in the Micro-Station core or in MDL applications. Objective MicroStation project files, however, persistently store schema that contain both data and executable code to operate on it. They continue to maintain the platform independence that has proven so valuable in design files.

By defining schema, Objective MicroStation developers create intelligent data types that are better able to represent the complex engineering models of modern users. Schema objects can be thought of as intelligent "applets" that capture the behavior of objects from an engineer's world. A pipe can now "understand" that it must not violate a structural member in a model of an oil refinery. It can complain if its maximum pressure rating is likely to be exceeded when it is connected to a particular pump. A beam can "know" what sort of a fixture is necessary to connect to a column and can add to a bill of materials appropriately.

Schema can be seen as the natural evolution of the idea of a feature. Feature-based modeling captures the shapes that are familiar to engineers: fillets, bosses, holes, etc.. Schema definitions capture both shape and behavior, enabling engineers finally to consider the function as well as the form of their designs. The benefit of such an approach is obvious; it represents a marriage of the design and analysis tasks that have so long been separated in the computer-aided engineering world.

Schema are developed using Bentley System's new ProActiveM™ technology. Building on the success of MDL, the ProActiveM environment provides a plat-form-inde-pendent means to define easily transferred data and executable code. In this respect, ProActiveM is similar to Sun Microsystem's Java language; hardware issues are not a concern to either Java developers or Objective MicroStation programmers. ProActiveM also provides support for distributed, persistent store of schema objects, functionality which is lacking from Java.

This approach enables enterprise-scale collaboration across a heterogeneous mix of machine hardware and operating systems. It guarantees data integrity across many platforms and for all time. Everyone working on a project need not standardize on one release of a single operating system, or worry about installing relevant system libraries and correct versions of dynamic link libraries (DLLs). Furthermore, objects never lose their intelligence as their executable methods are stored with them.

Objective MicroStation unites the extensibility and platform independence Bentley Systems has always maintained in its products with the object-oriented ideas that have evolved in the software industry. We believe it will prove to be the correct solution for managing large-scale engineering projects.

7   Conclusions

System extensibility is of crucial importance. As software developers we do not have expert knowledge of all design application areas. We have shown that by adopting an object-oriented view of a feature, defining it by the set of methods it must support, an extensible, parametric, feature-based modeling system may be created.

Acknowledgements. This work was performed by the following members of the Micro-Station Modeler design team: Raymond Bentley, Lu Han, R. Brien Bastings, Bhupinder Singh, and the author.

References

1. Pratt, M.J.: Synthesis of an Optimal Approach to Form Feature Modelling. ASME Computers Engineering.**1** (1988) 263–274
2. Roller, D.: An approach to computer-aided parametric design. Computer Aided Design **23** (1991) 385–391
3. Bentley Systems, Inc.: Home page URL. http://www.bentley.com
4. Mantyla, M.: An Introduction To Solid Modeling. Computer Science Press, Rockville, Maryland, 1988.
5. Shah, J.J.: Assessment of features technology. Computer Aided Design **23** (1991) 331–343
6. Wilson, P.R.: Multiple Representations of Solid Models. In: Wozny,M.J., McLaughlin, H.W., and Encarnaçao, J.L. (eds.) Geometric Modeling for CAD Applications. North-Holland, Amsterdam New York Oxford Tokyo, 1988, pp.99–114
7. Chen, X., Hoffmann, C.M.: On editability of feature-based design. Computer Aided Design **27** (1995) 905–914
8. Rappoport, A.: Geometric Modeling: a New Fundamental Framework and its Practical Implications. In: Hoffmann, C., Rossignac, J. (eds.) Proc. Third Symposium on Solid Modeling and Applications. ACM Press, New York, 1995, pp.31–42
9. Capoyleas, V, Chen, X., Hoffmann, C.M.: Generic naming in generative, constraint based design. Computer Aided Design **28** (1996) 17–26
10. Kripac, J.: A Mechanism for Persistently Naming Topological Entitiesin History-based Parametric Solid Models. In: Hoffmann, C., Rossignac,J. (eds.) Proc. Third Symposium on Solid Modeling and Applications. ACM Press, New York, 1995, pp.21–30

Contour Edge Analysis for Polyhedron Projections

Lutz Kettner and Emo Welzl*

Institut für Theoretische Informatik, ETH Zürich, CH-8092 Zürich, Switzerland

Abstract. Given a polyhedron (in 3-space) and a view point, an edge of the polyhedron is called *contour edge*, if one of the two incident facets is directed towards the view point, and the other incident facet is directed away from the view point. Algorithms on polyhedra can exploit the fact that the number of contour edges is usually much smaller than the overall number of edges. The main goal of this paper is to provide evidence for (and quantify) the claim, that the number of contour edges is small in many situations.

An asymptotic analysis of polyhedral approximations of a sphere with Hausdorff distance ε shows that while the required number of edges for such an approximation grows like $\Theta(1/\varepsilon)$, the number of contour edges in a random *orthogonal* projection is $\Theta(1/\sqrt{\varepsilon}\,)$.

In an experimental study we investigate a number of polyhedral objects from several application areas. We analyze the expected number of contour edges and the expected number of intersections of contour edges in a projection (a quantity relevant for line sweep algorithms). We conclude that, indeed, the number of contour edges is small and the number of intersections of contour edges appears to be even more favorable. As a specific application we describe the computation of the silhouette of a polyhedral object with a sweep line algorithm in object space.

1 Introduction

We are given a polyhedron (in 3-space) and a view point. Roughly speaking, an edge of the polyhedron is called *contour edge*, if one of the two incident facets is directed towards the view point, and the other incident facet is directed away from the view point, see Figure 1 for an illustration. Algorithms on polyhedra (or polyhedral scenes) can exploit the fact that the number of contour edges is usually much smaller than the overall number of edges. This was first pointed out by Appel [1] when describing an object space hidden surface removal algorithm.

The main goal of this paper is to provide evidence for (and quantify) the claim, that the number of contour edges is small in many situations. First we make an asymptotic analysis of polyhedral approximations of a sphere with Hausdorff distance ε. While the

* Part of this research has been carried out while the authors were at the Freie Universität Berlin, the first author with a scholarship from the Graduiertenkolleg "Algorithmische Diskrete Mathematik", DFG We 1265/2-1. Support from a Leibniz Award, DFG We 1265/5-1, and from the ESPRIT IV LTR Project No. 21957 (CGAL) is acknowledged.

Fig. 1. An example for contour edges: the hidden surface rendering[1], the complete line drawing, and the contour edges.

required number of edges for such an approximation grows like $\Theta(1/\varepsilon)$, the number of contour edges in a random *orthogonal* projection is $\Theta(1/\sqrt{\varepsilon}\,)$; here and below *random* means that the projection vector is uniformly distributed on the unit sphere in 3-space. We believe that this 'square-root' behavior of the number of contour edges as compared to the overall number of edges appears in general for approximations (of increasing quality) of smooth (and not necessarily convex) bodies.

The second part of the paper reports the results of an experimental study of a number of polyhedral objects from several application areas. We analyze the expected number of contour edges, $\overline{n_c}$, and the expected number of intersections of contour edges in a projection, $\overline{int_c}$, (a quantity relevant for line sweep algorithms). We compare these numbers with the number of edges, n, and the expected number of intersections of all edges in a projection, $\overline{int}$. For the 'mushroom' example as depicted in Figure 1, these values[2] are $n = 464$, $\overline{n_c} = 54$, $\overline{int} = 450$, $\overline{int_c} = 4$. Instead of sampling from and averaging over a certain number of random directions, we describe how to compute these expected quantities directly.

From the examples we conclude that, indeed, the number of contour edges is small and the number of intersections of contour edges appears to be even more favorable. The latter is particularly interesting for object space methods based on the line sweep algorithm, [7,18,15]. As a specific application we describe the computation of the silhouette of a polyhedral object.

This presentation is a preliminary report from a project called CEBaP(*Contour Edge Based Polyhedron Visualization*). A brief description of the project and its goals is given in Section 6.

2 Contour Edges and Probabilistic Analysis

We consider polyhedral surfaces P given by their vertices V, edges E, and facets F, together with their incidence relation (and geometry information $V \mapsto \mathbb{R}^3$). An edge is incident to one or two facets. A facet is bounded by a simple oriented planar polygon. An edge incident to two facets has opposite orientations in these facets. That is, we assume that the surface is orientable. We also assume that each facet has a non-empty area.

---

[1] The three-dimensional hidden surface renderings are done with `Geomview`, a software written at the Geometry Center, University of Minnesota.

[2] Throughout the paper, expectations are rounded to the next integer.

To each facet f we define its normal vector $\mathbf{n}_f$, such that the direction of $\mathbf{n}_f$ is consistent with the orientation of f following the right-hand rule. Edges incident to exactly one facet are called *border edges*.

Definition 1 (contour edge). For a given viewing direction d we make the following definitions:

1. A facet f is a *front face* if $\mathbf{n}_f \cdot d < 0$. Otherwise it is a *back face*.
2. An edge e is a *contour edge* if it is a border edge or if it is incident to both a front face and a back face.
3. An edge e is a *front edge* if it is a contour edge or if it is incident to two front faces.

The intuitive meanings of front face and back face assume that the polyhedral surface P encloses a solid, and the normal vectors associated with the facets point outside this solid. Although this intuition is not valid in our setting with a general polyhedral surface (also with boundaries), we want to distinguish between the inner (solid) side of a facet, and the outer side of a facet (where its normal vector points). The definition of a contour edge is sensitive to the choice of the orientation of a surface only if a facet normal is orthogonal to the viewing direction.

We want to analyze the expected number of contour edges of P with respect to a random viewing direction. Recall once more that a random viewing direction is a vector to a point uniformly distributed on the unit sphere in 3-space. A random projection is an orthogonal projection in such a random viewing direction. Because of linearity of expectation, it suffices to analyze for each edge e independently the probability that it is a contour edge. Clearly, for a border edge this probability is one.

Consider a sphere of infinitesimal radius centered at an interior point of an edge e with two incident facets f' and f''. Intersect the sphere with the wedge defined by the facets. (The facets define two complimentary wedges, and we refer to that one enclosing the smaller angle, a possible tie in case of coplanar facets broken arbitrarily). Let α_e, the normalized solid angle, be the ratio between the surface area of this intersection and the whole sphere itself. α_e can be computed as

$$\alpha_e = \frac{1}{2\pi} \arccos \frac{-\mathbf{n}_{f'} \cdot \mathbf{n}_{f''}}{|\mathbf{n}_{f'}||\mathbf{n}_{f''}|}.$$

We observe that α_e is the probability that both facets incident to e are front facets, and similarly, it is the probability that both facets are back facets. We conclude that

$$\Pr[e \text{ is contour edge}] = 1 - 2\alpha_e \ ,$$

and linearity of expectations yields

Lemma 2. *The expected number $\overline{n_c}$ of contour edges with respect to a random viewing direction is*

$$\overline{n_c} = \sum_{e \in E} 1 - 2\alpha_e \ .$$

We want to demonstrate in a concrete setting that $\overline{n_c}$ is much smaller than the overall number of edges. To this end we consider *convex* polyhedral approximations C of the unit sphere S. We call C an ε-approximation of S, if no point on S has distance larger than ε from C, and no point on C has distance larger than ε from S. This entails that the Hausdorff distance between C and S is upper bounded by ε.

Lemma 3. *If e is an edge of a convex ε-approximation C of the unit sphere with $\varepsilon < 1$ then $\alpha_e > \beta_\varepsilon/2\pi$, where β_ε is*

$$\beta_\varepsilon = \pi - 2 \arctan \frac{2\sqrt{\varepsilon}}{1-\varepsilon}$$

in radians. Moreover, $1 - 2\alpha_e < \frac{4\sqrt{\varepsilon}}{\pi(1-\varepsilon)} = \frac{4}{\pi}\sqrt{\varepsilon} + O(\varepsilon^{3/2})$.

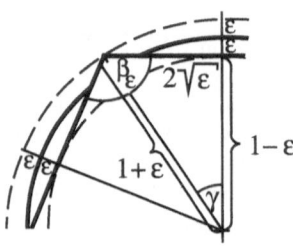

Fig. 2. The extremal configuration for planes supporting adjacent facets of an ε-approximation.

Proof. We exploit the facts that the supporting planes of the facets incident to an edge e must have distance at least $1 - \varepsilon$ from the center of S (because of the approximation property and convexity of C), and that they must have a common point in the ball of radius $1 + \varepsilon$ around the center of S (because their intersection carries an edge of C which must not have distance larger than ε from S). The extremal configuration, which allows the smallest possible angle β_ε between two such planes, is depicted in Figure 2. With the notation from that figure we have $\frac{\beta_\varepsilon}{2} = \frac{\pi}{2} - \gamma$ where $\tan\gamma = \frac{2\sqrt{\varepsilon}}{1-\varepsilon}$, β_ε and γ angles in radians. For the estimate of $1 - 2\alpha_e$ we use $\arctan x < x$ for $x > 0$. □

Lemma 4. *Let n be the number of edges of a convex ε-approximation of the unit sphere, and let $\overline{n_c}$ be the expected number of contour edges of this approximation. Then $\overline{n_c} = O(\sqrt{\varepsilon}\, n)$.*

Proof. For the set E of edges of an ε-approximation, we have $\overline{n_c} = \sum_{e \in E} 1 - 2\alpha_e = O(\sqrt{\varepsilon})|E|$ due to Lemmas 2 and 3. □

The number of edges required for an approximation of the unit sphere with Hausdorff distance ε is known to be $\Theta(1/\varepsilon)$ (see [8]). We conclude that

Theorem 5. *Let n be the number of edges of an optimal convex ε-approximation of the unit sphere and $\overline{n_c}$ the expected number of contour edges. Then we have $\overline{n_c} = \Theta(\sqrt{n})$.*

Proof. Since $n = \Theta(1/\varepsilon)$, we have $\overline{n_c} = O(1/\sqrt{\varepsilon})$ by Lemma 4. Simple considerations analogous to the proof of Lemma 3 show that every orthogonal projection of an ε-aproximation must have $\Omega(1/\sqrt{\varepsilon})$ edges. □

The asymptotic bounds obtained can be generalized to convex approximations of convex bodies of bounded curvature.

Here, and for the remainder of the paper, we analyze orthogonal projections (from a view point at infinity), simply because here a natural distribution of projections exists. Note, however, that it may very well be argued that orthogonal projections tend to maximize the number of contour edges. (Consider, for example, again polyhedral approximations of the unit sphere.)

3   Computation of Expected Number of Intersections

In the previous section we have seen how to compute $\overline{n_c}$. The complexity of the arrangement of the contour edges in the projection plane is determined by the number of projected contour edges and by the number of their intersections. For the sake of comparison and as a building component we present first how to compute the expected number $\overline{int}$ of such intersections of all edges (not just contour egdes) for random orthogonal projections. Then we condition on the event that the edges must be simultaneously contour edges while intersecting in the projection. This will be the expected number $\overline{int_c}$ of contour edge intersections. A known heuristic for hidden surface removal deletes all edges between two back faces before computing the visibility map since these edges can never be seen (at least, this is true for closed surfaces enclosing a solid). Only front edges remain for the visibility computation. For the sake of comparison with this approach, we compute the expected number $\overline{int_f}$ of intersections between front edges.

Intersection of Edges in the Projection

It suffices to consider all pairs of edges e' and e'' independently and to evaluate the probability that e' intersects e'' in a random projection. If e' and e'' intersect in space, then this probability is one. So let us assume that this is not the case, and when we refer to an intersection of two edges, we implicitly refer to the intersection of the projections of the edges in the projection plane.

Let T be the tetrahedron obtained as the convex hull of e' and e''. If e' and e'' are coplanar but disjoint, then they intersect with probability zero. So we may assume that T is a full dimensional simplex. We enumerate the edges of T as e_i, $i = 1, 2, \ldots 6$, so that $e_1 = e'$ and $e_2 = e''$. Vertices are enumerated as v_i, $i = 1, 2, 3, 4$. For the following analysis we refer to this tetrahedron when we talk about angles at edges, and about contour edges.

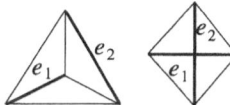

Fig. 3. Projections of tetrahedron T.

Observe that edges e_1 and e_2 intersect, if and only if none of the two edges is a contour edge of T. That is, we can derive:

$$\Pr[e_1 \text{ intersects } e_2] = \Pr[\text{none of } e_1, e_2 \text{ is contour edge}] =$$

$$\cdot \frac{1}{2} \left(\Pr[e_1 \text{ is not contour edge}] + \Pr[e_2 \text{ is not contour edge}] \right) \tag{1}$$

$$- \Pr[\text{exactly one of } e_1, e_2 \text{ is contour edge}] \right) .$$

The first two probabilities are $2\alpha_{e_1}$ and $2\alpha_{e_2}$, respectively. In order to calculate the third probability, note that exactly one of e_1, e_2 is a contour edge if and only if the tetrahedron projects to a triangle (see Figure 3). Equivalently, this means that one of the vertices does not appear on the boundary of the projection.

For the analysis of this event we define the normalized solid angle α_v at a vertex v similarly to the way we defined it for edges. That is, we center an infinitesimally small sphere at vertex v and take the ratio of the surface area in the tetrahedron compared to the full surface area. Consequently, $2\alpha_v$ is the probability that vertex v does not appear on the boundary of the projection. Since these events are disjoint for all vertices, $\sum_{i=1}^{4} 2\alpha_{v_i}$ is the probability that T projects to a triangle. Using the so-called Gram-Sommerville idendity [9,21] we can rewrite this sum as[3]

$$2\sum_{i=1}^{4} \alpha_{v_i} = 2\sum_{i=1}^{6} \alpha_{e_i} - 2 \; .$$

Plugging this into (1), we get

$$\Pr[e_1 \text{ intersects } e_2] = \frac{1}{2}\left(2\alpha_{e_1} + 2\alpha_{e_2} - \left(2\sum_{i=1}^{6} \alpha_{e_i} - 2\right)\right) \; .$$

Lemma 6. *Let e_1 and e_2 be two edges which do not lie in a common plane. Then e_1 and e_2 intersect in a random orthogonal projection with probability*

$$\Pr[e_1 \text{ intersects } e_2] = 1 - \sum_{i=3}^{6} \alpha_{e_i} \; ,$$

where e_3, e_4, e_5, and e_6 denote the remaining edges of the tetrahedron obtained as the convex hull of e_1 and e_2.

The expected number $\overline{int}$ of intersections of the set of edges E in a random projection is

$$\overline{int} = \sum_{e',e'' \in E, e' \neq e''} \Pr[e' \text{ intersects } e''] \; ,$$

which we can easily compute.

Intersection of Contour Edges

We must take a closer look at the geometric configuration of the two edges and their incident facets in 3-space. Consider an edge e with two incident facets f' and f''. Let h' be the positive open halfspace bounded by the supporting plane of f', that is, the halfspace where the normal vector $n_{f'}$ points to. It can be viewed as the set of view points for which f' is a front face. Similarly, define h'' for f''. Now let $W_c(e)$ be the symmetric difference of h' and h''. Intuitively speaking, $W_c(e)$ is the set of view points for which e is a contour edge (although we never formally defined a contour edge for a view point, we employ the obvious and natural extension for this illustration.) Observe that $W_c(e)$ is a double wedge (see Figure 4). For a border edge e, we define $W_c(e)$ to be the whole space.

If edges e' and e'' are contour edges and intersect in the projection plane, then the preimage of this intersection is a line in space intersecting e' and e'', and it is contained in $W_c(e')$ and $W_c(e'')$. It follows that only $e' \cap W_c(e'')$ can contribute to

---

[3] To see this directly, note that $\sum_{i=1}^{6} 1 - 2\alpha_{e_i}$ is the expected number of edges in the projection of T, $\sum_{i=1}^{4} 1 - 2\alpha_{v_i}$ is the expected number of vertices, and these two quantities have to be equal.

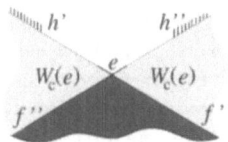

Fig. 4. The scenario around an edge e is shown when intersected with a plane orthogonal to e: The incident facets f' and f'', the halfspaces h' and h'', and the double wedge $W_c(e)$.

such an intersection, and, in fact, whenever the projection of $e' \cap W_c(e'')$ intersects the projection of e'', e'' has to be a contour edge. Because of symmetry, we conclude that e' and e'' intersect as contour edges if and only if $e' \cap W_c(e'')$ and $e'' \cap W_c(e')$ intersect.

Note that $e' \cap W_c(e'')$ need not be connected. It may be empty or consist of one or two line segments. Let S' denote the set of these line segments in $e' \cap W_c(e'')$, and let S'' be the segments in $e'' \cap W_c(e')$. Using this notation we get:

$$\Pr[e' \text{ and } e'' \text{ are contour edges and intersect}] = \sum_{g' \in S', g'' \in S''} \Pr[g' \text{ intersects } g'']$$

and it is now obvious how to calculate the expected number $\overline{int_c}$ of intersections of contour edges by summing these values for all pairs of edges in E.

Intersection of Front Edges

Some extra care has to be taken, because the definition of a front edge is sensitive to inverting the viewing direction. We denote the two remaining wedges apart from $W_c(e)$ at an edge e: The wegde $W_f(e)$ is the intersection of h' and h'', and $W_b(e)$ is the intersection of the complements of h' and h''. Intuitively speaking, $W_f(e)$ is the set of view points for which e is a *proper* front edge; that is, e is a front but not a contour edge. $W_b(e)$ is the set of view points where both facets incident to e are back facets. Both sets are defined to be empty for border edges.

Fig. 5. The wedges $W_c(e)$, $W_f(e)$, and $W_b(e)$ for an edge e with two incident facets.

Consider two edges e' and e''. The edge e' is subdivided in up to three segments by the regions $W_c(e'')$, $W_f(e'')$, and $W_b(e'')$. Let S' denote the set of these segments and let S'' denote the set of the segments obtained by intersecting e'' with $W_c(e')$, $W_f(e')$, and $W_b(e')$. The probability that two edges are front edges and intersect is

$$\Pr[e' \text{ and } e'' \text{ are front edges and intersecting}] = \sum_{g' \in S', g'' \in S''} p(g', g'') \,,$$

where

$$p(g', g'') = \begin{cases} \Pr[g' \text{ intersects } g''] & \text{for } g' \subseteq W_c(e'') \text{ and } g'' \subseteq W_c(e') \\ 0 & \text{for } g' \subseteq W_f(e'') \text{ and } g'' \subseteq W_f(e') \\ 0 & \text{for } g' \subseteq W_b(e'') \text{ and } g'' \subseteq W_b(e') \\ \frac{1}{2}\Pr[g' \text{ intersects } g''] & \text{otherwise} \,. \end{cases}$$

To show that this equation holds we look at the preimage of an intersection of the two edges in the projection plane. It is a line l in space intersecting both edges. Define the segment s as the segment on l between both edges. Figure 6 indicates all different cases (up to symmetry) of how s can lie relative to the wedges of the involved edges. The situation does only change if s is crossing the boundary of one of the wedges $W_c(e)$, $W_f(e)$, or $W_b(e)$. Hence each pair of segments $g' \in S'$ and $g'' \in S''$ belongs to exactly one case.

Fig. 6. The possible cases how a segment s can lie between two edges: **(a)** Both edges are contour edges in both viewing directions. **(b)** At least one edge is a *proper* front edge for one viewing direction and the other one is a front edge. **(c)** No viewing direction achieves that both edges are simultaneously front edges.

(a)   (b)   (c)

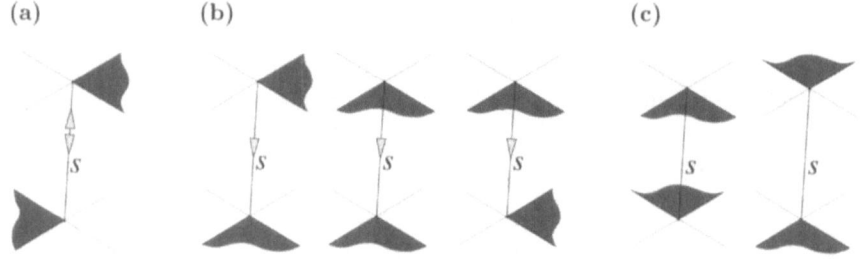

Case (a) depicts two intersecting contour edges. For the segment s holds $s \subseteq W_c(e')$ and $s \subseteq W_c(e'')$. It follows that $g'' \subseteq W_c(e')$ and $g' \subseteq W_c(e'')$. Therefore, in this case the probability that e' and e'' intersect as front edges is equal to the probability that g' and g'' intersect.

Case (b) depicts the other possibilities of how two front edges can intersect. At least one of the edges has to be a proper front edge, since otherwise we are in case (a). That is either $g'' \subseteq W_f(e')$ or $g' \subseteq W_f(e'')$. Note that the intersection depends on the orientation of the viewing direction. For the reverse viewing direction we get that either $g'' \subseteq W_b(e')$ or $g' \subseteq W_b(e'')$, respectivly. Here e' and e'' will not intersect as front edges. Summing up, in this case the probability that e' and e'' intersect as front edges is half the probability that g' and g'' intersect.

Case (c) depicts the situation where for any orientation of the viewing direction both edges cannot be front edges in the projection simultaneously. That is either $g'' \subseteq W_b(e')$ or $g' \subseteq W_b(e'')$. Thus, the probability of this case to contribute an intersection of front edges is zero.

4   Expectations Measured on Objects

We have collected three-dimensional objects from different application areas to evaluate their properties. The application areas are animation, architecture, terrain data from geographical information systems, mesh generation from CAD models, surface reconstruction, medical imaging, molecular modeling, and mathematics. We count the number $n = |E|$ of edges and the number $|\partial E|$ of border edges. We compute the the expected number $\overline{n_c}$ of contour edges where the border edges are included by definition.

Fig. 7. Examples from computer graphics. **head** is from the **Geomview** distribution, **honda**, **beethoven**, and **general** are from Viewpoint [20].

| head | honda | beethoven | general |

filename	n	$|\partial E|$	$\overline{n_c}$	$\% \frac{n_c}{n}$	$\overline{int}$	$\overline{int_f}$	$\overline{int_c}$	$\% \frac{\overline{int_c}}{\overline{int}}$
head	3106	58	203	6	2950	456	64	2
honda	13875	574	2998	21	36970	11918	2138	6
beethoven	5461	268	1003	18	9415	2983	617	7
general	25858	6	4350	16	93285	27014	3094	3

We compute the expected number $\overline{int}$ of intersections of all edges, $\overline{int_c}$ of intersections of contour edges and $\overline{int_f}$ of intersections of front edges.

In Figure 7 **head** is an example of an approximated curved surface in computer graphics. **honda** has the worst ratio of contour edges to edges under these examples due to the technical details. **beethoven** and **general** are handmade. Details like the faces and medals are modeled in great detail, other parts like the dress are not. For example, **general** consists of 118 components (hardly visible in the small picture). The details contribute a lot to the contour edges.

As for the intersections, observe that even $\overline{int_f}$ is a factor 5 to 10 larger than $\overline{int_c}$. The overall impression is that n seems to dominate all other quantities (apart from $\overline{int}$) in most examples.

The bad contour edge ratios of the examples in Figure 8 reveal that architecture might not be a good candidate for the contour edge approach. It is perhaps counter-intuitive to see that **powerlns** perform better than **pagota**. This is a consequence of the quality of representation of these models. **powerlns** consists of 20 components and has no border edges. **pagota** had originally 1086 edges with more than two facets incident. In order to prepare the data set for our analysis we have chosen to trivially break these edges apart resulting in the 4374 border edges. They count as contour edges for all viewing directions. The number of components is 1573 for the **pagota**. **epcot3** is our worst case example both with respect to contour edges and contour edge intersections.

The terrain data in Figure 9 perform quite well in the contour edge analysis. The terrains 1 to 3 are approximations of increasing density of the same terrain (defined by the limit of a fractal terrain generating process). We observe the decreasing ratio of contour edges with increasing approximation quality. Under the assumption that $\overline{n_c} = c_1\sqrt{n}$, we can compute the coefficient c_1 to be 4.5, 6.6, and 10.2 for **terrain** 1, 2, and 3, respectively.

Fig. 8. Architecture examples. `powerlns` and `skyscrpr` are from Avalon [2], `pagota` is from Viewpoint [20], `epcot3` is from the `Geomview` distribution.

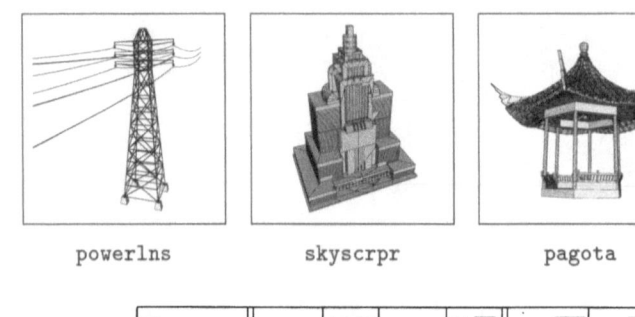

powerlns   skyscrpr   pagota   epcot3

filename	n	$|\partial E|$	$\overline{n_c}$	$\%\frac{n_c}{n}$	$\overline{int}$	$\overline{int_f}$	$\overline{int_c}$	$\%\frac{int_c}{int}$
powerlns	9214	0	2500	27	45162	17350	4530	10
skyscrpr	3711	0	1230	33	23462	10574	3796	16
pagota	44137	4374	16971	38	274119	131333	50317	18
epcot3	2304	0	928	40	5459	2594	1585	29

The first three examples in Figure 10 were generated by a mesh generation algorithm for objects with parametric face representations [14]. The number in the filename denotes the tolerated error bound in the approximation. The coefficient in $c_1\sqrt{n}$ is here 21.7 and 30 for `cola 0.05` and `cola 0.005`, respectively. `fandisk1` is an optimized mesh from surface reconstruction [13]. From that, a piecewise smooth surface has been com-

Fig. 9. Terrain data examples for geographical information systems. `kground` is from Avalon [2], the terrain data sets are fractal terrains.

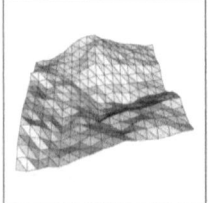
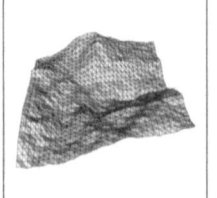

kground   terrain1   terrain2   terrain3

filename	n	$|\partial E|$	$\overline{n_c}$	$\%\frac{n_c}{n}$	$\overline{int}$	$\overline{int_f}$	$\overline{int_c}$	$\%\frac{int_c}{int}$
kground	10920	240	657	6	7604	2579	369	5
terrain1	800	64	130	16	623	230	49	8
terrain2	3136	128	371	11	3234	1124	201	6
terrain3	12416	256	1139	9	15807	5257	811	5

Fig. 10. Examples from mesh generation from CAD models and surface reconstruction. cola and pencil are from Tobias Hüttner and Reinhard Klein, WSI/GRIS, Universität Tübingen, Germany. fandisk and bunny are meshes from the Computer Graphics Group, University of Washington.

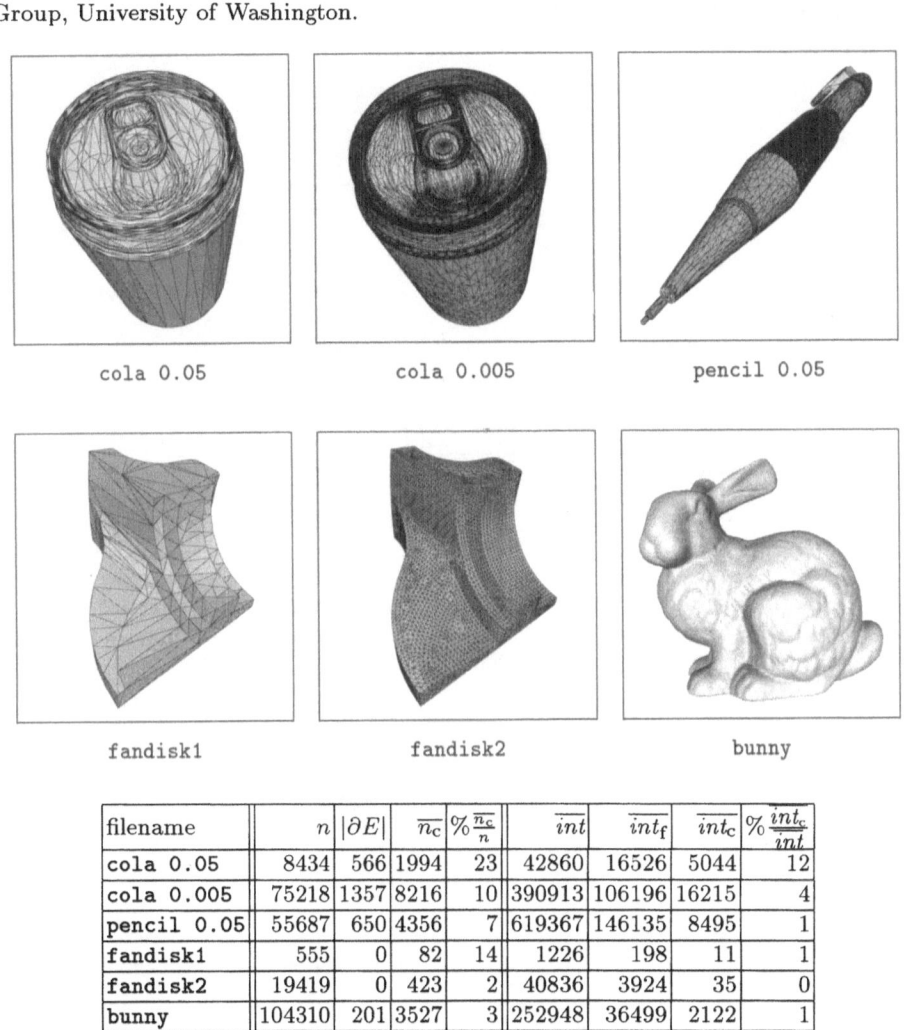

cola 0.05   cola 0.005   pencil 0.05

fandisk1   fandisk2   bunny

filename	n	$|\partial E|$	$\overline{n_c}$	$\%\frac{\overline{n_c}}{n}$	$\overline{int}$	$\overline{int_f}$	$\overline{int_c}$	$\%\frac{\overline{int_c}}{\overline{int}}$
cola 0.05	8434	566	1994	23	42860	16526	5044	12
cola 0.005	75218	1357	8216	10	390913	106196	16215	4
pencil 0.05	55687	650	4356	7	619367	146135	8495	1
fandisk1	555	0	82	14	1226	198	11	1
fandisk2	19419	0	423	2	40836	3924	35	0
bunny	104310	201	3527	3	252948	36499	2122	1

puted and fandisk3 is a piecewise linear approximation [12]. The coefficient c_1 is 3.5 and 3.0, respectively. bunny is a data set from a 3d-scanner. It has been used in [5,11].

face in Figure 11 is generated with a marching cube algorithm. Although this algorithm has produced a remarkable band structure in the reconstruction, the contour edge ratio is quite good. The reconstruction examples kopf, lunge, and becken from [22] have a significantly lower resolution and therefore medium contour edge ratios.

mol1 in Figure 12 is a relative crude approximation of the van der Waals surface of a molecule. mol2 is the molecular surface [17], also known as Conolly surface, of

Fig. 11. Examples from medical imaging. `face` was supplied by Herve Delingette, Project Epidaure, INRIA-Sophia-Antipolis, France. `kopf`, `lunge`, and `becken` were supplied by Barbara Wolfers, FU Berlin, Germany (reconstructed from data supplied by INRIA, Sopia Antipolis and Konrad-Zuse-Zentrum für Rechentechnik, Berlin.)

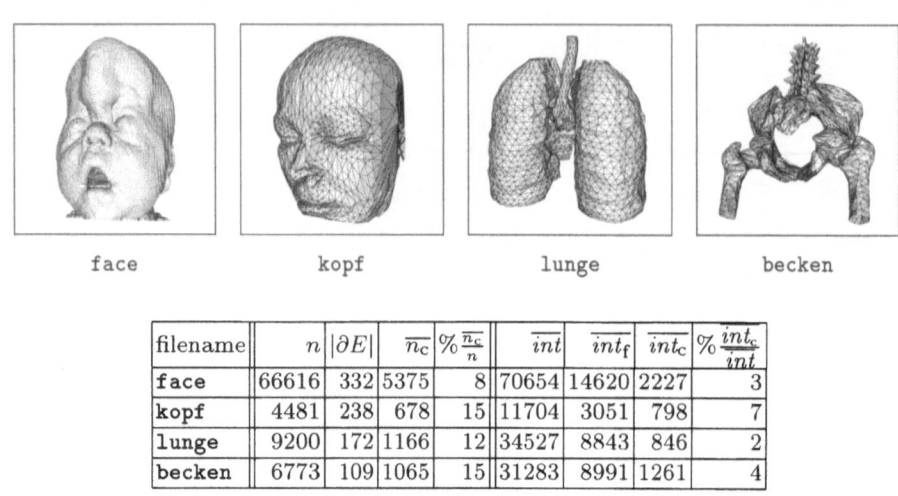

face   kopf   lunge   becken

filename	n	$\|\partial E\|$	$\overline{n_c}$	$\%\frac{\overline{n_c}}{n}$	$\overline{int}$	$\overline{int_f}$	$\overline{int_c}$	$\%\frac{\overline{int_c}}{int}$
face	66616	332	5375	8	70654	14620	2227	3
kopf	4481	238	678	15	11704	3051	798	7
lunge	9200	172	1166	12	34527	8843	846	2
becken	6773	109	1065	15	31283	8991	1261	4

Fig. 12. Examples from molecular modeling and mathematics. `mol1` and `mol2` were supplied by Nataraj Akkiraju and Herbert Edelsbrunner, University of Illinois at Urbana-Champaign. `hypersheet` is from the Computer Graphics Group, University of Washington. `tre_twist` is from Avalon [2].

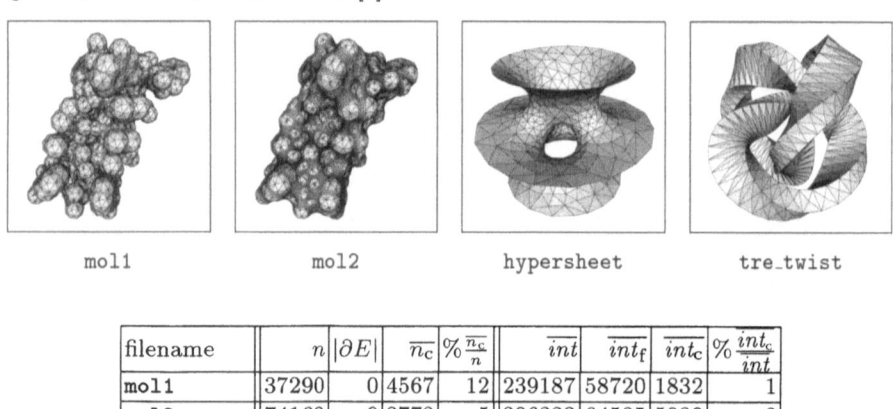

mol1   mol2   hypersheet   tre_twist

filename	n	$\|\partial E\|$	$\overline{n_c}$	$\%\frac{\overline{n_c}}{n}$	$\overline{int}$	$\overline{int_f}$	$\overline{int_c}$	$\%\frac{\overline{int_c}}{int}$
mol1	37290	0	4567	12	239187	58720	1832	1
mol2	74163	0	3772	5	306223	64585	5980	2
hypersheet	1407	63	154	10	2507	538	106	4
tre_twist	2400	0	375	15	13238	3313	490	4

the same molecule. It is obtained from the van der Waals surface by introducing so-called reentrant surface patches. Caveties are removed or flattened as can be seen from the silhouette of `mol1` and `mol2`. The genus of `mol1`, which is 37, is lowered to 1 for `mol2`. Therefore, not only the contour edge ratio is better for `mol2`, but also the absolute number of contour edges is smaller. `hypersheet` is again a reconstruction taken from [11]. `tre_twist` realizes surprisingly high values $\overline{int}$ and $\overline{int_f}$ for such a small object.

Fig. 13. All values for expected contour edges $\overline{n_C}$ drawn in one plot.

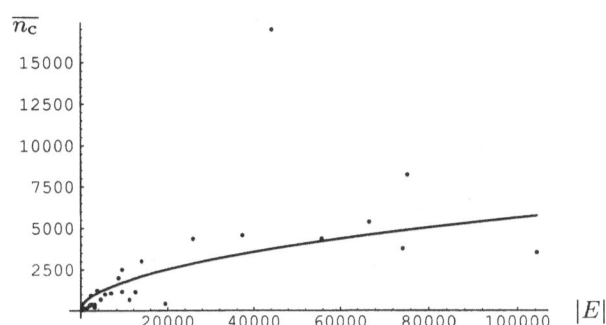

Figure 13 plots all values for the expected number $\overline{n_C}$ of contour edges that are mentioned in the table above. The absolute maximum is 16971 and is achieved by the **pagota** in figure 8. The overall average percentage of contour edges is 15%. Ignoring the obvious runaway **pagota**, we computed a least squares fitting curve $c_1\sqrt{n}$ with $c_1 = 17.8$ which is also depicted in Figure 13. If we choose $c_1\sqrt{n} + c_2 n$ as the fitting curve, the linear term turns out to be small with $c_2 = 0.0056$, while $c_1 = 16.5$.

5   The Silhouette of a Polyhedron

The *silhouette* is the boundary of the projected polyhedron. It separates the object from the background. This definition also includes boundaries of holes in the object through which one can see the background. As a data structure, the silhouette is a set of segments S. The preimage of each segment $s \in S$ is a piece e_s of an an edge e of the polyhedron.

The silhouette is useful in shadow casting. It defines the shadow volume as follows: The silhouette S is computed for a polyhedron P and a lighting direction d. The light source casts a shadow like a "curtain" from the preimage e_s of each segment s. This shadow is an unbounded facet of the shadow volume. They can be used to cast shadows onto other objects or even the object itself. The exact solution provided by an object space algorithm [19] avoids the drawback of aliasing effects of raster algorithms for shadow casting.

We use a line sweep algorithm to compute all intersections of the projected contour edges in the projection plane. It is sufficient to consider only contour edges since the silhouette edges must be a subset of the contour edges. It follows that the running time of this example application is intrinsically dependent on the number of contour edges and their intersections.

The sweep line stops at projected endpoints and intersections of contour edges. It stores all edges that are actually intersected by the sweep line as its status. To compute the silhouette we extend the sweep line status with visibility information as follows:

The notion of *quantitive invisibility* is defined for points in space as the number of facets hiding the point from the viewer [1]. The sweep line status can be extended by the quantitive invisibility of the background. It changes only at contour edges. Border edges make a difference of one in the counting, other contour edges make a difference of two. The sweep line starts with a count of zero. Each event point leads to local changes. E.g. a vertex with two border edges, one already intersecting the sweep line, changes nothing. A vertex with two edges, none of them intersecting the sweep line, opens a new region with a changed quantitive invisibility depending on whether the incident facets of the edges already intersect the sweep line or not (for details we refer to [10]). The silhouette is exactly the collection of those parts of the projected contour edges that separate regions of quantitive invisibility zero from those of non zero. They can be reported while the sweep line proceeds. Altogether, no additional effort is necessary for the silhouette computation except for the line sweep. Thus the expected running time (for a random projection) amounts to $O((\overline{n_c} + \overline{int_c}) \log n)$.

In Figure 14 we show a sequence of silhouettes obtained by rotating **honda** from Figure 7, computed by a recent implementation of this line sweep. The computation of a **honda**-silhouette takes approximately 0.15 seconds on a SUN Sparcstation 10-20, [10].

Fig. 14. A Sequence of four silhouettes of the **honda** example rotated.

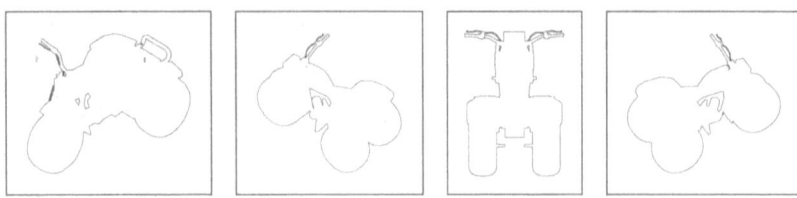

6   Conclusion

The work described here is part of the project CEBaP, an acronym for *Contour Edge Based Polyhedron Visualization*. In the project we study object space hidden surface removal algorithms [19]. It is intended to provide implementations for the CGAL project, a coordinated effort of seven research groups in Europe for constructing a *Computational Geometry Algorithms Library* [6,16,3].

The approach we are focussing on uses a line sweep algorithm for hidden surface elimination in the projection plane as in the example above for the silhouette computation. A similar line sweep considering all projected edges was studied by Hamlin and Gear [7]. They analyzed possible types of intersection between edges to reduce the number of depth comparisons during the sweep. Séquin and Wensley [18] extended this work to a broader class of input geometries including segments. Nurmi [15] studied the slightly different problem of hidden line elimination. He described a data structure for the sweep line status that supports the depth comparisons efficiently such that the running time is the well known bound for a Bentley-Ottmann line sweep, i.e. $(O(n + k) \log n)$ where n is the number of vertices and k the number of intersections of edges in the projection. A survey including other hidden surface removal algorithms can be found in [4].

Robustness and exactness are important issues for combinatorial algorithms such as the line sweep. We decided to implement the line sweep with exact arithmetic of bounded but sufficient precision. Thus, the projected contour edges are rounded to the resolution of the output device right before the sweep. At first, the remaining part of the algorithm used double arithmetic. This led to serious problems: The contour edges were extracted with the precision of the original coordinates in space, the sweep line on the other hand used the rounded coordinates. Thus, we extracted contour edges that were no longer contour edges and vice versa because the geometry had changed underway. We detected thin but longish triangles that reverse orientation when rounded. This resulted in cracks and holes in the silhouette computation. Figure 14 gives examples; three of the four pictures have cracks. Consequently, we are redesigning the algorithm to round the input data first and compute with bounded and exact arithmetic afterwards.

So the future directions of the project are towards a robust and exact implementation, for preprocessing, rounding, and complete hidden surface removal for polyhedra.

Acknowledgements

We would like to thank Bernd Gärtner, Michael Hoffmann, Sven Schönherr, and Michael Schutte for the many fruitful discussions in the CEBaP meetings. Thanks to Sven Schönherr for the ti_library that provides the basic data types for CEBaP, thanks to Michael Hoffmann and Michael Schutte for implementation work. Last but not least, we would like to thank all the people that provided us with numerous example objects for our analysis.

References

1. Arthur Appel. The Notion of Quantitive Invisibility and the Machine Rendering of Solids. In *Proc. ACM National Conf.*, pages 387–393, 1967.
2. Avalon. A server of public domain 3d objects: http://www.viewpoint.com/.
3. The CGAL home page: http://www.cs.ruu.nl/CGAL/.
4. Susan E. Dorward. A Survey of Object-Space Hidden Surface Removal. *Int. J. of Computational Geometry & Applications*, 4(3):325–362, 1994.
5. Matthias Eck, Tony DeRose, Tom Duchamp, Hugues Hoppe, Michael Lounsbery, and Werner Stuetzle. Multiresolution Analysis of Arbitrary Meshes. In *Computer Graphics (Proc. SIGGRAPH '95)*, volume 29, pages 173–182, 1995. Examples in file://ftp.cs.washington.edu/pub/graphics.
6. Andreas Fabri, Geert-Jan Giezeman, Lutz Kettner, Stefan Schirra, and Sven Schönherr. The CGAL Kernel: A Basis for Geometric Computation. In M. C. Lin and D. Manocha, editors, *ACM Workshop on Applied Computational Geometry*, Philadelphia, Pennsylvania, May, 27–28 1996. Lecture Notes in Computer Science 1148.

7. Jr. Griffith Hamlin and C. William Gear. Raster-Scan Hidden Surface Algorithm Techniques. In *Computer Graphics (Proc. SIGGRAPH '77)*, volume 11, pages 206–213, 1977.

8. Peter M. Gruber. Approximation of Convex Bodies. In Peter M. Gruber and Jörg M. Wills, editors, *Convexity and Its Applications*, pages 131–162. Birkhäuser, 1983.

9. Branko Grünbaum. *Convex Polytopes*. John Wiley & Sons, New York, 1967.

10. Michael Hoffmann. Line-Sweep auf einem Gitter, 1996. Diplomarbeit. Freie Univ. Berlin, Germany.

11. Hugues Hoppe. *Surface reconstruction from unorganized points*. PhD thesis, University of Washington, 1994.

12. Hugues Hoppe, Tony DeRose, Tom Duchamp, Mark Halstaed, Hubert Jin, John McDonald, Jean Schweitzer, and Werner Stuetzle. Piecewise Smooth Surface Reconstruction. In *Computer Graphics (Proc. SIGGRAPH '94)*, volume 28, pages 295–302, 1994. Examples and code in `file://ftp.cs.washington.edu/pub/graphics`.

13. Hugues Hoppe, Tony DeRose, Tom Duchamp, John McDonald, and Werner Stuetzle. Mesh Optimization. In *Computer Graphics (Proc. SIGGRAPH '93)*, volume 27, pages 19–26, 1993. Examples and code in `file://ftp.cs.washington.edu/pub/graphics`.

14. Reinhard Klein and Wolfgang Straßer. Mesh Generation from Boundary Models with Parametric Face Representations. In C. Hoffmann and J. Rossignac, editors, *Third Symposium on Solid Modeling and Applications*, pages 431–440. ACM Press, May 1995.

15. Otto Nurmi. A fast line-sweep algorithm for hidden line elimination. *BIT*, 25:466–472, 1985.

16. Mark H. Overmars. Designing the Computational Geometry Algorithms Library CGAL. In M. C. Lin and D. Manocha, editors, *ACM Workshop on Applied Computational Geometry*, Philadelphia, Pennsylvenia, May, 27–28 1996. Lecture Notes in Computer Science 1148.

17. Frederick M. Richards. Areas, Volumes, Packing and Protein Structure. *Annu. Rev. Biophys. Bioeng.*, 6:151–176, 1977.

18. Carlo H. Séquin and Paul R. Wensley. Visible Feature Return at Object Resolution. *IEEE Computer Graphics and Application*, 5:37–50, May 1985.

19. I. E. Sutherland, R. F. Sproull, and R. Schumaker. A Characterization of Ten Hidden-Surface Algorithms. *ACM Computing Surveys*, 6(1):1–55, 1974.

20. Viewpoint DataLabs, Orem UT, USA. Commercial 3d objects and the Avalon server of public domain 3d objects: `http://www.viewpoint.com/`.

21. Emo Welzl. Gram's Equation – A Probabilistic Proof. In *Results and Trends in Theoretical Computer Science*, pages 422–424. Springer Verlag, 1994. *Lecture Notes in Computer Science*, 812.

22. Emo Welzl and Barbara Wolfers. Surface Reconstruction between Simple Polygons via Angle Criteria. *J. Symbolic Computation*, 17:351–369, 1994.

Part IV
Automated Assembly

Atlas: An Automatic Assembly Sequencing and Fixturing System

Bruce Romney

Dept. of Computer Science
Stanford University
bruce@flamingo.stanford.edu

Abstract. We present a method to generate automatically both an assembly sequence and a fixture to hold all the intermediate subassemblies, given only a geometric description of the product. The main contribution here is that the sequence and fixture are generated concurrently rather than sequentially, unlike in previous assembly-planning work. Also, the representation of possible fixturing locations is more complete. This method has been implemented for the planar case, and we present experimental results below. We also discuss some of the key issues involved in the concurrent design of assembly sequences and fixtures. The long-term vision for this work is a CAD tool which can provide immediate feedback to a product designer about the ease of assembling the proposed product.

1 Introduction

Early mechanical CAD systems were little more than electronic draftboards, providing some part-interference information but leaving most advanced analysis to the designer. Over the years, though, as geometric modelling and related CAD technology have improved, CAD systems have been able to provide increasingly sophisticated feedback to the designer, not only about the product's functionality but also about its manufacturability and serviceability. These advances now allow the designer to quickly and automatically identify potential problems in these areas, and devote more attention to higher-level considerations. Recently, for example, a number of prototype systems have been presented which can automatically generate assembly sequences for a given product [1–5].

The purpose of the present research is to increase this level of sophistication still further, by automatically generating not only the assembly sequence but also the mechanical fixture required to hold the product in place during assembly. Assembly fixturing and stability have a crucial impact on the feasibility of assembly sequences, so a system to generate both a fixture and a sequence would be substantially more useful than a sequencer alone. Now, one could conceivably generate first one and then the other; but given the strong interdependence between sequencing and fixturing, it seems reasonable to expect that an approach which combines the two analyses could potentially produce better results. We refer to this technique as sequence/fixture *co-design*, and we employ it in Atlas, the system described below. Although the current implementation of Atlas is restricted to planar assemblies, all of its techniques extend to the three-dimensional case as well.

A sequencing and fixturing system such as Atlas, when fully developed and incorporated into a CAD tool, would be of tremendous value to a team of product designers. It could alert them to hidden problems in the assemblability of a product, or, conversely, present non-obvious solutions to known problems. This would be all accomplished "on line", without the need to construct a prototype.

In this paper, we begin by describing some of the past work in sequencing and fixturing in section 2. In section 3, we then formally define the problem and state our assumptions. Next, in section 4, we describe Atlas' basic operating procedures. In section 5, we explore Atlas' decision-making process from the perspective of wrench space. In section 6, we present some experimental results. Finally, in section 7, we discuss some directions of future research and conclude.

2 Related Work

The problem of automatically generating assembly sequences is, in its full generality, PSPACE-hard [6,7]. Even when every operation moves exactly two subassemblies into their final relative positions, the problem is still NP-hard [8]. As a result, much of the past and present work in this area focuses on even more restricted variants of the problem. For example, some systems insert parts only along the major axes [3,9-11], while others insert only one part at a time [2,4]. For the case of binary, monotone sequences using straight-line part insertions from any direction, Wilson and Latombe presented a complete polynomial-time algorithm in [12] using the *non-directional blocking graph (NDBG)* data structure.

One common thread that appears in much of the literature is the strategy of "assembly by disassembly", in which an assembly sequence is generated by starting with the completed product and working backwards through disassembly steps. We shall use this strategy as well.

The assembly-sequencing portion of Atlas is based on the STAAT system [1] and the related concept of the local translational freedom cone [13]. The approach has some similarities to techniques described by other authors [2,5,14,15]. However, little consideration is given in these papers to the problem of fixturing the intermediate subassemblies.

There is a great deal of literature on the fixturing or grasping of a single part. Much of it focuses on establishing form closure [16-21] or force closure [22-24] on that part. Some authors also propose figures of merit with which to compare different form- or force-closing configurations, often based on the necessary contact forces. However, in sequence/fixture co-design, even when treating an assembly as a single unit, we are not necessarily interested in a form-closing fixture; on the contrary, such a fixture would often be too restrictive to allow part removals or insertions. Instead, we merely need to stabilize the assembly and all subassemblies against gravity and the *insertion forces* (see below, section 3.1), while still admitting free insertion paths for the parts. Also, because of the difficulty of the combined sequencing/fixturing problem, this paper focuses on ways to find *one* feasible fixture and sequence, leaving the task of comparing multiple solutions to future efforts.

The field of *assembly* fixturing, in which all parts may move independently, is far less represented in the literature than that of part fixturing. Some early work was done by Fahlman [25] for structures of simple blocks. However, Fahlman himself acknowledged his stabilization techniques to be "a loose grouping of a very large number of essentially independent tricks". Later, Wolter and Trinkle [26] and Mattikalli [27] described a more systematic approach to analyzing and fixturing individual assemblies, by using linear programs to solve simultaneous force or velocity equations. Then, by scattering fixture elements all over the assembly and eliminating those that turned out to be unnecessary, these authors were able to stabilize assemblies with only a small number of fixture elements. Atlas' fixturing algorithm was originally inspired by these researchers' efforts.

However, to the author's knowledge, no one has previously addressed the problem of performing both assembly sequencing and assembly fixturing *concurrently*. This co-design approach raises a host of new issues, many of which were discussed in an earlier paper [28]. Atlas, then, represents the first attempt to both generate an assembly sequence and fixture that sequence, under a "unified" approach which recognizes and utilizes the interrelationships between fixturing and sequencing.

3   Problem Statement

The task we are trying to accomplish is as follows:

> Given a geometric description of a product, generate (1) a sequence of part insertions which will assemble the product from its parts, and (2) a fixture to hold the product in place during assembly, such that each intermediate subassembly is stable under gravity and all insertion forces.

3.1   Insertion Forces

Insertion forces are part-against-part forces which may arise from part insertions. For example, in Fig. 1, either friction or a slight misalignment in P_1's position could cause a force to be imparted on P_2 to the right. To allow for this possibility, we would need one or more additional fixture elements (or *fixels*) to the right of P_2.

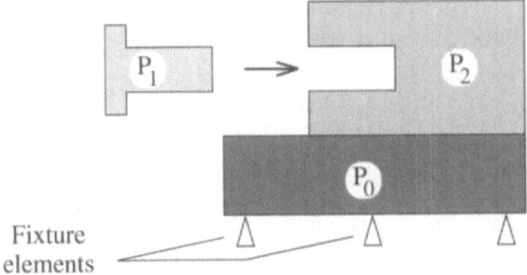

Fig. 1. An assembly operation which is unstable due to insertion forces.

To be conservative, we assume that these insertion forces could occur at any contact point(s), and in any combination of magnitudes. They could potentially be quite large if the insertions were performed by, say, heavy robotic machinery; therefore, we check for insertion-force stability in the presumed absence of gravity (recognizing that the insertion forces might overwhelm any gravitational resistance).

3.2   Assumptions

We assume that the assembly is *polyhedral, rigid*, and defined by *nominal geometry* (*i.e.,* no tolerances).

We assume that the assembly sequences are *monotone* (parts inserted into a sub-assembly do not later move with respect to that subassembly) and composed of *single-step translations*. Extensions to rotational and/or multi-step part motions remain a direction for future research.

We assume that the fixture is *firmly secured* and *non-adjustable*. The advantage of a single, non-adjustable fixture is that it can be constructed once and then left alone throughout the assembly process; upon completion, it can then be recycled again and again. (The techniques presented here, however, extend easily to allow the dynamic addition of fixture elements during the assembly process.) Note that the single-fixture model does pose substantial restrictions on the range of possible sequence/fixture solutions. Most notably, it implies *linearity* in the assembly sequence (that is, each step involves the insertion of exactly one part). This is because the insertion of a multi-part subassembly would imply the previous use of another fixture to create that subassembly.

We assume the fixture elements are *frictionless* and can only apply *nonnegative force* (that is, they can push but cannot pull). We also treat them as having *negligible size*, so we describe them only at their points of contact, as in Fig. 1. To be sure, the actual shape of the fixture elements could pose a significant assembly-sequencing constraint; and if there were, say, a library of predefined fixturing elements available on a CAD system, these constraints could be carefully considered. However, this analysis is beyond the scope of the present work.

Finally, sometimes (but not always), we assume that parts are *glued together* (or similarly attached) upon insertion, so that the entire assembly is rigid and unified for stability-checking purposes. This assumption is based on observations from tours of actual automobile factories. When parts are described as not glued together, we shall neglect friction between them in our stability-checking (in order to be conservative).

4   Description of System

4.1   Overview

Atlas' basic approach to sequencing and fixturing is to enclose the final assembly in a completely immobilizing fixture (a process called *overfixturing*), and then to remove parts one at a time, reducing the fixture as necessary. The major steps are shown in Fig. 2 and described in greater detail below. During the disassembly process, the requirements imposed by the evolving assembly sequence constrain the evolving fixture, and vice-versa. In general, when we modify the proposed fixture, we are modifying the *one* fixture which will hold *all* the subassemblies; therefore, we must ensure the stability not only of the current assembly, but also of all previously-seen superassemblies.

Naturally, a fixture must not be so constraining as to prevent the product's removal after it has been assembled (usually in the upward direction). So we must perform one additional step after overfixturing, to eliminate all portions of the fixture which would block an upward motion of the entire assembly. (Some of the figures in this paper may, however, neglect the need for such an "escape path", for clarity in illustration.)

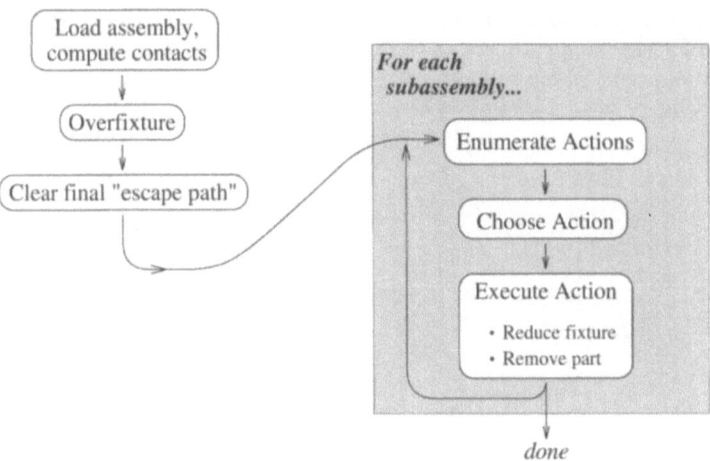

Fig. 2. Atlas' basic operating procedure.

4.2   The Fixture Continuum

Atlas represents the candidate fixture as a *continuum* of possible fixel locations, rather than as a finite set of discrete fixels. (See Fig. 3.) Although a discrete representation, such as was used in [27,28], is certainly valid, the continuous approach is a more complete representation of possible fixturing locations. This produces a potentially greater choice of stable disassembly steps. In essence, a continuous fixture can be thought of as an infinite set of differential fixture elements (the only caveat being that finite-sized forces can be applied at any point, unlike in the differential model).

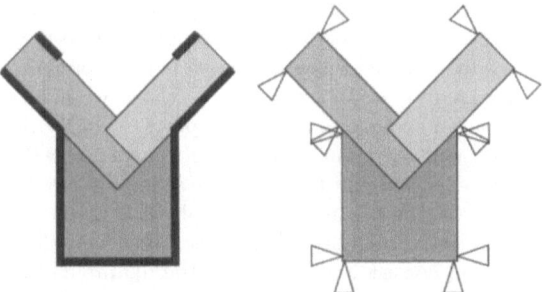

Fig. 3. A continuous fixture representation (left) describes possible fixturing locations more completely than a discrete fixture (right).

For purposes of analyzing stability, however, the fixture continuum is reduced to an equivalent, finite set of *virtual fixels* with which to balance the external forces. For planar assemblies, each "fixture segment" is reduced to two virtual fixels at the segment's endpoints, as shown in Fig. 4. Any net force and torque producible by the continuous segment can also be produced by these virtual fixels, and vice-versa; this makes virtual fixels useful for checking stability. In the three-dimensional case, the continuous fixture would be composed of two-dimensional "regions" rather than segments, and the virtual fixels would be at the corners of each region's convex hull.

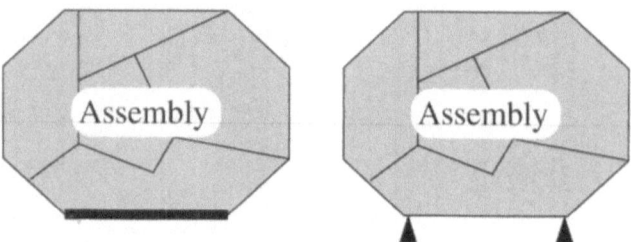

Fig. 4. To check the stability of a planar assembly, each continuous "fixture segment" (left) is reduced to two virtual fixels (right); together, these virtual fixels can produce any net force and torque that the original segment can produce.

This scheme resembles somewhat Wolter and Trinkle's use of two fixels at continuously-variable locations along each edge (in the planar case) [26]; however, the fixture continuum has the advantage of allowing segments to be broken up into arbitrarily many disconnected subsegments, as dictated by the evolving sequencing constraints.

4.3   The Disassembly Tree

In [29], Homem de Mello and Sanderson described a way to represent all possible disassembly sequences using an AND/OR graph. Each node in the graph represents a subassembly, and each AND'ed pair of children for that node represents two subassemblies into which the given subassembly can be decomposed. When there are multiple ways to break apart a subassembly, there are multiple pairs of children OR'd together.

Now, for linear disassembly sequences, such as those we are generating here, every assembly decomposition removes exactly one part. So one child node out of each AND'ed pair of children will represent an extracted part, and will therefore be a leaf node. For simplicity, we can suppress these nodes, reducing the AND/OR graph to a simple OR graph whose missing leaf nodes are implicit.

When fixture information is included in the nodes, however, the graph must become a tree in order to maintain consistency [28]. A partial example of such a tree is shown in Fig. 5. Note that the fixtures in most nodes represent partial solutions, admitting only the disassembly subsequence seen so far (at and above the given node). The final fixture solutions appear at the bottom, in the leaf nodes.

The arcs in the tree link a parent assembly and candidate fixture with each of its possible child subassemblies/fixtures. These arcs represent *actions*, of the form

Delete segments S from the fixture, and remove part P in direction D.

An action is said to be *feasible* if it involves no part collisions and the fixture modification violates no current or previous stability requirements.

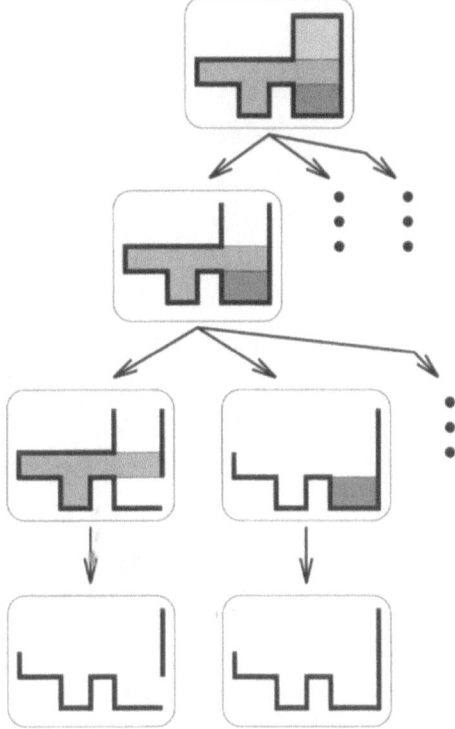

Fig. 5. A partial example of a disassembly tree.

Now, the continuous nature of the fixture implies that there may actually be an infinite number of distinct feasible children for a given node, each with at least a slightly different fixture. (See Fig. 6.) If we were able to somehow represent this continuum of children (and grandchildren, etc.) in our tree structure, and then exhaustively search the entire resulting tree, we would have a complete solution to the co-design problem; that is, given an assembly, we would be guaranteed to find a sequence and fixture for it when such a solution existed.

However, in practice, creating and managing such a tree would be difficult indeed. Equivalence classes of children are difficult to identify because the external forces to which they would respond equivalently are not all known until the sequence has been fully generated. Moreover, the tree would grow at least exponentially in the number of parts. To address the first problem, we shall discretize the continuum of children, by taking the continuum of possible actions and sampling from it a (hopefully) representative set of them. To address the exponential-growth problem, we shall explore only the most promising child which results; possible figures of merit for this are discussed at length below. To be sure, completeness is sacrificed by these measures, but the resulting problem becomes manageable, and by choosing our child nodes very carefully, we can make failures few and far between.

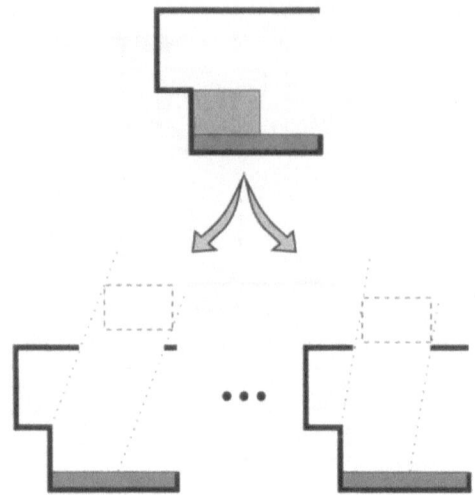

Fig. 6. Sometimes, there may be an infinite number of distinct feasible children, because the fixture is continuous.

4.4   Enumerating Possible Actions

A Complete Approach. In a complete algorithm, the set of all possible actions for an assembly can be found by considering each part in turn and generating its *extended translational freedom (ETF) cone*. An ETF cone is the subset of directions, from the entire sphere of directions, over which a given part can translate to infinity without colliding with another part. For example, in Fig. 7a, the possible motions of P_1 are described by the ETF cone of Fig. 7b. Note that in general, the ETF cone may not be cone-shaped at all; it may be non-convex, empty, degenerate, or composed of several disjoint regions.

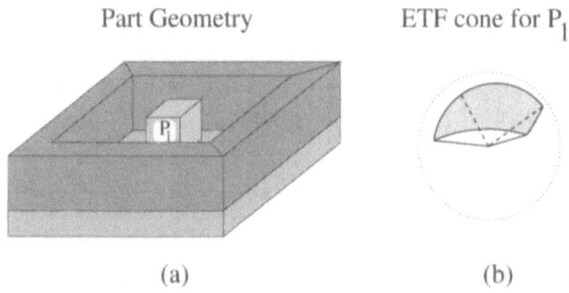

Part Geometry   ETF cone for P_1

(a)   (b)

Fig. 7. An example of an extended translational freedom (ETF) cone.

Not all directions in the ETF cone produce actions which are feasible. For any given direction, a part motion in that direction may require the elimination of crucial portions of the fixture, in terms of the stability either of the current assembly or of some previously-seen superassembly. The set of directions corresponding to feasible actions is a subset of the ETF cone. Conceivably, we could identify this subset by projecting the assembly stability constraints onto the ETF cone. Doing so would yield the (continuous) set of all feasible actions, and thus ensure the completeness of the algorithm.

Atlas' Algorithm. Again, however, due to the difficulty of such an operation, Atlas uses a simplified approach, illustrated in Fig. 8. The cone of directions is first discretized, and feasibility is checked on each of the resulting discrete directions. Furthermore, instead of the extended translational freedom cone, Atlas' method is based on the simpler *local translational freedom (LTF)* cone [13]. An LTF cone is the subset of directions over which a given part can move *infinitesimally* without colliding.[1]

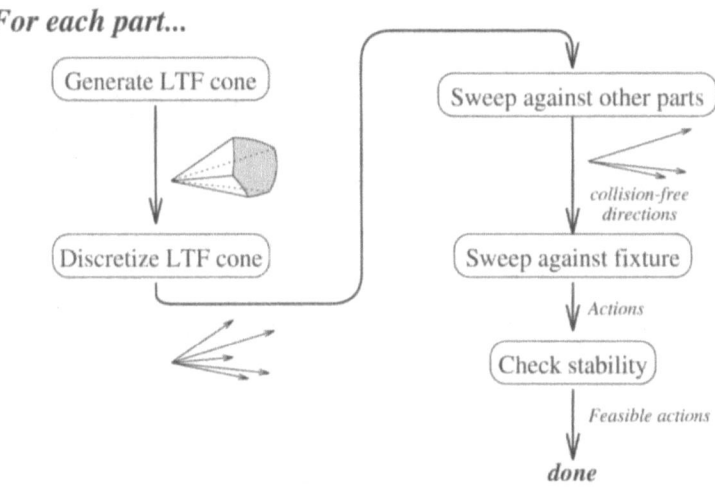

Fig. 8. Atlas' procedure for enumerating feasible actions.

For each part, Atlas generates its LTF cone with respect to all other parts, and from that cone selects a finite set of directions (such as the "corners" of the LTF cone). Atlas then "sweeps", or projects, the part out to infinity in each direction, to verify that it will not collide with more distant parts (see Fig. 9). For those directions that are indeed collision-free, a similar sweeping operation identifies the portions of the fixture which must be deleted. The result is a set of collision-free actions. Finally, for every such action, stability is checked for the current assembly and all previously-seen superassemblies, using linear programs. Each of these linear programs tries to find a combination of forces and torques from the virtual fixels and/or the interpart contacts to balance out the known external forces and torques. The formulas are presented briefly below, in section 5.1; for more information, the interested reader is referred to [26,27].

---

[1] For a more rigorous definition of LTF cones, the reader is referred to [30].

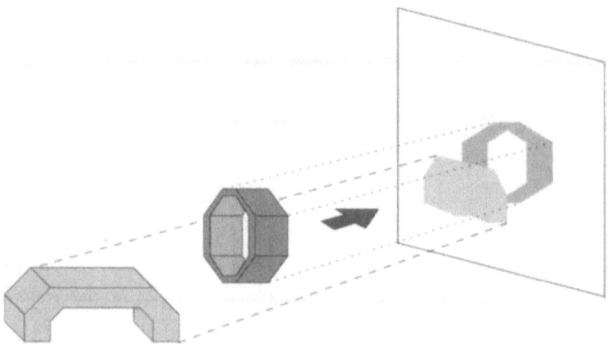

Fig. 9. By projecting, or "sweeping", parts in a proposed direction of motion, we can detect collisions between them. A similar operation detects part/fixture collisions.

One final point is worth mentioning about the enumeration of feasible actions. Sometimes it is possible to remove a part without reducing the fixture at all. In fact, such "fixture-preserving" actions may be preferred, since fixture reductions may reduce the ability of the fixture to stabilize future subassemblies. In order to try to find fixture-preserving actions when they exist, Atlas performs the preceding analysis a second time, this time with the fixture treated as unchangeable, like a big, immovable part. That is, in the second pass, the LTF cone includes constraints from the fixture as well as from other parts, and the sweeping module rejects any fixture-reducing actions entirely. Any feasible actions which emerge from this demanding process must be fixture-preserving, and are included in the list of feasible actions for consideration by the next phase of the co-design process.

4.5   Choosing an Action

Having enumerated a set of feasible actions, Atlas' next step is choosing among them. This is a crucial decision, because exhaustively exploring all possible assembly sequences is not practical. In [1], which considered sequencing but not fixturing, the choice of action was guided by the optimization of an objective function, such as removing a key part as quickly as possible. However, in sequence/fixture co-design, we must first and foremost ensure that our fixture successfully stabilizes the entire assembly sequence. Since this sequence is not fully known until the end, deciding which action will best accomplish that goal is a difficult task indeed.

Nonetheless, several strategies have been found which work well in practice. For example, one approach, used in [28], is to minimize the number of deleted fixels (or deleted virtual fixels) in each step. In the planar case this is equivalent to minimizing the number of deleted fixture segments. The assumption here is that a fixture with more remaining fixels will be more likely to stabilize future, as yet undetermined, subassemblies.

While this method is generally a good one (as evidenced by the results reported in [28]), it can sometimes lead to dead ends in the disassembly process. For example, consider the "tilted blender" example of Fig. 10a. If we delete as few virtual fixels as possible in each step, we eventually arrive at the subassembly of Fig. 10b. No further actions are feasible, primarily because of the need to counteract insertion forces. If we attempt to remove the lid to the left, we must delete the leftmost part of the fixture, which would leave nothing to counterbalance the insertions of the rightmost parts. Conversely, if we attempt to remove the jar to the right, we will be unable to counterbalance the insertion of the small "lid cap" from the left.

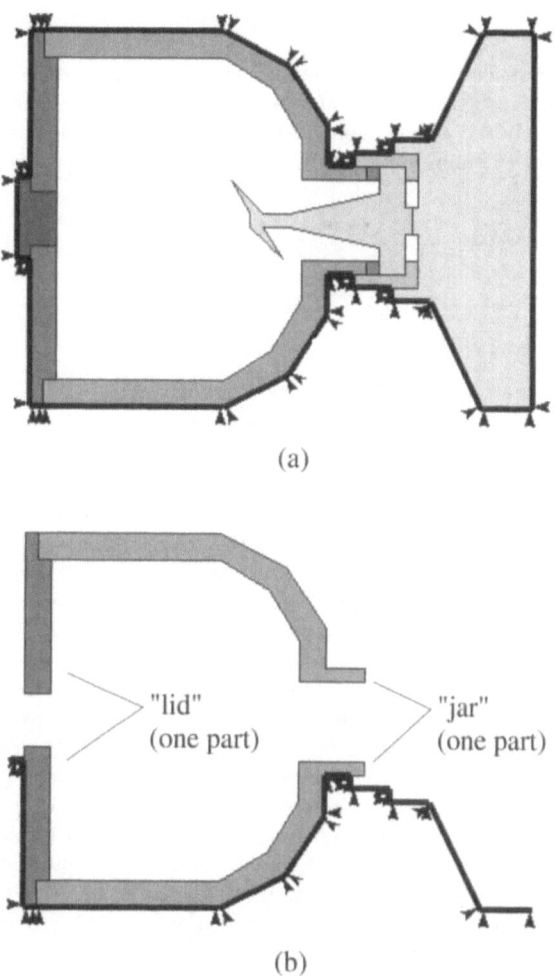

Fig. 10. The "tilted blender" example: (a) After overfixturing, with virtual fixels shown explicitly. (b) The dead-end situation that results from always deleting as few virtual fixels as possible.

We can gain additional insight into this problem, and infer a better heuristic for choosing actions, by considering the notion of *wrench space*.

5   Wrench Space and Convex Hulls

5.1   Theory

A *wrench* is a force and a torque together:

$$\mathbf{w} = \begin{pmatrix} \mathbf{F} \\ \tau \end{pmatrix}$$

Each fixel (or virtual fixel) imparts a wrench on the assembly, as do external forces such as gravity and insertion forces; the net sum of these wrenches must be zero to achieve stability. As noted earlier, none of the fixels can apply negative force. So if $\mathbf{w}_i$ is the unit-magnitude wrench imparted by fixel i, then there must be some $\alpha_i \in \Re$ such that [26,27]:

$$\alpha_0 \mathbf{w}_0 + \alpha_1 \mathbf{w}_1 + \cdots + \alpha_{m-1} \mathbf{w}_{m-1} + \mathbf{W}_{ext} = \mathbf{0}, \qquad \alpha_i \geq 0$$

or,

$$\alpha_0 \mathbf{w}_0 + \alpha_1 \mathbf{w}_1 + \cdots + \alpha_{m-1} \mathbf{w}_{m-1} = -\mathbf{W}_{ext}, \qquad \alpha_i \geq 0 \qquad (1)$$

For a planar assembly with glued parts, $\mathbf{w}_i \in \Re^3$. For a three-dimensional assembly, $\mathbf{w}_i \in \Re^6$. And if the n parts are not glued together, then $\mathbf{w}_i \in \Re^{6n}$, with six force/torque components acting on each part (most of which will be zero). In the unglued case, we must also include wrenches from the interpart contacts.

Atlas checks stability by seeking a solution $\{\alpha_i\}$ to eqn. (1) using a linear program. (For more information on these formulations, the reader is referred to [26,27].)

Graphically, what eqn. (1) implies is illustrated in Fig. 11. For the system to be stable, each negated external wrench $-\mathbf{W}_{ext,j}$ must be expressible as a nonnegative linear combination of the fixel wrenches $\mathbf{w}_i$, which means that each $-\mathbf{W}_{ext,j}$ is on or inside the "cone" of $\mathbf{w}_i$'s. If we were to project the wrenches onto a unit sphere, as Brost and Mason did in [31], the projected $-\mathbf{W}_{ext,j}$'s would have to lie on or inside the projected convex hull of the $\mathbf{w}_i$'s. Alternatively, the *ordinary* convex hull of the $\mathbf{w}_i$'s and the origin must contain $-\epsilon \mathbf{W}_{ext}$ for some suitably small ϵ. Note that in practice, this convex hull may not be cone-shaped; instead, it may take a variety of convex forms, such as a flat disc, or a "polyhedral sphere" completely surrounding the origin.

Over the course of the sequence/fixture generation, we will be deleting fixels (and their wrenches) to make way for part removals, and we will be adding new external wrenches corresponding to the gravity and insertion forces on different subassemblies. Throughout this process, we must ensure that all the external wrenches are contained in the fixel-wrench convex hulls in this manner.

5.2   Application

An example of an actual wrench-space configuration is shown in Fig. 12. This situation corresponds to the "tilted blender" example of Fig. 10, under the glued-parts assumption, and in particular to the dead-end case of Fig. 10b. As can be seen, the disassembly subsequence leading up to this point has produced a number of gravity and insertion-force wrenches which must be contained.

And here we see the difficulty of the current predicament: the external wrenches are so spread apart that maintaining containment of them requires the use of most of

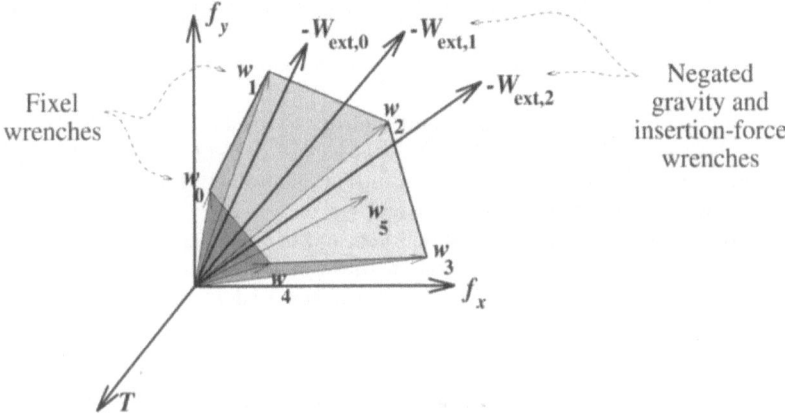

Fig. 11. The negated external wrenches $-\mathbf{W}_{ext,j}$ must be contained in the "cone" of fixel wrenches $\mathbf{w}_i$, in order to achieve stability.

the remaining fixels. As a result, Atlas does not have sufficient flexibility in deleting fixels. This observation suggests a disassembly strategy which keeps these wrenches more "clustered" together, or seeks redundancy in the spacing of the fixel wrenches, or both.

Clustering among the external wrenches can be measured in a variety of ways. If the parts are glued, the simplest might be to compute the average dot product among the external wrenches (after scaling the torques to bring their magnitudes in line with those of the forces). A more sophisticated approach might analyze the convex hull of the external wrenches, since containment of the wrenches on that convex hull implies containment of all externals.

For the case of unglued parts, the dot-product criterion would have to be modified, since any external wrenches ($\in \Re^{6n}$) affecting different parts would necessarily have a

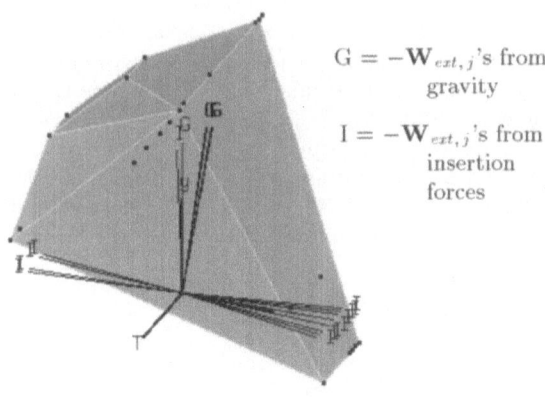

Fig. 12. The wrench-space representation of Fig. 10b.

zero dot product. One modification would be to first project these wrenches onto $\Re^3$ or $\Re^6$ before comparing them; in essence this treats the assembly as a single unit for purposes of comparing external wrenches, despite the assembly's many internal degrees of freedom.

Redundancy in the fixel wrenches is much harder to measure, and identifying useful measures of fixture redundancy remains the subject of current research. One possible approach, however, might be to find the minimum number of fixels whose removal would destabilize the assembly in some way.

Atlas currently selects the action which maximizes clustering, as measured by the dot-product test (using an $\Re^3$ projection in the case of unglued parts). This simple strategy was found to work well on all the examples tested.

5.3   Convex Hull Complexity

In general, a convex hull of m points in d dimensions has complexity $O(m^{\lfloor d/2 \rfloor})$. Therefore, if we assume the parts are glued together and act as a rigid body (so d=constant), the complexity is polynomial in the number of fixels m (but constant in the number of parts). If the n parts are not glued, however, the wrench space has dimension $d = 3n$ or $6n$, and the complexity is exponential in the number of parts. Therefore, for larger assemblies, an explicit convex hull computation is only practical under the glued-parts assumption. This may constrain the development of more sophisticated criteria for selecting actions.

Furthermore, a number of rather subtle complications exist to the convex hull structure described thus far. We now turn our attention to some of them.

5.4   Convex Hull "Shells"

When parts are removed, the fixels which touched them become "inactive"; that is, they are no longer useful in stabilizing the remaining subassemblies, although they are still valid for stabilizing the earlier subassemblies. For example, in Fig. 10b, the fixels on the lower right are inactive. Note that this situation is different from that of *deleted* fixels, which cannot be considered at all.

In wrench-space terms, inactive fixels may be used to contain external wrenches which affect the earlier subassemblies, but not wrenches affecting the later ones. Therefore, different convex hulls are applicable at different stages of the disassembly process. Since these hulls get progressively smaller as the disassembly progresses, we have a series of concentric convex-hull "shells". And the external wrenches do not all correspond to the same shell; our requirement for containment of an external wrench by the convex hull actually means containment by the particular shell corresponding to the subassembly which that wrench affects. This observation further complicates the assessment of a fixture's future usefulness in stabilizing the assembly sequence.

5.5   Disconnected Assemblies

Another complication arises when a subassembly becomes disconnected; that is, when its parts can be split into multiple connected components which do not touch each other. In this case, our assumption that the parts are glued together upon contact no longer ensures that the assembly will behave as a single rigid body (for indeed, components which do not touch cannot be glued together).

In this situation, the effect of a fixel on an assembly is properly described by a wrench of dimension $3c$ (for a planar assembly) or $6c$ (for a three-dimensional assembly), where c is the number of connected components. In wrench space, this is similar to an assembly of c unglued parts, but with two key differences. First, the absence of contacts between the components decouples the stability check into c independent subproblems. Second, the previously-generated subassemblies are still described by simple 3- or 6-dimensional wrenches (if parts are glued), and later subassemblies may be, too. So the wrench space is of non-constant dimensionality.

Fortunately, the dot-product "clustering" test can still be made to work. The $\Re^3$ projections of the $\Re^{3c}$ wrenches correctly describe the effect of each wrench on a particular component. And if the pivot point for the torques remains the same, then these projections are entirely comparable to the $\Re^3$ wrenches from the earlier subassemblies. Therefore, maximizing the dot product between the earlier $\Re^3$ external wrenches and the projections of the new $\Re^{3c}$ external wrenches will realize our goal of containing these wrenches with a relatively small number of fixel wrenches.

6   Experimental Results

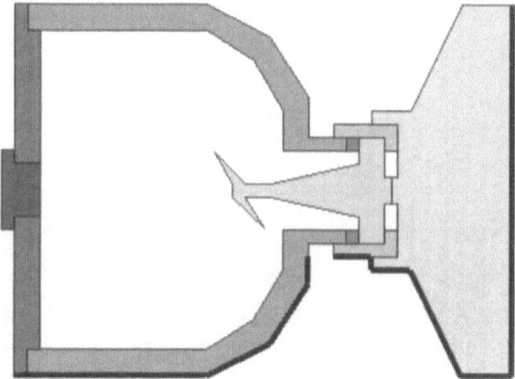

Fig. 13. The "tilted blender" assembly in its final (non-minimized) fixture, following a successful disassembly. The generated assembly sequence inserts the rightmost part downward, and then all other parts from left to right.

Using this dot-product "clustering" strategy, Atlas was able to successfully disassemble the blender example from Fig. 10. The final fixture is shown in Fig. 13. The corresponding assembly sequence inserts the rightmost part downward, and then all other parts from left to right. The final wrench-space configuration is shown in Fig. 14. For the case of glued parts, this solution was generated in 0.07 seconds of CPU time on a DEC alpha workstation. For the case of unglued parts, the same solution was generated in 0.84 CPU seconds. The higher time reflects the much greater complexity of the stability analysis in the unglued case.

In another example, Atlas was run on an 11-part cross-section of an "A' Discriminator", a safety device being developed at Sandia National Laboratories. It generated

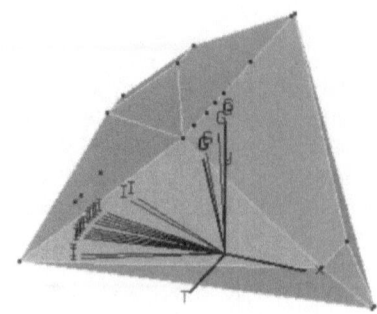

Fig. 14. The wrench-space representation of Fig. 13, showing all the external wrenches from the assembly process.

the fixture of Fig. 15 in 0.32 seconds (glued parts) or 3.00 seconds (unglued parts), along with a sequence that inserts all parts straight down. Again, the same solution was found for both cases. Although this solution seems fairly straightforward, the example is interesting nonetheless because one subassembly is disconnected (consisting specifically of the two largest parts), and therefore the projection approach described in section 5.5 had to be applied.

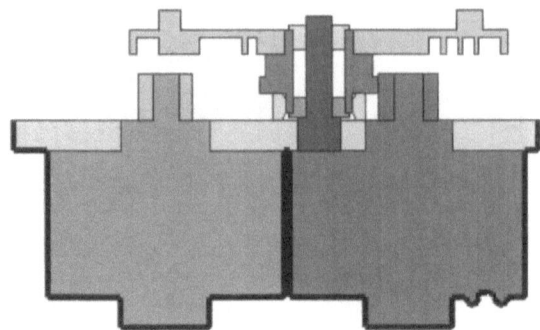

Fig. 15. The A' Discriminator example (courtesy Sandia National Laboratories), in its final fixture.

A third example is shown in Fig. 16. The corresponding sequence inserts the largest part straight down and then all other parts from left to right. This solution was generated in 0.02 CPU seconds (glued parts) or 0.14 CPU seconds (unglued parts).

7   Conclusion

A software system has been presented which can quickly generate an assembly sequence and a fixture for a product, given only a geometric description of that product. Stability against both gravity and insertion forces is guaranteed. To the author's knowledge,

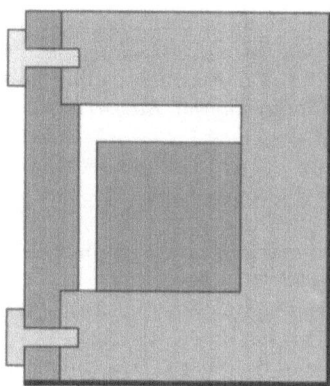

Fig. 16. The "box-lid-cargo" example, in its final fixture.

this system, and its predecessor in [28], represent the first attempt to address the problems of assembly sequencing and assembly fixturing in a unified fashion. The system is viewed as an important first step toward the realization of a CAD tool which can provide immediate feedback to a team of product designers about a product's assemblability.

However, in order to achieve that goal, much work remains to be done. One important subgoal will be a system which can return not just any sequence/fixture solution, but the "best" one according to some criterion. To accomplish this, some means will have to be found to search sequence/fixture trees effectively but non-exhaustively.

Another important direction for future research will be to release many of the assumptions described earlier, such as the restriction to polyhedral assemblies, the use of only one fixture, the linearity of the assembly sequence, and the ideality of the fixels. But it should be noted that Atlas' present application only to planar assemblies is merely a matter of implementation; the underlying theoretical basis for all of Atlas' techniques extends equally well to three-dimensional assemblies.

Finally, experiments with more example assemblies would be useful, in order to verify and/or refine the algorithm presented above.

Acknowledgement

This research was supported by a grant from the Stanford Integrated Manufacturing Association (SIMA), and by NSF/ARPA grant IRI-9306544.

References

1. B. Romney, C. Godard, M. Goldwasser, and G. Ramkumar, "An efficient system for geometric assembly sequence generation and evaluation," in *Proc. ASME International Computers in Engineering Conference*, (Boston), pp. 699–712, 1995.
2. J. Wolter, "On the automatic generation of assembly plans," in *Proc. IEEE International Conference on Robotics and Automation*, vol. 1, (Scottsdale, AZ), pp. 62–68, 1989.

3. R. Hoffman, "Automated assembly planning for B-rep products," in *Proc. IEEE International Conference on Systems Engineering*, (Pittsburgh), pp. 391–394, 1990.

4. T. Woo and D. Dutta, "Automatic disassembly and total ordering in three dimensions," *Journal of Engineering for Industry*, vol. 113, no. 2, pp. 207–213, May 1991.

5. A. Lin and T.-C. Chang, "3D MAPS: Three-dimensional mechanical assembly planning system," *Journal of Manufacturing Systems*, vol. 12, no. 6, pp. 437–456, 1993.

6. B. K. Natarajan, "On planning assemblies," in *Proc. 4th ACM Symp. on Computational Geometry*, pp. 299–308, 1988.

7. J. D. Wolter, *On the Automatic Generation of Plans for Mechanical Assembly*. PhD thesis, Univ. of Michigan, 1988.

8. L. Kavraki, J. Latombe, and R. Wilson, "On the complexity of assembly partitioning," in *Information Processing Letters*, vol. 48, pp. 229–235, 1993.

9. J. Miller and R. Hoffman, "Automatic assembly planning with fasteners," in *Proc. IEEE International Conference on Robotics and Automation*, vol. 1, (Scottsdale, AZ), pp. 69–74, 1989.

10. S. Lee and Y. Shin, "Assembly planning based on geometric reasoning," *Computers and Graphics*, vol. 14, no. 2, pp. 237–250, 1990.

11. A. Subramani and P. Dewhurst, "Automatic generation of product disassembly sequences," *Manufacturing Technology CIRP Annals*, vol. 40, no. 1, pp. 115–118, 1991.

12. R. Wilson and J. Latombe, "Geometric reasoning about mechanical assembly," *Artificial Intelligence*, vol. 71, no. 2, pp. 371–396, 1994.

13. L. Homem de Mello, *Task Sequence Planning for Robotic Assembly*. Ph.D. thesis, Carnegie Mellon University, Pittsburgh, 1989.

14. L. Homem de Mello and A. Sanderson, "A correct and complete algorithm for the generation of mechanical assembly sequences," in *Proc. IEEE International Conference on Robotics and Automation*, vol. 1, (Scottsdale, AZ), pp. 56–61, 1989.

15. A. Sanderson, "Automatic generation of mechanical assembly sequences," in *Geometric Modeling for Product Engineering*, (Amsterdam), pp. 461–482, North-Holland, 1990.

16. H. Asada and A. By, "Kinematic analysis of workpart fixturing for flexible assembly with automatically reconfigurable fixtures," *IEEE Journal of Robotics and Automation*, vol. RA-1, no. 2, pp. 86–94, 1985.

17. J. Bausch and K. Youcef-Toumi, "Kinematic methods for automated fixture reconfiguration planning," in *Proc. International Conference on Robotics and Automation*, vol. 2, (Cincinnati, OH), pp. 1396–1401, 1990.

18. R. Brost and K. Goldberg, "Complete algorithm for synthesizing modular fixtures for polygonal parts," in *Proc. IEEE International Conference on Robotics and Automation*, (San Diego), pp. 535–542, 1994.

19. X. Markenscoff and C. Papadimitriou, "Optimum grip of a polygon," *International Journal of Robotics Research*, vol. 8, no. 2, pp. 17–29, 1989.

20. B. Mishra, J. Schwartz, and M. Sharir, "On the existence and synthesis of multifinger positive grips," *Algorithmica*, vol. 2, no. 4, pp. 541–558, 1987.

21. Y. Xiong, D. Sanger, and D. Kerr, "Geometric modelling of boundless grasps," *Robotica*, vol. 11, pp. 19–26, 1993.

22. C. Ferrari and J. Canny, "Planning optimal grasps," in *Proc. International Conference on Robotics and Automation*, vol. 3, (Nice, France), pp. 2290–2295, 1992.

23. Z. Li and S. Sastry, "Task-oriented optimal grasping by multifingered robot hands," *IEEE Journal of Robotics and Automation*, vol. 4, no. 1, pp. 32–44, 1988.

24. V.-D. Nguyen, "Constructing force-closure grasps," *International Journal of Robotics Research*, vol. 7, no. 3, pp. 3–16, 1988.

25. S. Fahlman, "A planning system for robot construction tasks," *Artificial Intelligence*, vol. 5, pp. 1–49, 1974.

26. J. Wolter and J. Trinkle, "Automatic selection of fixture points for frictionless assemblies," in *Proc. IEEE International Conference on Robotics and Automation*, (San Diego), pp. 528–534, 1994.

27. R. Mattikalli, *Mechanics Based Assembly Planning*. Ph.D. thesis, Carnegie Mellon University, Pittsburgh, 1994.

28. B. Romney, "Issues in the co-design of assembly sequences and fixtures." Manuscript. Available at `http://robotics.stanford.edu/users/assembly`, 1995.

29. L. Homem de Mello and A. Sanderson, "AND/OR graph representation of assembly plans," *IEEE Transactions on Robotics and Automation*, vol. 6, no. 2, pp. 188–199, 1986.

30. R. Wilson, *On Geometric Assembly Planning*. Ph.D. thesis, Stanford University, Stanford, CA, 1992.

31. R. Brost and M. Mason, "Graphical analysis of planar rigid-body dynamics with multiple frictional contacts," in *Robotics Research: The Fifth International Symposium* (H. Miura and S. Arimoto, eds.), pp. 293–300, MIT Press, 1990.

Geometrical and Physical Reasoning for Stable Assembly Sequence Planning

F. Röhrdanz, H. Mosemann and F. M. Wahl

Technical University of Braunschweig, Institute for Robotics and Computer Control,
Hamburger Str. 267,
38114 Braunschweig, Germany

Abstract. We present a planning system generating sequences for the automated assembly of mechanical products by robots. Several physical and geometrical constraints have to be taken into account to compute efficient, robust and powerful assembly plans. Our assembly planning system automatically considers physical and geometrical constraints to generate stable robot assembly sequences. We propose a relational assembly model including a CAD description, the specification of features and symbolic spatial relations between the assembly components. The solid modeling system of a commercial robotics simulation system allows the user to define features of the assembly components and to specify symbolic spatial relationships. We developed an extended cycle finder which reduces the degrees of freedom defined by the symbolic spatial relationships and increases the efficiency of the assembly planning system. We use an optional specification of an arbitrary hierarchy of assemblies to speed up and guide the generation of sequences. Another important constraint is the stability of the generated (sub)assemblies. The presented system is the first assembly planning system which automatically determines the range of all stable orientations of an assembly for assembly plan generation.

1 Introduction

Assembly planning is an interesting area for automation, quality control and flexibility. The goal in robot assembly is to construct a mechanical product consisting of several parts which can be assembled by robots. A high level assembly planning system generates sequence plans specifying the order in which parts are to be assembled to form the desired product and computes the trajectories to bring the parts together. In addition to such sequence plans task plans for actually performing each assembly operation must be generated. The *high level assembly sequence plan* is the basis for such *lower level plans,* and taken together they ensure the efficient and flexible assembly of a mechanical product.

The main problem in assembly planning is to find a *good* sequence plan; the issues to be considered are very complex. An overview of typical constraints in automated assembly planning can be found in [15]. While fulfilling the criteria for *goodness* it is important to search for all valid plans satisfying those criteria. The number of all valid plans can be very large, even for small assemblies ([42]). Therefore, it is difficult to find good search heuristics based on geometrical reasoning only. The high level assembly planning system introduced in this paper uses for example an assembly hierarchy, an extended cycle finder and physical reasoning to simplify the search for an optimal plan. A very important internal physical constraint for the generation of sequence plans is the stability of (sub)assemblies. In this paper we introduce a new approach to integrate an algorithm for stability analysis of assemblies ([33]) into a high level assembly planning system.

Our system has a modular structure and an open architecture. It is a significant extension of the planning system introduced in [11]. The system covers all modern aspects of high level assembly planning and introduces new approaches for the generation of sequence plans like the stability analysis. CAD tools are used to specify assembly components. The description of the assembly components and an assembly specification using symbolic spatial relations are the input to the system.

2   Relation to Other Work

The first assembly planning systems strongly based on user interaction. They queried the user for information about e. g. geometric reasoning and precedence relations. Bourjault ([3]) uses a *subset* rule and its contrapositive the *superset* rule to reduce the number of questions to be asked to the user. De Fazio and Whitney ([6]) drastically reduce the user interaction by requiring each answer to state the situations in which one connection can be established.

One basic idea in most of the work on assembly planning is the *assembly by disassembly strategy*. Here, the assembly sequences are generated by starting with the completed mechanical product and decomposing it by disassembly operations. Much research work has been done to reduce the necessary user interaction. But even today it is very difficult for an assembly planning system to automatically generate assembly sequences in general. Kavraki et. al. ([17], [18]) proved, that the problem of automatically generating assembly sequences is NP-complete in two-dimensional and three-dimensional cases. As a matter of fact, past and present work in this research area make simplifications in order to cope with complexity.

In [44] *disassembly trees* are developed showing which subassemblies must be removed before other subassemblies. The system removes one subassembly at a time via single-step translations. In their work Woo and Dutta ([44]) showed how to find the optimal sequence to access and remove a given subassembly for a restricted class of assemblies. Besides this approach, many other assembly planning systems perform the *assembly by disassembly* strategy using translational motions along some major axes only ([12], [19], [24]). In [13] the restriction, that the translational motions are along major axis is dropped. A trial-and-error procedure guesses possible local motions to extend the analysis to translations and rotations of the participating subassemblies. To separate two polygons by a sequence of translations Pollack et al. ([29]) use computational geometry techniques. In [40] an approach is given to separate two three-dimensional polyhedra by multiple translations.

In [21] and [41] the computation of the directions of possible part removals is suggested by means of contact information between the subassemblies. Homem de Mello and Sanderson ([8]) introduce the *AND/OR*-graph representation of the space of assembly sequences. Goldwasser et al. ([35]) extended this approach by condensing the local analysis into a single compact data structure called the *non-directional blocking graph*. Kaufman et al. ([16]) introduce an assembly planning system which generates the optimized assembly sequence plan from a CAD model of the assembly and translates the plan into robot code for one specific assembly. By means of this code the mechanical product specified in the CAD model is assembled. In [15] Jones and Wilson give a list of constraints that have appeared in the assembly planning literature.

Other researchers focus their attention onto the physical reasoning needed to accomplish single steps in an assembly plan. A number of authors studied motion planning to assemble subassemblies ([22], [40]). Grasp ([28], [31], [36]) and fixture planning ([43]) are also important for a high level assembly planning system. And finally, Palmer ([26]) showed, that in general, deciding whether an assembly is stable, is a NP-complete problem.

3   Assembly Model

We use a relational assembly model similar to [7] to reason about the feasibility of assembly tasks . In contrast to [7] the system performs a powerful geometrical **and** physical reasoning for task sequence planning and plan evaluation with minimal user interaction. The assembly representation allows the planning system to automatically make deductions from the information contained in the representation in a straightforward and efficient manner. In the following, the assembly representation used by the algorithms presented in sections 4 to 6 is described.

3.1   Relational Assembly Model

A relational model $\mathcal{RM}$ of an n-component assembly is a tripel,
$\mathcal{RM} =< C, R, H >$ with:

- $C = \{c_1, c_2, \ldots, c_n\}$, $n \in \mathbf{N}$, is a set of tripels representing the assembly components.
    - $c_i =< B_i, \mathcal{F}_i, CSG_i >, i \in [1, \ldots, n]$
        * $B_i \in \mathbf{R}^{4 \times 4}$ is a homogeneous transformation matrix representing the base frame of the ith component
        * $\mathcal{F}_i = \{< F_{i_1}, f\_type >, \ldots, < F_{i_k}, f\_type >\}$ is a list of tupels describing the feature frames of the ith component. $F_{i_j} \in \mathbf{R}^{4 \times 4}, j \in [1, \ldots, k], k \in \mathbf{N}$, is a homogeneous transformation matrix and f_type $\in \{face, shaft, hole, edge\}$ corresponds to a feature type specification.
        * CSG_i is the CSG representation of the ith component.
- $R = \{r_1, r_2, \ldots, r_{max}\}$ is a set of four tuples representing relations between different components of the assembly, where $max = \binom{n}{2}$.
    - $r_l =< p, q,^P T_q, \mathcal{C}_{pq} >, l \in [1, \ldots, max]$
        * $c_p, c_q \in C, (p \neq q)$
        * $^P T_q \in \mathbf{R}^{4 \times 4}$ is a homogeneous transformation matrix representing the spatial relationship between components c_p and c_q.
        * $\mathcal{C}_{pq}$ is a list of tuples describing a set of symbolic spatial relations and the set of all physical contacts between c_p and c_q:

$$\mathcal{C}_{pq} = \begin{cases} \text{NULL,} & \text{if no relations exist between} \\ & \text{components } c_p \text{ and } c_q \\ < \mathcal{SSR}_{pq}, \mathcal{PC}_{pq} >, & \text{else} \end{cases}$$

- $\mathcal{SSR}_{pq} = \{< F_{p_1}, F_{q_1}, \text{r\_type} >, \ldots, < F_{p_k}, F_{q_k}, \text{r\_type} > \}$, r_type $\in$ {against, fits, aligned, coplanar}, describes a list of symbolic spatial relations between compatible features F_{p_i} and F_{q_i}, $i \in [1, \ldots, k], k \in$ **N**.

- $\mathcal{PC}_{pq}$ = { $<$ c_region$_{pq_1}$, $\mathbf{n}_{pq_1}$, c_type, a_type$>, \ldots$, $<$ c_region$_{pq_k}$, $\mathbf{n}_{pq_k}$, c_type, a_type$>$} describes all physical contacts between c_p and c_q. c_region$_{pq_i}$ lists the vertices of the ith contact region of the physical contact between c_p and c_q, $i \in [1, \ldots . k], k \in$ **N**, $\mathbf{n}_{pq_i} \in \mathbf{R}^3$ denotes the normal vector of the ith contact region pointing from c_p towards c_q, c_type $\in$ {face_face, face_edge, face_vertex, edge_edge, edge_vertex, vertex_vertex} describes the contact type and a_type $\in$ {glue, pressure_fit, screw, no_attach} corresponds to an attachment acting on c_region$_{pq_i}$.

- H is an optional specification of an arbitrary assembly hierarchy corresponding to the following regular expression:

$$
\begin{aligned}
assbly &= assbly\_with\_base \vee assbly\_no\_base; \\
assbly\_no\_base &= \{single\_part \cup assbly \,\}^+; \\
assbly\_with\_base &= base \cup \{single\_part \cup assbly \,\}^*; \\
base &= assbly;
\end{aligned}
\tag{1}
$$

The whole assembly is specified by $assbly$; a component c_i is denoted as $single\_part$. $assbly\_with\_base$ corresponds to an assembly with a special marked basepart ($base$) and $assbly\_no\_base$ denotes an assembly without any specified base part.

We distinguish between the following obligatory and optional attributes:

Obligatory Attributes (CSG_i, F_{i_j}, $\mathcal{SSR}_{pq}$) The assembly components c_i are represented as CSGs using the solid modeling system of the robotics simulation system $IGRIP$ ([38]). For each component feature frames F_{i_j} are defined and associated with the surface primitives using the $IGRIP$ three-tier menu system. A set of symbolic spatial relations $\mathcal{SSR}_{pq}$, each defined between two features of different components c_p and c_q, is used to model the composed assembly (see [1] for more details).

In Figure 1a a simple example consisting of 3 blocks is shown with some feature frames. Figure 1b shows the corresponding feature type specification and in Figure 1c the high level assembly specification is given.

Optional Attributes (H, a_type) One possibility to speed up and to guide the generation of valid assembly sequences (ref. to section 6.2) is the definition of an arbitrary hierarchy of subassemblies. We distinguish between assemblies with and without base part. Consider the assembly depicted in Figure 1a. The definition of $Block12$ (name of $assbly$) as assembly without basepart consisting of $Block1$ and $Block2$ and the definition of assembly $Block123$ with basepart $Block12$ and the component $Block3$ is specified as:

GROUP $Block12$ $Block1$ $Block2$ **END**
GROUP $Block123$ **BASE** $Block12$ $Block3$ **END**

In Figure 2a this hierarchy is illustrated corresponding to the regular expression given by equations (1). Taking into account the top level hierarchy the assembly is

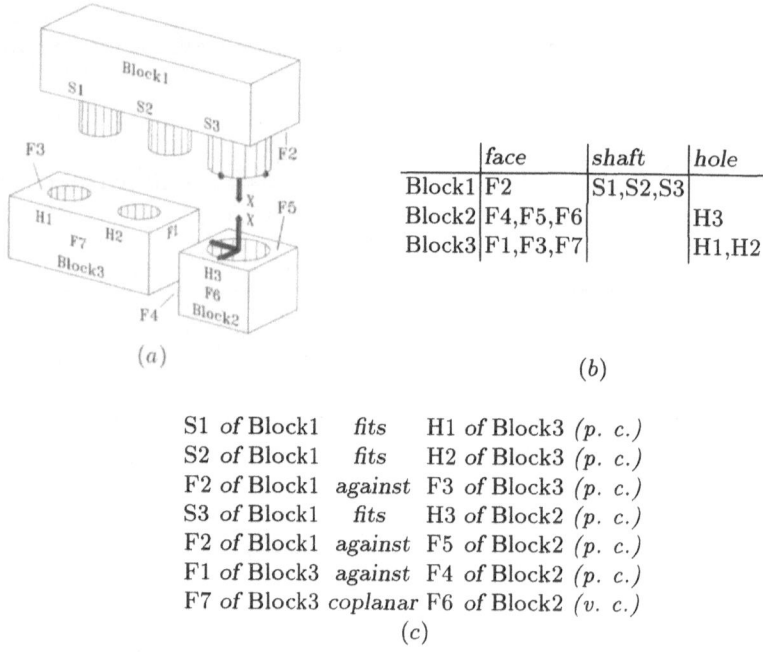

	face	shaft	hole
Block1	F2	S1,S2,S3	
Block2	F4,F5,F6		H3
Block3	F1,F3,F7		H1,H2

(a)

(b)

S1 *of* Block1   *fits*   H1 *of* Block3 *(p. c.)*
S2 *of* Block1   *fits*   H2 *of* Block3 *(p. c.)*
F2 *of* Block1 *against* F3 *of* Block3 *(p. c.)*
S3 *of* Block1   *fits*   H3 *of* Block2 *(p. c.)*
F2 *of* Block1 *against* F5 *of* Block2 *(p. c.)*
F1 *of* Block3 *against* F4 *of* Block2 *(p. c.)*
F7 *of* Block3 *coplanar* F6 *of* Block2 *(v. c.)*

(c)

Fig. 1. (a) Feature frames H3, F5 (coincident with H3) and S3 of an assembly, (b) feature type specification and (c) high-level assembly description. Note that some relations correspond to physical contacts *(p. c.)* and some relations describe virtual contacts *(v. c.)*. The system automatically determines these corresponding contact attributes.

managed as an assembly consisting of 2 objects ({*Block12*, *Block3*}). The removal of *Block3* splits the top level hierarchy resulting in the next hierarchy level for the subassembly {*Block1*, *Block2*}.

Attachments may act on physical contacts and eliminate all degrees of freedom for relative motion. If the user specifies a glue, pressure fit or screw attachment (a_type ∈ {*glue, pressure_fit, screw* }) the stability module of the planning system merges the attached components to a new compound object during the stability analysis (ref. to section 6.1). If no attachment attribute is specified the attachment attribute *no_attach* (a_type = *NO_ATTACH*) is used by default.

4   Extended Cycle Finder

The key idea of the cycle finder [30] is the reduction of the degrees of freedom defined by symbolic relationships SSR_{ik} by searching cycles consisting of two appropriate relationships $< F_{i_j}, F_{k_l}, r\_type >$, $< F_{i_n}, F_{k_m}, r\_type >$ between different components c_i, c_k $(i \neq k)$ (arbitrary features of these objects) and combine these relationships to a new one with equal or less degrees of freedom. The cycle finder runs until no further reduction can be made.

A sequence of reduction steps of the cycle finder for the assembly shown in Figure 1a is depicted in Figure 3. From Figure 3a to 3b the two *fits* relations between Block1 and

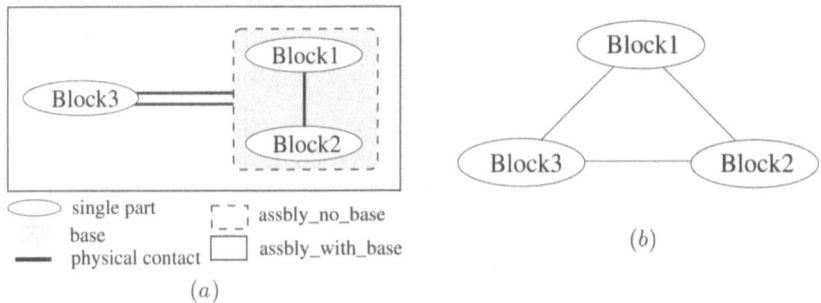

Fig. 2. (a) Optional assembly hierarchy and (b) assembly graph of the assembly depicted in Figure 2a.

Block3 in the depicted *relation graph* are reduced to a *lin* relation. *lin* is a relationship corresponding to one translational degree of freedom along one axis. In general, a *fits-fits* pair can be reduced to *fits* (if the two x-axes of the two *lin* relationships are collinear), to *lin* (the two x-axes are parallel, but not collinear) or to *fix* (the two x-axes aren't parallel). The system automatically chooses the correct case. In addition, the *coplanar-against* pair between Block2 and Block3 is substituted by *lin*. During the next step from 3b to 3c the *lin-against* pair is reduced to *fix*. The cycle finder published in [30] stops at this step. In contrast to this, our *extended cycle finder* has an additional kind of union feature. In our system all objects sharing a *fix* relationship at an arbitrary

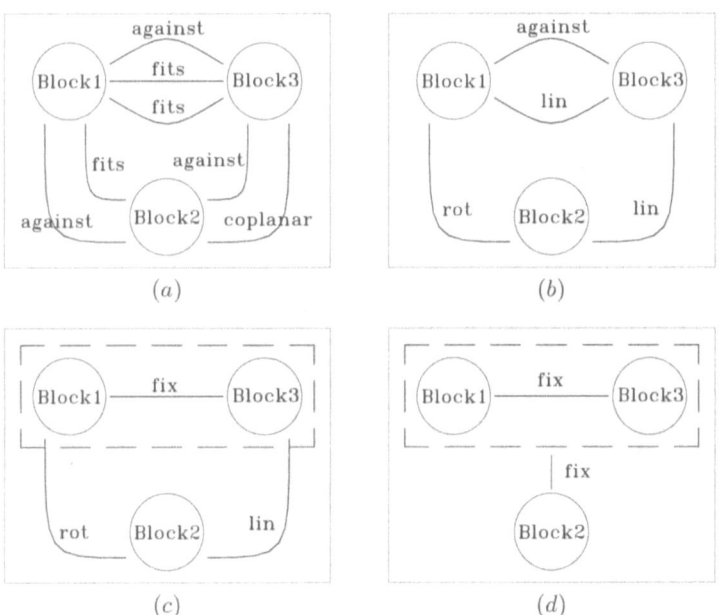

Fig. 3. Sequence of reduction steps by the cycle finder for the assembly specification depicted in Figure 2a.

reduction step are united to a new *compound object* (Figure 3c). Subsequently the reduction process in our example finds the *rot-lin* pair and reduces it to *fix*. Thus, the whole assembly is fixed and all homogeneous transformations automatically can be generated from the symbolic spatial relationships.

If the cycle finder can't reduce all relations between the assembly components to *fix* relations, two reasons are possible. First, the modeled assembly has some degrees of freedom in reality, like a shaft inside a hole. Second, there are cycles in the graph consisting of more then two nodes and two relationships; these cycles can't be further reduced by our proposed compound object technique. To increase the efficiency of the extended cycle finder, relations are automatically included by a *transitive rule*, if an assembly can't be reduced to *fix* using the above mentioned methods. The transitive rule inserts a new relation of type $\mathcal{T}$ between the features F_{i_j} and F_{k_l}, if two relations of type $\mathcal{T}$ exist between two feature pairs F_{i_j}, F_{m_n} and F_{m_n}, F_{k_l} with frames having the same orientation.

5   Contact Model

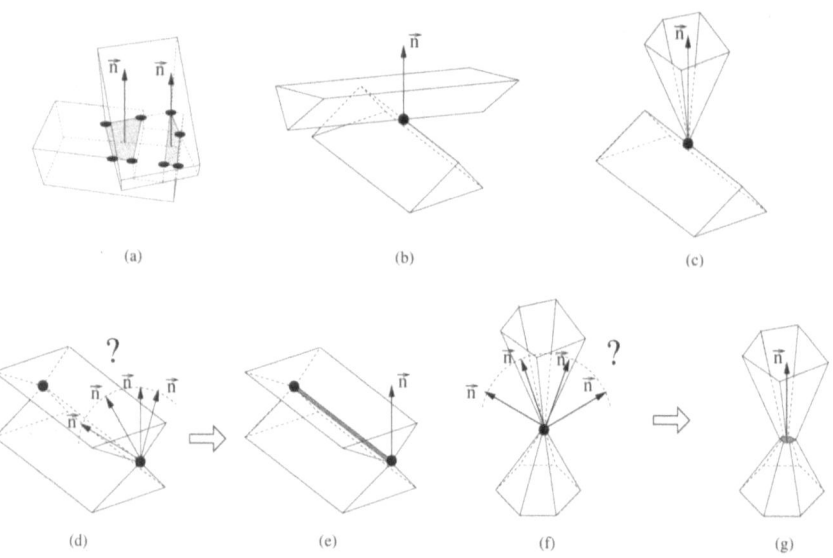

Fig. 4. Different types of contacts between polyhedral objects: (a) polygonal surface contact, (b) nonparallel edge-edge contact, (c) edge-vertex contact, (d) degenerate parallel edge-edge contact and (e) corresponding planar approximation, (f) degenerate vertex-vertex contact and (g) corresponding planar approximation.

Before planning, all physical contacts $\mathcal{PC}_{ij}$ $(i, j \in [1, \ldots, n], i \neq j)$ are determined. Input models include polyhedral, spherical and cylindrical contacts. At each contact point a surface normal is calculated. Figure 4 shows the normal vectors of some polyhedral contacts. For face-face (Figure 4a), face-edge and face-vertex contacts, the normal

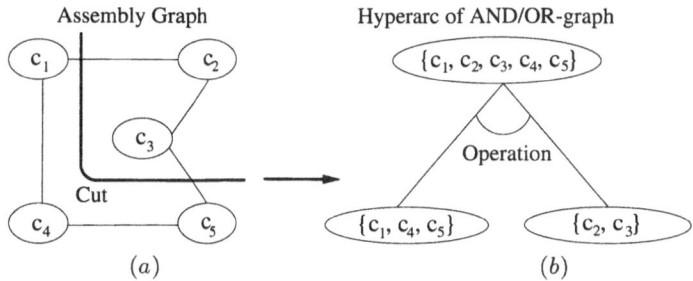

Fig. 5. Valid cut in the assembly graph (a) and the resulting hyperarc h=({c_1, c_2, c_3, c_4, c_5}, {c_1, c_4, c_5}, {c_2, c_3}) in the AND/OR-graph (b).

at any point of contact is well-defined. Figure 4b shows two nonparallel edges being in contact at one point. In this case the normal vector is also well-defined and is perpendicular to both edges. For the edge-vertex contact depicted in Figure 4c, a normal vector perpendicular to the corresponding edge lying inside the vertex cone is chosen. Figures 4d and 4e show a degenerate vertex-vertex resp. edge-edge contact with ill-defined contact normals. These contact situations make the stability problem NP-hard (ref. to section 6.1). Degenerate polyhedral contacts are modelled by planar approximations as indicated in Figures 4e and 4g.

In the following, any single action of merging components or subassemblies, or of moving components or subassemblies, is denoted as an *assembly operation*, and any valid assembly is identified by its corresponding set of components C.

6   Generation of Assembly Sequences

The generation of valid assembly sequences bases on the well-known assembly by disassembly philosophy[1] combined with a cut-set method (similar to [8], chapter 7). Our cut-set method uses the *assembly graph*, a simple undirected graph $G = (N, E)$ with $N = C$ representing the set of nodes and E representing the set of edges. An edge $e = (c_i, c_j)$ exists between a pair of components if $\mathcal{PC}_{ij} \neq \emptyset$. Figure 2b shows the assembly graph G for the assembly depicted in Figure 1a. An example of a cut in an arbitrary assembly graph is shown in Figure 5a. If a cut corresponds to a *feasible assembly operation* it is stored as a hyperarc of an AND/OR-graph which constitutes a compact representation of all feasible assembly sequences. Furthermore, time dependencies and independencies between assembly tasks can be represented (see e. g. [14] for more details about the representation of assembly sequences using AND/OR-graphs). Figure 5b shows the corresponding hyperarc of the valid cut depicted in Figure 5a.

A cut (C', C'') in G, $C' \cup C'' = C, C' \cap C'' = \emptyset$ corresponds to a feasible assembly operation if *local* and *global* assembly constraint are satisfied. Local constraints take into account the corresponding assembly operation; global constraints are applied during the whole assembly process (for a survey of constraints in automated assembly planning see [15]).

---

[1] As we assume, that there are no internal forces inside the assembly, the reverse of a feasible disassembly operation will be a feasible assembly operation.

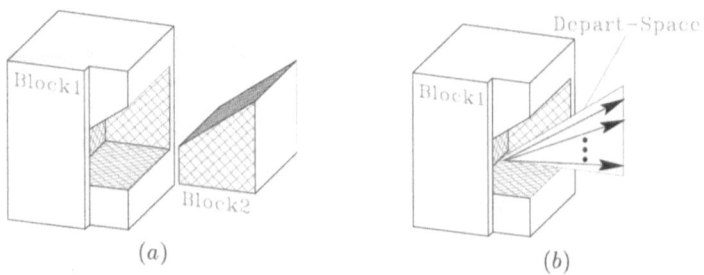

Fig. 6. Example for the depart space of type *sector_of_a_plane*.

6.1 Local Constraints

Connectivity This is a very common constraint being implicit in cut-set methods. It means, that in each step of an assembly sequence only two subassemblies are joined. This constraint results in a two-handed assembly sequence plan. The subassembly held and moved by the gripper is denoted as the *active subassembly*; the *passive subassembly* is placed on some work table or held by some fixture.

Geometric Feasibility Each sequence must be geometrically valid, i. e. none of the operations results in an interpenetration of components. The geometric feasibility of an operation is tested in two steps :

1. Step: The shape of the *local depart space* between C' and C'' for translational movements using the edges of G with the automatically generated transformations $^i T_j$ (section 4) is computed. For many types of contacts, only very few feasible motions between the components exist. For example, a pin in a hole (*fits*-relation) can translate along and rotate around the axis of the hole. Whenever a subassembly comprises such a constraining contact, the analysis of the local depart motion is conducted by checking the compatibility of the most restrictive contact with all other contacts. Similar to [7] we use the following types of local depart spaces : *no_depart_space, half_line, line, halfplane, sector_of_a_plane, plane, halfspace, infinite_wedge, polyhedral_cone, space*.

In Figures 6 and 7 examples for the depart spaces of type *sector_of_a_plane* and *polyhedral_cone* are shown. In both figures the contact faces which should be established or separated respectively are shown hatched. For the depart space of type *sector_of_a_plane* any vector indicated in Figure 6b is a valid local depart vector. Only the direction of this depart vector is of importance; its absolute position is arbitrary. Consequently, the depart space of Figure 6b can be translated to any other position. For the depart space of type *polyhedral_cone* all valid depart vectors are originating at the tip of the cone (Figure 7b) and completely are lying in the volume defined by the cone.

In order to decide whether a set of planar contacts does not completely constrain a subassembly, the corresponding depart spaces iteratively are intersected. If the resulting depart space is of type *no_depart_space*, then no local depart motion is possible and the cut into C' and C'' is invalid. Otherwise the shape and the position of the depart space is mathematically known.

2. Step: The *global feasibility* of depart motions is computed by a heuristically chosen set of directions lying in the local depart space, sweeping out one subassembly

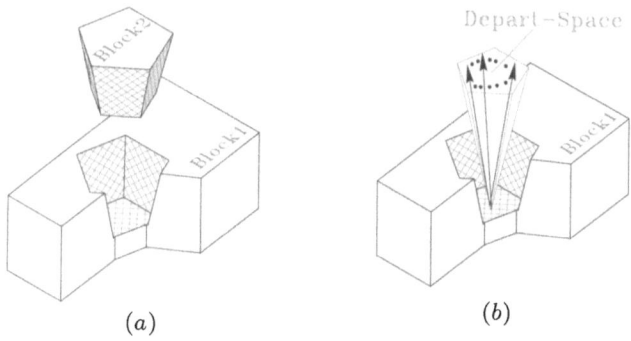

Fig. 7. Example for the depart space of type *polyhedral_cone.*

against the rest of the assembly. First, the optimal direction is chosen. Optimal in our case means, that all physical contacts are removed simultaneously, e.g. for the polyhedral cone all vectors inside the cone. In case of intersection, several valid local depart vectors lying inside the depart space are tested in a similar way. If no global depart vector can be computed, the user is asked for one, which automatically is tested by the system in terms of feasibility.

The mentioned connectivity and geometric feasibility criteria are corresponding to practical constraints often imposed on assembly sequences by manufacturing processes. Application of these constraints results in a plan in which the active subassembly is inserted immediately into its final position relative to the passive subassembly. Thus, a n-component assembly is executed in exactly $(n-1)$ operations[2]. I. e. , complicated motions and assembly operations making the execution of the assembly more difficult are avoided.

Assembly Stability At least one stable orientation must exist for the generated subassemblies. A (sub)assembly is stable if the rigid components are in static equilibrium under the influence of external and internal forces. External forces arise from gravitation and internal forces from the mutual contact of the objects.

For the determination of a stable orientation we take into account static friction and compute the net forces $\mathbf{F}_j = (f_{n_j}, f_{t_{x_j}}, f_{t_{y_j}})^T$ and net torques $\boldsymbol{\tau}_j = (\tau_{x_j}, \tau_{y_j}, \tau_{z_j})^T$ acting on the jth component of the assembly (see [33] for more details about the computation of the net forces and net torques). We assume that n components contact at finite many points, which are indexed from 1 to m. If we define the vectors $\mathbf{r}$, $\mathbf{e}$ $\in \mathbf{R}^{6n}$ and $\mathbf{f} \in \mathbf{R}^{3m}$ as the collections

---

[2] Such a plan is denoted as a *monotone assembly plan.*

$$\mathbf{r} = \begin{pmatrix} \mathbf{F_1} \\ \boldsymbol{\tau_1} \\ \vdots \\ \mathbf{F_n} \\ \boldsymbol{\tau_n} \end{pmatrix} \quad \text{and } \mathbf{e} = \begin{pmatrix} M_1\mathbf{g} \\ 0 \\ \vdots \\ M_n\mathbf{g} \\ 0 \end{pmatrix} \quad \text{and } \mathbf{f} = \begin{pmatrix} f_{n_1} \\ f_{t_{x_1}} \\ f_{t_{y_1}} \\ \vdots \\ f_{n_m} \\ f_{t_{x_m}} \\ f_{t_{y_m}} \end{pmatrix} \quad (2)$$

we can write $\mathbf{r} = \mathbf{Af} + \mathbf{e}$ where $\mathbf{A}$ is a $6n \times 3m$ matrix whose coefficients are given by the contact geometry of the assembly, M_i denotes the mass of the ith component, $\mathbf{g} \in \mathbf{R}^3$ is the gravity vector, f_{n_i} describes the force components acting normal to the contact surface of the ith contact point and $f_{t_{x_i}}, f_{t_{y_i}}$ describe the force components lying in the contact plane of the ith contact point. An assembly is said to be *potentially stable* if there exist contact forces such that the net force and net torque on every body is zero. If we define the vector $\mathbf{f_N} \in \mathbf{R}^m$ as the collection $\mathbf{f_N} = (f_{n_1}, f_{n_2}, \dots, f_{n_m})^T$ a given gravity vector $\mathbf{g}$ induces potential stability if there exists $\mathbf{f}$ such that[3]

$$\mathbf{r} = \mathbf{Af} + \mathbf{e} = \mathbf{0} \quad \text{and} \quad \mathbf{f_N} \geq \mathbf{0} \quad \text{and}$$
$$\mathbf{C}\left(f_{t_{x_i}}, f_{t_{y_i}}, \mu\right)^T \leq \mathbf{0}, \quad i \in [1, \dots, m] \quad (3)$$

Whether such a vector $\mathbf{f}$ exists can be decided with *linear programming* (see e. g. [25]). We introduce the components of the gravity vector as variables in the linear program (3) and search among all possible directions of the vector $\mathbf{g}$. For that purpose the constraint $\|\mathbf{g}\|_\infty = 1$ is applied (for a vector $\mathbf{g} = (g_0, g_1, g_2)^T \in \mathbf{R}^3$, $\|\mathbf{g}\|_\infty = max_i \mid g_i \mid$). This constraint forms a unit cube around the origin. The following linear program describes the search for a stable orientation in the jth facet , $j \in [0, \dots, 5]$, of the unit cube[4]:

Minimize:
$$z = \sum_{i=1}^m (f_{t_{x_i}} + f_{t_{y_i}}) \quad (4)$$

subject to:
$$\mathbf{Af} + \mathbf{e} - \mathbf{r} = \mathbf{0}$$
$$\mathbf{f_N} \geq \mathbf{0}$$
$$\mathbf{f_N} \leq \mathbf{f}_{n,max}$$
$$\mathbf{C}(f_{t_{x_i}}, f_{t_{y_i}}, \mu)^T \leq \mathbf{0}, \quad i \in [1, \dots, m]$$
$$g_{(j+1)mod3} \leq 1, \quad g_{(j+1)mod3} \geq -1$$
$$g_{(j+2)mod3} \leq 1, \quad g_{(j+2)mod3} \geq -1$$
$$g_{jmod3} = (-1)^j$$

---

[3] The $l \times 3$ matrix $\mathbf{C}$ approximates the friction cones at the i contact points by $l = 2^n$ facets, $n \geq 2$, and μ describes the coefficient of static friction. For further details see [33].

[4] The choice of the objective function z results in an orientation $\mathbf{g}$ with minimal emerging tangential forces. Such an orientation is classified as the most frictional stable orientation of an assembly.

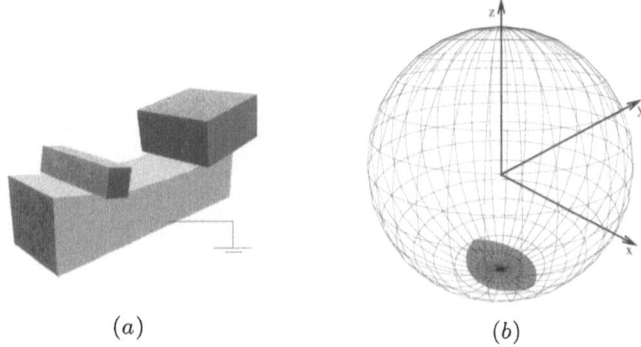

(a) (b)

Fig. 8. (a) Example assembly with a chamfered component of different slopes. The chamfered component is fixed by a supporting surface or a gripper, as indicated by the ground symbol. For this assembly no frictionless stable orientation is found; considering friction results in the frictional stable range shown shaded in Figure (b).

We search in the $(j+1)$th facet for a possible frictional stable orientation if no stable orientation in any previous facet has been found.

Simply using potential stability as the criterion for computing the set of stable orientations results in solutions which are not guaranteed to be stable in reality. In contrast to [23] we restrict the set of physically incorrect solutions introducing the constraint $\mathbf{f}_N \leq \mathbf{f}_{n,max} = (\sum_{i=1}^{m} M_i)\mathbf{g}$ in the linear program (4) as an upper bound for the magnitudes of possible normal force components.

For each AND/OR-graph node the whole *set of frictional stable orientations* of the corresponding (sub)assembly is taken into account: Each of the solution sets P_j, $j \in [0, \dots, 5]$, to the problem (4) is considered. The vertices of the solution sets P_j are found by the algorithm for vertex enumeration of polyhedra proposed in [2] using a dictionary enumeration algorithm (see [33] for more details). Having found all vertices of the solution sets, the g_x, g_y, g_z components of the gravity vector are transformed to the unit sphere to mark the stable region on the sphere. Figure 8 shows an example of an assembly with the corresponding set of frictional stable orientations.

6.2   Global Constraints

Assembly Hierarchy To avoid blind recursion of the cut-set generation in the assembly graph, a *guided recursive search* of connected subgraphs is implemented, which automatically generates the hyperarcs of the AND/OR-graph.

The key ideas of the guided search are :

- A valid cut-set of an assembly C into two connected subgraphs (C', C'') with $C' \cap C'' = \emptyset$ and $C' \cup C'' = C$ is modified by adding one $c'' \in C''$ to C' ($C' = C' \cup \{c''\}$, $C'' = C'' - \{c''\}$), if the two new subgraphs remain connected and additional constraints are satisfied (s. below). Then the *calculate_cut* procedure is started with (C', C''), $(\emptyset, C')$ and $(\emptyset, C'')$. The last two calls are only executed, if these cuts aren't already known! At the very beginning *calculate_cut* is started with $(\emptyset, C)$.

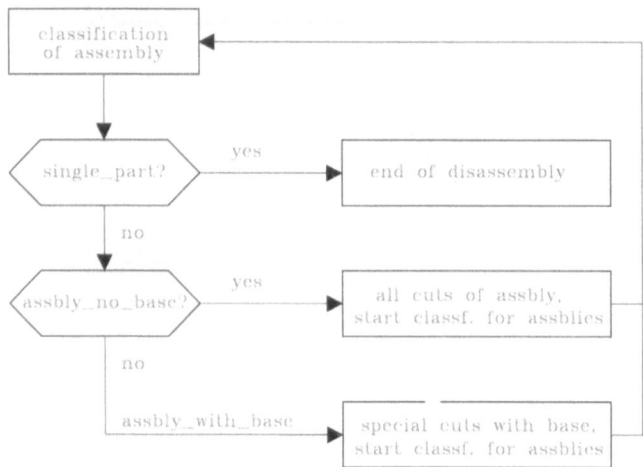

Fig. 9. Disassembling using hierarchies of subassemblies

- Each pair (C', C'') is computed by *calculate_cut* only, if $|C'| \leq |C''|$

The optional declaration of a hierarchy of assemblies (equations (1)) influences the *calculate_cut* procedure. As shown in Figure 9, the *calculate_cut* procedure sees only the top level of the subassembly hierarchy inside the actual assembly. Thus, at this decomposition level each subassembly is treated as a single part and is handled as complete union. If *calculate_cut* is called with $(\emptyset, C)$ the top level of the assembly hierarchy is splitted and the next level of the hierarchy is seen, and so on. On each disassembly level, called with $(\emptyset, C)$, we distinguish three types of assemblies (see regular expression given by equations (1)):

- **single part** $=$ *single_part*: End of recursion because single parts can't be decomposed further.
- **assembly without base part** $=$ *assbly_no_base*: Inside the assembly all assemblies (*single_part* or *assbly*) are treated equally. Consequently all cut combinations are computed.
- **assembly with base part** $=$ *assbly_with_base*: Inside the assembly the specially marked base part ($=$ *base*) is the starting part for the assembly process of this level and all other assemblies (*single_part* or *assbly*) are iteratively added to this *virtual base*. The *virtual base* is permanently updated and corresponds to the union of the *base* with all parts already iteratively added.

The generation of a special cut of C into (C', C'') will be only blocked, if the previous mentioned local constraints are violated.

7 Results

The planning system generates an AND/OR-graph using the cut-set method with the mentioned constraints representing all possible assembly sequences. Due to the large number of feasible assembly sequences it is desirable to select a few best sequences

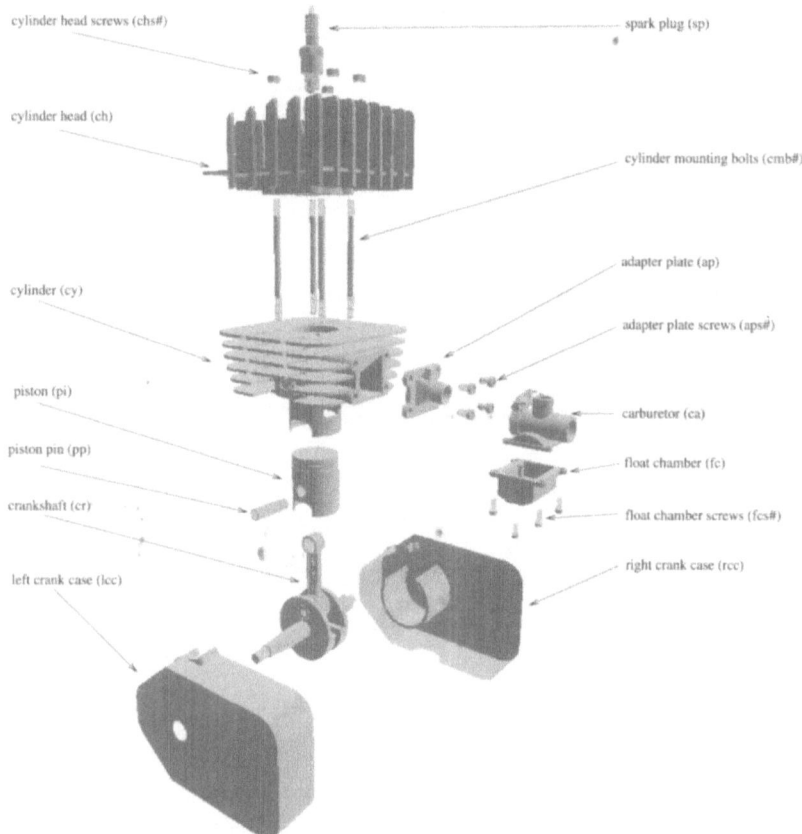

Fig. 10. Exploded view of a Yamaha RD80 motorcycle engine.

or the best assembly sequence. However, this has been hampered by the difficulty of selecting proper performance criteria ([14], [20], [37], [41]) and relating them directly to assembly cost. The planning system assigns weights to the nodes and hyperarcs of the AND/OR-graph to select an assembly plan minimizing the costs from the initial nodes up to the goal nodes (see [32] for more details about the AND/OR-graph evaluation).

To illustrate the efficiency of the planning system we describe an assembly planning example of a Yamaha RD80 motorcycle engine. Figure 10 depicts an exploded view of the assembly consisting of 27 components. For ease of notation in the following, the different components will be denoted by the corrresponding abreviations as indicated in the braces in Figure 10. The following hierarchy is specified by the user:

GROUP *ca_group* **BASE** *ca ap fc fcs1 fcs2 fcs3 fcs4* **END**
GROUP *cr_group* **BASE** *rcc lcc pi pp cr* **END**

The cut-set generation is influenced by this specification (section 6.2) disassembling the *ca_group* by removing the four adapter plate screws. The best (dis)assembly

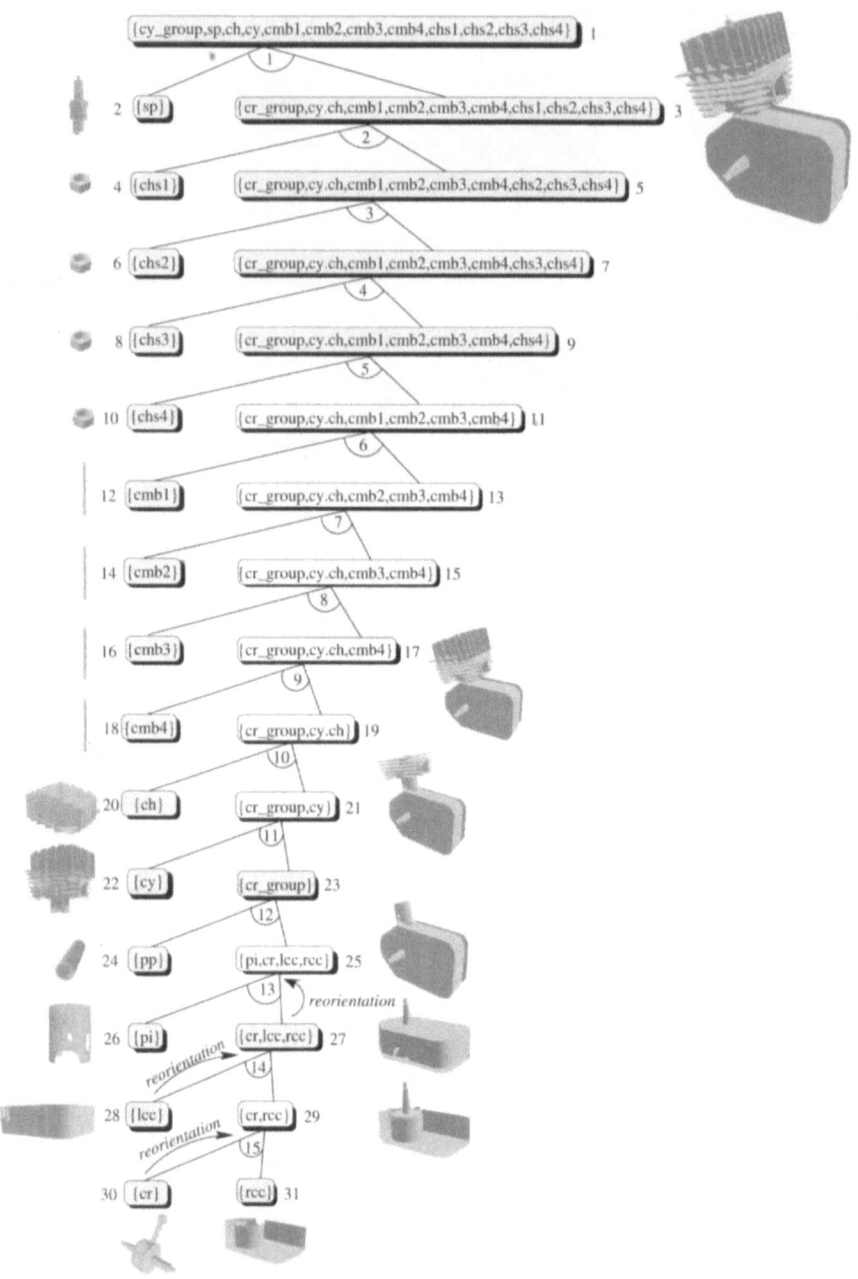

Fig. 11. AND/OR-graph of the best assembly sequence for the {cr_group, ch, cy, sp, cmb1, cmb2, cmb3, cmb4, chs1, chs2, chs3, chs4}-assembly. Shaded nodes indicate assemblies being stable under any orientation. The initial orientations of the assembly components are specified by the user leading to the indicated reorientations during plan execution.

sequence for the remainder of the assembly ($\{cr\_group, ch, cy, sp, cmb1, cmb2, cmb3, cmb4, chs1, chs2, chs3, chs4\}$) is depicted in Figure 11.

For each hyperarc $h = (C' \cup C'', C', C'')$ the set of stable orientations for assembling the subassemblies C' and C'' is calculated by intersecting the corresponding sets of frictional stable orientations. If there exists a non-empty intersection the planning system chooses an assembly orientation corresponding to the center of the intersection. The user can affect the calculation of the assembly orientations by specifying orientations of components/subassemblies which may be located in part feeders or fixtures or by choosing a fixed orientation for the goal assembly. The former results in a bottom-up strategy for the calculation of successive assembly orientations, while the latter results in a top-down procedure. For this planning example the user has specified the initial orientations of the assembly components as indicated in Figure 11. Therefore, three reorientations must be performed (hyperarcs 13, 14, 15) as indicated in the Figure by the arcs.

8   Conclusions and Future Work

In this paper we presented a sophisticated assembly planning system capable to generate stable robot assembly sequences based on geometrical and physical reasoning. The system takes the CAD description of the assembly components and a user-friendly high level assembly specification as input. A relational assembly model allows the planning system to automatically make deductions from the information contained in the representation in a straightforward and efficient manner. All valid assembly sequences are generated and evaluated with minimal user interaction.

In the future we will improve and speed up the geometric reasoning algorithms making use of the graphics hardware of the applied Silicon Graphics Indigo2 with Extreme Graphics. We will use, e. g. ,the z-buffer to check collision free movements of global depart motions. Furthermore, we plan the animation of workcells performing the assembly with robots, tooling and fixturing. Therefore, external constraints must be taken into account, e. g. fixture requirements, tool requirements, tool accessibility. Grasp planning will also be crucial to the further development of a practical assembly planner. We plan the integration of the grasp planning system [31] applied in [34]. Moreover, a fixture planning synthesis algorithm will be integrated taking into account the complete solution set (forces and orientations) of the range of frictional stable orientations of an assembly. It is planned that components with maximal emerging tangential forces will be fixed. Another important point is the introduction of clearances to compensate for tolerances and tolerance propagation as the generated plans are to be executed by robots. Tolerance representation is a well studied problem in the field of geometric modeling as well as robotics. However, it has proved itself to be a very tough problem because no single representational method has been found that can be used to represent a part that is "within tolerances" as opposed to one that is not (see e. g. [4], [39]). Furthermore, we will use the output of the planning system to automatically generate complex sensor guidance information and the task structure of the necessary assembly operations. These task structures and sensor guidance information will be implemented and used in our parallel robot programming environment [27].

References

1. A. P. Ambler and R. J. Popplestone. Inferring the Positions of Bodies from Specified Spatial Relationships. *Artificial Intelligence*, 6:157 – 174, 1975.
2. D. Avis and K. Fukuda. A pivoting algorithm for convex hulls and vertex enumeration of arrangements and polyhedra. *Discrete and Computational Geometry*, 8:295–313, 1992.
3. A. Bourjault. Methodology of assembly automation: A new approach. pages 37 – 45. International Society for Productivity Enhancement, Springer-Verlag, July 1987.
4. M. Boyer and N. F. Stewart. Imperfect tolerancing on manifold objects: a metric approach. *International Journal of Robotics Research*, 11(5):482–490, 1992.
5. CPLEX Optimization, Inc. *Using the CPLEX Callable Library*, 1994.
6. T. L. De Fazio and D. E. Whitney. Simplified Generation of All Mechanical Assembly Sequences. *IEEE Journal of Robotics and Automation*, RA-3(6):640 – 658, December 1987.
7. L. Homem de Mello. *Task Sequence Planning for Robotic Assembly*. PhD thesis, Carnegie Mellon University, May 1989.
8. L. S. Homem de Mello and S. Lee, editors. *Computer-Aided Mechanical Assembly Planning*. Kluwer Academic Publishers, 1991.
9. H. Goldstein. *Classical Mechanics*. Addison-Wesley, Reading, Massachusets, 1983.
10. R. Gould. *Graph theory*. Benjamin/Cummings Publishing, 1988.
11. R. Gutsche, F. Röhrdanz, and F. M. Wahl. Assembly Planning Using Symbolic Spatial Relationships. In W. Strasser and F. M. Wahl. editors, *Graphics And Robotics*. Springer-Verlag, 1994.
12. R. Hoffman. Automated assembly planning for b-rep products. *IEEE International Conference on Systems Engineering*, pages 391–394, 1990.
13. R. Hoffman. *Computer-Aided Mechanical Assembly Planning*, chapter A common sense approach to assembly sequence planning. Kluwer Academic Publishers, 1991.
14. L. S. Homem de Mello and A. C. Sanderson. AND/OR-Graph Representation of Assembly Plans. *IEEE Trans. Robotics and Automation*, 6(2):188 – 199, April 1990.
15. R. E. Jones and R. H. Wilson. A survey of constraints in automated assembly planning. *IEEE International Conference on Robotics and Automation*, pages 1525–1532, 1996.
16. S. G. Kaufman, R. H. Wilson, R. E. Jones, and T. L. Calton. The archimedes 2 mechanical assembly planning system. *IEEE International Conference on Robotics and Automation*, pages 3361–3368, 1996.
17. L. Kavraki and M. Kolountzakis. Partitioning a planar assembly into two connected parts is np-complete. *Information Processing Letters*, 55:159–165, 1995.
18. L. Kavraki, J. C. .Latombe, and R. W. Wilson. On the complexity of assembly partitioning. *Information Processing Letters*, 48:229–235, 1993.
19. S. Lee and Y. G. Shin. Assembly Planning based on Geometric Reasoning. *Computers and Graphics*, 14(2):237 – 250, 1990.
20. S. Lee and Y. G. Shin. Assembly Planning based on Subassembly Extraction. In *Proc. IEEE International Conf. on Robotics and Automation*, pages 1601–1611, May 1990.
21. A. C. Lin and Y. G. Shin. 3d maps: Three-dimensional mechanical assembly planning system. *Journal of Manufacturing Systems*, 12(6):437–456, 1993.

22. T. Lozano-Pérez. Spatial Planning: A Configuration Space Approach. *IEEE Trans. Computers*, C-32(2):26–38, February 1983.
23. R. Mattikalli, D. Baraff, P. Khosla, and B. Repetto. Finding All Stable Orientations of Assemblies with Friction. *IEEE Trans. Robotics and Automation*, 12(2):290–301, 1996.
24. J. M. Miller and R. L. Hoffman. Automatic assembly planning with fasteners. *IEEE International Conference on Robotics and Automation*, 1:69–74, 1989.
25. K. Murty. *Linear Programming*. Wiley, 1983.
26. R. S. Palmer. *Computational Complexity of Motion and Stability of Polygons*. PhD thesis, Cornell University, January 1987.
27. C. Pelich, H. Mosemann, and F. M. Wahl. A Powerful, Flexible and Process-Modular Robot Control Environment. In *"IASTED International Conference Robotics and Manufacturing"*, August 1996.
28. J. Pertin-Troccaz. *Grasping: A State of the Art*, chapter Programming, Planning, and Learning, pages 71–98. The Robotics Review. MIT Press, 1989.
29. R. Pollack, M. Sharir, and S. Sifrony. Separating two simple polygons by a sequence of translations. *Discrete and Computational Geometry*, 3(2):123–136, 1988.
30. R. J. Popplestone. Specifying manipulation in terms of spatial relationships. Technical Report 117, Department of Artificial Intelligence, University of Edinburgh, 1979.
31. F. Röhrdanz, R. Gutsche, and F. M. Wahl. Assembly Planning and Geometric Reasoning for Grasping. In I. Plander, editor, *Sixth Int. Conference on Artificial Intelligence and Information-Control Systems of Robots*, pages 93–106. World Scientific, September 1994.
32. F. Röhrdanz, H. Mosemann, and F. M. Wahl. Highlap: A high level system for generating, representing, and evaluating assembly sequences. In *IEEE International Joint Symposia on Intelligence and Systems*, November 1996.
33. F. Röhrdanz, H. Mosemann, and F. M. Wahl. Stability Analysis of Assemblies Considering Friction. Technical Report 5-1996-1, Institute of Robotics and Computer Control, Braunschweig, Germany, May 1996.
34. F. Röhrdanz and F. M. Wahl. Incremental Free Space Acquisition and Representation for Automatic Grasping. In *International Conference on Automation, Robotics and Computer Vision*, Singapore, December 1996.
35. B. Romney, C. Godard, M. Goldwasser, and G. Ramkumar. An efficient system for geometric assembly sequence generation and evaluation. *ASME International Computers in Engineering Conference*, pages 699–712, 1995.
36. K. B. Shimoga. Robot Grasp Synthesis Algorithms: A Survey. *International Journal of Robotics Research*, 15(3):230–266, June 1996.
37. C. K. Shin, D. S. Hong, and H. S. Cho. Disassemblability Analysis for Generating Robotic Assembly Sequences. *Proc. IEEE International Conf. on Robotics and Automation*, 2:1284 – 1289, May 1995.
38. *User Manual and Tutorials, IGRIP Version 3.0*. Deneb Robotics, 3285 Lapeer Road West, P.O. Box 214687, Auburn Hills, MI 48321-4687, 1994.
39. J. U. Turner, S. Subramaniam, and S. Gupta. Constraint representation and reduction in assembly modeling and analysis. *IEEE Journal of Robotics and Automation*, 8(6):741–750, 1992.
40. J. M. Valade. Geometric reasoning and automatic synthesis of assembly trajectory. *International Conference on Advanced Robotics*, pages 43–50, 1985.
41. J. Wolter. On the Automatic Generation of Assembly Plans. In *Proc. IEEE International Conf. on Robotics and Automation*, pages 62–68, 1989.

42. J. D. Wolter. A combinatorial analysis of enumerative data structures for assembly planning. *IEEE International Conference on Robotics and Automation*, pages 611–618, 1991.
43. J. D. Wolter and J. C. Trinkle. Automatic selection of fixture points for frictionless assemblies. *IEEE International Conference on Robotics and Automation*, pages 528–534, 1994.
44. T. C. Woo and D. Dutta. Automatic disassembly and total ordering in three dimensions. *Transactions of the ASME*, 113(2):207–213, 1991.

Focus on Computer Graphics
(Formerly EurographicSeminars)

Data Structures for Raster Graphics. Edited by L. R. A. Kessener, F. J. Peters, M. L. P. van Lierop. VII, 201 pages, 80 figs., 1986

Advances in Computer Graphics I. Edited by G. Enderle, M. Grave, F. Lillehagen. XII, 512 pages, 168 figs., 1986

Advances in Computer Graphics II. Edited by F. R. A. Hopgood, R. J. Hubbold, D. A. Duce. X, 186 pages, 96 figs., 1986

Advances in Computer Graphics Hardware I. Edited by W. Straßer. X, 147 pages, 76 figs., 1987

Intelligent CAD Systems I. Theoretical and Methodological Aspects. Edited by P. J. W. ten Hagen, T. Tomiyama. XIV, 360 pages, 119 figs., 1987

Advances in Computer Graphics III. Edited by M. M. de Ruiter. IX, 323 pages, 247 figs., 1988

Advances in Computer Graphics Hardware II. Edited by A. A. M. Kuijk, W. Straßer. VIII, 258 pages, 99 figs., 1988

Intelligent CAD Systems II. Implementational Issues. Edited by V. Akman, P. J. W. ten Hagen, P. J. Veerkamp. X, 324 pages, 114 figs., 1989

Advances in Computer Graphics IV. Edited by W. T. Hewitt, M. Grave, M. Roch. XVI, 248 pages, 138 figs., 1991

Advances in Computer Graphics V. Edited by W. Purgathofer, J. Schönhut. VIII, 223 pages, 101 figs., 1989

User Interface Management and Design. Edited by D. A. Duce, M. R. Gomes, F. R. A. Hopgood, J. R. Lee. VIII, 324 pages, 117 figs., 1991

Advances in Computer Graphics Hardware III. Edited by A. A. M. Kuijk. VIII, 214 pages, 88 figs., 1991

Advances in Object-Oriented Graphics I. Edited by E. H. Blake, P. Wisskirchen. X, 218 pages, 74 figs., 1991

Advances in Computer Graphics Hardware IV. Edited by R. L. Grimsdale, W. Straßer. VIII, 276 pages, 124 figs., 1991

Advances in Computer Graphics VI. Images: Synthesis, Analysis, and Interaction. Edited by G. Garcia, I. Herman. IX, 449 pages, 186 figs., 1991

Intelligent CAD Systems III. Practical Experience and Evaluation. Edited by P. J. W. ten Hagen, P. J. Veerkamp. X, 270 pages, 116 figs., 1991

Graphics and Communications. Edited by D. B. Arnold,
R. A. Day, D. A. Duce, C. Fuhrhop, J. R. Gallop, R. Maybury, D. C. Sutcliffe.
VIII, 274 pages, 84 figs., 1991

Photorealism in Computer Graphics. Edited by K. Bouatouch, C. Bouville.
XVI, 230 pages, 118 figs., 1992

Advances in Computer Graphics Hardware V. Rendering, Ray Tracing and
Visualization Systems. Edited by R. L. Grimsdale, A. Kaufman.
VIII, 174 pages, 97 figs., 1992

Advances in Scientific Visualization. Edited by F. H. Post, A. J. S. Hin.
X, 212 pages, 141 figs., 47 in color, 1992

Computer Graphics and Mathematics. Edited by B. Falcidieno, I. Herman,
C. Pienovi. VII, 318 pages, 159 figs., 8 in color, 1992

Rendering, Visualization and Rasterization Hardware. Edited by A. Kaufman.
VIII, 196 pages, 100 figs., 1993

Visualization in Scientific Computing. Edited by M. Grave, Y. Le Lous,
W. T. Hewitt. XI, 218 pages, 120 figs., 1994

Photorealistic Rendering in Computer Graphics. Edited by P. Brunet,
F. W. Jansen. X, 286 pages, 175 figs., 1994

From Object Modelling to Advanced Visual Communication. Edited by
S. Coquillart, W. Straßer, P. Stucki. VII, 305 pages, 128 figs., 38 in color, 1994

Photorealistic Rendering Techniques. Edited by G. Sakas, P. Shirley, S. Müller.
X, 448 pages, 155 figs., 16 color plates, 1995

Interactive Systems: Design, Specification, and Verification.
Edited by F. Paternó. X, 447 pages, 176 figs., 1995

Object-oriented and Mixed Programming Paradigms. New Directions in Computer
Graphics. Edited by P. Wisskirchen.
X, 196 pages, 68 figs, 1996

Geometric Modeling: Theory and Practice. Edited by W. Strasser, R. Klein, R. Rau.
IX, 434 pages, 236 figs., 1997

Springer
and the
environment

At Springer we firmly believe that an international science publisher has a special obligation to the environment, and our corporate policies consistently reflect this conviction.

We also expect our business partners – paper mills, printers, packaging manufacturers, etc. – to commit themselves to using materials and production processes that do not harm the environment. The paper in this book is made from low- or no-chlorine pulp and is acid free, in conformance with international standards for paper permanency.